Essentials of Human Anatomy and Physiology

Second Edition

Elaine N. Marieb, R.N., Ph.D.

Holyoke Community College

The Benjamin/Cummings Publishing Company, Inc.

Menlo Park, California • Reading, Massachusetts
Don Mills, Ontario • Wokingham, U.K. • Amsterdam • Sydney
Singapore • Tokyo • Madrid • Bogota • Santiago • San Juan

Sponsoring Editor: Connie Spatz
Production Supervisor: Mary Shields
Cover and Book Design: John Edeen
Artist (new art): Sara Lee Steigerwald
Reese Thornton, Folium
Developmental Editor: Janet Wagner
Copy Editor: Diane Denny
Composition, Second Edition: Graphic Typesetting Service

Photo credits follow the index.

Library of Congress Cataloging in Publication Data

Marieb, Elaine Nicpon, 1936–
    Essentials of human anatomy and physiology.

    Includes index.
    1. Human physiology.   2. Anatomy, Human.   I. Title.
[DNLM: 1. Anatomy.   2. Physiology.   QS 4 M334e]
QP34.5.M455   1988   612   87–27735
ISBN 0–8053–6739–X

    EFGHIJ-DA-89

Benjamin/Cummings Publishing Company, Inc.
2727 Sand Hill Road
Menlo Park, California 94025

Preface to the Instructor

The second edition of *Essentials of Human Anatomy and Physiology,* like the first, is designed to introduce students pursuing careers in the allied health fields to the structure and function of the human body. Some sections have been updated to ensure that explanations are both accurate and current, and new material has been added to even better meet the needs of both students and instructors.

The style of writing and presentation, that, in the first edition, helped students with limited backgrounds in the sciences grasp the fundamental concepts of human anatomy and the inner workings of the body has been retained and employed again in a new chapter on immunology. Because this difficult subject is receiving increasing attention, both in the media and in clinical agencies, a creative and clear explanation of the fundamentals of immunology is of growing importance to students in the allied health fields. Also new to this edition is an appendix on chemistry, included in response to the desire expressed by many instructors that such an introduction appear in subsequent editions.

Organization

Instructors often prefer different topic sequences in their presentations and lectures on body systems. Extensive reviews by anatomy and physiology instructors helped me decide on a particular presentation. My choice was to construct a sequence of topics that encourages learning. The first chapter orients students to the new world of learning in which they will soon become enmeshed. Important anatomic terms, named gross body areas, and an overview (kept intentionally brief) of all body organ systems set the stage for future learning.

Cells, tissues, and the first organ system (skin) are treated in succession. Most students find that the skin is an interesting system, and thus the transition to the organ system discussions is easily accomplished. The subsequent chapters include systems requiring a good deal of anatomic terminology (for example, skeletal, muscular, and nervous systems). Each organ system is approached from simple to increasingly complex levels. Building-in an understanding of the concepts, rather than rote memorization, is emphasized.

Organizational Flexibility

A good textbook should meet the needs of both instructors and students. Because every instructor has a personal style and philosophic approach to teaching and course organization, I have written each chapter to allow for flexibility in topic presentation. All chapters are self-contained. An instructor may choose a chapter sequence that is personally preferred. Topic omission may be necessary in programs in which less than one semester is devoted to anatomic and physiologic concepts.

Presentation of Anatomy, Physiology, and Pathology

The content presents an equal balance of anatomic and physiologic concepts. Complex topics are explained consistently by the use of analogies to foster, rather than discourage, student interest. In instances requiring more than usual amounts of memorization of terminology and facts, the material is presented in both a text discussion and tabular form to reinforce learning.

Pathologic examples and discussions are chosen based on conditions seen most often by those working in the allied health fields. Pathologic examples are introduced not as an end to themselves, but to familiarize students to possible consequences when body structures are damaged or malformed. A student must understand normal functions and structures in order then to comprehend abnormal processes.

Special Features

Much thought was given to the learning process when designing and incorporating the unique features in this text. These include:

- **Exceptional Art Program.** The art program uses color to highlight and distinguish important structures and concepts. The quality of the art, labeling, and text and the figure-referencing system are designed to retain student attention.

- **Topic Boxes.** Each chapter incorporates topic boxes that present timely and interesting concepts. Such topics as "The Shrinking Brain" and "Monoclonal Antibodies" are presented in the hope that students will then develop curiosity in the sciences.

- **Developmental Aspects.** Each chapter ends with a section on developmental aspects of that particular system. The section presents the formation in the embryo or fetus through old age. Important health problems unique to that system are introduced, and emphasis is placed on problems that accompany the aging process. Because many health workers are intricately involved in care of elderly people, this emphasis should be valuable to today's student.

- **Writing Style.** The writing style is intentionally informal. New terminology is set in boldface or italic type, followed by phonetic pronunciation. Terms are defined within the text and again in the main glossary.

- **Pedagogical Features.** Several pedagogical devices are used to ensure that students are introduced to important terminology and concepts. Each chapter begins with a chapter outline and objectives. A summary of text contents and review questions are at the end of each chapter. Important terms are listed, and the student is referred to the main glossary.

Teaching Package

Teaching structure and function is a challenging and rewarding endeavor. To assist instructors, we are providing the most comprehensive teaching package available. This package includes an instructor's guide and a transparency package.

Related Texts

- **A & P Coloring Workbook: A Complete Study Guide to Anatomy and Physiology, 2nd edition.** Available for student use, this unique combination of a coloring book and study guide will provide a learning/testing tool of multiple choice, matching, diagram-labeling, and coloring exercises. These exercises complement and enhance the learning and retention of textual material.

- **Brief Laboratory Manual.** For those instructors who have a lab component, a laboratory manual is available. The manual includes in-class dissection techniques for demonstration by instructor or student.

Acknowledgments

Many people contributed to my efforts in the creation of the first and second editions of this book. I would like to thank the many students, and particularly my former student, Ann Allworth, who allowed me to use them as "sounding boards" to try out my analogies and ideas.

I also would like to thank the following reviewers of this work for their thoughtful critiques: Kay Brashear, El Centro College, Dallas, TX; Annette Brown, Broometioga Board of Cooperative Educational Services, Broometioga, NY; Ray Canham, Ph.D., Richland College, Dallas, TX; Elizabeth Carl, Nassau Technological Center (BOCES), Westbury, NY; Jean Cons, College of San Mateo, San Mateo, CA; Luci Constant, W.M.L. Dawson School of Practical Nursing, Chicago, IL; Sue D. Hall, R.N., Rogue

Community College, Grants Pass, OR; John P. Harley, Ph.D., Eastern Kentucky University, Richmond, KY; Carol Holley, San Jacinto College, Pasadena, TX; Mary Hopkins, R.N., B.S., Kiamichi Voc-Tech, Wilburton, OK; Drucilla B. Jolly, Forsyth Technical College, Winston-Salem, NC; Fred Klaus, Ph.D., East Texas State University, Commerce, TX; Bonnie Kroemmelbein, M.Ed., Upper Bucks County AVTS Practical Nursing Program, Perkasie, PA; John Lammert, Gustavus Adolphus College, St. Peter, MN; Jeri Lindsey, Ph.D., Tarrant County Jr. College, Hurst, TX; Bennie Marshal, STOP/CETA School of Practical Nursing, Norfolk, VA; William Matthai, Tarrant County Jr. College, Hurst, TX; Walter Matulis, Ph.D., Mid-Michigan Community College, Harrison, MI; Roxine McQuitty, Milwaukee Area Technical College, Milwaukee, WI; Lew Milner, Ph.D., North Central Technical College, Mansfield, OH; Bridget Price, R.N., Vocational Nursing School of California, Anaheim, CA; Ann Senisi, R.N., M.A., Nassau Technological Center (BOCES), Westbury, NY; Ann Welch, Indiana Vocational/Technical College, Indianapolis, IN; and Eileen Williams, R.N., B.S.N., M.Ed., Nassau Technological Center (BOCES), Westbury, NY.

The staff of Benjamin/Cummings Publishing Company contributed immensely in the form of support and guidance. I would especially like to thank Connie Spatz and Langdon Faust, my editors, and Mary Shields, my production supervisor. Finally, I wish to thank my patient and loving husband for always being there at the end of the day.

Elaine Nicpon Marieb

Preface to the Student

This book is written with you, the student, in mind. Human anatomy and physiology is more than just interesting—it is fascinating. To help get you involved in the study of this subject, a number of special features are incorporated throughout the book.

The *informal writing style* invites you to learn more about anatomy and physiology without the intimidation other such texts often present. We want you to enjoy reading this book.

Topic boxes and *tables* are designed with you in mind. The topic boxes present scientific information that can be applied to your daily life. When reading the topic boxes, you will probably find yourself saying, "I didn't know that," or "Now I understand why. . . ." The tables are summaries of important information in the text. You should be able to use the tables when studying for an exam or reviewing an important topic.

Important terms are defined within the text as they are introduced and are listed at the ends of the chapters. An extensive *glossary* is provided at the back of the book to help you review these terms.

Phonetic spellings are provided for many of these important terms, especially those which are likely to be unfamiliar to you. To read these, you will need to remember the following rules:

1. Accent marks follow stressed syllables. The primary stress is shown by ′, and the secondary stress by ″.

2. Unless otherwise noted, assume that vowels at the ends of syllables are long; vowels followed by consonants are short.

For example, the phonetic spelling of "thrombophlebitis" is *throm″bo-fle-bi′tis*. The next-to-the-last syllable (*bi′*) receives the greatest stress, and the first syllable (*throm″*) gets the secondary stress. The vowel in the second syllable comes at the end of the syllable and is long.

Any exam causes anxiety. Exams in anatomy and physiology are no exception. To help you better prepare for an exam or comprehend the material you have just read, extensive *review questions* and *summaries* are found at the ends of the chapters.

The art program is designed to help you learn the different structures and functions of the human body. All figures are referred to within the text discussion. The best way to make use of the extensive art program is to carefully study each illustration when it is discussed in the text.

We hope that you enjoy *Essentials of Human Anatomy and Physiology* and that this book makes learning about intricate structures and functions of the human body a fun and rewarding process.

Elaine Nicpon Marieb

Contents

CHAPTER 4
Skeletal System   55

CHAPTER 5
Muscular System   87

CHAPTER 6
Nervous System   115

CHAPTER 12

Respiratory System   269

CHAPTER 13

Digestive System   287

CHAPTER 14

Urinary System   319

TOPIC BOXES AND TABLES

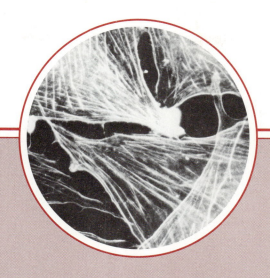

CHAPTER 1

The Human Body— an Orientation

Chapter Contents

After completing this chapter, you should be able to:

- Define anatomy, physiology, and homeostasis.

- Describe how anatomy and physiology are related.

- Describe the anatomical position verbally or demonstrate it.

- Use proper anatomical terminology to describe body directions, planes, surfaces, and cavities.

- Name the major organ systems and state the major function of each.

- Classify the organs discussed by organ system.

- Identify the organs discussed on a diagram or a dissectible torso.

ANATOMY AND PHYSIOLOGY

Most of us have a natural curiosity about our bodies; we want to know what makes us tick. This fact is demonstrated in infants who can keep themselves happy for long periods of time staring at their own hands or pulling at their mother's nose. Older children wonder where food goes when they swallow it, and some believe that they will grow a watermelon in their "belly" if they swallow the seeds. They scream loudly when approached by any medical personnel (fearing body mutilation) and play "doctor." Adults become upset when their heart pounds, they have uncontrollable "hot flashes," or they cannot keep their weight down.

Anatomy and physiology, subdivisions of biology, explore and describe how the body is put together and how it works.

Anatomy

Anatomy (ah-nat′o-me) is the study of the structure and shape of the body and body parts and their relationship to one another. Whenever we look at our own body, we are observing *gross* anatomy, that is, we are studying large, easily observable structures. If a microscope or magnifying instrument is used to see very small structures in the body, we are studying *microscopic* anatomy.

Physiology

Physiology (fiz″e-ol′o-je) is the study of how the body and its parts work or function. For example, cardiac physiology is the study of the function of the heart, which acts as a muscular pump to keep blood flowing throughout the body.

Relationship Between Anatomy and Physiology

It is difficult to separate anatomy and physiology because they are always related. The parts of your body are combined and arranged to form a well-organized unit, and each of those parts has a job to do to see that the body operates as a whole. Structure determines what functions can happen. For example, the lungs are not muscular chambers like the heart and cannot pump blood through the body, but because the walls of their air sacs are very thin, they *can* exchange gases and provide oxygen to the body. Most specialized body functions act to maintain **homeostasis** (ho″me-o-sta′sis), which can be most simply defined as a relatively stable condition of the body's internal environment. For example, one area of your brain acts as a thermostat to keep your body temperature within a narrow range—97° to 99°F. If you go out on a cold day, your body temperature will begin to drop rapidly, but the activation of the thermostat will soon cause shivering which

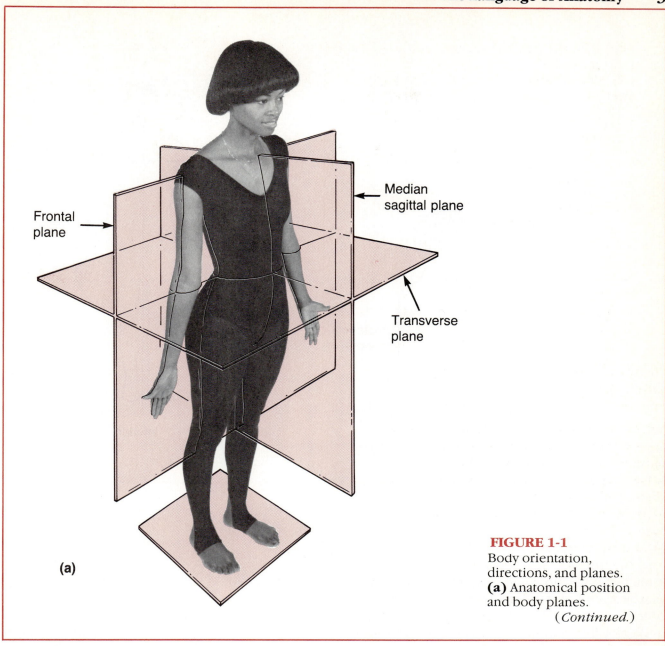

Frontal plane

Median sagittal plane

Transverse plane

(a)

FIGURE 1-1
Body orientation, directions, and planes. **(a)** Anatomical position and body planes.
(*Continued.*)

increases heat production and brings body temperature back up. Not all homeostatic mechanisms act in the same way, but all of them have the same goal—preventing sudden, severe changes within the body. A relatively constant, internal environment is absolutely necessary for survival. When the ability of the body to maintain homeostasis is overwhelmed, disease, illness, and even death occur. As we grow older, the efficiency of body organs decreases and internal conditions become less stable, leading to the changes we recognize as the aging process.

THE LANGUAGE OF ANATOMY

It is easy to be confused by the language of anatomy because many of its special terms are also part of your everyday vocabulary. But without this special terminology, problems are bound to occur because people in health sciences must be able to describe the areas of the body precisely so that there is no misunderstanding. For example, what do <u>over</u>, <u>on top of</u>, <u>above</u>, and <u>behind</u> mean in reference to the human body? Do they mean the

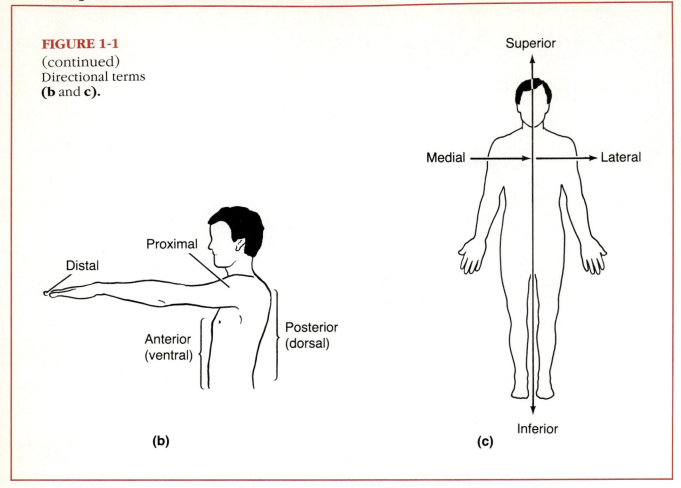

FIGURE 1-1
(continued)
Directional terms
(b and **c).**

same thing when you are standing on your head as when you are standing upright? There is an accepted set of reference terms that are universally understood and allow body structures to be located and identified clearly with just a few words. All parts of the body are described in relation to other parts, so to avoid confusion it is always assumed that the body is in the **anatomical position.** It is important to understand this position because most body terminology used in this book refers to this body positioning *regardless* of the position the body happens to be in. Figure 1-1 illustrates the anatomical position. The body is erect with feet together, and arms are hanging at the sides with the palms facing forward. Stand up and assume the anatomical position. Notice that it is not very comfortable. This is because the hands are held unnaturally forward rather than hanging partially cupped toward the thighs.

Body Orientation and Direction

When reading through the descriptions of the terms given next, keep referring to Figure 1-1*b* and *c* to make sure you understand the terms clearly.

SUPERIOR/INFERIOR

Superior/inferior means *above/below.* These terms refer to the position of a body structure along the body's long axis. Thus, superior structures are always above other structures, and inferior structures are always below other structures. For example, the nose is superior to the mouth, and the abdomen is inferior to the chest region.

CEPHALAD/CAUDAD

Cephalad/caudad means *toward the head/toward the tail.* These terms are used interchangeably with superior and inferior in human beings. Still another synonym for cephalad is *cranial.*

ANTERIOR/POSTERIOR

Anterior/posterior means *toward the front of the body/toward the back of the body.* In humans the most anterior surfaces are those most forward—the face, chest, and abdomen. Posterior surfaces are those on the backside of the body—the buttocks and shoulder blades. The anterior and posterior surfaces of our bodies may also be described by the words *ventral* (literally, bellyside) and *dorsal* (backside).

MEDIAL/LATERAL

Medial/lateral means *toward the body midline/away from the body midline.* For example, the breastbone is medial to the ribs, and the ear is lateral to the eye. Notice that in the anatomical position, the great toes are on the medial aspects of the feet whereas the thumbs are on the lateral side of the hands.

PROXIMAL/DISTAL

Proximal/distal means *nearer the body trunk or attached end/farther from the body trunk or attached end.* These two terms are usually used to refer to the extremities or limbs and to locate various regions on them. For example, the fingers are distal to the elbow. This means that the fingers are farther from the attachment point of the arm to the body (the shoulder region) than the elbow. It is much easier to say that the fingers are distal to the elbow.

SUPERFICIAL/DEEP

Superficial/deep means *on or near the body surface/deeper inward or away from the body surface.* For example, a superficial injury is one that involves the skin (a paper cut or bruise), but a lacerated liver is definitely a deep injury. Synonyms for superficial and deep are *external* and *internal,* respectively.

Before continuing, take a minute to check your understanding of what you have just read. Give the relationship between the following body parts using the correct anatomical terms.

The wrist is _____ to the hand.

The breastbone is _____ to the spine.

The brain is _____ to the spinal cord.

The lungs are _____ to the stomach.

The bridge of the nose is _____ to the eyes.

Body Planes and Sections

When preparing to look at the internal structures of the body, medical students find that it is necessary to make a **section** or cut. When the section is made through the body wall or through an organ, it is made along an imaginary line called a **plane.** Since the body is three-dimensional, we can refer to three types of planes or sections that lie at right angles to one another (see Figure 1-1*a*).

SAGITTAL SECTION

A **sagittal** (saj'i-tal) **section** is a cut made along the lengthwise, or longitudinal, plane of the body that divides the body into right and left parts. If the cut is made down the median plane of the body and the right and left parts are equal in size, it is called a **midsagittal section.**

FRONTAL SECTION

A **frontal section** is a cut made along a lengthwise plane that divides the body into anterior and posterior parts. It is also called a **coronal** (ko-rŏ'nal) **section.**

TRANSVERSE SECTION

A **transverse section** is a cut made along a transverse, or horizontal, plane that divides the body into superior and inferior parts; it may also be called a **cross section.**

Surface Anatomy

There are many visible landmarks on the surface of the body. Once you know their proper anatomical names, you can be specific in referring to different areas of the body.

FIGURE 1-2
Surface anatomy.
(a) Anterior body landmarks, and
(b) posterior body landmarks.

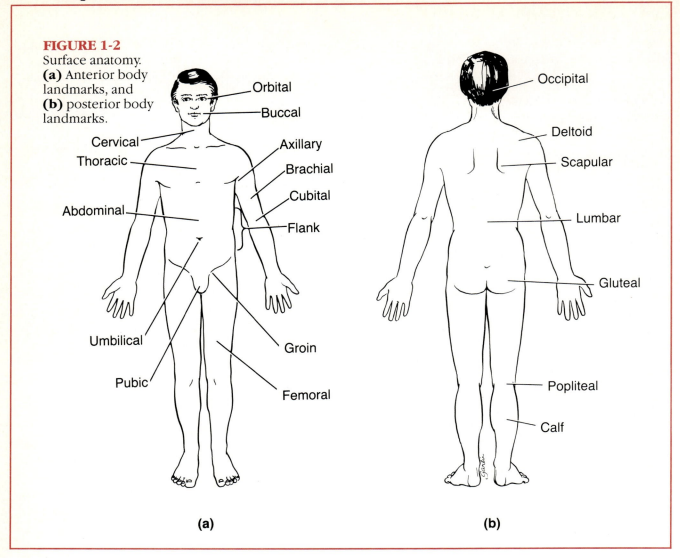

(a) (b)

ANTERIOR BODY LANDMARKS

Look at Figure 1-2*a* to find the following body regions. Once you have identified all the anterior body landmarks, cover the labels that describe what the structures are, and again go through the list pointing out the specific areas on your own body.

abdominal anterior body trunk

axillary *(ak'sĭ-lar"e)* armpit

brachial *(bra'ke-al)* arm

buccal *(buk'al)* cheek area

cervical neck region

cubital *(ku'bĭ-tal)* anterior surface of the elbow

femoral *(fem'or-al)* thigh

flank lateral surface of the body trunk from the rib cage to the hip

groin area where the thigh meets the body trunk

orbital *(or'bĭ-tal)* eye area

pubic *(pu'bik)* genital region

thoracic *(tho-ras'ik)* chest

umbilical *(um-bill'ĭ-kal)* navel

POSTERIOR BODY LANDMARKS

Identify the following body regions in Figure 1-2*b* and then locate them on yourself without referring to this book.

calf the posterior surface of the lower leg

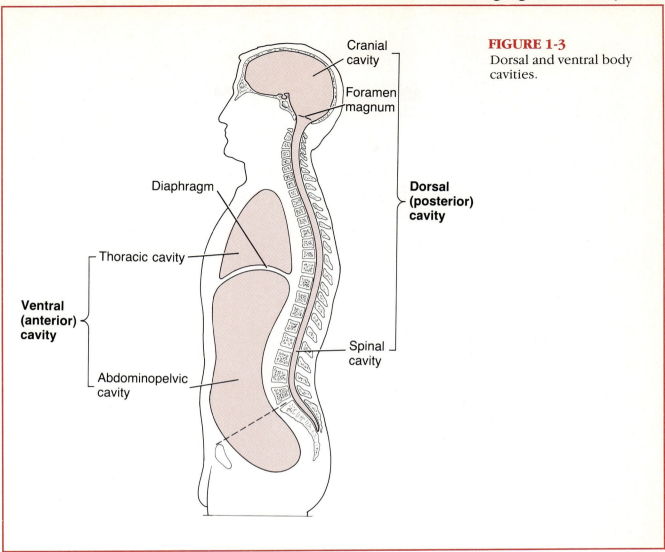

Cranial cavity

Foramen magnum

Diaphragm

Thoracic cavity

Dorsal (posterior) cavity

Ventral (anterior) cavity

Abdominopelvic cavity

Spinal cavity

FIGURE 1-3
Dorsal and ventral body cavities.

deltoid *(del'toid)* curve of the shoulder formed by the large deltoid muscle

gluteal *(gloo'te-al)* buttocks

lumbar area of the back between the ribs and hips

occipital *(ok-sip'i-tal)* posterior surface of the head

popliteal *(pop-lit'e-al)* knee area

scapular *(skap'u-lar)* shoulder blade area

Body Cavities

The body has two sets of internal cavities that provide different degrees of protection to the organs within them (Figure 1-3).

DORSAL BODY CAVITY

The **dorsal body cavity** has two subdivisions, which are continuous with each other. The **cranial cavity** is the space inside the bony skull. The brain is well protected because it occupies the cranial cavity. The **spinal cavity** extends from the cranial cavity nearly to the end of the vertebral column. The spinal cord, which is a continuation of the brain, is protected by the vertebrae, which surround the spinal cavity.

VENTRAL BODY CAVITY

The **ventral body cavity** is much larger than the dorsal cavity. It contains all of the structures within the chest and abdomen. Like the dorsal cavity, the ventral body cavity is subdivided. The superi-

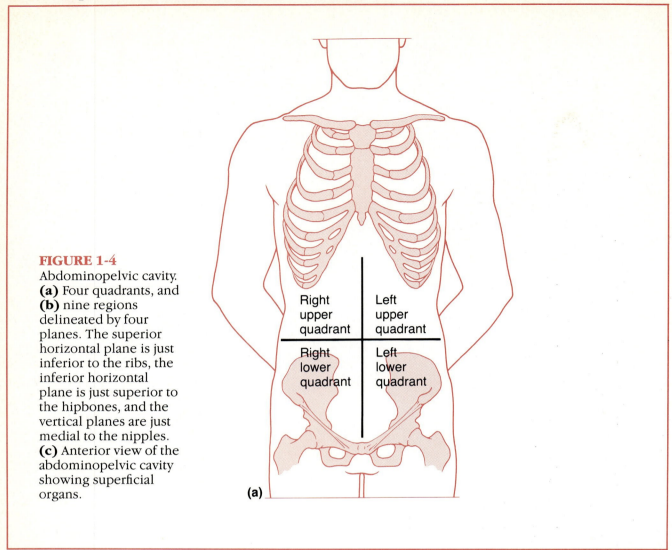

FIGURE 1-4
Abdominopelvic cavity.
(a) Four quadrants, and
(b) nine regions
delineated by four
planes. The superior
horizontal plane is just
inferior to the ribs, the
inferior horizontal
plane is just superior to
the hipbones, and the
vertical planes are just
medial to the nipples.
(c) Anterior view of the
abdominopelvic cavity
showing superficial
organs.

Right upper quadrant

Left upper quadrant

Right lower quadrant

Left lower quadrant

(a)

or **thoracic cavity** is separated from the rest of the ventral cavity by a dome-shaped muscle, the **diaphragm** (di′ah-fram). The organs in the thoracic cavity (lungs, heart, and others) are somewhat protected by the rib cage. The cavity inferior to the diaphragm is the **abdominopelvic** (ab-dom″ĭ-no-pel′vik) **cavity.** (Some prefer to subdivide it into a superior abdominal cavity containing the stomach, liver, intestines, and other organs, and an inferior pelvic cavity with the reproductive organs, bladder, and rectum; however, there is no physical structure dividing the abdominopelvic cavity.) If you look carefully at Figure 1-3, you will see that the pelvic cavity is not continuous with the abdominal cavity in a straight plane, but that it tips away from it in the posterior direction. There is no bony protection for the abdominal portion of the ventral body cavity.

Because the abdominopelvic cavity is quite large and contains many organs, it is helpful to divide it up into smaller areas for study. One scheme divides the abdominopelvic cavity into four more or less equal regions called *quadrants;* the quadrants are then simply named according to their relative position, that is, right upper quadrant, right lower quadrant, left upper quadrant, and left lower quadrant (Figure 1-4*a*).

Another commonly used system divides the abdominopelvic cavity into nine separate regions by four planes as shown in Figure 1-4*b*. Although the names of the nine regions are unfamiliar to you now, with a little patience and studying they will become easier to remember. As you locate these regions in the figure, notice the organs they overlie by referring to Figure 1-4*c*.

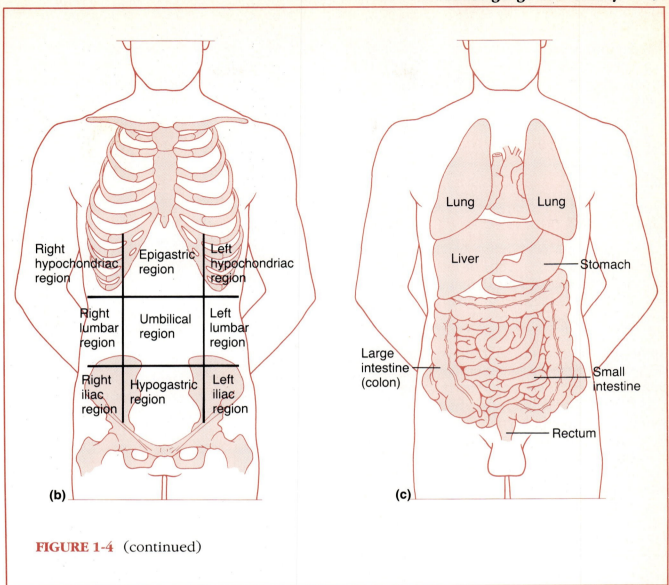

Right hypochondriac region

Epigastric region

Left hypochondriac region

Right lumbar region

Umbilical region

Left lumbar region

Right iliac region

Hypogastric region

Left iliac region

(b)

Lung

Lung

Liver

Stomach

Large intestine (colon)

Small intestine

Rectum

(c)

FIGURE 1-4 (continued)

umbilical region The central region, which includes the navel or umbilicus.

epigastric *(ep″y-gas′trik)* **region** The area just above the umbilical region, which contains most of the stomach.

hypogastric *(hi″-po-gas′trik)* **region** The area just inferior to the umbilical region and often called the pubic area.

iliac (inguinal) *(ing′gwĭ-nal)* **regions** The areas lateral to the hypogastric region on each side and overlying the inferior parts of the hip bones.

lumbar regions The areas lateral to the umbilical region on each side and overlying the superior flaring parts of the hip bones.

hypochondriac *(hi″po-kon′dre-ak)* **regions** The areas lateral to the epigastric region on each side and overlying the lower ribs.

The ventral cavity wall is lined with a smooth serous membrane called the **parietal serosa** (pah-ri′ĕ-tal se-ro′sah). This membrane is continuous with a similar membrane **(visceral** [vis′er-al] **serosa)** that covers the organs contained in the cavity. These membranes produce a lubricating fluid that allows the body organs to slide over one another or to rub against the body wall without friction. The specific names of the serous membranes are dependent on their location. The serosa lining the abdominal cavity and covering its organs is the **peritoneum** (per″i-to-ne′um), that around the lungs is the **pleura** (ploor′ah), and that around the heart is the **pericardium** (per″ĭ-kar′de-um).

ORGAN SYSTEM OVERVIEW

Levels of Organization

The smallest unit or building block of all living things is the **cell.** The cells of the body can be grouped into one of four different **tissue** types according to similarities in their structure and function (see Chapter 2, p. 31).

An **organ** is a structure composed of two or more of these tissue types that performs a specific function for the body. For example, the small intestine, which digests and absorbs food, is composed of all four tissue types. Finally, all organs of the body are organized so that a number of organ systems are formed. An **organ system** is a group of organs that work together to perform a specific body function. For example, the digestive system is made up of the esophagus, stomach, and small and large intestines (to name a few of its organs). They each have their own job to do, and working together, they keep food moving through the digestive system so that it is properly broken down and absorbed into the blood, providing fuel for all the body's cells. In all, there are ten organ systems. The major organs of each of the organ systems are shown in diagrammatic form in Box 1-1. Refer to the figures as you read through the descriptions of the organ systems next.

Integumentary System

The **integumentary** (in-teg-u-men′tar-e) **system** is the external covering of the body or the skin. It waterproofs the body and cushions and protects the deeper tissues from injury. It also functions in excretion of salts and urea, which are lost in perspiration, and aids in the regulation of body temperature.

Skeletal System

The **skeletal system** consists of bones, cartilages, tendons, and joints. Its major roles are to support the body and provide a framework that the skeletal muscles can use to cause movement. Obviously it also has a protective function (for example, the skull encloses and protects the brain). **Hemopoiesis** (he″mo-poi-e′sis), or formation of blood cells, goes on within the cavities of the skeleton.

Muscular System

The muscles of the body have only one function—to *contract* or *shorten.* When this happens, movement occurs. Thus, as the heart (cardiac muscle) contracts, it pumps blood through blood vessels. When the large, fleshy muscles attached to the bones (skeletal muscles) contract, you are able to walk, run, grasp, and smile. When the smooth muscle layers in the walls of hollow organs contract, the organs change shape and substances are moved through organs along definite pathways (for example, the movement of food through the digestive tract organs, or the passage of urine from the bladder to the outside of the body).

Nervous System

The **nervous system** is the body's fast-acting control system. It consists of the brain, spinal cord, nerves, and sensory receptors. The body must be able to respond to irritants or stimuli coming from outside the body (such as light, sound, and changes in temperature) and from inside the body (such as decreases in oxygen and stretching of tissue). The sensory receptors detect these changes and send messages to the central nervous system (brain and spinal cord) so that it is constantly informed about what is going on. The central nervous system then assesses this information and responds by activating the appropriate body muscles or glands.

Endocrine System

Like the nervous system, the **endocrine** (en′do-krin) **system** controls body activities and maintains homeostasis. The endocrine glands produce chemical molecules called *hormones* and release them into the blood to travel to relatively distant target organs.

(Text continues on p. 15.)

BOX 1-1
Organ Systems

(a) Integumentary system
(b) Skeletal system

(a)

(b)

BOX 1-1
Organ Systems
(continued)

The organ systems of
the body.
(c) Muscular system
(d) Nervous system

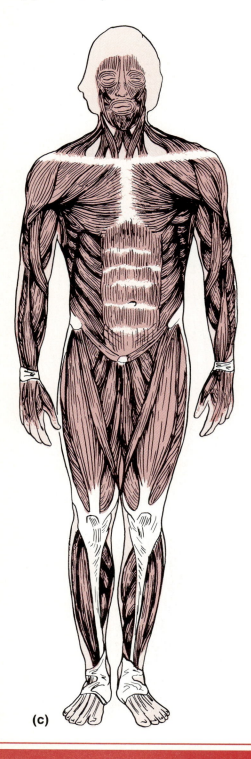

(c)

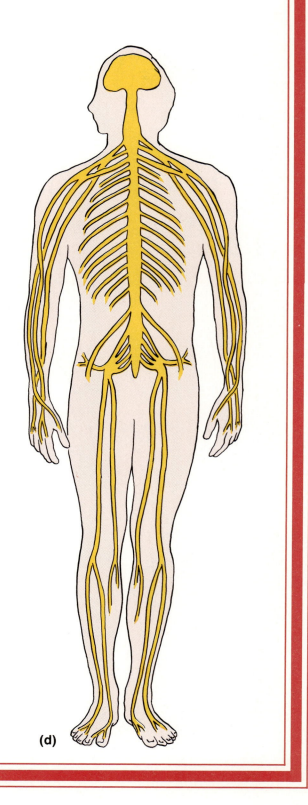

(d)

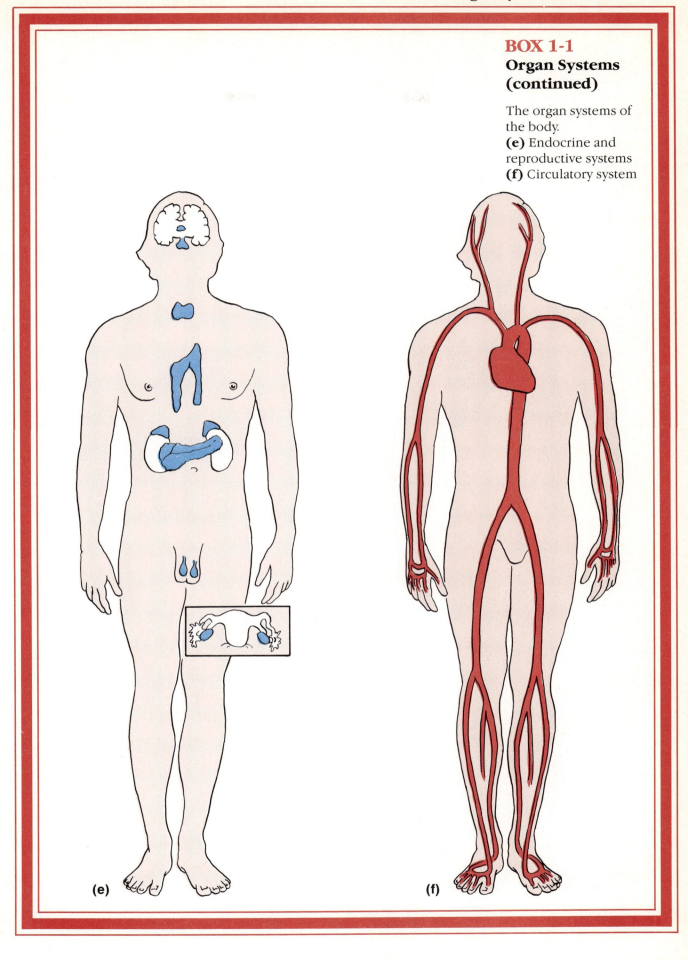

BOX 1-1
Organ Systems (continued)

The organ systems of the body.
(e) Endocrine and reproductive systems
(f) Circulatory system

(e)   (f)

BOX 1-1
**Organ Systems
(continued)**

The organ systems of
the body.
(g) Respiratory and
urinary systems
(h) Digestive system

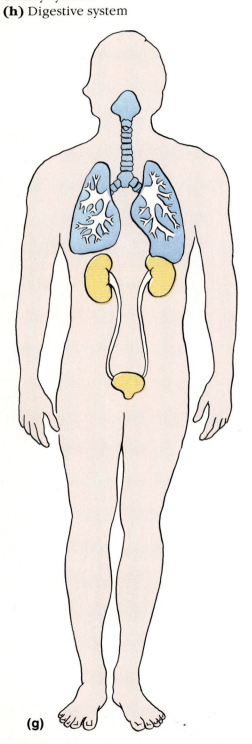

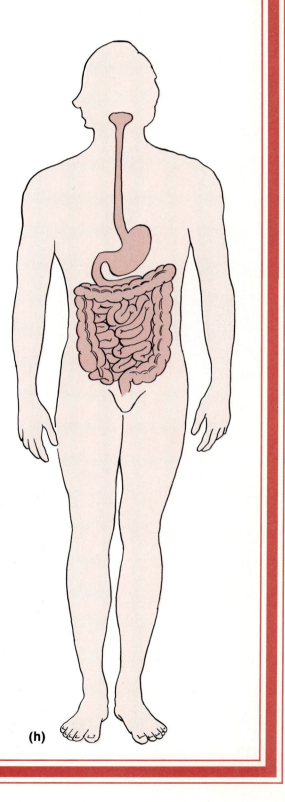

(g)

(h)

The endocrine glands include the pituitary, thyroid, parathyroids, adrenals, thymus, pancreas, pineal, ovaries (in the female), and testes (in the male). The endocrine glands are not connected anatomically in the same way parts of the other organ systems are. What they have in common is that they all secrete hormones, which regulate other structures. The body functions controlled by hormones are many and varied. They involve every cell in the body. Growth, reproduction, sexual maturation, and food use by the cells are all controlled (at least in part) by hormones.

Circulatory System

The **circulatory system** is a transport and delivery system. It consists primarily of the heart, blood vessels, and blood. It acts to carry oxygen, carbon dioxide, nutrients, metal ions, hormones, and other substances to and from the tissue cells where exchanges are made. White blood cells and antibodies in the blood protect the body from such "foreign invaders" as bacteria, toxins, and tumor cells. The heart acts as the "blood pump," propelling blood through the blood vessels to all body tissues.

Respiratory System

The job of the **respiratory system** is to keep the body constantly supplied with oxygen and remove carbon dioxide. The respiratory system consists of the nasal passages, pharynx, larynx, trachea, bronchi, and lungs. Within the lungs are tiny air sacs. It is through the thin walls of these air sacs that gas exchanges are made, to and from the blood.

Digestive System

The **digestive system** is basically a tube running through the body from mouth to anus. The organs of the digestive system include the mouth, esophagus, stomach, small and large intestines, and rectum. Their role is to break down food and deliver the products to the blood for dispersal to the body cells. The undigested food that remains in the tract at the rectum leaves the body through the anus as feces. Both mechanical food breakdown (chewing and churning) and chemical digestion, which uses *enzymes,* occur. The breakdown activities begin in the mouth, continue in the stomach, and are completed in the small intestine. From that point on, the major function of the digestive system is to remove water. The liver is considered to be a digestive organ, because the bile it produces helps in the breakdown of fats. The pancreas, which delivers digestive enzymes to the small intestine, also is functionally a digestive organ.

Urinary System

As it functions, the body produces wastes, which must be disposed. Carbon dioxide is one of these wastes; its disposal is handled by the respiratory system. Another type of waste is nitrogen-containing waste (primarily in the form of urea, uric acid, and ammonia), which results from the breakdown of proteins and nucleic acids by the body cells.

The **urinary system** removes nitrogenous wastes from the blood and flushes them from the body in urine. This system, often called the excretory system, is composed of the kidneys, ureters, bladder, and urethra. Other important functions of this system include maintaining the body's water balance and maintaining the acid-base balance of the blood.

Reproductive System

The **reproductive system** exists primarily to produce offspring. Sperm are produced by the *testes* of the male. Other male reproductive system structures are the scrotum, penis, accessory glands, and the duct system, which carries sperm to the outside of the body. The *ovary* of the female produces the eggs, or ova; the female duct system consists of the fallopian tubes, uterus, and vagina. The uterus provides the site for the development of the fetus (immature infant) once fertilization has occurred.

SUMMARY

A. ANATOMY AND PHYSIOLOGY

1. Anatomy is the study of structure. Observation is used to see size and relationships of body parts.

2. Physiology is the study of how a structure (which may be a cell, an organ, or an organ system) functions or works.

3. Structure determines what functions can occur; therefore, if the structure changes, the function must also change. This partially explains aging.

4. Body functions act to maintain homeostasis, or a relatively stable internal environment, in the body. Homeostasis is necessary for survival and good health.

IMPORTANT TERMS*

Anatomy *(ah-nat'o-me)*

Physiology *(fiz"e-ol'o-je)*

Homeostasis *(ho"me-o-sta'sis)*

Cranial cavity

Spinal cavity

Thoracic *(tho-ras'ik)* **cavity**

Abdominopelvic *(ab-dom"ĭ-no-pel'vik)* **cavity**

Cell

Tissue

Organ system

Organism

Integumentary *(in-teg-u-men'tar-e)* **system**

Skeletal system

Nervous system

Endocrine *(en'do-krin)* **system**

Circulatory system

Respiratory system

Digestive system

Urinary system

Reproductive system

---

*For definitions, see Glossary.

B. THE LANGUAGE OF ANATOMY

1. All anatomical terminology is relative and relates to the body in the anatomical position (erect, palms facing forward).

2. Body orientation
   a. Superior (cephalad): above something else, toward the head.
   b. Inferior (caudad): below something else, toward the tail.
   c. Anterior (ventral): toward the front of the body or structure.
   d. Posterior (dorsal): toward the rear or back of the body.
   e. Medial: toward the midline of the body.
   f. Lateral: away from the midline of the body.
   g. Proximal: closer to the point of limb attachment.
   h. Distal: farther from the point of limb attachment.
   i. Superficial (external): at or close to the body surface.
   j. Deep (internal): below or away from the body surface.

3. Body planes and sections
   a. Sagittal section: separates the body longitudinally into right and left parts.
   b. Frontal (coronal) section: separates the body on a longitudinal plane into anterior and posterior parts.
   c. Tranverse (cross) section: separates the body on a horizontal plane into superior and inferior parts.

4. Surface anatomy. Visible landmarks on the body surface may be used to specifically refer to a body part or area. See p. 6 for terms referring to anterior and posterior surface anatomy.

5. Body cavities
   a. Dorsal: well protected by bone; has two subdivisions.
      (1) Cranial: contains the brain.
      (2) Spinal: contains the spinal cord.
   b. Ventral: less protection than dorsal cavity; has two subdivisions.
      (1) Thoracic. The superior cavity that extends inferiorly to the diaphragm; contains the heart and lungs, which are protected by the rib cage.
      (2) Abdominopelvic. The cavity inferior to the diaphragm that contains digestive,

urinary, and reproductive organs. Abdominal portion has poor protection since it is protected only by the trunk muscles. There is some protection of the pelvic portion by the bony pelvis. The abdominopelvic cavity is often divided into four quadrants or nine regions (see p. 8 for ease of study).

The ventral cavity wall and organs are covered with a serous membrane, which decreases friction.

C. ORGAN SYSTEM OVERVIEW

  1. There are five levels of organization. The unit of life is the cell. Cells are grouped into tissues which in turn are arranged in specific ways to form organs. A number of organs form an organ system, which has a specific function to perform for the body (which no other organ system can do). All of the organ systems together form the organism or living body.

  2. For a description of organ systems naming the major organs and functions, see pp. 10–15.

REVIEW QUESTIONS

  1. Define anatomy, physiology, and homeostasis.

  2. You would have a hard time trying to learn and understand physiology if you did not study anatomy also. Why?

  3. Describe the anatomical position.

  4. On what body surface are the following located? Nose, calf of leg, ears, umbilicus, and fingernails.

  5. Several pairs of structures are given next. Choose the one in each case that meets the condition given.
     a. Distal—the knee or the foot
     b. Lateral—the cheekbone or the nose
     c. Superior—the neck or the chin
     d. Anterior—the heel or the toenails
     e. External—the skin or the skeletal muscles

  6. What kind of a section would have to be made to cut the brain into anterior and posterior parts?

  7. A nurse informed John that she was about to take blood from his *cubital* region. What part of his body should he have held out? Later,

she came back and said that she was going to give him an antibiotic shot in the *deltoid* region. Did he take off his shirt or drop his pants to get the shot? Before John left the office, the nurse noticed that he was badly bruised on his left *flank*. What part of his body was bruised?

  8. List the ten organ systems of the body, briefly describe the function of each, and then name two organs of each system.

  9. Which of the following organ systems (digestive, respiratory, reproductive, circulatory, urinary, or muscular) are found in both subdivisions of the ventral body cavity? Which are found in the thoracic cavity only? Which in the abdominopelvic cavity only?

  10. Make a drawing of the nine abdominopelvic regions and identify each region by name.

  11. Name the five levels of organization, *in order*, beginning with the cell and ending with the organism (the body).

  12. What is the function of the serous membrane lining the ventral body cavity and covering its organs?

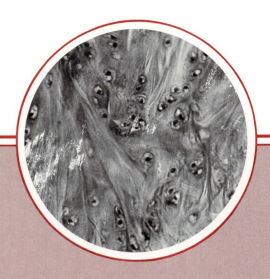

CHAPTER 2

Cells and Tissues

Chapter Contents

Function of cells • to carry out all the chemical activities needed to sustain life

Function of tissues • to provide for a division of labor among body cells

After completing this chapter, you should be able to:

- Name the four elements making up the bulk of living matter and list several trace elements.

- Define cell, organelle, and inclusion.

- Identify on a cell model or diagram the major cell regions (nucleus, cytoplasm) and organelles and discuss briefly the major function of each.

- Describe the makeup of genes and explain their function.

- Define selective permeability, diffusion (including dialysis and osmosis), active transport, passive transport, permease system, phagocytosis, pinocytosis, hypertonic, hypotonic, and isotonic.

- Explain how the various transport processes account for the directional movements

of specific substances across the cell membrane.

- Explain briefly the process of mitosis and the importance of mitotic cell division.

- Name the four major tissue types and the major subcategories of each, and explain how the four major tissue types are different structurally and functionally.

- Give the important locations of the various tissue types in the body.

- Define neoplasm and distinguish between benign and malignant neoplasms.

- Explain the importance of the fact that some tissue types (muscle and nerve) are amitotic after the growth stages are over.

CELLS

In the late 1600s, Robert Hooke was looking through a primitive microscope at plant tissue—cork. He saw some cubelike structures that reminded him of the long rows of monk's rooms (or cells) at the monastery, so he named these structures **cells.** The living cells, which had formed the cork, were long since dead; however, the name stuck and is used to describe the unit, or the building block, of all living things—plants and animals alike.

Perhaps the most striking thing about this microscopic unit of life is its organization. If we chemically analyze cells, we find that they are made up primarily of four elements—carbon, oxygen, hydrogen, and nitrogen, plus trace amounts of several other elements (such as iron, sodium, and potassium). Although the four major elements build most of the cell's structure (which is largely protein), the trace elements are very important for certain cell functions. For example, calcium is needed for blood clotting (among other things). Iron is necessary to make hemoglobin, which carries oxygen in the blood; iodine is needed to

make the thyroid hormone, which controls metabolism. Many of the metals (such as calcium, sodium, and potassium) can carry an electrical charge; when they do they are called **ions** (i'ons), or **electrolytes** (e-lek'tro-līts). Sodium and potassium ions are essential if nerve impulses are to be transmitted and muscles are to contract.

Strange as it may seem, especially when feeling firm muscles, living cells are about 60% water, which is one of the reasons water is essential for life. In addition to containing large amounts of water, all cells of the body are constantly bathed in a dilute saltwater solution (something like sea water) called **tissue fluid.** All exchanges between cells and blood are made through this tissue fluid.

Cells vary tremendously in size. The smallest cells (*bacteria*) are about 1/1000th of an inch (2.5 mm) in diameter. The largest may extend for 4 inches (10 cm) or more across (for example, an ostrich egg), or 3 feet (1 m) or more in length (some nerve cells). Cells also come in many different shapes. For example, some are disk-shaped (red blood cells), some have many threadlike extensions (nerve cells), others are like toothpicks

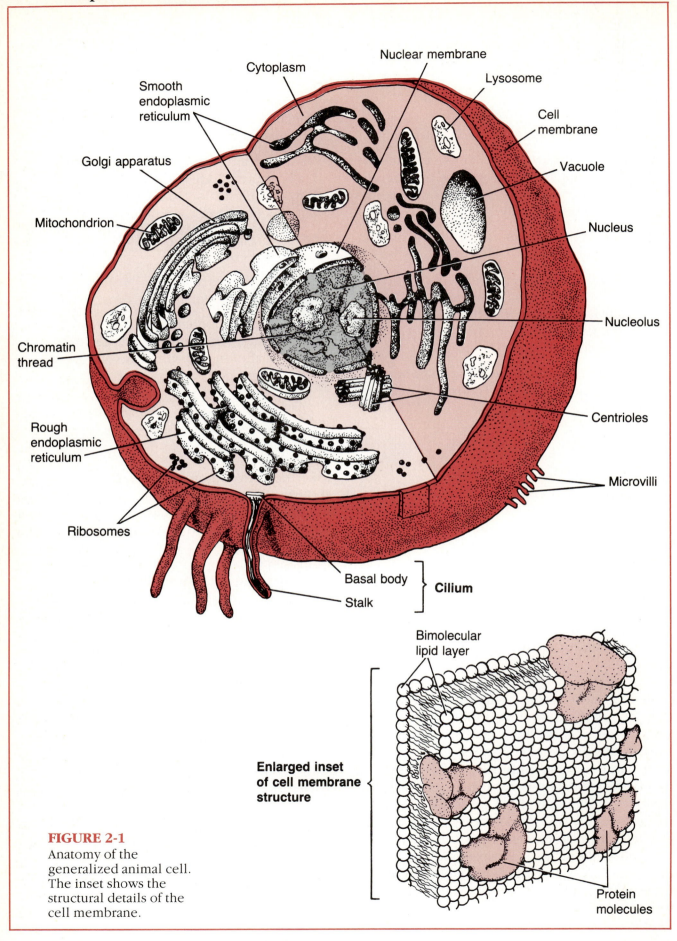

Cytoplasm

Nuclear membrane

Lysosome

Smooth
endoplasmic
reticulum

Cell
membrane

Golgi apparatus

Vacuole

Mitochondrion

Nucleus

Nucleolus

Chromatin
thread

Centrioles

Rough
endoplasmic
reticulum

Microvilli

Ribosomes

Basal body

Cilium

Stalk

Bimolecular
lipid layer

**Enlarged inset
of cell membrane
structure**

Protein
molecules

FIGURE 2-1
Anatomy of the
generalized animal cell.
The inset shows the
structural details of the
cell membrane.

that are pointed at each end (smooth muscle cells), or cubelike (some types of epithelial cells).

Cells also vary dramatically in the functions, or roles, they play in the body. For example, white blood cells wander freely through the body tissues and protect the body by destroying bacteria and other foreign substances. Some cells make hormones, chemicals that regulate other body cells. Still others take part in gas exchanges in the lungs or cleanse the blood (kidney tubule cells). A cell's structure often reflects its function; this will become clear when you read the discussion of tissues later in this chapter.

Anatomy of a Generalized Cell

The cells of the multicellular human body are very different in their sizes, shapes, and functions. Thus no one cell type represents all others. Yet cells do have many common structural features, and there are some functions that *all* must perform to maintain life. Here we will talk about the "generalized" cell, which demonstrates many typical structures and functions.

In general all cells have two main regions or parts—a nucleus (nu'kle-us) and cytoplasm (si'to-plazm") The nucleus is usually a round or oval structure near the center of the cell. It is surrounded by the semifluid cytoplasm, which is in turn enclosed by the cell membrane. Since the invention of the electron microscope, even smaller structures called organelles (or"gan-els') have been identified. Figure 2-1 shows the structure of the generalized cell as revealed by the electron microscope.

NUCLEUS

The **nucleus** is the headquarters, or control center, of the cell. It not only governs the activities occuring in the cytoplasm, but also is absolutely necessary for cell reproduction. A cell that has lost or ejected its nucleus (for whatever reason) is literally "programmed to die." The nucleus has its controlling role because it is the site of genetic material or *deoxyribonucleic* (de-ok"se-ri"bo-nu'-kle-ik) *acid* (DNA). DNA is much like a blueprint that contains all of the instructions needed for building the whole body; so as one might expect,

human DNA is different from that of a frog. More specifically, DNA has the instructions for building **proteins,** and each gene carries the code for building one protein.

Proteins are key substances for all aspects of cell life. Certain types of proteins, called *structural* proteins, are the major building materials for constructing the cells. Other proteins, called *functional* proteins, do things other than build structures. For example, hemoglobin, which carries oxygen in the blood, is a functional protein as are many hormones which regulate growth, development, and reproduction. All **enzymes,** which regulate chemical reactions in the cells, are functional proteins. Enzymes are *catalysts* (kat'ah-lists), which means that they are able to increase the chemical reaction rates without being used up in those reactions. There is not a single chemical reaction that goes on in the body that does not require an enzyme. So it follows that enzymes control or direct the chemical reactions in which carbohydrates, fats, other proteins, and even DNA are made and broken down. Chemical reactions in which molecules are made are called *anabolic* reactions, while those reactions that are basically destructive or break molecules down are called *catabolic* reactions. *Metabolism* is the total of all the anabolic and catabolic reactions that take place in the body. Metabolism is considered in detail in Chapter 13.

DNA is a very complex molecule (Figure 2-2). It is a ladderlike molecule which is coiled into a spiral staircase form called a *helix.* The uprights of the ladder are formed of alternating subunits of phosphate and sugar, and the rungs of the ladder are made of pairs of nitrogen-containing bases. The bases are of four types, and the binding together of the bases to form the rungs is very specific— that is, *adenine* (A) can only bind to *thymine* (T), and *cytosine* (C) can only bind to *guanine* (G). A and T fit well together and can interact; thus, they are said to be *complementary bases.* The same is true of C and G. The information that DNA carries is in the sequence of the bases along each side of the ladderlike molecule. Each sequence of three bases determines the code for a particular *amino acid.* (Amino acids are the building blocks of proteins. They are linked together during protein synthesis.) Just as different arrangements of notes on sheet music are played as different melodies,

FIGURE 2-2
Structure of DNA. The uprights of the ladderlike molecule are formed from alternating sugar and phosphate units. The rungs are formed by nitrogen bases (A, G, C, T) bound together by hydrogen bonds.

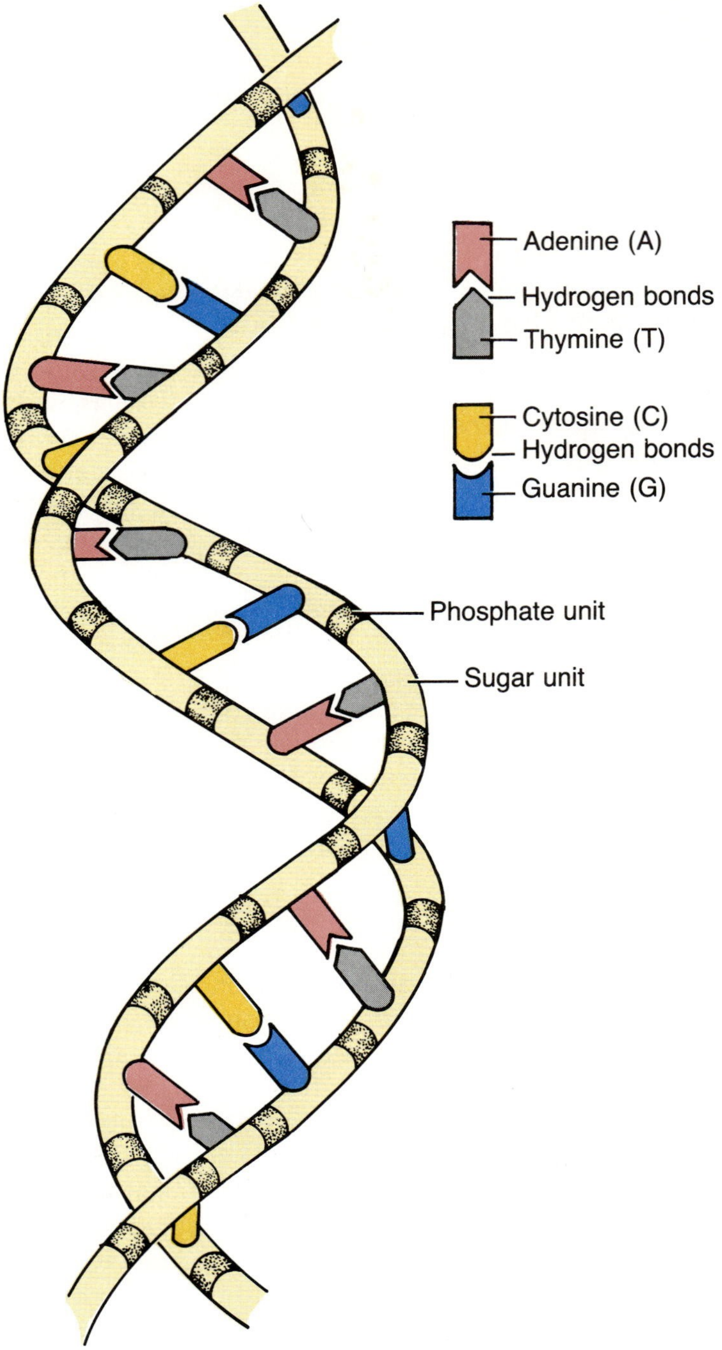

Adenine (A)
Hydrogen bonds
Thymine (T)

Cytosine (C)
Hydrogen bonds
Guanine (G)

Phosphate unit

Sugar unit

variations in the arrangements of A, C, T, and G in each gene allow body cells to make all the different kinds of protein needed. It has been estimated that a single *gene,* which usually carries instructions for building one protein, has between 500 and 2000 base pairs in sequence. The instructions for building the proteins are carried out into the cytoplasm to the sites of protein synthesis by other molecules called *ribonucleic acid* (RNA), which are molecular slaves for the DNA. The actual building of the proteins happens at the **ribosomes** (ri′bo-soms) (the cell's protein "factories") and is overseen by enzymes.

When a cell is getting ready to divide, all of its DNA molecules are replicated, or duplicated, so that for a short time the cell nucleus contains a double dose of genes. When the nucleus does divide, each daughter cell ends up with exactly the same genetic information as the original mother cell. For DNA replication to occur, the DNA molecule uncoils to form the straight ladderlike structure, and enzymes break the bonds between the base pairs. Each of the two single strands now acts as a model for building another whole DNA molecule. Since DNA always replicates *before* a cell divides, all the cells in your body (except sperm or eggs, which are special cases) have exactly the same genetic information as each other and as the fertilized egg from which you began.

When a cell is not dividing, DNA is combined with protein and is loosely scattered throughout the nucleus in a granular or threadlike form called **chromatin** (krow'mah-tin). When a cell is in the process of dividing to form two daughter cells, the chromatin coils and condenses to form dense rodlike bodies called **chromosomes**— much the way a stretched spring becomes shorter and thicker when relaxed. (Cell division is discussed later in this chapter.)

The nucleus also contains one or more small, round bodies called **nucleoli** (nu-kle'o-li). Nucleoli are believed to form ribosomes, which eventually migrate into the cytoplasm and serve as the actual sites of protein synthesis.

The nucleus is bound by a double-layered membrane with fairly large pores, the **nuclear membrane** (or nuclear envelope). Each layer of the nuclear membrane is similar to other cell membranes in its makeup.

Cell Membrane

The **cell membrane,** or plasma membrane, separates the cell contents from the surrounding environment. It is made up of protein and lipid (fat) and appears to have a core of two fat layers in which protein molecules float (see inset, Figure 2-1). The cell membrane is believed to have very tiny pores that allow the passage of only very small molecules. Besides forming a protective barrier for the cell, the cell membrane determines the substances that can enter or leave the cell. In some cells (for example, cells lining the small intestine) the cell membrane is thrown into tiny fingerlike projections, or folds, called **microvilli**

(mi"kro-vil'i). The microvilli greatly increase the surface area available for absorbing or transporting materials so that the process can occur more quickly.

Cytoplasm

The **cytoplasm** consists of the cell material outside the nucleus and inside the cell membrane. It is the major site of most activities that have to be done by the cell, so it can probably be referred to as the "factory area" of the cell. Within the cytoplasm are a large number of "specialists," each with its own job to do to maintain the life of the cell. The proper term for these specialized structures is **organelles.** The organelles, which are the metabolic machinery of the cell, are described next.

Ribosomes. Ribosomes are tiny, round, dark bodies that are the actual sites of protein production in the cell. Some ribosomes float free in the cytoplasm, and others attach to membranes. When ribosomes are attached to membranes, the whole ribosome–membrane combination is called the granular, or rough, endoplasmic reticulum.

Endoplasmic Reticulum. Endoplasmic reticulum (en"do-plas'mik re-tik'u-lum) (ER) is a system of tubules, or canals, that coils and twists through the cytoplasm. It serves as a minicirculatory system for the cell because it provides a network of channels for carrying substances (primarily proteins) from one part of the cell to another. There are two forms of ER; a particular cell may have both or only one, depending on its specific functions.

Granular, or *rough, ER,* as mentioned earlier is studded with ribosomes. It is believed to store proteins made on the ribosomes and deliver them to other areas of the cell. In general the amount of rough ER a cell has is a good clue to the amount of protein that cell makes. Rough ER is especially abundant in cells that produce protein products for export—for example, pancreas cells, which produce digestive enzymes that will be delivered to the small intestine.

Agranular, or *smooth, ER* is present in large amounts in cells that produce steroid-based hormones—for example, the interstitial cells of the testes of the male, which produce testosterone. Smooth ER is also thought to have something to do with fat metabolism and is found in large amounts in liver cells, which handle a large part of the body's fat metabolism.

GOLGI APPARATUS.   The **Golgi** (gol′je) **apparatus,** a stack of flattened sacs with swollen ends, is generally found close to the nucleus. The Golgi apparatus modifies and packages proteins for export (proteins are delivered to it by the rough ER). As proteins accumulate in the Golgi apparatus, the sacs swell; the sac ends, filled with protein, pinch off and form *vesicles* (ves′i-k′ls), which then travel to the cell membrane. When the vesicles reach the cell membrane, they fuse with it; the membrane then ruptures, and the contents of the sac are ejected to the outside of the cell. Mucus is packaged this way, as are digestive enzymes made by the pancreas cells.

LYSOSOMES.   Lysosomes (li′so-sōms), which appear in different sizes, are membrane sacs containing powerful digestive enzymes. The enzymes are capable of digesting worn-out cell structures and foreign substances that enter the cell. Since the lysosomes have the ability to totally destroy the cell, they are often called the "suicide sacs" of the cell.

MITOCHONDRIA.   Mitochondria   (mi″-to-kon′ dre-ah) are sausage-shaped bodies with a double-membrane wall. Enzymes on or inside the mitochondria carry out the reactions of cellular *oxidation* in which oxygen is used to break down foods. As the foods are broken down, energy is released. Much of this energy escapes as heat, but some is captured and used to form *adenosinetriphosphate* (ah-den″o-sin-tri-fos′fāt) (ATP) molecules. ATP is used to provide the energy for all cell work that has to be done. Every living cell requires a constant supply of ATP for its many activities. As the mitochondria provide most of this ATP, they are referred to as the "powerhouses" of the cell.

CENTRIOLES.   The paired **centrioles** (sen′tri-ols) lie close to the nucleus. They are rod-shaped bodies, which lie at right angles to each other; internally they are made up of fine *tubules*. During cell division, the centrioles form the *mitotic spindle* (see p. 29–30).

In addition to these cell structures, some cells have projections called cilia or flagella. **Cilia** (sil′e-ah) act to move substances past the cell surface—for example, the ciliated cells of the respiratory-system lining move mucus up and away from the lungs. Where they appear, there are usually many cilia projecting from the cell surface. **Flagella** (flah-jel′lah) act to propel a cell. The tail of a sperm is a flagellum.

The cytoplasm contains various other substances and structures, including stored foods, pigment granules, crystals of various types, and water *vacuoles* (vak′u-ōls). But these are not part of the active metabolic machinery of the cell, and they are therefore called **inclusions.**

Cell Physiology

As mentioned earlier, each of the cell's internal parts is organized to perform a specific function for the cell. Most cells have the ability to *metabolize* (to use nutrients to build new cell material, break down substances, make ATP, and dispose of wastes), *grow* and *reproduce,* and *respond* to a stimulus (irritability). Most of these functions are considered in detail in later chapters. For example, metabolism is covered in Chapter 12, and the ability to react to a stimulus is covered in the chapter on the nervous system (Chapter 6). Here we will only consider the functions of cell reproduction (cell division) and membrane transport (the means by which substances get through cell membranes).

TRANSPORT OF SUBSTANCES

The cell membrane is selective about what passes through it. It allows nutrients to enter the cell but keeps many undesirable substances out. At the same time, valuable cell proteins and other substances are kept within the cell, and wastes are allowed to pass to the exterior. This property is known as **selective permeability** and is typical only of healthy unharmed cells. When a cell dies or is badly damaged, its membrane can no longer be selective and becomes permeable to nearly everything. This is why burn patients become dehydrated and lose precious ions and proteins from their bodies. The burned cells are no longer able to retain their internal contents so fluid keeps oozing from the burned surfaces.

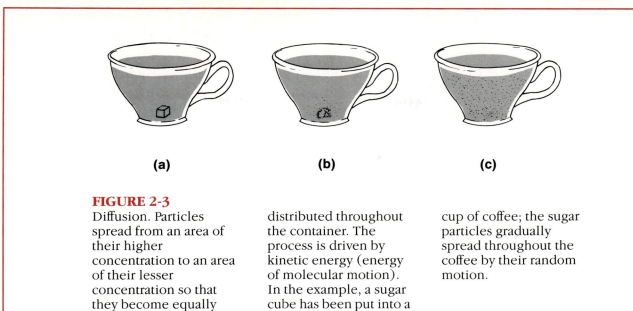

(a)   **(b)**   **(c)**

FIGURE 2-3
Diffusion. Particles spread from an area of their higher concentration to an area of their lesser concentration so that they become equally distributed throughout the container. The process is driven by kinetic energy (energy of molecular motion). In the example, a sugar cube has been put into a cup of coffee; the sugar particles gradually spread throughout the coffee by their random motion.

Movement of substances through the cell membrane happens basically in two ways—passive transport and active transport. In **passive transport** processes, substances are moved across the membrane without any help from the cell. In **active transport,** the cell itself becomes involved and provides the energy (ATP) for the transport process.

PASSIVE TRANSPORT—DIFFUSION. All molecules possess *kinetic energy* (energy of motion), which causes them to move randomly at very high speeds. In general, the smaller the particle, the faster it moves. Also, the higher the temperature, the faster the particles move. **Diffusion** is the process by which randomly moving particles scatter themselves evenly throughout the available space. In diffusion, the particles move along a *concentration gradient,* that is, from a region of their higher concentration (where they are more numerous) to a region of their lower concentration (where there are fewer of them). The driving force is the kinetic energy of the molecules themselves. Two examples should help you to understand diffusion. Picture yourself pouring a cup of hot coffee, and then adding (but not stirring) 2 teaspoons of sugar. After you have just added the sugar, the phone rings, and you are called in to work; you never drank any of the coffee. Upon returning that evening, you find that the coffee tastes sweet even though it was never stirred. This is because the sugar molecules had been moving around all day and eventually, as a result of their activity, had become equally distributed throughout the coffee (Figure 2-3). Or try this example: You walk into a room and after a few minutes, your co-worker says "I like your cologne." Obviously she (or he) did not smell the cologne until it began to diffuse out into the air of the room and away from you (the area of highest concentration).

The cell membrane is a physical barrier to diffusion, and yet needed substances (and wastes) must be able to pass through it if cell activity and homeostasis are to be maintained. Molecules will diffuse *passively* through the cell membrane if they are small enough to pass through its pores (as does water) or if they can dissolve in the fat portion of the membrane (as in the case of other fats, oxygen, and carbon dioxide) (Figure 2-4). The diffusion of *solutes* (particles dissolved in water) through a selectively permeable membrane is called **dialysis** (di-al'ĭ-sis); the diffusion of *water* through a selectively permeable membrane is called **osmosis** (oz-mo'sis). Both dialysis and osmosis are examples of diffusion (a passive transport process); substances that pass into and out of cells by diffusion save the cell a great deal of energy. When considering that two of these substances, water and oxygen, are vitally important to cells, it becomes apparent just how necessary

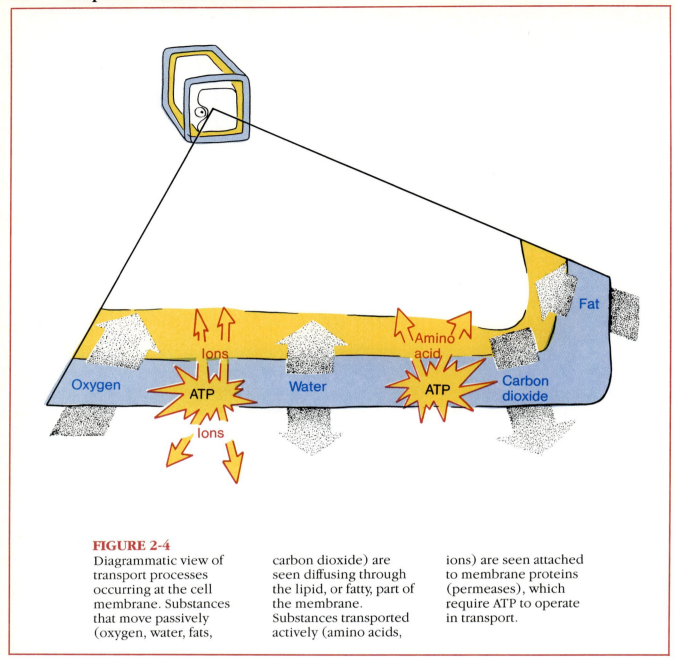

FIGURE 2-4

Diagrammatic view of transport processes occurring at the cell membrane. Substances that move passively (oxygen, water, fats, carbon dioxide) are seen diffusing through the lipid, or fatty, part of the membrane. Substances transported actively (amino acids, ions) are seen attached to membrane proteins (permeases), which require ATP to operate in transport.

these passive transport processes really are. Oxygen continually moves into the cells (where it is in lower concentration because the cells keep using it up), and carbon dioxide (a waste product) continually moves out of the cells into the blood where it is in lower concentration.

Osmosis into and out of cells is occurring all the time as water moves according to its concentration gradient. Because osmosis occurs very quickly and a cell can easily become dehydrated or rupture due to excessive water entry, it is very important that medical personnel give only the proper intravenous, or into-the-vein, solutions to patients. If red blood cells are exposed to a **hypertonic** (hi″per-ton′ik) **solution** (a solution that contains more solutes, or dissolved substances, than there are inside the cells), the cells will begin to shrink, or become **crenated** (kre′nat-ed). This is because water is in higher concentration inside the cell than it is outside the cell, so it follows its concentration gradient and leaves the cell (Figure 2-5*a*). Hypertonic solutions are sometimes given to patients who have *edema* (swollen feet and hands because of fluid retention). Such solutions draw water out of the tissue spaces into the blood stream so that the excess fluid can be eliminated by the kidneys.

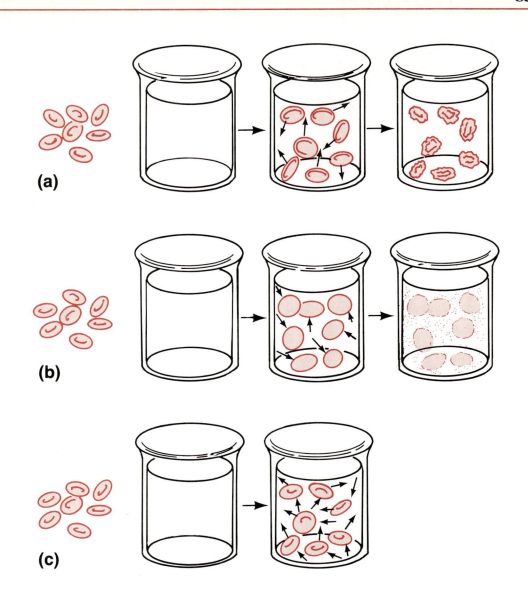

FIGURE 2-5
The effect of solutions of varying concentrations on living cells (red blood cells [RBCs] are used in this example). **(a)** When suspended in a hypertonic solution (containing more solutes than are present inside the RBCs), the cells lose water and become crinkled or crenated. **(b)** When suspended in a hypotonic solution (containing fewer solutes than are present inside the cell), the cells take in water by osmosis until they become very bloated or burst (hemoloysis). **(c)** In isotonic solutions (same solute/water concentrations as inside cells) no change is noted in the cells, which retain their normal size and shape.

When a solution contains fewer solutes (and therefore more water) than are found inside the cell, it is said to be **hypotonic** (hi-po-ton′ik) to the cell. In hypotonic solutions, red blood cells first "plump up" and then suddenly start to disap-pear. This phenomenon is **hemolysis** (he-mol′i-sis); the red blood cells are bursting as the water floods into them (Figure 2-5b). Distilled water is an example of a hypotonic substance. Various types of hypotonic solutions are sometimes administered to patients who are extremely dehydrated. **Isotonic** (i″so-ton′ik) **solutions** have the same solute and water concentrations as the cells do. Isotonic solutions cause no visible changes in the cells, and the cells retain their normal shape (Figure 2-5c). As you might guess, most intravenous solutions are isotonic solutions.

PASSIVE TRANSPORT—FILTRATION.   Filtration is the process by which water and solutes are forced through a membrane by fluid, or *hydrostatic pressure*. In the body, hydrostatic pressure is usually exerted by the blood. Like diffusion, filtration is a passive process, and a gradient is involved. In filtration, the gradient is a *pressure gradient,* and substances move from the higher pressure area to the lower pressure area. Filtration is necessary for the kidneys to do their job properly. In the kidneys, water and small solutes filter out of the capillaries into the kidney tubules because the blood pressure in the capillaries is greater than the fluid pressure in the tubules. Proteins and blood cells are too large to pass through the pores, and so they remain in the capillaries. Part of the *filtrate* (water and solutes) formed in this way eventually becomes urine. Filtration is not a very selective process, and only molecules too large to pass through the membrane pores are held back. The amount of filtrate formed almost entirely depends on the difference in pressure on the two sides of the membrane and the size of the membrane pores.

**ACTIVE TRANSPORT.   **Whenever a cell uses its energy from ATP to move substances across the membrane, the process is referred to as being active. Substances moved by active means are usually unable to pass in the desired direction by diffusion. They may be too large to pass through the pores, they may not be able to dissolve in the fat core, or they may have to move *against* rather than with a concentration gradient.

The **permease** (per′me-ās) **system** accounts for one type of active transport. Many important substances move across the membrane by combining with a protein carrier called a *permease,* which is located in the membrane. ATP is used and in many cases the substances move *against* a concentration gradient (see Figure 2-4). Amino acids, some sugars, and most ions are transported by permeases. Amino acids and sugars are too large

to pass through the pores and are not lipid soluble. Sodium ions (Na^+) are moved out of cells by permeases. There are more sodium ions outside the cells than there are inside, so they tend to remain in the cell unless the cell uses ATP to force or pump them out. This sodium pump is absolutely necessary for normal transmission of impulses by nerve cells. Each of the permeases in the cell membrane transports only specific substances; thus, this provides a way for the cell to be *very* selective (that is, no permease—no transport) in cases where substances cannot pass by diffusion.

Phagocytosis (fag″o-si-to′sis) and **pinocytosis** (pi″no-si-to′sis) also require ATP. In phagocytosis (cell-eating) parts of the cell membrane flow around some relatively large or solid material (bacteria, cell debris, and the like) and engulf it, enclosing it within a sac (Figure 2-6a). The sac or pouch, which is formed, then detaches from the cell membrane and forms a vacuole inside the cytoplasm. In many cases, this vacuole is combined with a lysosome so that the material in the phagocytic vacuole can be digested or broken down. Phagocytosis is a way for the cell to bring in fairly large objects that could not get in any other way.

If we say that a cell can eat, we can also say that they drink, and pinocytosis is cell-drinking. In pinocytosis the cell membrane does not engulf the material, but instead sinks beneath the material to form a small sac that pinches off into the cell interior (Figure 2-6b). Pinocytosis is commonly used to take in liquids that contain proteins or fats. It is seen most often in cells that have microvilli (for example, in kidney tubule cells and mucosa cells of the small intestine), which increase the surface area for absorption.

CELL DIVISION

Cell division, in all cells other than bacteria and some cells of the reproductive system, consists of two events. **Mitosis** (mi-to′sis), division of the nucleus, occurs first. The second event is division of the cytoplasm, which begins after mitosis is nearly completed. Although mitosis and division of the cytoplasm usually go hand-in-hand, in some cases the cytoplasm is not divided. This leads to the formation of *binucleate* (two nuclei) or *multinucleate* cells. This is fairly common in the liver.

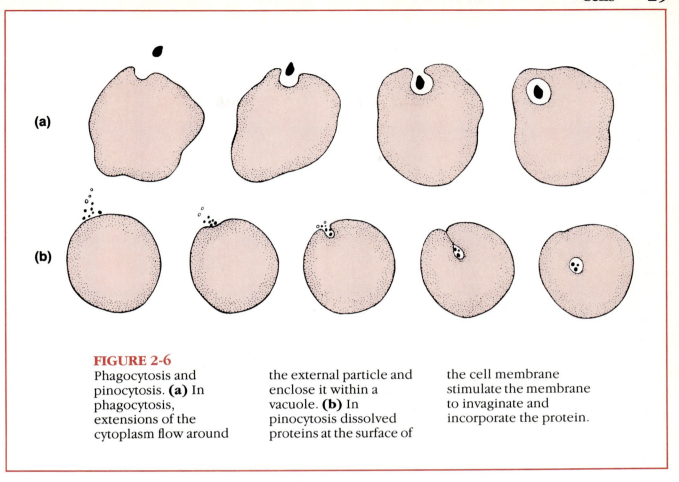

(a)

(b)

FIGURE 2-6
Phagocytosis and pinocytosis. **(a)** In phagocytosis, extensions of the cytoplasm flow around the external particle and enclose it within a vacuole. **(b)** In pinocytosis dissolved proteins at the surface of the cell membrane stimulate the membrane to invaginate and incorporate the protein.

Mitosis results in the formation of two daughter nuclei, which have exactly the same genes as the mother nucleus. The function of cell division is to increase the number of cells in the body for growth and repair processes and to make sure that all daughter cells have the same genetic material. An important event *always precedes* cell division. The genetic material (the DNA molecules that form part of the chromatin) is duplicated exactly. This occurs during the part of the cell's life called *interphase* (in'ter-fāz). Although some people refer to interphase as the cell's resting period, it is a misleading description; the cell is very active in its normal ongoing activities and is resting *only* from cell division.

The stages of mitosis, which are diagrammed in Figure 2-7, include the following events:

- **Prophase** (pro'fāz) As cell division begins, the chromatin threads begin to coil and shorten so that barlike bodies called *chromosomes* appear. Since DNA replication has already occurred, each chromosome is actually made up of two strands (each called a **chromatid** [kro'mah-tid]) held together by a small buttonlike body called a *centromere* (sen'tro-mēr). The centrioles separate from each other and begin to move toward opposite sides of the cell, spinning out the **mitotic spindle** (composed of thin tubules) between them as they move. The spindle provides a scaffolding for the attachment and movement of the chromosomes during the later mitotic stages. By the end of prophase, the nuclear membrane and the nucleoli have broken down and disappeared, and the chromosomes have become attached randomly to the spindle fibers by their centromeres.

- **Metaphase** (met'ah-fāz) In this short stage, the chromosomes become aligned at the center of the spindle midway between the centrioles so that a straight line of chromosomes is seen.

- **Anaphase** (an'ah-fāz) During anaphase, the centromeres holding the chromatids together break, and the chromatids (now called chromosomes again) begin to move slowly apart toward opposite ends of the cell. The chromosomes

FIGURE 2-7
Stages of mitosis.

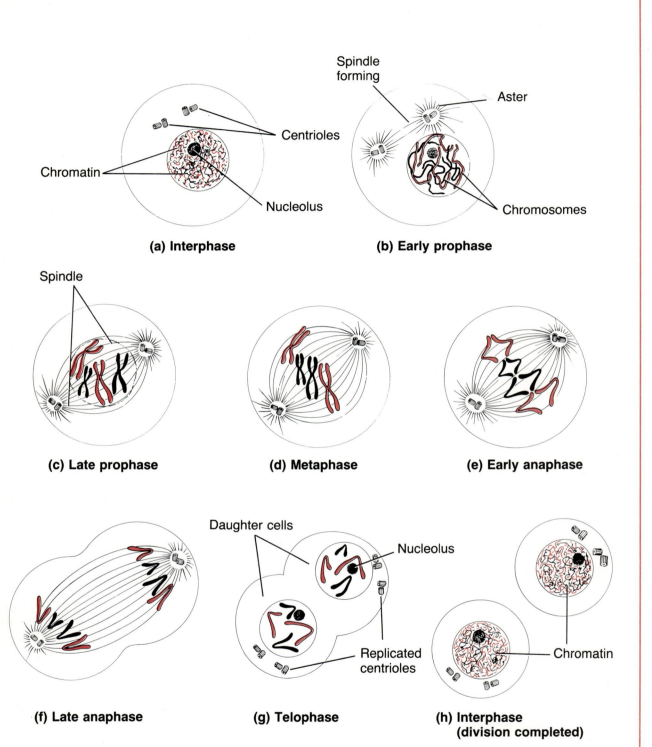

(a) Interphase

(b) Early prophase

(c) Late prophase

(d) Metaphase

(e) Early anaphase

(f) Late anaphase

(g) Telophase

(h) Interphase (division completed)

seem to be pulled by their half-centromeres with their "arms" dangling behind them. Anaphase is over when chromosome movement ends.

- **Telophase** (tel'o-fāz) Telophase is essentially prophase in reverse. The chromosomes at each end of the cell uncoil to become threadlike chromatin again. The spindle breaks down and disappears, a nuclear membrane forms around each chromatin mass, and nucleoli appear in each of the daughter nuclei.

Mitosis is basically the same in all animal cells. Depending on the type of tissue, it takes from 5 minutes to several hours to complete. Centriole replication is deferred until late interphase of the next cell cycle when DNA replication begins prior to the onset of mitosis.

The division of the cytoplasm begins and ends during telophase. A furrow begins to form over the midline of the spindle, and it eventually splits or pinches the original cytoplasmic mass into two parts. Thus, at the end of cell division two daughter cells exist. Each daughter cell is smaller and has less cytoplasm than the mother cell, but it is genetically identical to it. The daughter cells grow and carry out normal cell activities until it is their turn to divide. As mentioned earlier, mitosis is the basis of body growth in youth and is necessary for repair of body tissue all through life. Also, mitosis "gone wild" is the basis of tumors or cancers.

BODY TISSUES

The human body, complex as it is, starts out as a single cell—the fertilized egg—which divides almost endlessly. The thousands of cells that result become specialized for particular functions. Some become muscle cells, others the transparent lens of the eye, still others skin cells, and so on. Thus, there is a division of labor in the body with certain groups of highly specialized cells to perform functions that benefit the organism as a whole and contribute to homeostasis.

Cell specialization carries with it certain dangers or hazards. When a small group of cells is indispensable, its loss of function can paralyze or destroy the entire body. For example, the action of the heart depends on a very specialized group of cells in the heart muscle that control its contractions. If they are damaged or stop functioning, the heart will no longer work efficiently, and the whole body will suffer or die from lack of oxygen.

Groups of cells that have similarities in structure and function are called **tissues.** The four primary tissue types—epithelium, connective tissue, nervous tissue, and muscle—have recognizable structures, patterns, and functions. As explained in Chapter 1, tissues are organized into organs such as the heart, kidneys, and lungs. Most organs contain several tissue types, and the arrangement of the tissues determines the organ's structure and what it is able to do. Thus, a study of tissues should be helpful to your later studies of the body's organs and how they work.

Here we are concerned with becoming familiar with the major similarities and differences in the primary tissues. Because epithelium and some types of connective tissue are not to be considered again, they are emphasized more in this section than muscle, nervous tissues, and bone (a connective tissue) which are covered in more depth in later chapters.

Epithelial Tissue

Epithelial tissue, or **epithelium** (ep"ĭ-the'le-um), which covers all body surfaces, contains versatile cells. One type of epithelium forms the outer part of the skin and protects the body's interior from what is outside. Epithelium also lines body cavities—it is found lining the mouth and the whole length of the digestive tract, passages of the respiratory system, and blood vessels. Epithelium also forms various glands in the body. Its functions include *protection, absorption, filtration,* and *secretion.* For example, the epithelium covering the body protects against bacterial and chemical damage, and the epithelium lining the respiratory tract has cilia, which sweep dust and other debris away from the lungs. Epithelium specialized to absorb substances lines the stomach and small intestine where food is absorbed into the body. In the kidneys, epithelium both absorbs and filters. Secretion is a specialty of the glands, which produce such substances as perspiration, oil, digestive enzymes, and mucus (which pre-

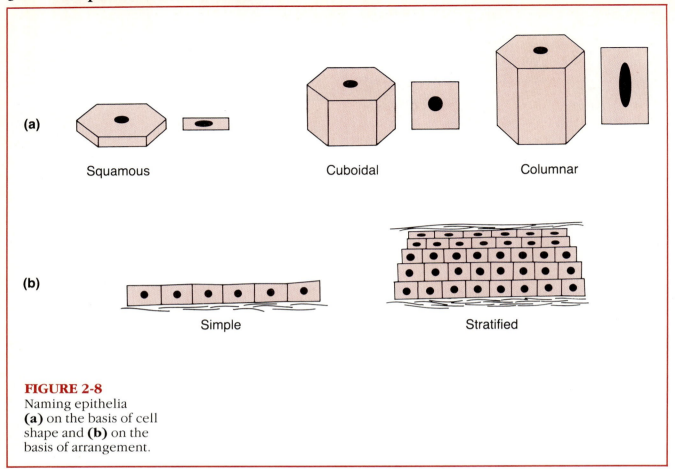

(a) Squamous   Cuboidal   Columnar

(b) Simple   Stratified

FIGURE 2-8
Naming epithelia
(a) on the basis of cell
shape and **(b)** on the
basis of arrangement.

vents the body from drying out and acts as a lubricant).

Epithelium generally has the following characteristics:

- Cells fit closely together to form membranes or sheets of cells.

- The membranes always have one free (unattached) surface or edge.

- The cells are attached to a **basement membrane,** a structureless material secreted by the cells.

- Epithelial tissues have no blood supply of their own (that is, they are *avascular*) and depend on diffusion from the capillaries in the underlying connective tissue for a supply of food and oxygen.

- If well nourished, epithelial cells regenerate themselves easily.

The covering and lining epithelia are classified according to cell shape and arrangement (Figure 2-8). Squamous (skway'mus) cells are flattened like fish scales, cuboidal (cue-boi'dal) cells are cube-shaped like dice, and columnar cells are shaped like columns. On the basis of arrangement, there is simple epithelium (one layer of cells) and stratified epithelium (more than one layer of cells). The terms describing the shape and arrangement are then combined to describe the epithelium fully. Stratified epithelia are named for the cells at the *surface* of the epithelial membrane, not those resting on the basement membrane.

SIMPLE SQUAMOUS EPITHELIUM

Simple squamous epithelium is a single layer of thin squamous cells resting on a basement membrane. The cells fit closely together much like tiles on a kitchen floor. This type of epithelium usually forms membranes where exchanges occur. It is in the air sacs of the lungs where oxygen and carbon dioxide are exchanged, and it forms the walls of capillaries where exchanges of foods, gases, and the like are made between the tissue cells and the blood in the capillaries. Simple squamous epithelium also forms the **serous**

membranes, or **serosae** (se-ro′sā), which line body cavities *closed* to the exterior and cover the organs in those cavities. The serous membranes produce a thin lubricating fluid that decreases friction as the organs work inside the body. These membranes were introduced in Chapter 1 as the *pleura, peritoneum,* and *pericardium*—names that indicate the organs with which the membranes are associated.

SIMPLE CUBOIDAL EPITHELIUM

Simple cuboidal epithelium, which is one layer of cuboidal cells resting on a basement membrane, is common in glands and their ducts (for example, salivary glands and pancreas). It also forms the walls of the kidney tubules.

SIMPLE COLUMNAR EPITHELIUM

Simple columnar epithelium is made up of a single layer of tall cells that fit closely together. **Goblet cells,** which produce mucus, are often seen in this type of epithelium. Simple columnar epithelium lines the entire length of the digestive tract from the stomach to the anus. Epithelial membranes that line body cavities, which are open to the body exterior, are called **mucosae** (mu-ko′sā) or **mucous membranes.**

STRATIFIED SQUAMOUS EPITHELIUM

Stratified squamous epithelium is the only common stratified epithelium in the body. It usually consists of several layers of cells; the cells at the free edge are squamous cells whereas those close to the basement membrane are cuboidal or columnar. Stratified squamous epithelium is found in sites, such as the esophagus or the outer portion of the skin, that receive a good deal of abuse or friction.

STRATIFIED CUBOIDAL EPITHELIUM AND STRATIFIED COLUMNAR EPITHELIUM

Stratified cuboidal epithelium is found in sweat glands while **stratified columnar epithelium** makes up the mucosa of the male urethra.

Figure 2-9 diagrams the most common epithelia and shows their typical locations in the body.

Epithelial cells, which form glands, are specialized to remove materials from the blood and use them to make new materials, which they then se-crete. There are two major types of glands which develop from epithelial sheets. The **endocrine glands** lose their connection to the surface (duct), thus they are often called *ductless* glands. Their secretions (all hormones) diffuse directly into the blood vessels that weave through the glands. Examples of endocrine glands include the thyroid, adrenals, and pituitary. The **exocrine glands** retain their ducts, and their secretions empty through the ducts to the epithelial surface. The exocrine glands, which include the sweat and oil glands, liver, and pancreas, are both internal and external. They are discussed with the organ systems to which their products are related.

Connective Tissue

Connective tissue, as its name suggests, connects body parts. It is found everywhere in the body; it is the most abundant and widely distributed of the tissue types.

The characteristics of connective tissue include the following:

- Most connective tissues are well *vascularized* (that is, have a good blood supply), but there are exceptions. Cartilages, tendons, and ligaments, all of which have a poor blood supply, heal very slowly when injured. (This is why many people say they would rather have a broken bone than a torn ligament.)

- Connective tissues are made up of many different types of cells, plus a substance found outside the cells called the **matrix.**

The nonliving matrix deserves a bit more explanation primarily because it is what makes connective tissue so different from the other tissue types. The matrix is produced within the connective tissue cells and then secreted to their exterior. Depending on the connective tissue type, the matrix may be liquid, semisolid or gel-like, or very hard. The matrix is responsible for the strength associated with structures consisting of connective tissue, but there is variation. At one extreme, fat tissue is composed mostly of cells, and the matrix is soft. At the opposite extreme, bone and cartilage have very few cells and large amounts of hard matrix, which makes them extremely strong. Various types and amounts of fibers are deposited in and form a part of the matrix material. They include

FIGURE 2-9
Types of epithelia
showing their common
locations in the body.

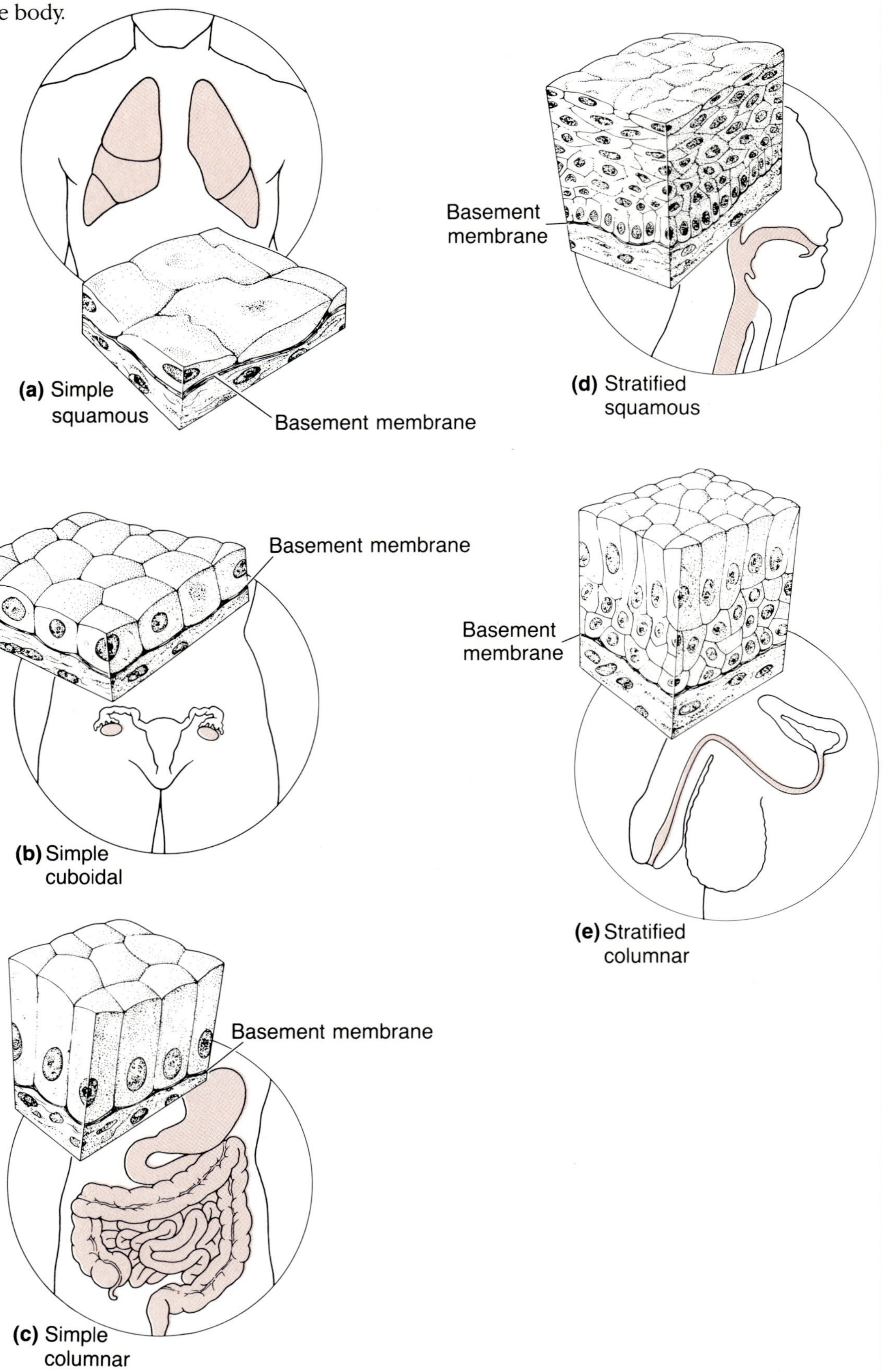

(a) Simple
squamous

Basement membrane

Basement
membrane

(d) Stratified
squamous

Basement membrane

(b) Simple
cuboidal

Basement
membrane

(e) Stratified
columnar

Basement membrane

(c) Simple
columnar

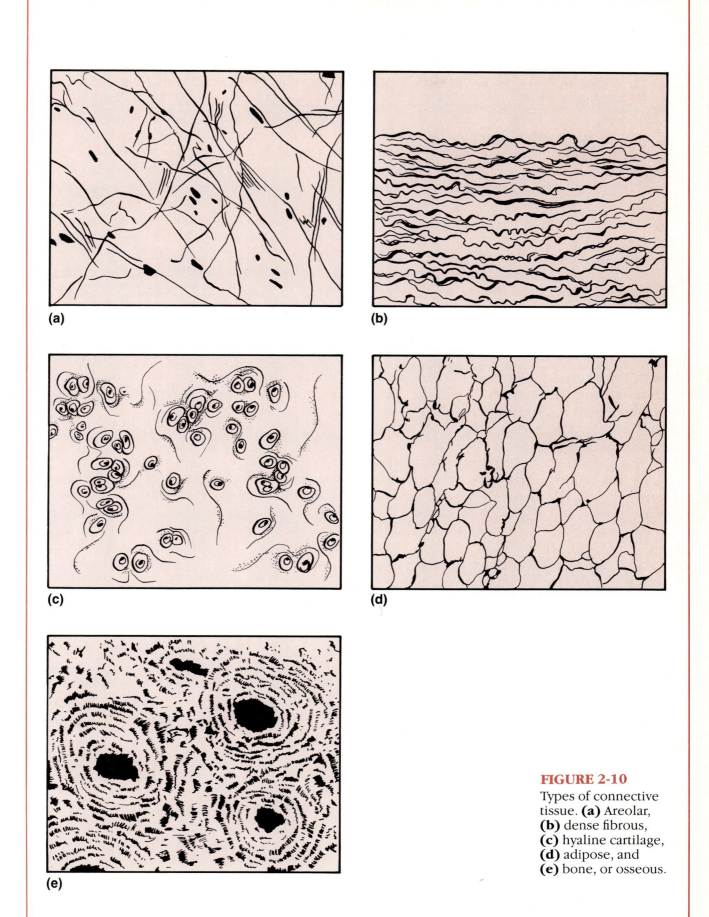

FIGURE 2-10
Types of connective tissue. **(a)** Areolar, **(b)** dense fibrous, **(c)** hyaline cartilage, **(d)** adipose, and **(e)** bone, or osseous.

collagen (white) fibers, *elastic* (yellow) fibers, and *reticular* (fine collagen) fibers. Like the matrix, the fibers are made by connective tissue cells and then secreted to lie in the matrix outside the cells.

Connective tissues perform many functions, but they are primarily involved in *protecting, supporting,* and *binding together* other tissues of the body. Find the various types of connective tissues in Figure 2-10 as you read through their descriptions.

Osseous Tissue

Osseous tissue, commonly called *bone,* is composed of bone cells surrounded by layers of a very hard matrix that contains calcium salts. The bones protect and support other body organs (for example, the skull protects the brain).

Dense Fibrous Tissue

Dense fibrous tissue forms strong cordlike structures such as tendons and ligaments. **Tendons,** which attach skeletal muscles to bones, are made up mostly of the tough collagen fibers. **Ligaments,** which connect bones to bones (at joints), are more stretchy and contain more elastic fibers than tendons. Another type of dense fibrous tissue makes up the lower layers of the skin (dermis).

Hyaline Cartilage

Hyaline (hi′ah-līn) **cartilage** is rubbery and smooth. Its matrix is somewhat hard but not nearly as hard as bone matrix. It is found only in a few places in the body. It forms the supporting structure of the larynx, or voicebox; attaches the ribs to the breastbone; and covers the ends of bones where they form joints. The skeleton of a fetus is made of hyaline cartilage, but most of the cartilage has been replaced by bone by the time the baby is born.

Areolar Tissue

Areolar (ah-re′o-lar) **tissue** is a soft, packaging tissue that cushions and protects the body organs it wraps. It functions as a connective tissue "glue" since it helps to hold the internal organs together and in their proper positions. Its matrix is semifluid and contains all types of fibers that form a loose network. Many types of *phagocytes* wander through this tissue scavenging for bacteria, dead cells, and other "debris," which they destroy.

Adipose Tissue

Adipose (ad′i-pōs) **tissue** is commonly called fat tissue. Forming the subcutaneous tissue beneath the skin, it acts as an insulation for the body and protects it from both extreme heat and cold. Adipose tissue also protects some organs individually—for example, the kidneys are surrounded by a capsule of fat, and adipose tissue cushions the eyeballs in their sockets. There are also fat "depots" in the body, such as the hips and breasts, where fat is stored and available as an energy source if needed. Adipose cells are often called "signet ring cells" because of the way they look. These cells contain a large central mass of stored fat, which looks clear. Because their nuclei are pushed to one side by the fat, the cells look like the seal of a signet ring.

Hemopoietic Tissue

Hemopoietic (he″mo-poi-et′ik) **tissue** is blood-forming tissue. It is found within the bone cavities where it continually replenishes the body's supply of red blood cells.

Muscle Tissue

Muscle tissues are highly specialized to *contract* or *shorten* to produce movement. Although the cells of the different types of muscle tissue do differ, they all are elongated to provide a long axis for contraction. The three types of muscle are illustrated in Figure 2-11. Notice their similarities and differences.

Skeletal Muscle

Skeletal muscle, the flesh of the body, is attached to the skeleton. It is controlled *voluntarily* (or consciously), and when it contracts it generally moves the limbs and other external body parts. The cells of skeletal muscle are long, cylindrical, and multinucleate; they have obvious *striations* (stripes).

Cardiac Muscle

Cardiac muscle is found only in the heart. As it contracts, the heart acts as a pump and propels blood through the blood vessels. Like skeletal muscle, cardiac muscle has striations, but cardiac cells are branching cells that fit tightly together (like clasped fingers) at junctions called **intercalated disks.** These structures allow the cardiac muscle to act as a unit. Cardiac muscle is under

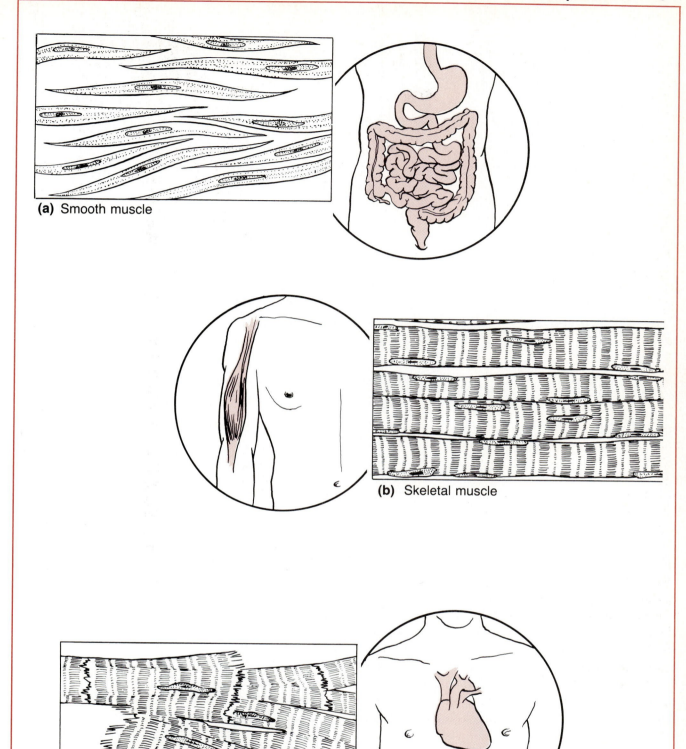

(a) Smooth muscle

(b) Skeletal muscle

(c) Cardiac muscle

FIGURE 2-11

Types of muscle tissue showing their common locations in the body.

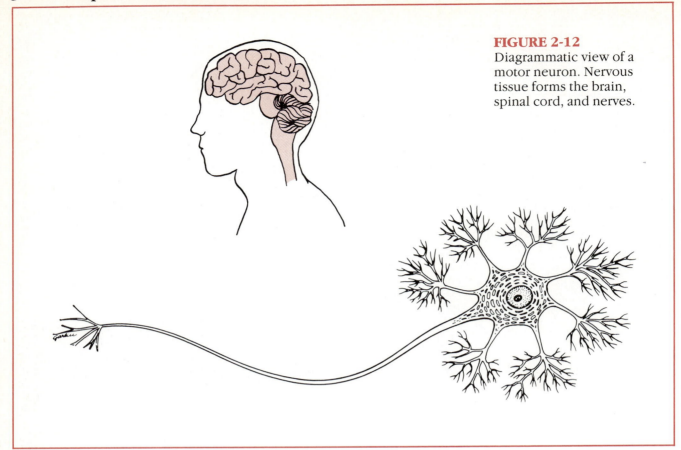

FIGURE 2-12
Diagrammatic view of a motor neuron. Nervous tissue forms the brain, spinal cord, and nerves.

involuntary control, which means that we cannot consciously control the activity of the heart. (There are, however, rare individuals who claim they have such an ability.)

SMOOTH MUSCLE

Smooth, or **visceral,** muscle, is found in the walls of hollow organs such as the stomach, bladder, uterus, and blood vessels. When smooth muscle contracts, the cavity of an organ alternately becomes smaller (constricts) or enlarges (dilates) so that substances are propelled through the organs along a specific pathway. Smooth muscle cells look quite different from skeletal or cardiac muscle cells—no striations are visible. In addition, the cells have only one nucleus and are spindle-shaped (pointed at each end). Smooth muscle contracts much more slowly than the other two muscle types. *Peristalsis* (per"i-stal'sis), a wavelike motion which keeps food moving through the small intestine, is typical of its activity.

Nervous Tissue

Nervous tissue is composed of cells called **neurons.** All neurons receive and conduct electro-chemical impulses from one part of the body to another; thus *irritability* and *conductivity* are the two major functional characteristics. The structure of neurons is unique (Figure 2-12). The cytoplasm is drawn out into long extensions (as much as 3 feet or more, as in the leg), which allows a single neuron to conduct a stimulus over long distances in the body. Neurons make up the structures of the nervous system—the brain, spinal cord, and nerves.

DEVELOPMENTAL ASPECTS OF CELLS AND TISSUES

As said earlier, we all begin life as a single cell, which divides thousands of times to form our multicellular embryonic body. Very early in embryonic development the cells begin to specialize to form the primary tissues, and by the time of birth, most organs are well formed and functioning. The body continues to grow and enlarge by forming new tissue throughout childhood and adolescence.

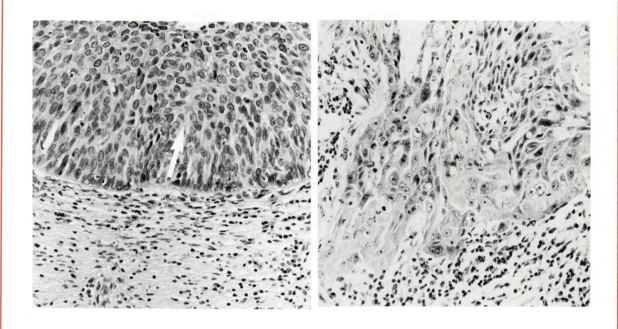

BOX 2–1

Non-Age Related Modifications in Cells and Tissues

In addition to tissue changes associated with aging, which are accelerated during the later years of life, there are other modifications of cells and tissues that may occur at any time. For example, **neoplasms** (ne'o-plazms) are growths or cell masses that result from the loss of normal controls on cell division. Neoplasms may be either benign or malignant. *Benign* (be-nin'), or innocent, neoplasms tend to be confined to one area and surrounded by a capsule. *Malignant* (mah-lig'nant), or *cancerous,* neoplasms tend to invade other body tissues and spread to distant parts of the body through the blood where new neoplasms are begun. Various factors (such as viruses, continual irritation, or some chemicals) seem to encourage the formation of neoplasms, but in most cases the cause is not really known. The left photomicrograph shows an early cancerous neoplasm of the uterus. Except for the slight whitened area (indicated by *arrow*), the tissue appears normal. Right view shows a malignant uterine neoplasm that has become invasive. Notice how different the invasive cells are in appearance and how they are spreading "crablike" into adjacent tissue areas.

The three major cancer killers in the United States are lung, colon, and breast cancer. Skin cancer is the single *most common* type of cancer. However, most skin cancers are seen and treated early before they *metastasize* (me-tas'tah-sīz), or spread, to invade other body organs. Thus skin cancers are usually much less dangerous.

(*Continued.*)

BOX 2-1
(continued)

Neoplasms are diagnosed by a *biopsy* (bī′ŏp-se). To do a biopsy, samples of the questionable tissue are removed surgically (or scraped off a surface) and examined under the microscope. In a cancerous growth the cells lose the specialized look of that particular tissue or organ and resemble embryonic cells. The treatment of choice for either type of neoplasm is surgical removal. If surgery is not possible—as in cases where the cancer has spread widely or is inoperable—radiation and drugs (chemotherapy) are used.

Not all increases in cell number involve neoplasms. Certain body tissues (or organs) may enlarge because there is some local irritant or condition that stimulates the cells. This is called *hyperplasia* (hī″per-plā′ze-ah). For example, when one is anemic, the bone marrow undergoes hyperplasia so that red blood cells may be produced at a faster rate. When (and if) the anemia is remedied, the enlargement or overproliferation of cells in the marrow disappears. Another example is when a woman's breasts enlarge during pregnancy in response to increased hormones; this is a normal situation, which doesn't have to be treated. On the other hand, *atrophy* (at′ro-fe), or decrease in size of an organ or body area, can occur if it loses its normal stimulation. For example, muscles not used or that have lost their nerve supply begin to atrophy and waste away rapidly.

Cell division is extremely important during the body's growth period. Most cells (except neurons) undergo mitosis until the end of puberty when adult body size is reached and overall body growth ends. After this time only certain cells will routinely carry out cell division—for example, cells exposed to abrasion such as skin and gut cells. Other cell groups, such as liver cells, stop dividing; however, they still have this ability should some of them die or become damaged and have to be replaced. Still other cell groups (for example, skeletal muscle and nervous tissue) completely lose the ability to divide (that is, they become *amitotic*) (am″ĭ-tot′ik). Amitotic tissues are severely handicapped by injury because the lost cells cannot be replaced. Repair *will* occur as a result of scar-tissue formation by connective tissue cells, but the scar tissue does not have the capabilities of the original tissue. This is why the heart of an individual who has had several severe heart attacks becomes weaker and weaker. Since the damaged cardiac muscle (another amitotic tissue) cannot be regenerated and is replaced by scar tissue which cannot contract, the heart becomes less and less capable of acting as an efficient pump.

The aging process begins once maturity has been reached. (Some believe it begins at birth.) No one has been able to explain just *what* causes aging, but there have been many suggestions. Some believe it is a result of little "chemical insults," which occur continually throughout life—for example, the presence of substances (such as alcohol, certain drugs, or carbon monoxide) in the blood that are toxic to cells; or the temporary absence of needed substances such as glucose or oxygen. Perhaps the effect of these chemical insults is cumulative and finally succeeds in upsetting the delicate chemical balance of the body cells. Others think that external physical factors such as radiation (X rays or cosmic waves) contribute to the aging process. Several believe that the aging "clock" is genetically programmed or built into our genes. This last suggestion is very believable—we all know of cases like the "radiant woman of 50 that looks about 35" or the "barely-out-of-adolescence woman of 24 that looks 40," and that such traits can run in families. There is no question that certain events are part of the aging process. For example, the glands of the body (epithelial tissue) become much less productive as we age, which is why we begin to "dry out'' as less oil, mucus, and sweat are produced. The endocrine glands also produce decreasing amounts of hormones, and the body processes that they control (such as metabolism and reproduction) become less efficient or stop altogether.

Connective tissue structures also show changes with age. Bones tend to become porous and weaken. Repair of tissue injuries becomes slower, and the elasticity of the skin is lost, allowing the skin to sag.

IMPORTANT TERMS*

Electrolytes *(e-lek′tro-līts)*

Nucleus *(Nu′kle-us)*

Enzymes

Ribosomes *(ri′bo-sōms)*

Chromatin *(kro′mah-tin)*

Nucleoli *(nu-kle′o-li)*

Microvilli *(mi″kro-vil′i)*

Cytoplasm *(si′to-plazm″)*

Organelles *(or″gan-els′)*

Endoplasmic reticulum *(en″do-plas′mik rĕ-tik′u-lum)*

Golgi *(gol′jē)* **apparatus**

Lysosomes *(li′so-sōms)*

Mitochondria *(mi″to-kon′dre-ah)*

Centrioles *(sen′trĭ-ols)*

Cilia *(sil′e-ah)*

Selective permeability

Diffusion

Filtration

Permease *(per′me-ās)* **system**

Phagocytosis *(fag″o-si-to′sis)*

Pinocytosis *(pi″no-si-to′sis)*

Mitosis *(mi-to′sis)*

Epithelial *(ep″ĭ-the′le-al)* **tissue**

Connective tissue

Matrix

Muscle tissue

Nervous tissue

*For definitions, see Glossary.

SUMMARY

A. CELLS

1. Anatomy of the Generalized Cell
   a. A cell is composed primarily of four elements—carbon, hydrogen, oxygen, and nitrogen, plus many trace elements (such as calcium, sodium, iron, potassium). Living matter is over 60% water. The major building substance of the cell is protein.
   b. Cells vary in size from being microscopic to over 3 feet in length. Shape often reflects function; for example, muscle cells have a long axis to allow shortening, and nerve cells have long cell extensions to allow them to conduct impulses over long distances through the body.
   c. Cells have two major regions—nucleus and cytoplasm
      (1) The nucleus, control center of the cell, directs cell activity and is necessary for reproduction. The nucleus contains genetic material (DNA), which carries instructions for the synthesis of proteins. Enzymes (protein catalysts) direct all chemical reactions in the cells.
      (2) The cytoplasm is the area of the cell where most cellular activities occur. It contains highly specialized bodies called organelles, each of which has a specific function. For example, mitochondria are sites of ATP synthesis, ribosomes are sites of protein synthesis, and the Golgi apparatus packages substances for export from the cell. Inclusions are nonliving stored, or inactive, materials in the cytoplasm such as fat globules, water vacuoles, crystals, and the like.

The cell membrane limits and encloses the cytoplasm and acts as a selective barrier to the movement of substances into and out of the cell.

2. Cell Physiology
   a. All cells exhibit irritability and are able to reproduce, grow, and metabolize.
   b. Transport of substances through the cell membrane
      (1) Passive transport processes
         (a) Diffusion is the movement of a substance from an area of its higher concentration to an area of its lower concentration. It occurs because of kinetic energy (energy of motion) of the molecules themselves. The diffusion of water through the cell membrane is called osmosis; the diffusion of dissolved solutes through the cell membrane is called dialysis.
         (b) Filtration is the movement of substances through a membrane from an area of high hydrostatic pressure to an area of lower fluid pressure. In the body, the driving force is blood pressure.
      (2) Active transport processes: Energy (ATP) provided by the cell
         (a) In the permease system substances are moved across the membrane by a protein carrier (permease) against a concentration gradient (that is, from lower to higher concentration). Accounts for the transport of amino acids, some sugars, and most ions.
         (b) Phagocytosis is the engulfment of solid materials by the cell.
         (c) Pinocytosis is the intake of fluids from the cell exterior by vesicle formation.
   c. Effect of hypertonic, hypotonic, and isotonic solutions on cells.
      (1) Hypertonic solutions contain more solutes (and less water) than is contained inside the cells. Cells lose water by osmosis and crenate.
      (2) Hypotonic solutions contain fewer solutes (and more water) than do the cells. Cells swell and may rupture (hemolysis) as water rushes in by osmosis.
      (3) Isotonic solutions have the same solute-to-solvent ratio as contained by

the cell. These solutions cause no changes in the size of cells.
   d. Cell division has two parts, mitosis (nuclear division) and division of the cytoplasm. Mitosis begins to occur after DNA has been replicated and consists of four stages—prophase, metaphase, anaphase, and telophase. The result is two daughter nuclei each identical to the mother nucleus. Division of the cytoplasm occurs during telophase and pinches the cytoplasm in half. Mitotic cell division provides an increased number of cells for growth and repair.

B. BODY TISSUES

1. Epithelial tissue is the covering, lining, and glandular tissue. Functions include protection, absorption, and secretion. They are named according to cell shape (squamous, cuboidal, or columnar) and arrangement (simple or stratified).

2. Connective tissue is the supportive, protective, and binding tissue. It is characterized by the presence of a nonliving matrix, which is produced and secreted by the cells. The matrix varies in amount and consistency. Fat, ligaments and tendons, bones, and cartilages are all connective tissues or connective tissue structures.

3. Muscle tissue is specialized to contract or shorten, which causes movement. There are three types—skeletal (attached to the skeleton), cardiac (forms the heart), and smooth (in the walls of hollow organs).

4. Nervous tissue is composed of cells called neurons, which are highly specialized to receive and transmit nerve impulses. Neurons are important for control of the body (as a whole). They are located in nervous system structures—brain and spinal cord.

C. DEVELOPMENTAL ASPECTS OF CELLS AND TISSUES

1. Growth through cell division continues until the end of puberty. Cell populations (such as epithelium) that are exposed to friction continue to form replacement cells throughout life. Connective tissue remains mitotic to form repair (scar) tissue. Muscle tissue becomes amitotic by the end of

puberty, and nervous tissue becomes amitotic shortly after birth. Each is severely handicapped by injury.

2. The cause of aging is unknown, but chemical and physical insults and genetic programming are suggested.

3. Neoplasms, both benign and cancerous, represent situations (that is, abnormal cell masses) in which the normal controls on cell division are not working. Atrophy (decrease in size) of a tissue or organ occurs when the organ is no longer stimulated normally; hyperplasia (increase in size of the tissue or organ) occurs, in some cases, when the tissue is strongly stimulated or irritated.

REVIEW QUESTIONS

1. Name the four elements making up the bulk of living matter.

2. Why is water important to the body?

3. Define organelle and cell.

4. Although cells have differences that reflect their special functions in the body, what functional abilities do all cells exhibit?

5. Describe the general function of the nucleus. Describe the special function of DNA found in the nucleus. What nuclear structures contain DNA?

6. Describe the general function of the cytoplasm.

7. Name the cellular organelles and explain the function of each.

8. What is the value of ATP to the body?

9. Define diffusion, osmosis, dialysis, filtration, permease transport, phagocytosis, and pinocytosis.

10. What is the difference between an active and passive transport process?

11. What two structural characteristics of cell membranes determine if substances can pass through them passively? What determines whether or not a substance can be actively transported through the membrane?

12. Explain the effect of the following solutions on living cells: hypertonic, hypotonic, and isotonic.

13. Define mitosis. Why is mitosis important?

14. What is the role of the spindle in mitosis?

15. Why can an organ be permanently damaged if its cells are amitotic?

16. Define tissue. List the four major types of tissue. Which of the four major tissue types is most widely distributed in the body?

17. Describe the general characteristics of epithelial tissue. List the most important functions of epithelial tissues and give examples.

18. How are epithelial tissues classified?

19. Where is ciliated epithelium found, and what role does it play?

20. How do the endocrine and exocrine glands differ in structure and function?

21. What are the general structural characteristics of connective tissues? What are the functions of connective tissues? How are its functions reflected in its structure?

22. Name a connective tissue with (a) a liquid matrix, (b) a soft matrix, and (c) a stoney hard matrix.

23. What is the function of muscle tissue?

24. Name the three types of muscle tissue and tell where each would be found in the body.

25. What is meant by "Smooth muscles are involuntary in action"? Which muscle type is voluntary in action?

26. What two functional characteristics are highly developed in neurons?

27. In what ways are neurons similar to other cells? In what ways are they different?

28. Define neoplasm, atrophy, and hyperplasia.

29. How are benign neoplasms different from cancerous neoplasms?

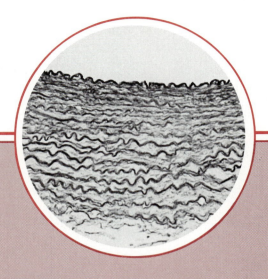

CHAPTER 3

Skin and Body Membranes

Chapter Contents

Function of skin • to protect against injuries of many types as the outermost boundary of the body

Function of body membranes • to line, protect, and lubricate body surfaces

After completing this chapter, you should be able to:

- List several important functions of the integumentary system and explain how these functions are accomplished.

- Recognize and name the following skin structures: epidermis, dermis (papillary and reticular layers), hair and hair follicle, sebaceous gland, sudoriferous gland—when provided with a model or diagram of the skin.

- Name the uppermost and deepest layers of the epidermis and describe the characteristics of each.

- Describe the distribution and function of the epidermal derivatives—sebaceous glands, sudoriferous glands, and hair.

- Name the factors that determine skin color and describe the function of melanin.

- Name several types of sensory receptors in the skin and explain their importance.

- List the general functions of each membrane type—cutaneous (epidermal), mucous, serous, and synovial—and give its location in the body.

- Compare the structure and function of the major membrane types and tell which are epithelial and which are connective tissue membranes.

Body membranes, which cover surfaces, line body cavities, and form protective (and often lubricating) sheets around organs, fall into two major groups. There are (a) epithelial membranes (consisting of the epidermis of the skin and the mucous and serous membranes), and (b) connective tissue membranes represented by synovial membranes. The epidermis is introduced in this chapter in conjunction with a broader topic—the skin or integumentary system. The other membranes are considered briefly at the end of the chapter.

INTEGUMENTARY SYSTEM (SKIN)

Basic Structure and Function

The **integument** (in-teg′u-ment), or skin, is often thought of as an organ system because it is large and complex. It is much more than an external body covering. It is an absolute essential because it keeps water in and keeps water (and other things) out. (This is why one can swim for hours without becoming waterlogged.) Structurally the skin is a marvel. It is pliable and yet tough, which

allows it to take constant punishment from external agents.

The skin has many functions; most (but not all) are protective (Table 3-1). It insulates and cushions the deeper body organs and protects the entire body from mechanical damage (bumps and cuts), chemical damage (such as acids and bases), thermal damage (heat and cold), and bacteria that live on it. The uppermost layer of the skin is *cornified,* or hardened, and prevents water loss from the body surface. The skin's rich capillary network (controlled by the nervous system) plays an important role in regulating heat loss from the body surface. The skin acts as a mini-excretory system; urea, salts, and water are lost when we sweat. The skin also synthesizes vitamin D. Modified cholesterol molecules located in the skin are converted to vitamin D by sunlight. Finally, the *cutaneous* (ku-ta′ne-us) *sense organs* (touch, temperature, pressure, and pain receptors) are located in the skin.

The skin is composed of two kinds of tissue. The outer **epidermis** (ep″ĭ-der′mis) is made up of stratified squamous epithelium that is capable of *keratinizing* (ker′ah-tin-īz-ing), or of becoming hard and tough. The underlying **dermis** is made up of dense fibrous connective tissue. The epider-

TABLE 3-1 Functions of the Skin

Functions	How Accomplished
Protects deeper tissues from:	
• Mechanical damage (bumps)	Physical barrier contains pressure receptors, which alert the nervous system to possible damage
• Chemical damage (acids and bases)	Has relatively impermeable keratinized cells; contains pain receptors, which alert the nervous system to possible damage
• Bacterial damage	Has an unbroken surface and "acid mantle" (skin secretions are acidic, and thus inhibit bacteria)
• Ultraviolet radiation (damaging effects of sunlight)	Has melanin protection by melanocytes
• Thermal (heat or cold) damage	Contains heat/cold/pain receptors
• Desiccation (drying out)	Contains waterproofing keratin
Aids in body heat loss or heat retention (controlled by the nervous system)	Heat loss: By activation of sweat glands and allowing blood to flush into skin capillary beds. Heat retention: By not allowing blood to flush into skin capillary beds
Aids in excretion of urea and uric acid	Contained in perspiration produced by sweat glands
Synthesizes vitamin D	Modified cholesterol molecules in skin converted to vitamin D by sunlight

mis and dermis are firmly cemented together. However, friction (like the rubbing of a poorly fitting shoe) may cause them to separate, which results in a blister. Under the dermis is the **subcutaneous tissue,** which contains many fat cells; it is not considered to be part of the skin. The subcutaneous tissue serves as a shock absorber and insulates the deeper tissues from extreme temperature changes occurring outside the body. It is also responsible for the curves that are more a part of a woman's anatomy than a man's. The main skin areas and structures are described next. As you read, locate the described areas or structures on Figure 3-1.

EPIDERMIS

The epidermis is *avascular,* that is, it has no blood supply of its own. This explains why a man can

shave daily and not bleed even though he is cutting off many cell layers each time he shaves.

The deepest cell layer of the epidermis is called the **stratum germinativum** (stra'tum jer″mĭ-na-tiv'um). It lies closest to the dermis and contains the only epidermal cells that receive enough nourishment (through diffusion of nutrients from the dermis). These cells are constantly undergoing cell division. Millions of new cells are produced daily. The daughter cells are pushed upward, away from the source of nutrition, to become part of the epidermal layers closer to the skin surface. As they move away from the dermis, they become increasingly keratinized and gradually die.

The uppermost layer is the **stratum corneum** (kor'ne-um); it is sometimes called the *horny lay-*

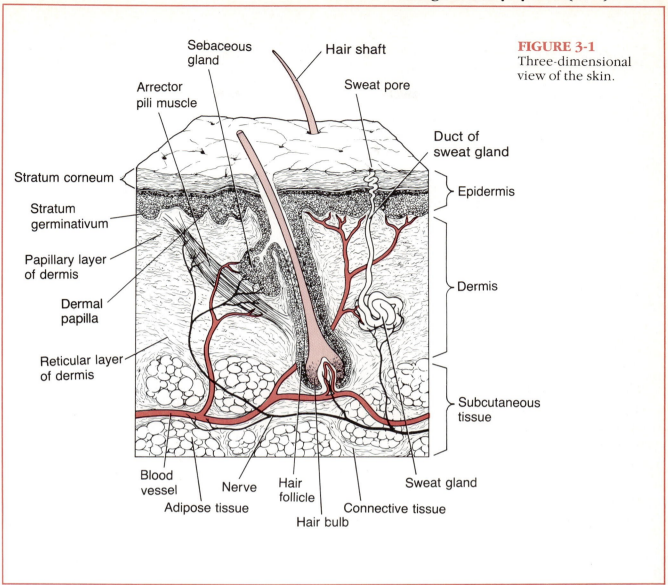

FIGURE 3-1
Three-dimensional view of the skin.

er because it consists of shingle-like, fully keratinized cells. *Keratin* is a protein with water-proofing properties. Thus, this layer provides a natural raincoat for the body and is important in preventing water loss from the deeper tissues. The stratum corneum cells are dead. They are constantly rubbing and flaking off in a slow steady way and being replaced by the division of the deeper cells.

Melanin (mel′ah-nin), a brown pigment, is produced by specials cells (*melanocytes* [mel′ah-no-sīts]) found in the stratum germinativum. Tanning of the skin is due to an increase in melanin production by the melanocytes when the skin is exposed to sunlight. Melanin, which somehow gets into the stratum germinativum cells (some say that melanocytes inject it into them; others say

that the stratum germinativum cells phagocytize the pigment), forms a protective pigment "umbrella" over their nuclei that shields genetic material (DNA) from the damaging effects of ultraviolet radiation in sunlight. *Freckles* and *moles* are seen where melanin is concentrated in one spot.

Skin color is a result of two major factors—the amount of melanin produced and the amount of oxygen in the blood. People who produce a lot of melanin have brown-toned skin. In light-skinned people, who have less melanin, the dermal blood supply flushes through the transparent cell layers above and gives the skin a rosy glow. When the blood is poorly oxygenated the skin takes on a bluish color; this condition of the skin is called *cyanosis* (si″ah-no′sis).

DERMIS

The dermis is your "hide." It is a strong stretchy envelope that helps to hold the body together. When you purchase leather goods (bags, shoes, and the like), you are buying the treated dermis of animals.

The connective tissue making up the dermis consists of two major regions—the papillary and the reticular areas. Like the epidermis, the dermis varies in thickness. For example, it is particularly thick on the palms of the hands and soles of the feet and is quite thin on the eyelids.

The **papillary layer** is the upper dermal region. It is uneven and has conelike projections from its superior surface, the **dermal papillae** (pah-pil'e), which attach it to the epidermis above. These projections result in our fingerprints—unique patterns of ridges, which can be used for identification and remain unchanged throughout life. A rich blood supply in the papillary layer furnishes nutrients to the epidermis and allows heat to radiate to the skin surface. Pain and touch receptors are also found here.

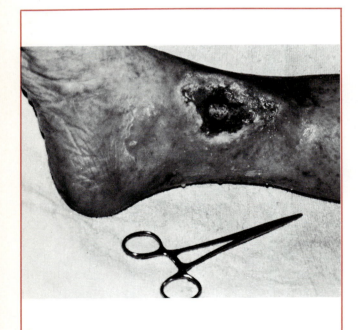

FIGURE 3-2
A decubitus ulcer on the ankle of a patient.

The **reticular layer** is the deepest skin layer. It contains blood vessels, sweat and oil glands, and pressure receptors. Many phagocytes are found here (in fact throughout the dermis). They act to prevent bacteria, which have managed to get through the epidermis, from penetrating any deeper into the body.

Both *collagen* and *elastic fibers* are found throughout the dermis. Collagen fibers are responsible for the toughness of the dermis. Elastic fibers give skin its elasticity when we are young. As we age, the number of elastic fibers decreases, and the subcutaneous tissue loses fat. As a result the skin becomes less elastic and sags and wrinkles.

The dermis is abundantly supplied with blood vessels that play a role in regulating body temperature. When body temperature is high, the capillaries of the dermis become engorged, or swollen, with heated blood. This allows body heat to radiate from the skin surface. If the environment is cool and body heat must be conserved or saved, blood bypasses the dermis capillaries temporarily causing internal body temperature to stay high. Any restriction of the normal blood supply to the skin results in cell death and, if severe or prolonged enough, skin ulcers. *Decubitus* (de-ku'bĭ-tus) *ulcers* (bedsores) occur in bedridden patients who are not turned regularly. The weight of the body puts pressure on the skin, especially over bony projections. This restricts the blood supply, and the cells die. A photograph of a decubitus ulcer is shown in Figure 3-2.

The dermis also has a rich nerve supply. As mentioned earlier, many of the nerve endings have specialized receptor organs that transmit messages to the central nervous system for interpretation when they are stimulated by environmental factors (pressure, temperature, and the like). These cutaneous receptors are discussed in more detail in Chapter 6.

Appendages

The appendages of the skin—hair, nails, and glands—all arise from the epidermis but are found in the dermis. As they are formed by the cells of the stratum germinativum, they push downward into the deeper skin regions.

CUTANEOUS GLANDS

The cutaneous glands fall into two groups: sebaceous glands and sudoriferous glands. The **sebaceous** (se-ba'shus) **glands** are found all over the skin except for the palms of the hands and the soles of the feet. They are **exocrine** (ek'so-krin) **glands.** Their ducts usually empty into a hair follicle, but some open directly onto the skin surface.

The product of the sebaceous glands, **sebum** (se'bum), is a mixture of oily substances and fragmented cells. For this reason, the sebaceous glands are sometimes called *oil glands*. Sebum is a lubricant that keeps the skin soft and moist and prevents the hair from becoming brittle. The sebaceous glands become very active when male sex hormones are produced in increased amounts (as in both sexes during adolescence); thus, the skin tends to become oilier during this period of life. *Blackheads* are accumulations of dried sebum and bacteria in the gland and its duct. *Acne* is caused by an active infection of the sebaceous glands.

Epithelial openings called **pores** are the outlets for the **sudoriferous** (su"do-rif'er-us) **glands,** or *sweat glands*. These exocrine glands are widely distributed in the skin, and their number is staggering—approximately 2 million per person. There are two types of sudoriferous glands based on the contents of their secretions. The **eccrine** (ek'rin) **glands** are found all over the body. They produce a clear perspiration which is primarily water, plus some salts (sodium chloride) and urea. The **apocrine** (ap'o-krin) **glands,** largely found in the axillary and genital areas of the body, secrete a milky substance containing proteins as well as all the substances in the eccrine secretion. Bacteria, which live on the skin, use the apocrine gland secretion as a source of nutrients for their growth.

Sweat glands are an important part of the body's heat-regulating equipment. They are richly supplied with nerve endings that cause them to secrete perspiration when the external temperature or body temperature is high. When perspiration evaporates off the skin surface, it carries large amounts of body heat with it. On a hot day it is possible to lose 7 quarts of body water in this way. The heat-regulating functions of the body are important—if internal temperature changes more than a few degrees from the normal 98.6°F, life-threatening changes occur in the body.

HAIR

There are millions of hairs scattered all over the body, but for all practical intents, hairs are one of the few parts of the body without a purpose. They served cavemen (and still serve hairy animals) by providing warmth in the cold weather, but modern man has other means of keeping warm.

A hair, enclosed in a **follicle**, is also an epithelial structure. That part of the hair enclosed in the follicle is called the **root;** the part projecting from the surface of the scalp or skin is called the **shaft** (see Figure 3-1). A hair is formed by cell division of the well-nourished germinal epithelial cells at the inferior end of the follicle (the hair bulb). As the daughter cells are pushed farther away from the growing region, they become keratinized and die. Thus the bulk of the hair shaft, like the bulk of the epidermis, is dead material and almost entirely protein. A hair consists of a central region (**medulla** [mĕ-dul'ah]) surrounded by a protective **cortex**. Abrasion or harsh treatment of the cortex results in "split ends" in which the medulla is no longer bound together by the cortex. Hair color reflects the amount of pigment (generally melanin) in the cortex. The more pigment there is, the darker the hair color.

Hairs come in a variety of sizes and shapes. They are short and stiff in the eyebrows, long and flexible on the head, and usually nearly invisible almost everywhere else. When the hair shaft is oval, the person has straight hair; when it is flat and ribbonlike, the hair is curly or kinky. Hairs are found all over the body surface except for the palms of the hands, soles of the feet, and the lips. Humans are born with as many hair follicles as they will ever have, and hairs are among the fastest growing tissues in the body. Hormones account for the development of "hairy" regions—the scalp, and in the adult, the pubic and axillary areas.

Look carefully at the structure of the hair follicle in Figure 3-1, and note that it is slanted. Small bands of smooth muscle cells—**arrector pili** (ah-rek'tor pi'li)—connect each side of the hair follicle to the papillary layer of the dermis. When

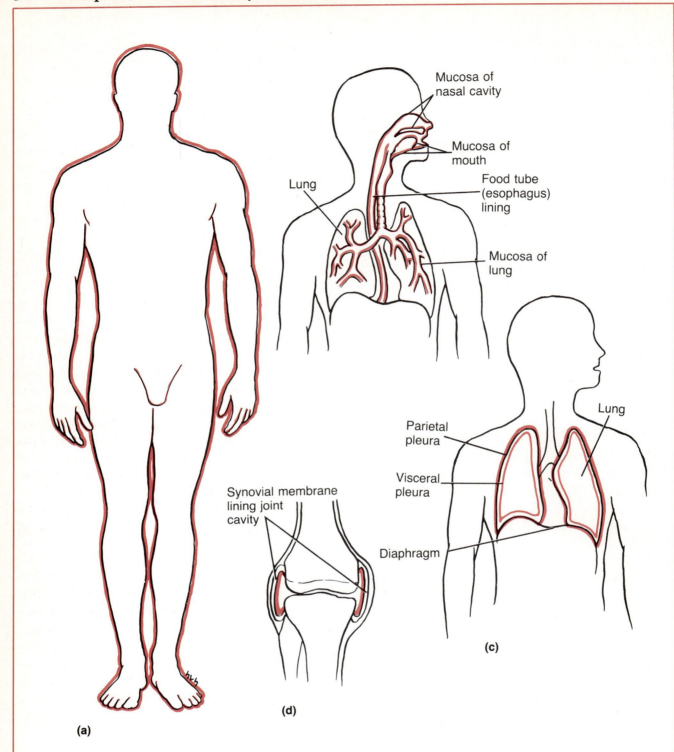

Mucosa of
nasal cavity

Mucosa of
mouth

Lung

Food tube
(esophagus)
lining

Mucosa of
lung

Lung

Parietal
pleura

Visceral
pleura

Diaphragm

Synovial membrane
lining joint
cavity

(a)

(d)

(c)

FIGURE 3-3

Classes of body membranes. **(a)** Cutaneous membrane, or epidermis of the skin (epithelial membrane); **(b)** mucous membrane lines body cavities that are *open* to the exterior (epithelial membrane); **(c)** serous membrane lines body cavities that are *closed* to the exterior (epithelial membrane); **(d)** synovial membrane lines joint capsules (connective tissue membrane).

these muscles contract (as when we are cold or frightened), the hair is pulled upright, dimpling the skin surface with "goose bumps." There is no question that this helps keep animals warm in the winter by adding a layer of insulating air in the fur. It is especially dramatic in a scared cat whose fur actually stands on end to make it look larger to hopefully scare off its enemy. (However, this phenomenon is not very useful to human beings.)

NAILS

Nails are hornlike structures produced by the epidermis. They consist of a **root**, which adheres to the nail bed, and a **body**. As the germinal cells in the nail root divide, the nail grows. Like the hair shaft, the nail is mostly made up of nonliving material. Nails are transparent and nearly colorless but appear pink because of the blood supply in the nail bed. When someone is cyanotic due to a lack of oxygen in their blood, the nails take on a blue cast.

CLASSIFICATION OF BODY MEMBRANES

Body membranes fall into two major groups according to the type of tissue forming them. The epithelial membrane types were introduced in Chapters 1 and 2; thus the purpose here is simply to provide a summary of the body membranes. Figure 3-3 shows the locations of these membranes in the body.

Epithelial Membranes

Epithelial membranes include the epidermis of the skin, the mucous membranes, and the serous membranes. Since the epidermis has just been discussed in some detail earlier in this chapter, it will not be considered here except to list it as a subcategory of the epithelial membranes.

CUTANEOUS MEMBRANE

The **cutaneous membrane** is the epidermis of the skin, which is composed of a stratified squamous keratinizing epithelium.

MUCOUS MEMBRANE

As described earlier, a **mucous membrane** (*mucosa*) is composed of epithelial cells resting on a connective tissue membrane. This membrane type lines all body cavities that open to the body exterior—the respiratory, digestive, and urinary tracts. Notice that the term *mucosa* refers only to the location of the epithelial membranes, not the cell makeup of the membrane. The types of epithelium composing mucous membranes *do* differ; for example, that of the esophagus (food tube) is stratified squamous, and that lining the trachea (wind pipe) is a special type of ciliated columnar epithelium. Although mucous membranes often secrete mucus, this is *not* a requirement. The mucosae of both the digestive and respiratory tracts secrete mucus, but that of most of the urinary tract does not.

SEROUS MEMBRANE

A serous membrane (*serosa*) is composed of two layers—a layer of simple squamous epithelium on a basement membrane. Serous membranes occur in pairs. The *parietal layer* lines a body cavity, and the *visceral layer* covers the outside of the organs in that cavity. In contrast to mucous membranes, which line open body cavities, serous membranes line body cavities closed to the exterior (except for the dorsal body cavity and joint cavities). The serosae secrete a thin fluid (serous fluid), which lubricates the organs and reduces friction as they slide across one another and the cavity walls. Special names are given to the serosae surrounding the organs in the ventral body cavity: the *pleura* surrounds the lungs, the *pericardium* the heart, and the *peritoneum* the abdominal organs. A serous membrane also lines the blood vessels *(endothelium)* and the heart *(endocardium)*. The entire wall of capillaries is composed of a serous membrane that acts as a selectively permeable membrane between the blood and the tissue fluid of the body. Because it is very thin, exchanges are made easily.

Connective Tissue Membrane

Synovial (sĭ-no′ve-al) **membrane** is composed of connective tissue and contains no epithelial cells at all. These membranes line the fibrous capsules surrounding joints where they provide a smooth surface and secrete a lubricating fluid. They also line small sacs of connective tissue called *bursae* (ber′se) and tendon sheaths. Both of these structures cushion organs moving against each other, as during muscle activity.

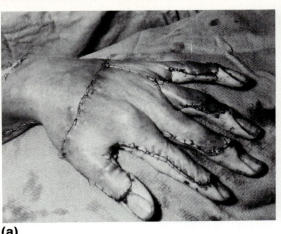

(a)

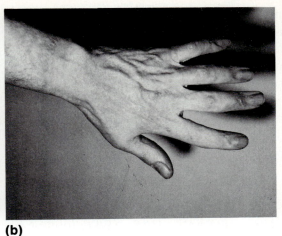

(b)

BOX 3-1
Burns

There are few threats to skin more serious than burns. When the skin is burned and its cells are destroyed, two major problems result. First, since the protective mantle is no longer there, *pathogens* (path'o-jens) (bacteria, fungi, and the like) can easily invade the body. Second, the body loses its precious supply of water, ions, and proteins as these flow freely from the burned surface. Burns are classed according to the amount of damage. In *first degree burns,* only the epidermis is damaged, and the area becomes red and swollen. Except for the temporary discomfort, first degree burns are not usually serious. In *second degree burns*, there is damage to both the epidermis and dermis. The skin is red and painful, and blisters containing tissue fluid form. Although more serious than first degree burns, regrowth of the epithelium can occur, and ordinarily no permanent scars result. In *third degree burns,* both the epidermal and dermal layers are completely destroyed in that area. The skin appears blanched (a grayish color), and since the nerve endings in the area have been destroyed, there is no pain. In third degree burns regeneration is not possible, and skin grafting must be done to cover the underlying exposed tissues.
(a) Skin from the inside of the thigh is grafted into a hand with third-degree burns. **(b)** The same hand healed about 10 weeks later.

DEVELOPMENTAL ASPECTS OF SKIN AND BODY MEMBRANES

When a baby is born, its skin is covered with *vernix caseosa* (ver'niks ka-se-o'sah). This white substance, produced by the sebaceous glands, protects the baby's skin while it is floating in its water-filled sac inside the mother. The newborn's skin is very thin, and blood vessels are easily seen through it. Commonly, there are small white spots called *milia* (mil'e-ah), which are accumulations in the sebaceous glands, on the baby's nose and forehead. These normally disappear by the third week after birth. As the baby grows, its skin becomes thicker and moist, and more subcutaneous fat is formed.

As we age the skin on the exposed parts of our body is continually insulted by abrasion, chemicals, wind, sun, and other irritants; and its pores become clogged with air pollutants and bacteria. As a result, pimples, scaling, and various kinds of *dermatitis* (der″mah-ti′tis), or skin inflammations, begin to become common.

As old age is approached, the amount of subcutaneous tissue begins to decrease, leading to an intolerance for cold, which is common in the elderly person. The skin also becomes drier (due to decreased oil production), and as a result it becomes itchy and bothersome. Thinning of the skin is another result of the aging process, causing it to become easily bruised and more susceptible to other types of injuries. The decreasing elasticity of the skin (along with the loss of the subcutaneous fat tissue) allows bags to form under our eyes, and our jowls begin to sag. This loss of elasticity is speeded up by sunlight, so perhaps one of the best things you can do for your skin is to protect it from the sun. In so doing, you will also be decreasing (as much as you possibly can) the chance of skin cancer. Although there is no way to avoid the aging of the skin, good nutrition, plenty of fluids, and cleanliness may help to delay the process somewhat.

Many men become bald as they age. This loss of hair is called *alopecia* (al″o-pe′she-ah). A bald man is not really hairless—he has hairs in the bald area. Because those hair follicles have begun to degenerate, the hairs are colorless and very tiny (so much so they may not even emerge from the follicle). Such hairs are called "velis hairs." Another phenomenon of aging is the graying of hair. This, like balding, is usually genetically controlled by a "delayed action" gene. Once the gene

becomes effective, the amount of melanin deposited in the hair decreases or becomes entirely absent, which results in gray-to-white hair. Certain events can cause hair to gray or fall out prematurely. For example, many people have claimed that they turned gray nearly overnight because of some emotional crisis in their life. In addition, we know that anxiety, therapy with certain chemicals (chemotherapy), radiation, excessive vitamin A, and certain fungus diseases can cause both graying and hair loss. However, when the cause of these conditions is not genetic, hair loss is usually not permanent.

IMPORTANT TERMS*

Integument *(in-teg′u-ment)*

Epidermis *(ep″i-der′mis)*

Keratinizing *(ker′ah-tin-iz-ing)*

Dermis

Subcutaneous tissue

Melanin *(mel′ah-nin)*

Sebaceous *(se-ba′shus)* **glands**

Exocrine *(ek′so-krin)* **glands**

Sudoriferous *(su″do-rif′er-us)* **glands**

Follicle

Nails

Cutaneous membrane

Mucous membrane

Serous membrane

Synovial *(si-no′ve-al)* **membrane**

Keratinizing *(ker′ah-tin-iz-ing)*

---

*For definitions, see Glossary.

SUMMARY

A. INTEGUMENTARY SYSTEM (SKIN)

1. Functions include protection of the deeper tissue from chemicals, bacteria, bumps, and drying out; regulation of body temperature through radiation and perspiring; and formation of vitamin D. The skin is the location of the cutaneous sensory receptors.

2. The epidermis, superior part of the skin, is formed of stratified squamous keratinizing epithelium and is avascular. Cells at its surface are dead and continually flake off.

They are replaced by the cell division of the basal cell layer. As the cells move away from the basal layer, they begin to accumulate keratin and gradually die. Melanin is produced by melanocytes. It protects the epithelial cells from the damaging rays of the sun.

3. The dermis is composed of dense fibrous connective tissue. It is the site of blood vessels, nerves, and the epidermal appendages. It has two regions, the papillary and reticular layers. The papillary layer has ridges, which produce fingerprints.

4. The appendages are formed from the epidermis, but reside in the dermis.
   a. Sebaceous glands produce an oily product (sebum), which is usually ducted into a hair follicle. Sebum keeps the skin and hair soft.
   b. Sudoriferous (sweat) glands, under the control of the nervous system, produce perspiration which is ducted to the epithelial surface. These glands are part of the heat-regulating apparatus of the body. There are two types: eccrine (most numerous) and apocrine (product includes protein, which skin bacteria metabolize).
   c. Hair is primarily dead keratinized cells and is produced by the hair bulb. The root is enclosed in a sheath called the hair follicle.
   d. Nails are hornlike derivatives of the epidermis. Like hair, nails are primarily dead keratinized cells.

B. CLASSIFICATION OF BODY MEMBRANES

1. Epithelial
   a. Cutaneous: epidermis of skin; protects body surface.
   b. Mucous: lines body cavities open to the exterior.
   c. Serous: lines body cavities closed to the exterior.

2. Connective tissue: synovial—lines joint cavities.

C. DEVELOPMENTAL ASPECTS OF THE SKIN AND BODY MEMBRANES

1. The skin is thick, resilient, and well hydrated in youth but loses its elasticity and thins as aging occurs. Skin cancer is a major threat in skin exposed heavily to sunlight.

2. Balding and/or graying occurs with aging. Both are genetically determined but other factors (drugs, emotional stress, and so on) can result in either.

3. Because they interfere with skin's protective functions, burns represent a major threat to the body. The result is loss of body fluids and invasion of bacteria.

REVIEW QUESTIONS

1. What primary tissues are destroyed when the skin is damaged?

2. From what types of damage does the skin protect the body?

3. A nurse tells the doctor that a patient is cyanotic. What is cyanosis? What does its presence indicate?

4. Explain *how* we become tanned as a result of sitting in the sun.

5. What is a decubitus ulcer? Why does it occur?

6. Name two different categories of skin secretions and the glands that manufacture them.

7. How does the skin help in regulating body temperature?

8. What is a blackhead?

9. What are arrector pili? What do they do?

10. How does a mucosa differ from a serosa?

11. What is the name of the connective tissue membrane found lining joint cavities?

12. Imagine yourself without any cutaneous sense organs. Why might this be very dangerous?

13. Why does hair turn gray?

14. Name three changes that occur in the skin as one ages.

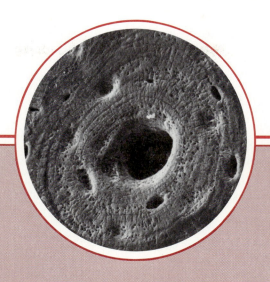

CHAPTER 4

Skeletal System

Chapter Contents

Functions of the skeletal system • to provide an internal framework for the body, protect organs by enclosure, and anchor skeletal muscles so that muscle contraction can cause movement

The above electron micrograph from *Tissues and Organs: A Text-Atlas of Scanning Electron Microscopy* by Richard G. Kessel and Randy H. Kardon. W. H. Freeman and Co. Copyright © 1979.

After completing this chapter, you should be able to:

- List at least three functions of the skeletal system.

- Name the four main kinds of bones.

- Identify the major anatomical areas of a long bone.

- Explain the role of bone salts and the organic matrix in making bone both flexible and hard.

- Identify the subdivisions of the skeleton as axial or appendicular.

- Identify and name the bones of the skull when provided with a skull or diagram.

- Describe how the skull of a newborn infant (or fetus) differs from that of an adult and explain the function of fontanels.

- Name the parts of a typical vertebra and explain in general how the cervical, thoracic, and lumbar vertebrae differ from one another.

- Discuss the importance of the intervertebral disks and spinal curvatures.

- Explain how the abnormal spinal curvatures (scoliosis, lordosis, and kyphosis) differ from one another.

- Identify on a skeleton or diagram the bones of the shoulder and pelvic girdles and their attached limbs.

- Explain important differences between a male and a female pelvis.

- Name the three major categories of joints and compare the amount of movement allowed by each.

- Name and explain the various types of fractures.

- Identify some of the causes of bone and joint problems.

FUNCTION

Our skeleton is shaped by an event that happened more than a million years ago—when a human being first stood erect on hind legs. It is a tower of bones arranged so that we can run, walk, and balance ourselves. It anchors skeletal muscles, allowing them to move the body parts, and protects vital organs. Bones, the "steel girders" and "reinforced concrete" of the body, form its internal framework. No other animal has such relatively long legs (compared to the arms or forelimbs) or such a strange foot, and few have such remarkable grasping hands. Even though the infant's backbone is like an arch, it soon changes to a "swayback" (S-shaped structure), which is required for the upright posture.

Besides supporting and protecting the body as an internal framework, the skeleton provides a storage area for lipids (fats) and calcium. A small amount of calcium in its ion form (Ca^{++}) must be present in the blood at all times for the nervous system to transmit messages, muscles to contract, and blood to clot. Because most of the body's calcium is deposited in the bones as calcium salts, the bones are a convenient place to get more calcium ions for the blood as they are used up. Problems occur not only when there is too little calcium in the blood, but also when there is too much (leading to kidney stones and problems with heart activity). Hormones control the movement of calcium to and from the bones and blood according to the needs of the body at any particular time. Finally, the red marrow cavities of bones provide a site for *hemopoiesis,* or blood cell formation.

The skeleton is formed from two of the strongest and most supportive tissues in the body—cartilage and bone. In embryos, the skeleton is primarily made of cartilage, but in the young child most of the cartilage has been replaced by bone, which is much more rigid. Cartilage remains only in isolated areas such as the bridge of the nose, parts of the ribs, and the joints.

The skeleton is subdivided into two divisions: the **axial skeleton** (bones that form the longitudinal axis of the body) and the **appendicular skeleton** (bones of the limbs and girdles). In addition to bones, the skeletal system also includes joints and ligaments (fibrous cords that bind the bones at joints). The joints give the body flexibility and allow movement to occur.

Before beginning a study of the skeleton, imagine that your bones have turned to putty. What if you were running when this change took place? Now imagine your bones forming a continuous metal framework inside your body, somewhat like a system of plumbing pipes. What problems could there be with this arrangement? These images should help you understand how well the skeletal system provides support and protection while allowing movement to occur.

BONES: AN OVERVIEW

Bone Markings

Even when looking casually at bones, one can see that the surfaces are not smooth but scarred with bumps, holes, and ridges. These bone markings reveal where muscles, tendons, and ligaments were attached, and where blood vessels and nerves passed. There are two categories of bone markings: (a) *projections,* or *processes,* which grow out from the bone surface, and (b) *depressions,* or cavities, which are indentations in the bone.

A few of the terms used to describe the different types of bone markings are listed next. These terms do not have to be learned now, but hopefully they will help you remember some of the specific markings on bones to which you will be introduced later in this chapter.*

Projections that are sites of muscle attachment:

- **crest**—narrow ridge of bone
- **spine**—sharp slender projection

---

*There is a little trick for remembering the bone markings listed here: All the terms beginning with *T* are projections; all the terms beginning with *F* are depressions.

- **trochanter** (tro-kan′ter)—very large, irregularly shaped process
- **tubercle** (tu′ber-k′l)—small rounded projection
- **tuberosity**—large rounded projection or roughened area

Projections that help form joints:

- **condyle** (kon′dīl)—rounded projection
- **epicondyle**—raised area on a condyle
- **head**—extension carried on a narrow neck
- **ramus** (ra′mus)—armlike bar

Depressions commonly seen are:

- **fissure**—narrow slitlike opening
- **foramen** (fo-ra′men)—opening through a bone, often round/oval
- **fossa** (fos′ah)—shallow depression in a bone
- **meatus** (me-a′tus)—canal-like passageway
- **sinuses**—depressions, filled with air and lined with mucous membranes within bones

Classification of Bones

The 206 bones of the adult skeleton are composed of two basic types of osseous, or bone, tissue. **Compact bone** is dense and looks smooth and homogeneous. **Spongy bone** is composed of small needlelike pieces of bone and lots of open space. Bones may by classified on the basis of their gross anatomy into four groups: long, short, flat, and irregular.

Long bones are generally longer than wide. They have a shaft with heads at both ends. Long bones are mostly compact bone. All bones of the limbs, except wrist and ankle bones, are long bones.

Short bones are generally cube-shaped. They contain more spongy bone than compact. The bones of the wrist and ankle are short bones.

Flat bones are thin and flattened like pancakes. They have two thin layers of compact bone sandwiching a layer of spongy bone between them.

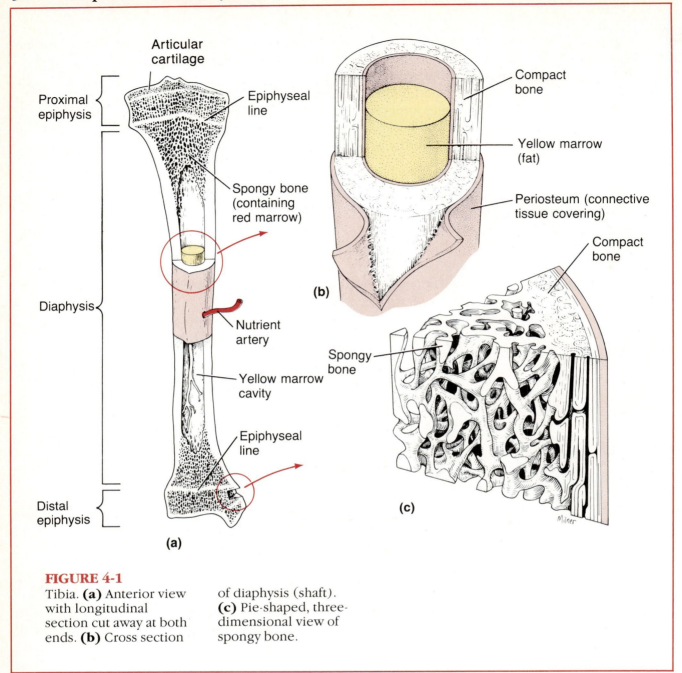

FIGURE 4-1

Tibia. **(a)** Anterior view with longitudinal section cut away at both ends. **(b)** Cross section of diaphysis (shaft). **(c)** Pie-shaped, three-dimensional view of spongy bone.

The bones of the skull and the sternum (breast-bone) are flat bones.

Bones that do not fall into one of the preceding categories are called **irregular bones.** The vertebrae, which make up the spinal column, fall into this group.

Anatomy of a Long Bone

The structure of a long bone is shown in Figure 4-1. The **diaphysis** (di-af′i-sis), or shaft, makes up most of the bone's length and is composed of compact bone. The diaphysis is covered and protected by a fibrous connective tissue membrane, the **periosteum.** The **epiphyses** (ĕ-pif′ĭ-sēs) are the ends of the long bone. Each epiphysis is composed of a thin layer of compact bone filled with spongy bone. **Articular cartilage** covers its external surface. Because the articular cartilage is glassy hyaline cartilage, it provides a smooth slippery surface to prevent friction at joint surfaces.

The adult bone has a thin line of bone spanning

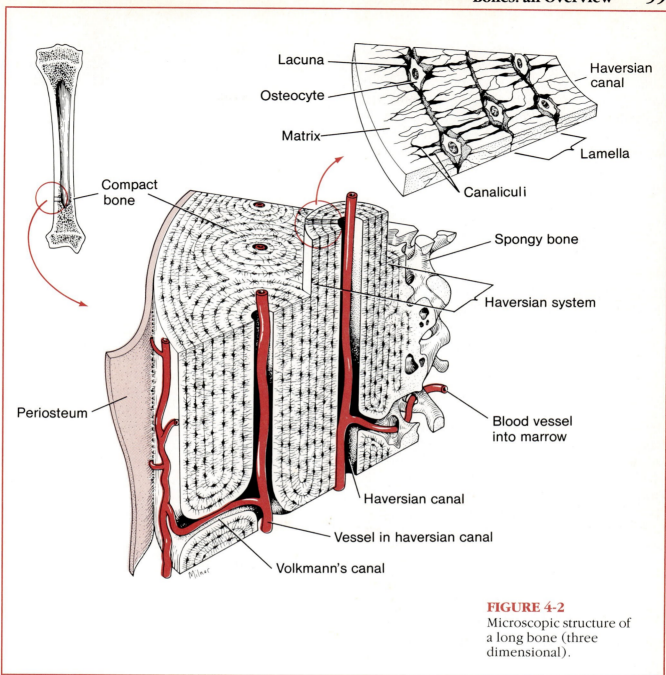

Lacuna

Osteocyte

Matrix

Haversian canal

Lamella

Canaliculi

Compact bone

Spongy bone

Haversian system

Periosteum

Blood vessel into marrow

Haversian canal

Vessel in haversian canal

Volkmann's canal

Milner

FIGURE 4-2
Microscopic structure of a long bone (three dimensional).

the epiphysis, which looks different from the rest of the bone in that area. This is the **epiphyseal line.** The epiphyseal line is a remnant of the **epiphyseal disk** (a flat plate of hyaline cartilage) seen in a young growing bone. Epiphyseal disks cause the lengthwise growth of the long bone. By the end of puberty, when hormones stop long bone growth, the epiphyseal disks have been replaced by bone.

In adults the cavity of the shaft is primarily a storage area for adipose tissue (fat). However, in in-fants this area forms blood cells, and **red marrow** is found there. In adult bones, red marrow is confined to the spongy bone of flat bones and the epiphyses of long bones.

Spongy bone has a spiky, openwork appearance whereas compact bone appears to be very dense. However, when looking at compact bone tissue through the microscope, one can see that it has a complex structure (see Figure 4-2). It is riddled with passageways carrying nerves, blood vessels, and the like, which provide the living bone cells with nutrients and a route for waste disposal. The mature bone cells, *osteocytes* (os′te-o-sīts″), are found in tiny cavities, or chambers, called **lacunae** (lah-ku′ne), which are arranged in circles (**lamellae** [lah-mel′e]) around central canals (Haversian canals). **Haversian canals** run lengthwise through the bony matrix, carrying blood vessels and nerves to all areas of the bone. Tiny canals, **canaliculi** (kan″ah-lik′u-li), radiate outward from the Haversian canals to all lacunae. Thus, the canaliculi form a transportation system, which connects all the bone cells to the nutrient supply, through the hard bone matrix. Because of this very elaborate network of canals, bone cells are well nourished in spite of the hardness of the matrix, and bone injuries heal quickly and well. The communication pathway from the outside of the bone to its interior (and the Haversian canals) is completed by **Volkmann's canals,** which run into the compact bone at right angles to the shaft.

Bone is one of the hardest materials in the body and, although relatively light in weight, it has a remarkable ability to resist tension and other forces acting on it. Nature has given us an extremely strong, exceptionally simple (almost crude), supporting system without giving up mobility. The calcium salts deposited in the matrix give bone its hardness, whereas the organic parts (especially the collagen fibers) provide flexibility.

Bone Remodeling

Many people mistakenly think that bones are lifeless structures that never change once long bone growth has ended. Nothing could be further from the truth. Bones are remodeled continually in response to changes in two factors:

1. Blood calcium levels
2. The pull of gravity and muscles on the skeleton

The manner in which these factors influence bones is outlined next.

When blood calcium levels drop below homeostatic levels, the parathyroid glands (located in the throat) are stimulated to release parathyroid hormone (PTH) to the blood. PTH activates *osteo-*

clasts (giant bone-destroying cells) in the bones to break down bone matrix and release calcium ions to the blood. On the other hand, when blood calcium levels are too high (*hypercalcemia* [hi″per-kal-se′me-ah]), calcium is deposited in bone matrix as hard calcium salts.

As long as we remain active, bones tend to remain healthy and strong. In fact, bones become thicker and form larger projections to increase their strength in areas where bulky muscles are attached and in areas of greatest stress. At such sites bone-forming cells, *osteoblasts,* lay down new matrix and become trapped in it. Once they are trapped, they become osteocytes, or mature bone cells.

To explain the interaction between these two controlling mechanisms as simply as possible, PTH determines *when* (or *if*) bone is to be broken down or formed in response to the need for more or fewer calcium ions in the blood. On the other hand, the stresses of muscle pull and gravity acting on the skeleton determine *where* this is to occur so that the skeleton can remain as strong and vital as possible.

AXIAL SKELETON

As noted earlier, the skeleton is divided into two parts, the axial and appendicular skeletons. The axial skeleton, which forms the longitudinal axis of the body, is shown as the colored portion of Figure 4-3. It can be divided into three parts—the skull, the vertebral column, and the thorax.

Skull

The **skull** is formed by two sets of bones. The cranium encloses and protects the fragile brain tissue; the facial bones hold the eyes in an anterior position and allow the facial muscles to show our feelings through smiles or frowns. All but one of the bones of the skull are joined together by *sutures,* which are interlocking joints. The mandible (jawbone) is attached to the rest of the skull by a freely movable joint.

CRANIUM

The boxlike **cranium** is composed of eight, large, flat bones. With the exception of two paired

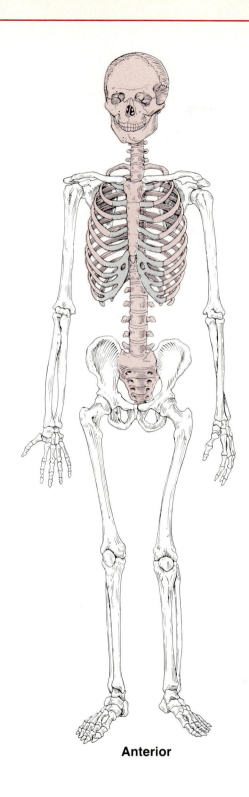
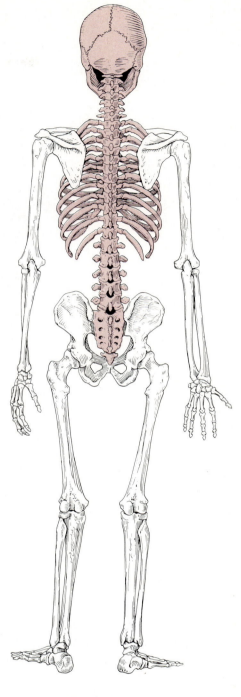

Anterior   **Posterior**

FIGURE 4-3
The human skeleton.
The colored area is the
axial skeleton; the
unshaded area is the
appendicular skeleton.

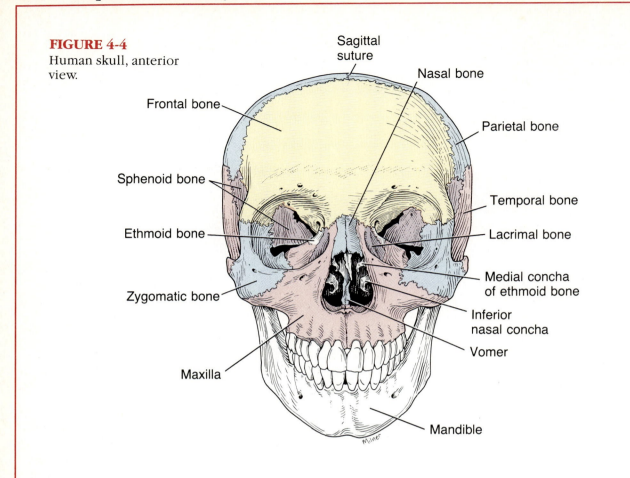

Sagittal suture

Nasal bone

Frontal bone

Parietal bone

Sphenoid bone

Temporal bone

Ethmoid bone

Lacrimal bone

Medial concha of ethmoid bone

Zygomatic bone

Inferior nasal concha

Vomer

Maxilla

Mandible

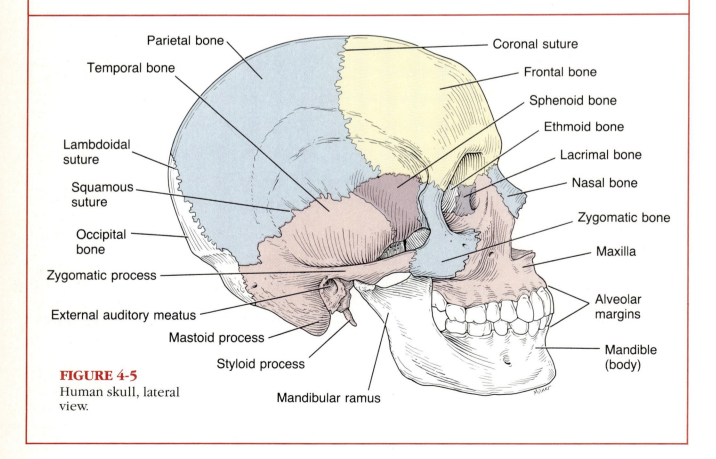

Parietal bone

Coronal suture

Temporal bone

Frontal bone

Sphenoid bone

Ethmoid bone

Lambdoidal suture

Lacrimal bone

Squamous suture

Nasal bone

Occipital bone

Zygomatic bone

Zygomatic process

Maxilla

External auditory meatus

Alveolar margins

Mastoid process

Styloid process

Mandible (body)

FIGURE 4-5
Human skull, lateral view.

Mandibular ramus

bones (the parietal and temporal bones), they are all single bones.

FRONTAL BONE.   The frontal bone forms the forehead, the bony projections under the eyebrows, and the superior part of each eye's orbit (Figure 4-4).

PARIETAL BONES.   The paired parietal (pah-ri′ĕ-tal) bones form most of the superior and lateral walls of the cranium (Figure 4-5). They meet in the midline of the skull to form the *sagittal suture,* and they form the *coronal suture* where they meet the frontal bone.

TEMPORAL BONES.   As Figure 4-5 illustrates, the temporal bones lies inferior to the parietal bones; they join them at the *squamous sutures.* Several important bone markings appear on the temporals.

1. The **external auditory meatus** is a canal that leads to the eardrum and the middle ear.

2. The **styloid process,** a sharp needlelike projection, is just inferior to the external auditory meatus. Many muscles use the styloid process as an attachment point.

3. The **zygomatic** (zi″go-mat′ik) **process** is a thin bridge of bone that joins with the cheekbone (zygomatic bone) anteriorly.

4. The **mastoid** (mas′toid) **process** is a rough projection posterior and inferior to the external auditory meatus; it provides an attachment site for some neck muscles. The mastoid process is full of air cavities (mastoid sinuses). It is so close to the middle ear (a high-risk spot for infections) that it may become infected too, a condition called *mastoiditis.* In addition, because this area is so close to the brain, mastoiditis can spread to the brain coverings and inflame them; this condition is known as *meningitis* (men″in-ji′tis). Using yourself as a model, see if you can find your mastoid process.

OCCIPITAL BONE.   If you look at Figures 4-5 and 4-6, you can see that the occipital (ok-sip′ĭ-tal) bone is the most posterior bone of the cranium. It forms the floor and back wall of the skull. The occipital bone joins the parietal bones anteriorly at the *lambdoidal suture.* The base of the occipital bone encloses a large opening, the **foramen magnum** (literally, large hole). The foramen magnum surrounds the lower part of the brain and allows the spinal cord to connect with the brain. Lateral to the foramen magnum on each side are the rockerlike **occipital condyles,** which rest on the first vertebra of the spinal column.

SPHENOID BONE.   The butterfly-shaped sphenoid (sfe′noid) bone spans the width of the skull and forms part of the cranial floor (Figure 4-7). In the midline of the sphenoid is a small depression, the **sella turcica** (sel′ah tur′si-cah), or *Turk's saddle,* which holds the pituitary gland in place in the living person. Parts of the sphenoid can be seen exteriorly forming both part of the eye orbits and the lateral part of the skull (see Figure 4-4). The central part of the sphenoid bone is riddled with air cavities, the *sphenoid sinuses.*

ETHMOID BONE.   The ethmoid (eth′moid) bone is very irregular-shaped and lies anterior to the sphenoid (see Figures 4-4, 4-5, and 4-7). It forms the roof of the nasal cavity and part of the medial walls of the orbits. Projecting from its superior surface is the **crista galli** (cock's comb); the outermost covering of the brain attaches to this projection. On each side of the crista galli, the ethmoid has an area with many small holes. These holey areas, the **cribriform** (krib′rĭ-form) **plates,** allow fibers carrying impulses from the olfactory receptors (receptors for smell) of the nose to reach the brain.

FACIAL BONES
There are 14 bones composing the face, and 12 are paired. Only the mandible and vomer are single. Figures 4-4 and 4-5 show most of the facial bones.

MANDIBLE.   The mandible, or lower jaw, joins the temporal bones on each side of the face. These joints are the only freely movable joints in the skull. You can find these joints on yourself by placing your fingers over your cheekbones and opening and closing your mouth. The horizontal part of the mandible (the **body**) forms the chin. Two upright bars of bone (the **rami**) extend from the body to connect the mandible with the temporal bone. The lower teeth lie in *alveoli* (sockets) on the superior edge of the body.

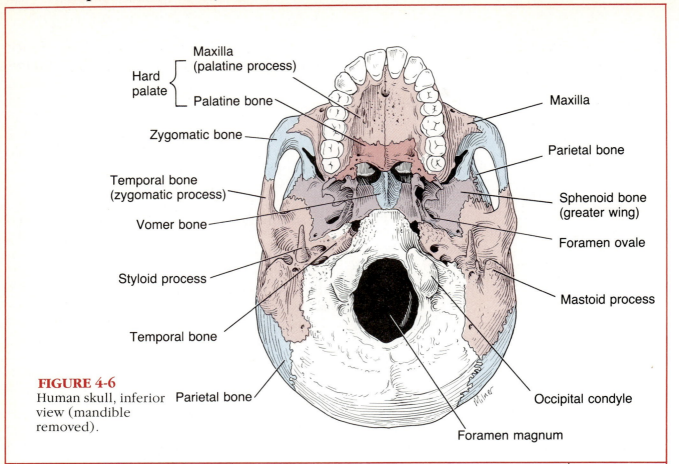

FIGURE 4-6
Human skull, inferior
view (mandible
removed).

Maxilla
(palatine process)

Hard
palate

Palatine bone

Zygomatic bone

Temporal bone
(zygomatic process)

Vomer bone

Styloid process

Temporal bone

Parietal bone

Maxilla

Parietal bone

Sphenoid bone
(greater wing)

Foramen ovale

Mastoid process

Occipital condyle

Foramen magnum

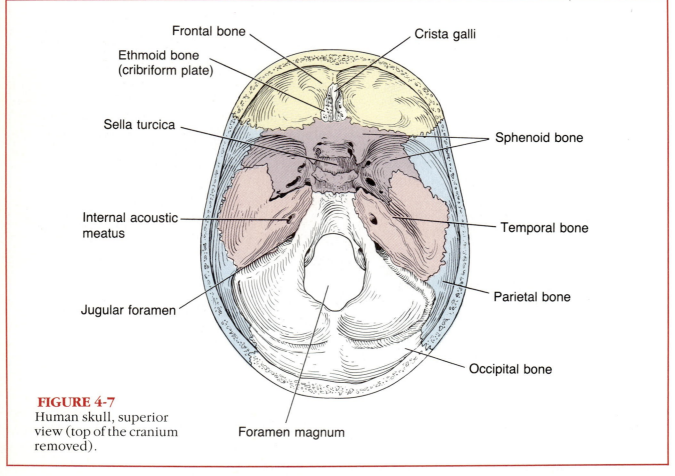

FIGURE 4-7
Human skull, superior
view (top of the cranium
removed).

Frontal bone

Crista galli

Ethmoid bone
(cribriform plate)

Sella turcica

Sphenoid bone

Internal acoustic
meatus

Temporal bone

Jugular foramen

Parietal bone

Occipital bone

Foramen magnum

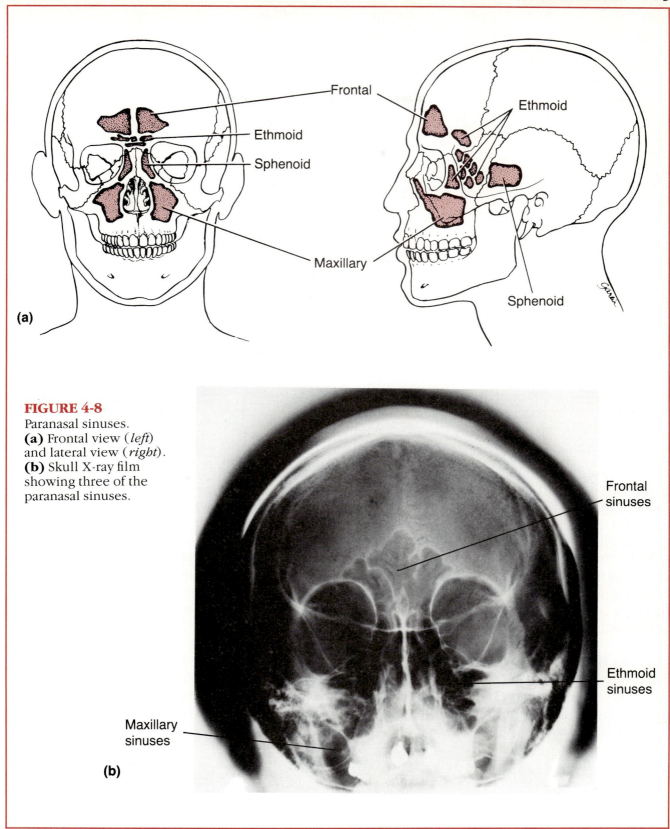

FIGURE 4-8
Paranasal sinuses.
(a) Frontal view (*left*) and lateral view (*right*).
(b) Skull X-ray film showing three of the paranasal sinuses.

MAXILLAE. The two maxillae are fused to form the upper jawbone. All facial bones except the mandible join the maxillae, thus they are the main, or "keystone," bones of the face. The maxillae carry the upper teeth in the alveolar margin. Extensions of the maxillae, the **palatine** (pal′ah-tīn) **processes,** form the anterior part of the hard palate of the mouth (see Figure 4-6). Like many other facial bones, the maxillae contain sinuses, which drain into the nasal passages (Figure 4-8).

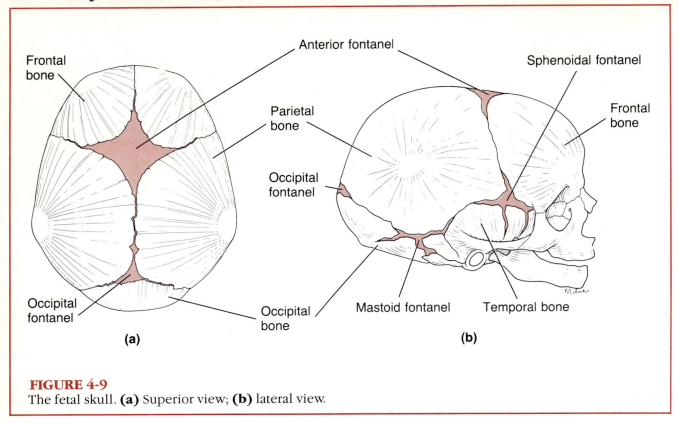

Frontal bone

Anterior fontanel

Sphenoidal fontanel

Parietal bone

Occipital fontanel

Frontal bone

Occipital fontanel

Occipital bone

Mastoid fontanel

Temporal bone

(a)

(b)

FIGURE 4-9
The fetal skull. **(a)** Superior view; **(b)** lateral view.

These paranasal (literally, around the nasal cavity) sinuses lighten the skull bones and probably act to amplify sounds we make during speech. They also cause many people a great deal of misery. Since the mucosa lining these sinuses is continuous with that in the nasal passages and throat, infections in these areas tend to migrate into the sinuses causing *sinusitis*. Depending on which sinuses are infected, a headache or upper jaw pain is the usual result.

PALATINE BONES.   The paired palatine bones are found posterior to the palatine processes of the maxillae. They form the posterior part of the hard palate. Failure of these or the palatine processes to fuse medially results in *cleft palate.*

ZYGOMATIC BONES.   Zygomatic bones are commonly referred to as the cheekbones. They also form a good sized portion of the lateral walls of the orbits, or eye sockets.

LACRIMAL BONES.   Lacrimal (lak'rĭ-mal) bones are tiny fingernail-size bones forming part of the medial walls of each orbit. Each lacrimal bone has a groove that serves as a passageway for tears (*lacrima* means tear).

NASAL BONES.   The small rectangular bones forming the bridge of the nose are the nasal bones. (The lower part of the nose is made up of cartilage.)

VOMER BONE.   The single bone in the median line of the nasal cavity is the vomer. (Vomer means plow, which refers to the bone's shape.) The vomer forms most of the nasal septum.

INFERIOR CONCHAE.   The inferior conchae (kong'ke) are thin curved bones projecting out from the lateral walls of the nasal cavity. (The superior and medial conchae are similar but are parts of the ethmoid bone.)

FETAL SKULL
The skull of a fetus or newborn infant is different in many ways from an adult skull. As Figure 4-9 illustrates, the infant's face is very tiny compared to the size of its cranium, and yet the skull as a whole is large when compared to the infant's total body length. The adult skull represents only one-eighth of the total body length whereas that of a newborn infant is one-fourth as long as its entire body. When a baby is born its skeleton is still unfinished; many areas of cartilage remain to be os-

sified, or converted, to bone. The skull also has fibrous cartilage areas between the cranial bones. These membranous areas are called **fontanels** (fon"tan-nels'), commonly known as "soft spots." The largest fontanels are the diamond-shaped anterior fontanel and the smaller triangular occipital fontanel. The fontanels allow the fetal skull to be compressed slightly during birth. In addition, because they are flexible, they allow the infant's brain to grow during the later part of pregnancy. (This would not be possible if the cranial bones were fused in sutures as in the adult skull.) These areas are gradually converted to bone during the early part of infancy, and the fontanels can no longer be felt by age 20–22 months.

Vertebral Column (Spine)

Some people think that the **vertebral column,** or **spine,** is a rigid supporting rod, but this picture is far from accurate. The vertebral column is formed from 24 single bones (vertebrae) and 2 fused bones (sacrum and coccyx). These are connected in such a way that a flexible curved structure is formed (Figure 4-10). The different areas of the vertebral column are the cervical, thoracic, lumbar, sacral, and coccyx regions.

The single vertebrae are separated by pads of fibrous elastic cartilage—**intervertebral disks**—which cushion the vertebrae and absorb shocks. In a young person, the disks have a high water content (about 90%) and are spongy and compressible. But as a person ages, the water content of the disks decreases (as it does in other tissues throughout the body), and the disks become harder and less compressible. This situation, along with a weakening of the ligaments of the vertebral column, predisposes older people to "slipped disks."

The disks and the S-shaped structure of the vertebral column work together to prevent shock to the head when walking or running. They also make the body trunk flexible. The spinal curvatures in the thoracic and sacral regions are referred to as *primary curvatures,* since they are present when we are born. Later, the *secondary curvatures* are formed. The cervical curvature appears when a baby begins to raise his or her head, and the lumbar curvature develops when the baby

begins to walk. Figure 4-11 shows three abnormal spinal curvatures—*scoliosis* (sko"le-o'sis), *kyphosis* (ki-fo'sis), and *lordosis* (lor-do'sis). These abnormalities may be congenital (present at birth) or result from disease or poor posture. As you look at these diagrams, try to pinpoint how each of the conditions differs from the normal healthy spine.

Most vertebrae have some common structural features as shown in Figure 4-12:

- **body** (or **centrum**): disklike central part of the vertebra facing anteriorly in the vertebral column
- **vertebral arch**: arch formed from the joining of all posterior extensions from the vertebral body
- **vertebral foramen**: canal through which the spinal cord passes
- **transverse processes**: two lateral projections from the body
- **spinous process**: single posterior projection from body
- **superior and inferior articular processes**: paired projections lateral to the vertebral foramen allowing a vertebra to form joints with adjacent vertebrae

In addition to the common features just defined, vertebrae in the different regions of the spine have very specific structural characteristics. These unique characteristics of the vertebrae of each region of the spine are described next.

CERVICAL VERTEBRAE
The seven **cervical vertebrae** (referred to as C_1 to C_7) form the neck region of the spine. The first two vertebrae (atlas and axis) are different because they perform functions not shared by the other cervical vertebrae. As you can see in Figure 4-13a (p. 70), the **atlas** (C_1) has no body. The superior surfaces of its transverse processes contain large depressions that receive the occipital condyles of the skull. This joint allows you to nod "yes." The **axis** (C_2) acts as a pivot for the rotation of the atlas (and skull) above. It has a large upright process, the **odontoid** (o-don'toid) **process,** or **dens,** which acts as the pivot point. The joint between C_1 and C_2 allows you to rotate your head from side to side to indicate "no."

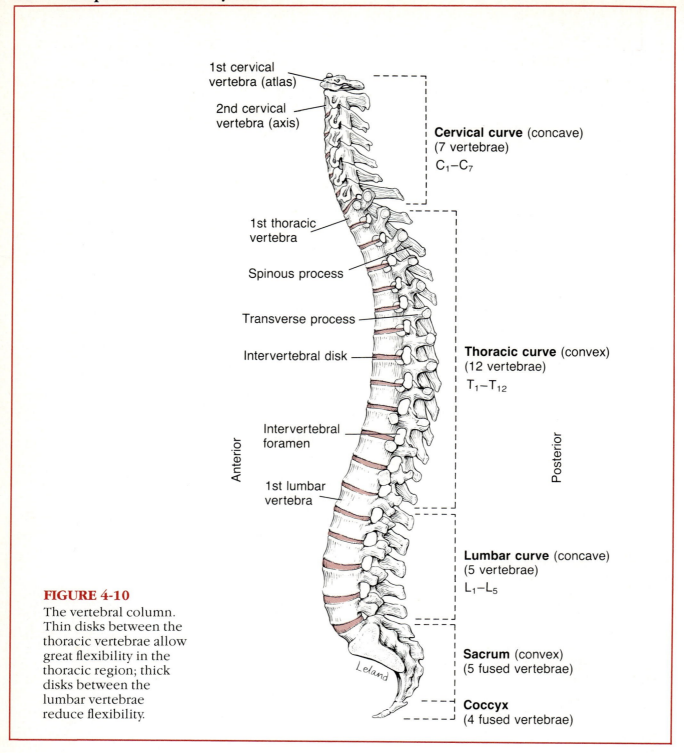

1st cervical
vertebra (atlas)

2nd cervical
vertebra (axis)

Cervical curve (concave)
(7 vertebrae)
C_1–C_7

1st thoracic
vertebra

Spinous process

Transverse process

Intervertebral disk

Thoracic curve (convex)
(12 vertebrae)
T_1–T_{12}

Anterior

Posterior

Intervertebral
foramen

1st lumbar
vertebra

Lumbar curve (concave)
(5 vertebrae)
L_1–L_5

Leland

Sacrum (convex)
(5 fused vertebrae)

Coccyx
(4 fused vertebrae)

FIGURE 4-10
The vertebral column.
Thin disks between the
thoracic vertebrae allow
great flexibility in the
thoracic region; thick
disks between the
lumbar vertebrae
reduce flexibility.

The "typical" cervical vertebrae (C_3 through C_7) are shown in Figure 4-13b. They are the smallest, lightest vertebrae, and most often the spinous process is short and divided into two branches. The transverse processes of the cervical vertebrae contain foramina (openings) through which the vertebral arteries pass on their way to the brain above. Any time you see these foramina in a verte-bra, you should know immediately that it is a cervical vertebra.

THORACIC VERTEBRAE
The 12 **thoracic vertebrae** (T_1–T_{12}) are all typi-cal. As seen in Figure 4-13c, they have a larger body than the cervical vertebrae. Their body is somewhat heart-shaped and has two costal facets

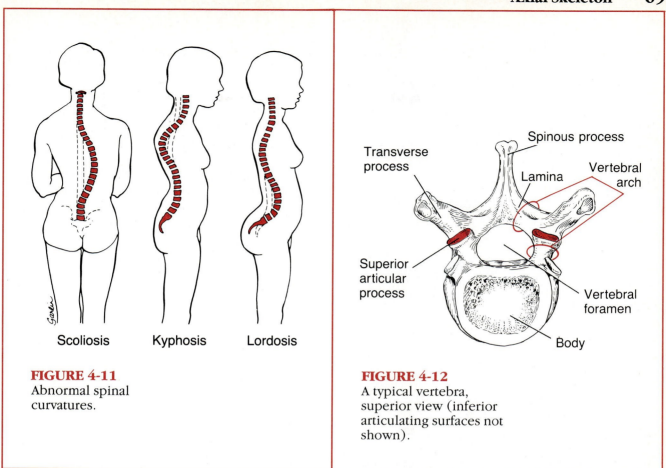

FIGURE 4-11
Abnormal spinal curvatures.

Scoliosis Kyphosis Lordosis

FIGURE 4-12
A typical vertebra, superior view (inferior articulating surfaces not shown).

Transverse process

Spinous process

Lamina

Vertebral arch

Superior articular process

Vertebral foramen

Body

(articulating surfaces) on each side. The spinous process is long with a sharp downward hook.

LUMBAR VERTEBRAE

The five **lumbar vertebrae** (L_1–L_5) have large blocklike bodies and short, thick spinous processes (Figure 4-13*d*). Since most of the stress on the vertebral column occurs in the lumbar region, these are the sturdiest of the vertebrae.

The spinal cord ends at the superior edge of L_2, but the outer covering of the cord, filled with cerebrospinal fluid, extends well beyond. Thus, a *lumbar puncture* (for examination of the cerebrospinal fluid) or "saddle block" anesthesia administration for childbirth is normally done between L_3 and L_5 where there is no chance of injuring the delicate spinal cord.

SACRUM

The **sacrum** (sa′krum) is formed by the fusion of five vertebrae (Figure 4-14). Superiorly it articulates with L_5, and inferiorly it connects with the coccyx. The winglike *alae* articulate laterally with the hipbones, forming the sacroiliac joints. The

sacrum forms the posterior wall of the pelvis. The vertebral canal continues inside the sacrum as the **sacral canal.**

COCCYX

The **coccyx** is formed from the fusion of three to five small, irregularly shaped vertebrae (Figure 4-14). It is the human "tailbone," a remnant of the tail that other vertebrate animals have.

Bony Thorax

The sternum, ribs, and thoracic vertebrae make up the **bony thorax.** The bony thorax is often called the thoracic cage because it forms a protective cone-shaped cage of slender bones around the organs of the thoracic cavity (heart, lungs, and major blood vessels). The bony thorax is shown in Figure 4-15.

STERNUM

The **sternum** (breastbone) is a typical flat bone and the result of the fusion of three bones—the **manubrium** (mah-nu′bre-um), **body,** and **xiphoid** (zif′oid) **process.** It is attached to the first

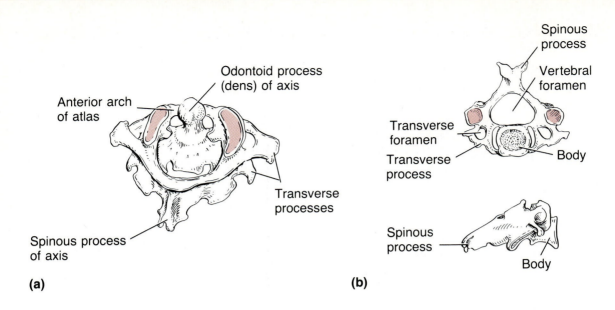

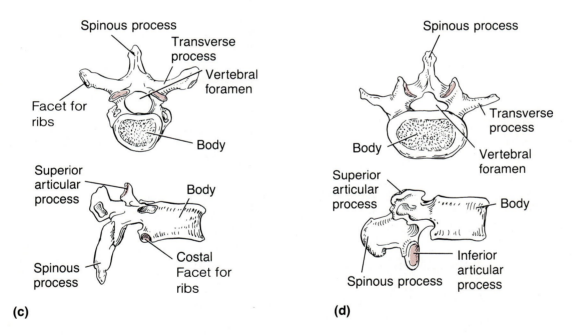

FIGURE 4-13
Cervical vertebrae.
(a) Superior view of the articulated atlas and axis. **(b)** Cervical vertebrae; superior view above, lateral view below. **(c)** Thoracic vertebrae; superior view above, lateral view below. **(d)** Lumbar vertebrae; superior view above, lateral view below.

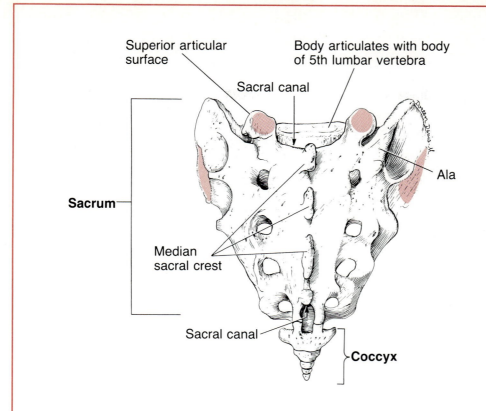

FIGURE 4-14
Sacrum and coccyx, posterior view.

Superior articular surface

Body articulates with body of 5th lumbar vertebra

Sacral canal

Sacrum

Ala

Median sacral crest

Sacral canal

Coccyx

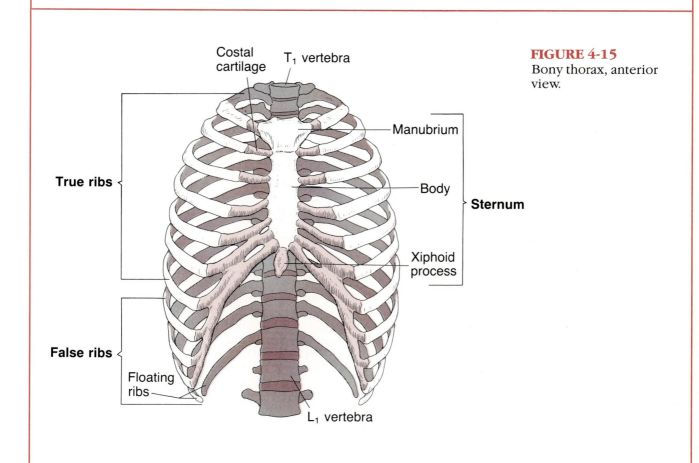

FIGURE 4-15
Bony thorax, anterior view.

Costal cartilage

T_1 vertebra

Manubrium

True ribs

Body

Sternum

Xiphoid process

False ribs

Floating ribs

L_1 vertebra

seven pairs of ribs. Because the sternum is so close to the body surface, it is easy to obtain samples of blood-forming (hemopoietic) tissue for the diagnosis of suspected blood diseases from this bone. A needle is inserted into the marrow of the sternum, and the sample is withdrawn; this is called a *sternal puncture.*

RIBS

The 12 pairs of **ribs** form the walls of the thoracic cage. (Contrary to a popular misconception, males do *not* have one less rib than females.) All of the ribs articulate with the vertebral column posteriorly and then curve downward and toward the anterior body surface. The *true ribs,* the first 7 pairs, attach directly to the sternum by costal cartilages. *False ribs,* the next 5 pairs, have cartilage attachments to the sternum that are indirect or are not attached to the sternum at all. The last 2 pairs of false ribs lack sternal attachment, and so they are also called *floating ribs.* Intercostal spaces (spaces between the ribs) are filled with the intercostal muscles that aid in breathing. Take a deep breath to expand your chest. Notice how your ribs seem to move outward and how your sternum rises.

APPENDICULAR SKELETON

The appendicular skeleton is shown by the unshaded part of Figure 4-3. It is composed of 126 bones of the limbs (appendages) and the pectoral and pelvic girdles, which attach the limbs to the axial skeleton.

Bones of the Shoulder Girdle

The **shoulder girdle,** or **pectoral girdle,** consists of four bones—two clavicles and two scapulae (Figure 4-16).

The **clavicle** (klav′ĭ-k′l), or collarbone, is a slender doubly curved bone. It attaches to the manubrium of the sternum medially and to the scapula laterally (where it helps to form the shoulder joint). The clavicle acts as a brace to hold the arm away from the top of the thorax and helps prevent shoulder dislocation. When the clavicle is broken

the whole shoulder region caves in medially, which shows how important it is as a brace.

The **scapulae** (skap′u-le), or shoulder blades, are triangular and are commonly called "wings." Each scapula has a flattened body and two important processes—the **acromion** (ah-kro′me-on) **process,** which is the enlarged end of the spine of the scapula, and the beaklike **coracoid** (kor′-ah-koid) **process.** The acromion process connects with the clavicle laterally. The coracoid process points over the top of the shoulder and provides a site of attachment for some of the muscles of the upper arm. The scapula is not directly attached to the axial skeleton; it is loosely held in place by trunk muscles. The scapula has three angles (superior, inferior, and lateral) and three borders (superior, vertebral, and axillary). The **glenoid fossa,** a shallow socket that receives the head of the upper arm bone, is located in the lateral angle.

The shoulder girdle is very light and allows the upper limb to have exceptionally free movement. This is due to the following factors:

1. Each shoulder girdle attaches to the axial skeleton at only one point—the sternoclavicular joint.

2. The loose attachment of the scapulae allows them to slide back and forth against the thorax as muscles act.

3. The glenoid fossa is shallow, and the shoulder joint is poorly reinforced with ligaments.

    However, this exceptional flexibility also has a drawback; the shoulder girdle is very easily dislocated.

Bones of the Upper Limbs

UPPER ARM

The upper arm is formed by a single bone, the **humerus** (hu′mer-us), which is a typical long bone (Figure 4-17*a* and *b*). At its proximal end is a rounded head that fits into the shallow glenoid fossa of the scapula. Opposite the head are two bony projections—the **greater** and **lesser tubercles,** which are sites of muscle attachment. In the midpoint of the shaft is a roughened area called

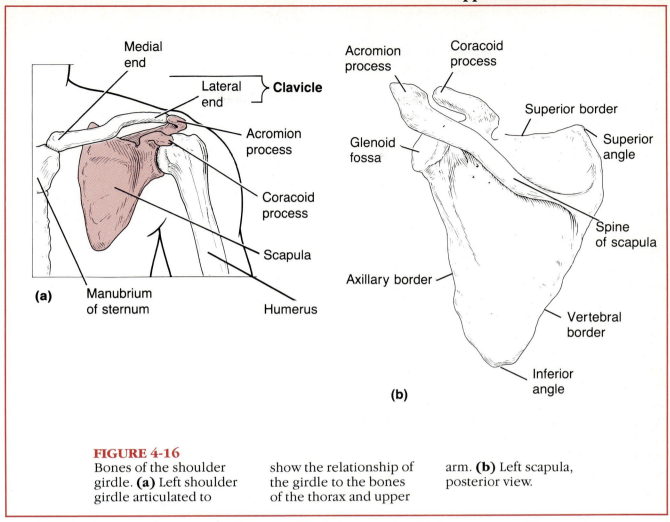

FIGURE 4-16
Bones of the shoulder girdle. **(a)** Left shoulder girdle articulated to show the relationship of the girdle to the bones of the thorax and upper arm. **(b)** Left scapula, posterior view.

the **deltoid tuberosity,** where the large, fleshy deltoid muscle attaches. At the distal end of the humerus is the medial **trochlea** (trok′le-ah), which looks rather like a spool, and the lateral **capitulum** (kah-pit′u-lum); both of these processes articulate with the bones of the lower arm (forearm). Above the trochlea anteriorly is a depression, the **coronoid fossa;** on the posterior surface is the **olecranon** (o-lek′rah-non) **fossa.** These two depressions allow the corresponding processes of the ulna to move freely when the elbow is bent and extended.

FOREARM
Two bones, the radius and the ulna, form the skeleton of the forearm, or lower part of the arm (Figure 4-17c). When the body is in the anatomical position, the **radius** is the lateral bone, that is, it is on the thumb side of the arm. When the hand is rotated so that the palm faces backward, the distal end of the radius ends up medial to the ulna.

Proximally, the disk-shaped head of the radius forms a joint with the capitulum of the humerus. Just below the head is the **radial tuberosity,** the point where the tendon of the biceps muscle attaches.

When the arm is in the anatomical position, the **ulna** is the medial bone (on the little finger side) of the forearm. On its proximal end are the anterior **coronoid process** and the posterior **olecranon process,** which are separated by the **semilunar notch.** Together these two processes grip the trochlea of the humerus in a plierslike joint.

WRIST AND HAND
The wrist (*carpus*) is composed of eight **carpal bones** that are arranged in two irregular rows of four bones each. The carpals are bound closely together by ligaments that restrict movements between them (Figure 4-18).

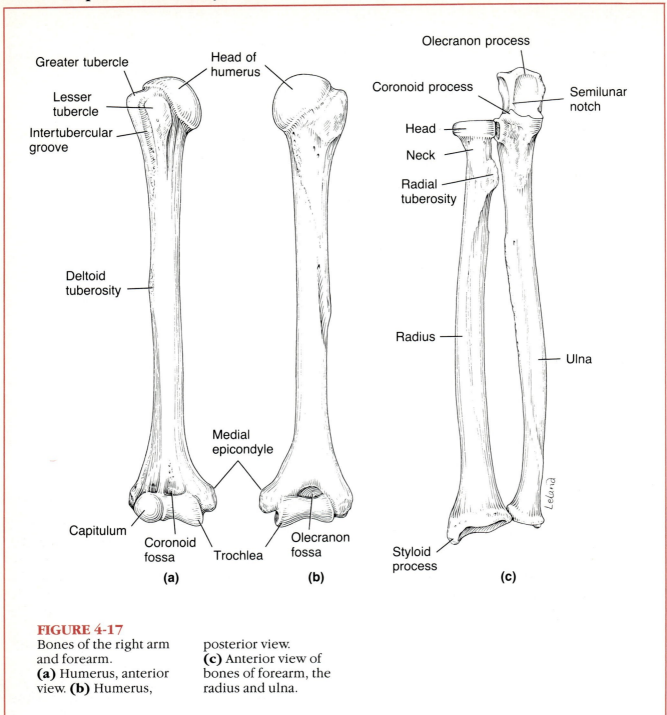

Greater tubercle

Head of humerus

Olecranon process

Lesser tubercle

Coronoid process

Semilunar notch

Intertubercular groove

Head

Neck

Radial tuberosity

Deltoid tuberosity

Radius

Ulna

Medial epicondyle

Capitulum

Coronoid fossa

Trochlea

Olecranon fossa

Styloid process

(a) **(b)** **(c)**

FIGURE 4-17
Bones of the right arm and forearm.
(a) Humerus, anterior view. **(b)** Humerus, posterior view.
(c) Anterior view of bones of forearm, the radius and ulna.

The hand consists of the **metacarpals** (bones of the palm) and the **phalanges** (fah-lan′jēz), or bones of the fingers. The metacarpals are numbered 1 to 5 from the thumb side of the hand toward the little finger. When the fist is clenched, the heads of the metacarpals become obvious as the "knuckles." Each hand contains 14 phalanges. There are 3 in each finger (proximal, middle, and distal) except the thumb, which has only 2 (proximal and distal).

Bones of the Pelvic Girdle

The **pelvic girdle,** or **pelvis,** is formed by two **os coxae** (os kok′se), or hipbones, plus the sacrum and the coccyx (Figure 4-19). The bones of the pelvic girdle are large and heavy, and the pelvis is attached securely to the axial skeleton. The sockets, which receive the thighbones, are deep and heavily reinforced by ligaments to hold the leg

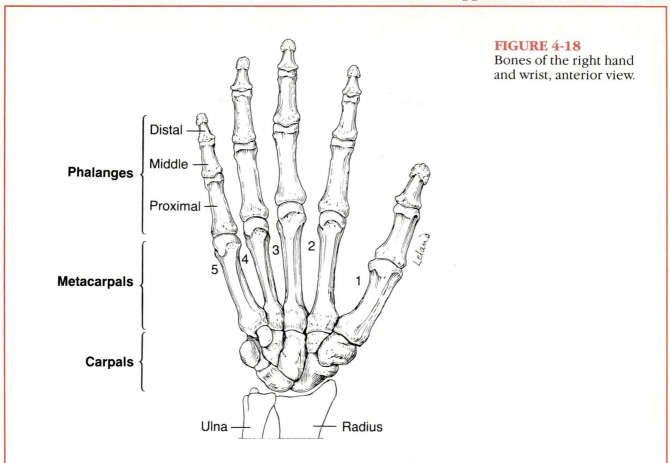

FIGURE 4-18
Bones of the right hand
and wrist, anterior view.

Phalanges — Distal, Middle, Proximal

Metacarpals — 5 4 3 2 1

Carpals

Ulna — Radius

firmly attached to the girdle. Bearing weight is the most important function of this girdle, and the total weight of the upper body rests on the pelvis. The reproductive organs, urinary bladder, and part of the large intestine lie within and are protected by the pelvis.

Each os coxa is formed by the fusion of three bones: the ilium, ischium, and pubis. The **ilium** (il'e-um), which connects posteriorly with the sacrum at the **sacroiliac** (sa"kro-il'e-ak) **joint,** is a large flaring bone that forms most of the os coxa. When you put your hands on your hips, they are resting over the ilia. The upper edge of the ilium, the **iliac crest,** is an important anatomical landmark that is always remembered by those who give injections.

The **ischium** (is'ke-um) is the "sitdown bone," since it forms the most inferior part of the os coxa. The **ischial tuberosity** is a roughened area that receives body weight when sitting. The **ischial spine,** superior to the tuberosity, is an important anatomical landmark particularly in the pregnant

woman because it narrows the outlet of the pelvis through which the baby must pass during the birth process. Another important structural feature of the ischium is the **greater sciatic notch,** which allows blood vessels and the large sciatic nerve to pass from the pelvis posteriorly into the thigh; medical personnel giving injections in the buttock should always stay well away from this area.

The **pubis** (pu'bis) is the most anterior part of the os coxa. Fusion of the rami of the pubic bone anteriorly and the ischium posteriorly forms a bar of bone enclosing the **obturator** (ob'tu-ra"tor) **foramen,** an opening which allows blood vessels and nerves to pass into the anterior part of the thigh. The pubic bones of each hipbone fuse anteriorly to form a cartilage joint, the **pubic symphysis** (pu'bik sim'fĭ-sis).

The ilium, ischium, and pubis fuse at the deep socket called the **acetabulum** (as"ĕ-tab'u-lum), which means "vinegar cup." The acetabulum receives the head of the thighbone.

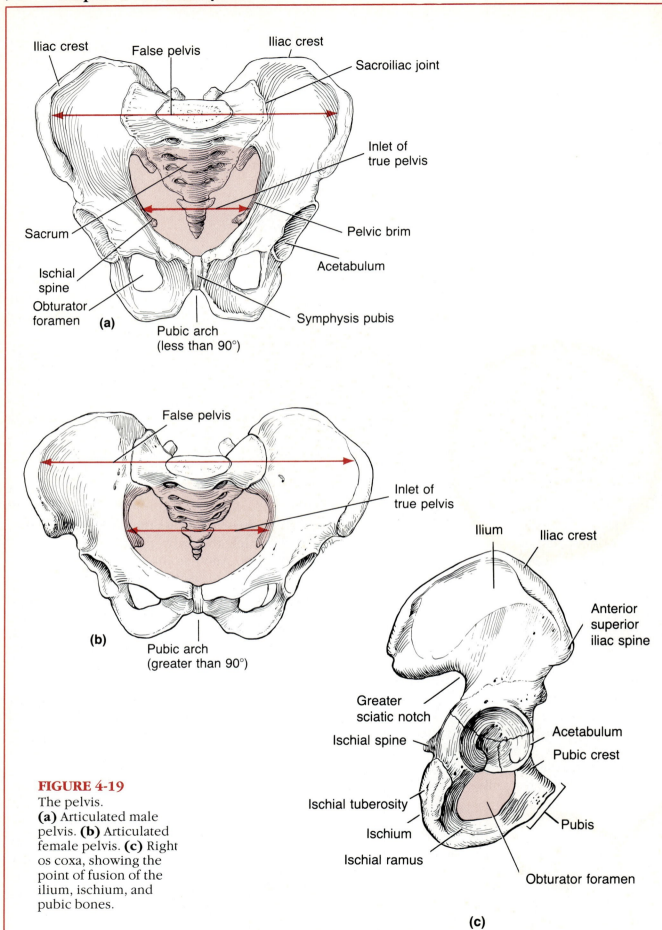

FIGURE 4-19
The pelvis.
(a) Articulated male
pelvis. **(b)** Articulated
female pelvis. **(c)** Right
os coxa, showing the
point of fusion of the
ilium, ischium, and
pubic bones.

The pelvis is divided into two regions. The *true pelvis* is surrounded by bone and lies inferior to the flaring parts of the ilia (the pelvic brim). The *false pelvis* is superior to the true pelvis; it is the area medial to the flaring portions of the ilia. The dimensions of the true pelvis of a woman are very important because they must be large enough to allow the infant's head (the largest part of the infant) to pass during childbirth. The dimensions of the cavity, the outlet (the inferior opening of the pelvis) and the inlet (superior opening) are critical, and they are carefully measured by the obstetrician.

Of course, individual pelvic structure varies, but there are differences between a male's and a female's pelvis that are fairly constant. Look at Figure 4-19 again and note the following characteristics that differ in the male and female pelves.

- The female inlet is larger and more circular.

- The female pelvis as a whole is more shallow, and the bones are lighter and thinner. The ilia flare more laterally.

- The female sacrum is broader and less curved, and the pubic arch is more rounded.

- The female ischial spines are shorter, thus the outlet is larger.

Bones of the Lower Limbs

THIGH

The **femur** (fe'mur), or thighbone, is the only bone of the thigh or upper leg (Figure 4-20a and b). It is the heaviest, strongest bone in the body. Its proximal end has a ball-like head, a neck, and **greater** and **lesser trochanters** (separated posteriorly by the **intertrochanteric crest**). The head of the femur articulates with the acetabulum of the hipbone in a deep, secure socket. However, the neck of the femur is a common fracture site, especially in old age.

The femur slants medially as it runs downward to join with the lower leg bones; this brings the knees in line with the body's center of gravity.

The medial course of the femur is more noticeable in females because of the wider female pelvis.

Distally on the femur are the **lateral** and **medial condyles,** which articulate with the tibia below. The trochanters, intertrochanteric crest, and the **gluteal tuberosity,** located on the shaft, all serve as sites for muscle attachment.

LOWER LEG

Two bones, the tibia and fibula, form the lower leg (see Figure 4-20c). The **tibia,** or shinbone, is larger and more medial. At the proximal end, the **medial** and **lateral condyles** (separated by the intercondylar eminence) articulate with the distal end of the femur to form the knee joint. The patellar (kneecap) ligament attaches to the **tibial tuberosity,** a roughened area on the anterior tibial surface. Distally the **medial malleolus** (mal-le'o-lus) process forms the inner bulge of the ankle. The anterior surface of the tibia is a sharp ridge (anterior crest) that is unprotected by muscles; thus, it is easily felt beneath the skin.

The **fibula,** which lies alongside the tibia, has no part in forming the knee joint. The fibula is thin and sticklike. Its distal end, the **lateral malleolus,** forms the outer part of the ankle.

ANKLE AND FOOT

The **tarsus,** or ankle, is composed of seven tarsal bones (Figure 4-21). Body weight is mostly carried by the two largest tarsals, the **calcaneus** (kal-ka'ne-us), or heelbone, and the **talus** (ta'lus) that lies between the tibia and the calcaneus.

The bones of the foot include 5 **metatarsals,** which form the instep, and 14 **phalanges,** which form the toes. Like the fingers of the hand, each toe has 3 phalanges except the great toe, which has 2.

The bones in the foot are arranged to form three strong arches: two longitudinal (medial and lateral) and one transverse (Figure 4-22). **Ligaments,** which bind the foot bones together, and **tendons** of the foot muscles help to hold the bones firmly in the arched position but still allow a certain amount of give or springiness. Weak arches are referred to as fallen arches or flat feet.

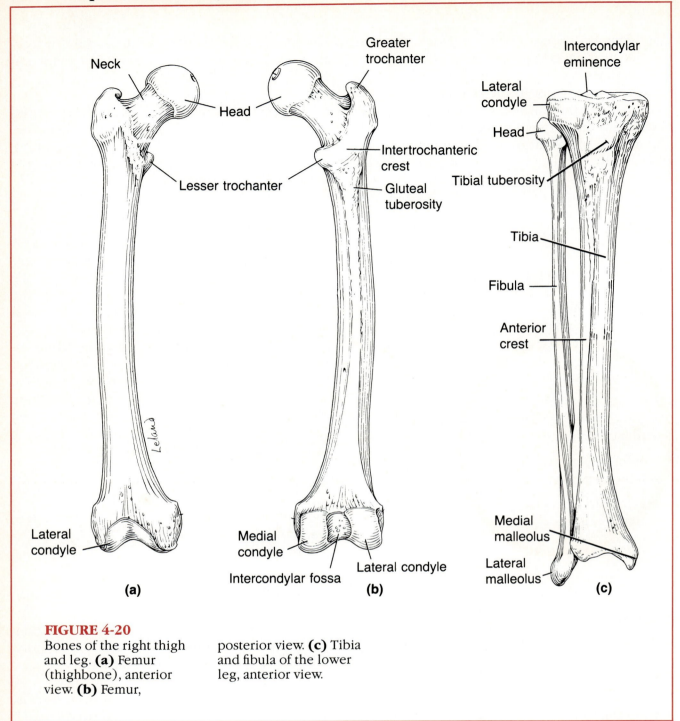

Neck

Head

Lesser trochanter

Lateral condyle

(a)

Greater trochanter

Intertrochanteric crest

Gluteal tuberosity

Medial condyle

Intercondylar fossa

Lateral condyle

(b)

Intercondylar eminence

Lateral condyle

Head

Tibial tuberosity

Tibia

Fibula

Anterior crest

Medial malleolus

Lateral malleolus

(c)

FIGURE 4-20
Bones of the right thigh and leg. **(a)** Femur (thighbone), anterior view. **(b)** Femur, posterior view. **(c)** Tibia and fibula of the lower leg, anterior view.

ARTICULATIONS, OR JOINTS

With one exception (the hyoid bone of the neck), every bone in the body forms a joint with at least one other bone. **Articulations,** or joints, have two functions: They hold the bones together and allow the rigid skeletal system to become somewhat flexible.

All joints consist of bony regions separated by cartilage or fibrous connective tissue, and they are classified into three groups according to the amount of movement they allow. The joint types are shown in Figure 4-23 and described next.

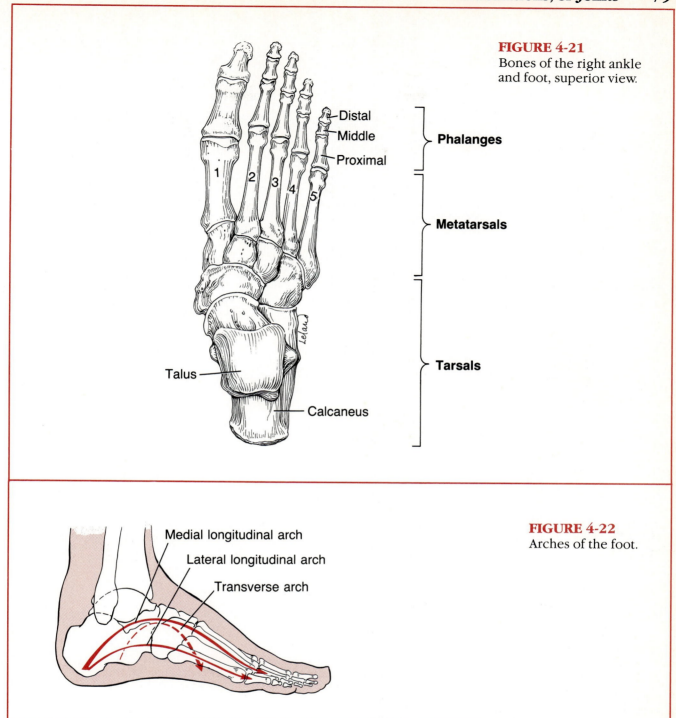

FIGURE 4-21
Bones of the right ankle and foot, superior view.

Distal
Middle
Proximal
Phalanges

Metatarsals

Talus

Calcaneus

Tarsals

FIGURE 4-22
Arches of the foot.

Medial longitudinal arch
Lateral longitudinal arch
Transverse arch

Synarthroses

Synarthroses (sin″ar-thro′sēs) are joints that allow essentially no movement, thus they are also called *immovable joints*. The bone ends forming the joint are connected by fibrous tissue. The best examples of this type of joint are the sutures of the skull. In sutures, the irregular edges of the bones interlock and are bound tightly together by fibrous connective tissue.

Amphiarthroses

Amphiarthroses (am-fe-ar-thro′sēz) are joints that are *slightly movable*. Basically they are joints in which bones are connected by a cartilage disk.

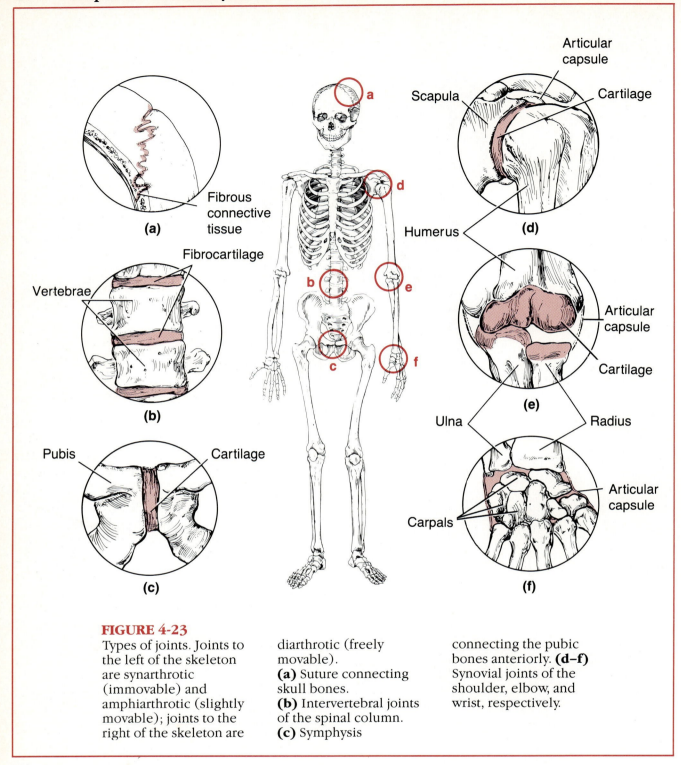

(a) Fibrous connective tissue

Fibrocartilage

Vertebrae

(b)

Pubis   Cartilage

(c)

Scapula

Articular capsule

Cartilage

Humerus

(d)

Articular capsule

Cartilage

(e)

Ulna   Radius

Carpals

Articular capsule

(f)

FIGURE 4-23
Types of joints. Joints to the left of the skeleton are synarthrotic (immovable) and amphiarthrotic (slightly movable); joints to the right of the skeleton are diarthrotic (freely movable).
(a) Suture connecting skull bones.
(b) Intervertebral joints of the spinal column.
(c) Symphysis connecting the pubic bones anteriorly. **(d–f)** Synovial joints of the shoulder, elbow, and wrist, respectively.

Examples of this joint type include the pubic symphysis of the pelvis and intervertebral joints of the spinal column.

Diarthroses

Diarthroses (di″ar-thro′sēs) have much more freedom or flexibility than the other two joint types, and they are commonly referred to as *freely movable* joints. However, their flexibility does vary. Some diarthroses can move in only one plane—for example, the hingelike elbow joint or the pivot joint between the atlas and axis. Others can move in two planes (biaxial joints like the wrist joints and knuckles), and still others are multiaxial joints that move in all planes such as the ball and socket joints (shoulder and hip joints). All joints of the limbs are diarthrotic. The various types of movements that occur at diarthrotic joints are discussed in detail on pp. 94–96, because they relate to muscle activity.

All diarthrotic joints (also called synovial joints) have the following structural plan (Figure 4-24):

1. The joint surfaces are enclosed by a sleeve or capsule of fibrous connective tissue.

2. The capsule is lined with a smooth synovial membrane, which secretes a lubricating fluid to reduce friction in the joint. (For this reason these joints are also called synovial joints.)

3. Hyaline (articular) cartilage covers the ends of the bones forming the joint.

4. The fibrous capsule is usually reinforced with ligaments, and bursae (fluid-filled synovial membrane sacs) are often found cushioning tendons where they cross bone.

Most of us don't think too much about our joints until something goes wrong with them. Joint pains and problems may be caused by many things. For example, a hard blow to the knee can cause *bursitis* or "water on the knee" due to inflammation of bursae or synovial membrane. Sprains and dislocations are other types of joint problems. In a *sprain,* the ligaments or tendons reinforcing a joint are damaged by excessive stretching, or they are torn away from the bone. Since both tendons and ligaments are cords of dense fibrous connective tissue with a poor blood supply, sprains heal slowly and are painful. *Dislocations* happen when bones are forced out of their normal position in the joint cavity. The process of returning the bone to its proper position should be done only by a physician. Attempts by an untrained person to "snap the bone back into its socket" are usually more harmful than helpful.

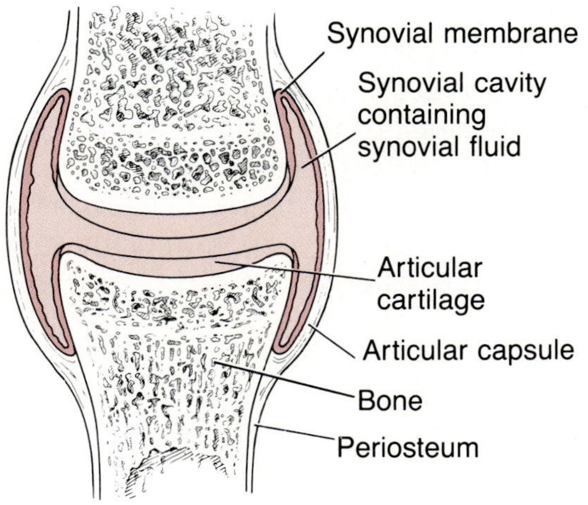

DEVELOPMENTAL ASPECTS OF THE SKELETON

In the very young fetus the long bones are formed of hyaline cartilage, and the flat bones of the skull are formed of fibrocartilage membranes. A fetal skeleton is illustrated in Figure 4-25. As the fetus continues to develop and grow, both flat and long bones are converted to bone. At birth, small areas of the fibrocartilage membrane (fontanels) still remain in the skull to allow for brain growth. These areas are usually fully ossified by age 2 years. In the ends of the long bones are platelike areas of hyaline cartilage called epiphyseal disks (see p. 59), which provide for long bone growth in childhood. By the end of puberty, the epiphyseal disks become fully ossified, and long bone growth ends.

Most cases of abnormal spinal curvatures, such as scoliosis and lordosis (see pp. 67 and 69), are congenital but can result from injuries. The abnormal curvatures are usually treated by surgery, braces, or casts when diagnosed. Generally speaking, young healthy people have no skeletal problems assuming that their diet is nutritious and they stay reasonably active.

Rickets is a disease of children that results from the failure of bones to calcify. As a result, the bones soften and a definite bowing of the weight-bearing bones of the legs occurs. Rickets is usually due to a lack of calcium in the diet or a lack of vitamin D, which is needed to absorb calcium

Synovial membrane

Synovial cavity containing synovial fluid

Articular cartilage

Articular capsule

Bone

Periosteum

FIGURE 4-24
Structure of a diarthrotic joint.

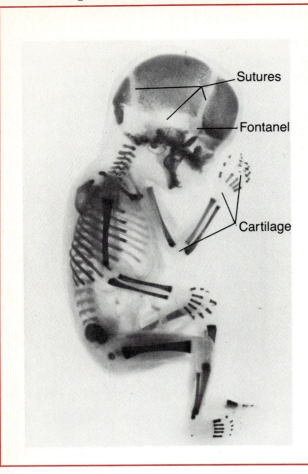

Sutures

Fontanel

Cartilage

FIGURE 4-25
Skeleton of a 12-week fetus. Darker areas of the skeleton represent ossified, or bony, areas whereas the light areas indicate cartilage that has not yet been replaced by bone. The fontanels, soft membranous areas between the sutures, are still clearly visible. Bone ossification is not complete until puberty.

into the blood stream. Rickets is not seen very often in the United States where great stress is put on good nutrition. Milk, bread, and other foods are fortified with vitamin D, and most children drink enough calcium-rich milk. However, it remains a problem in other parts of the world.

It cannot be emphasized strongly enough that bones have to be physically stressed to remain healthy. The more you use them, the stronger they get. When we remain active physically and muscles and gravity pull on the skeleton, the bones respond by becoming stronger. On the other hand, if we are totally inactive, they begin to atrophy, waste away, and become thin and fragile. Bone wasting *(osteoporosis)* is common in elderly people for two reasons. First, hormonal changes occur that lead to the withdrawal of calcium from the bones and a generalized bone thinning. This is particularly rapid in women after menopause, or "change of life." Second, many elderly people feel that they are helping themselves by "saving their strength" and not doing anything too physical. Their reward for this is pathologic fractures (spontaneous breaks without apparent injury),

which increase dramatically with age and are the single most common skeletal problem for this age group.

Advancing years also take their toll on joints. Weightbearing joints in particular begin to degenerate and become painful, a condition called *osteoarthritis,* or *degenerative arthritis. Adhesions* (a growing together of the bony areas) may form between the bone surfaces, and extra bone tissue *(spurs)* may grow along the joint edges. Such degenerative changes lead to the complaint so often heard from the aging person: "My joints are getting so stiff . . .". There are many types of arthritis (which means joint inflammation). Another familiar type is *rheumatoid arthritis,* which affects both young and old alike. In rheumatoid arthritis, the joints swell and the synovial membranes become very inflamed and are gradually destroyed, leading to crippling. Usually there is bilateral joint involvement, that is, if it is in the wrists, *both* wrists will affected. The cause of rheumatoid arthritis is not really known, but it is believed to be an autoimmune disease (a situation in which the body destroys its own tissues).

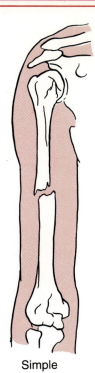
Simple

Compound

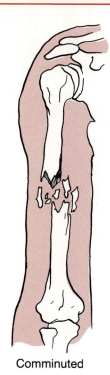
Comminuted

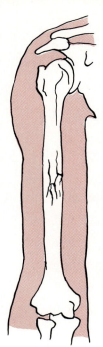
Greenstick

BOX 4-1
Fractures

Despite their remarkable strength, bones are susceptible to breaks or **fractures** all through life. During youth most fractures result from trauma or external forces, which act to twist or compress the bone. Sports activities such as skiing and football often jeopardize the bones. In old age bone is lost faster than it is replaced, which leads to its thinning and general loss of strength.

There are five common types of fractures. In *simple fractures* the bone breaks but does not penetrate the skin. Simple fractures are not nearly as dangerous as *compound fractures* in which the soft tissues are also torn and the broken ends of the bone

protrude through the skin. Exposure of the broken bone ends to bacteria often results in a very severe bone infection, *osteomyelitis* (os″te-o-mi″ĕ-li′tis). This condition requires massive doses of antibiotics for its control. *Comminuted* (kom′i-nut″ed) *fractures* occur when there is more than one line of break and several bone fragments are formed. *Compression fractures,* in which the bone is crushed, are a common type of vertebral column fracture seen in elderly people. A type of fracture typical of the way children's bones break is a *greenstick fracture*. In this type of fracture the bone breaks incompletely much in the same way a green branch breaks. This is because children's bones are more flexible

and have less calcium salts (which make bones hard and brittle) than do adult's bones. Some of these common fracture types are illustrated here.

A fracture is treated by *reduction,* which is the realignment of the broken bone ends. Reduction may be *closed*—bone ends are coaxed back into their normal position by the physician with his or her hands—or *open.* In open reductions surgery is performed, and the bone ends are held together with pins or wires. Once surgery is completed, the broken area is immobilized by a cast or traction. The healing time of a simple fracture is 6–8 weeks, but it is often much longer for elderly persons because of their poorer circulation.

IMPORTANT TERMS*

Axial skeleton	Red marrow
Appendicular skeleton	Skull
Compact bone	Vertebral column
Spongy bone	Intervertebral disks
Long bones	Bony thorax
Short bones	Shoulder girdle
Flat bones	Pelvic girdle
Irregular bones	Ligaments
Diaphysis *(di-af″i-sis)*	Articulations
Epiphyses *(ĕ-pif″i-sēs)*	Synarthroses *(sin″ar-thro′sēs)*
Articular cartilage	Amphiarthroses *(am″fe-ar-thro′sēs)*
Epiphyseal disk	Diarthroses *(di″ar-thro′sēs)*

*For definitions, see Glossary.

SUMMARY

A. FUNCTION

1. Bones provide a hard internal support for the body organs and protect some organs by enclosing them.

2. Bones serve as levers for the muscles to pull on to cause movement at joints.

3. Bones store calcium, fats, and other substances for the body.

4. Bones are the location for red marrow, the site of blood cell production.

B. BONES, AN OVERVIEW

1. Bone markings, projections and depressions are important anatomical landmarks that reveal where muscles attach(ed) and where blood vessels and nerves pass(ed).

2. Bones are classified into four groups—long, short, flat, and irregular—on the basis of their shape and amount of compact or spongy bone they contain.

3. A long bone is composed of a shaft (diaphysis) with two ends (epiphyses). The shaft is compact bone, and its cavity contains yellow marrow. The epiphyses are covered with hyaline cartilage, and they contain spongy bone (where red marrow is found). In a growing bone epiphyseal disks are the growth areas for longitudinal growth.

4. The organic parts of the matrix make bone flexible; calcium salts deposited in the matrix make bone hard.

C. AXIAL SKELETON

1. The skull is formed by cranial and facial bones. The 8 cranial bones protect the brain. They include the frontal, occipital, ethmoid, and sphenoid bones, and the paired parietal and temporal bones. The 14 facial bones are all paired (maxillae, zygomatics, palatines, nasals, lacrimals, and inferior conchae) except for the vomer and mandible.

2. Skulls of newborn infants contain *fontanels* (membranous areas), which allow brain growth. The facial bones are very small compared to the size of the infant's cranium.

3. The vertebral column is formed from 24 vertebrae, the sacrum, and the coccyx. There are 7 cervical vertebrae, 12 thoracic vertebrae, and 5 lumbar vertebrae, which have common as well as unique features. The vertebrae are separated by fibrous disks that allow the vertebral column to be flexible. The vertebral column is **S**-shaped to allow for the upright posture. Spinal curvatures present at birth are the thoracic and sacral curvatures; the secondary curvatures (cervical and lumbar) form after birth.

4. The bony thorax is formed from the sternum and 12 pairs of ribs. All ribs attach posteriorly to thoracic vertebrae; anteriorly the first 7 pairs attach directly to the sternum (true ribs). The last 5 pairs attach indirectly or not at all (false ribs). The bony thorax encloses the lungs, heart, and other thoracic cavity organs.

D. APPENDICULAR SKELETON

1. The shoulder girdle, composed of two bones—the scapula and clavicle—attaches the arm to the axial skeleton. It is a light, poorly reinforced girdle that allows the arm a great deal of freedom.

2. The bones of the upper limb include the humerus (upper arm), radius and ulna (forearm), eight carpals (wrist bones), and the metacarpals and phalanges of the hand.

3. The pelvic girdle is formed by the two heavy os coxae. Each os coxa is the result of fusion of the ilium, ischium, and pubis bones. The pelvic girdle is securely attached to the axial skeleton, and the socket for the thighbone is deep and heavily reinforced. This girdle receives the weight of the upper body and transfers it to the legs. The female pelvis is lighter and broader than the male's, and its inlet and outlet are larger. All of these modifications reflect the childbearing function of the female.

4. The bones of the lower limb include the femur (thighbone), tibia and fibula (leg bones), seven tarsal or ankle bones, and metatarsals and phalanges of the foot.

E. ARTICULATIONS, OR JOINTS

1. Joints hold bones together and allow movement of the skeleton.

2. Joints fall into three categories: synarthroses (immovable joints), amphiarthroses (slightly movable joints), and diarthroses (freely movable joints).

3. Most joints of the body are diarthrotic, or synovial. In synovial joints, the articulating bone surfaces are covered with articular cartilage and enclosed in a fibrous capsule lined with a synovial membrane.

F. DEVELOPMENTAL ASPECTS OF THE SKELETON

1. The fetal skeleton is cartilage most of which is replaced by bone by birth. Cartilage areas persist for a brief period in the skull (fontanels) and until the end of puberty in the long bones (epiphyseal disks).

2. Abnormal spinal curvatures are usually congenital and are treated by surgery or casts. During youth, most other bone problems are due to trauma (fractures) or poor nutrition (rickets). Bones require stress to remain healthy.

3. Bone fractures are the most common bone problem in elderly persons. Common types of fractures include: simple, compound, compression, comminuted, and greenstick. Bone fractures must be reduced to heal properly. Osteoporosis, a condition of bone wasting that results from hormone imbalance or inactivity, is also very common in elderly individuals.

4. The most common joint problem is arthritis, or inflammation of the joints. Rheumatoid arthritis occurs in both young and older adults; it is believed to be an autoimmune disease. Osteoarthritis, or degenerative arthritis, is a result of the "wear and tear" on joints over many years and is a common affliction of the aged.

REVIEW QUESTIONS

1. Name three functions of the skeletal system.

2. Name the four major classifications of bones. Name two examples of each type.

3. What is the proper anatomical name for the shaft of a long bone? For its ends? What is yellow marrow? How do spongy bone and compact bone look different?

4. Why do bone injuries heal much more rapidly than injuries to cartilage?

5. What is the function of the organic part of bone matrix? The inorganic part (bone salts)?

6. Name the three major parts of the axial skeleton.

7. Name the eight bones of the cranium.

8. What bones are connected by the frontal suture? By the sagittal suture?

9. With one exception, all skull bones are joined by sutures. Name the exception.

10. Name the four bones containing the paranasal sinuses.

11. What facial bone forms the chin? The cheekbone? The upper jaw? The bony eyebrow ridges?

12. Name two ways in which the fetal skull differs from the adult skull.

13. Name the five major regions of the vertebral column.

14. Diagram the normal spinal curvatures and then the curvatures seen in scoliosis and lordosis.

15. What is the function of the intervertebral disks?

16. Name the major components of the thorax.

17. What is a true rib? A false rib? Is a floating rib a true rib or false rib? Why do you think floating ribs are easily broken?

18. What is the general shape of the thoracic cage?

19. Name the bones of the shoulder girdle.

20. Name all the bones with which the ulna articulates.

21. The major function of the shoulder girdle is flexibility. What is the major function of the pelvic girdle?

22. What bones make up each hipbone or os coxa? Which of these is the largest? Which has tuberosities that we sit on? Which is most anterior?

23. List three differences between the male and female pelves.

24. Name the bones of the lower limb superiorly to inferiorly.

25. What is the function of joints?

26. Compare the amount of movement possible in synarthrotic, amphiarthrotic, and diarthrotic joints.

27. What is a fracture? What is reduction?

28. List two factors that keep bones healthy. Now list two factors that can cause bones to become soft or atrophy.

29. Define arthritis. What type of arthritis is most common in the elderly?

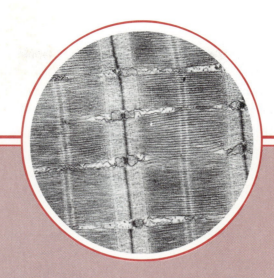

CHAPTER 5

Muscular System

Chapter Contents

Functions of the muscular system • to provide for movement of the body and its parts (as muscles shorten), and to generate heat

After completing this chapter, you should be able to:

- Describe similarities and differences in the three types of muscle tissue and note where they are found in the body.

- Describe the structure of skeletal muscle from gross to microscopic levels.

- Define and explain the role of the following: endomysium, perimysium, epimysium, tendon, and aponeurosis.

- Describe how an action potential is initiated in a muscle cell.

- Describe the events of muscle cell contraction.

- Define graded response, tetanus, muscle fatigue, isotonic and isometric contractions, and muscle tone.

- Define the terms: origin, insertion, prime mover, antagonist, synergist, and fixator as they relate to muscles.

- Demonstrate or identify the different types of body movements.

- List some criteria used in naming muscles.

- Name and locate the major muscles of the human body (on a torso model, muscle chart, or diagram) and state the action of each.

- Explain the importance of a nerve supply and exercise in keeping muscles healthy.

- Describe the changes that occur in muscle with aging.

The only function of muscle is contraction or shortening—a unique characteristic that sets it apart from any other body tissue. As a result of this ability, the muscular system is responsible for all body movement. The bulk of the body's muscle is voluntary, and its contraction is subject to our command. It is also called *skeletal muscle,* because it is attached to the skeleton (or underlying connective tissues). This type of muscle accounts for about 40% of body weight. Skeletal muscle helps form the contours of the body, provides a means of locomotion (swimming, running, and cross-country skiing, for example), and allows one to manipulate the environment. The 600-odd skeletal muscles are attached to bones like cables; when they contract, the muscles pull on bones causing movement. The most obvious examples of the action of muscles on bones are the movements that occur at the arm or leg joints. However, other less freely movable bones are also tugged into motion by the muscles; for example, movements that take place between the vertebrae when the torso is bent to the side. Muscles also pull on other muscles and the skin. When they pull the skin of your face, you can look interested or bored depending on your disposition at that particular time.

The balance of the body's muscle, which forms most of the walls of hollow organs (smooth muscle) and the heart (cardiac muscle), is involved in the transport of materials inside the body. These muscles are separate and distinct from the skeletal muscles. They have no relationship to the bony skeleton and operate without our conscious control.

Each of the three muscle types has a structure and function well suited for its job in the body. But since the term *muscular system* applies specifically to skeletal muscle, we will be concentrating on this muscle type in this chapter. The structure of the other two muscle types is considered briefly and only for comparison.

STRUCTURE OF MUSCLE TISSUE

Skeletal Muscle

Skeletal muscle is also known as *voluntary muscle* (because it is under our conscious control) and as *striated muscle* (because it appears to be striped). Skeletal muscle has some very special characteris-

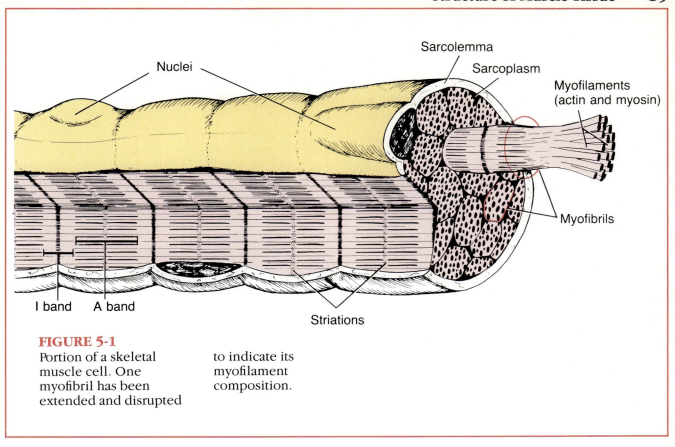

Nuclei

Sarcolemma

Sarcoplasm

Myofilaments
(actin and myosin)

Myofibrils

I band   A band

Striations

FIGURE 5-1

Portion of a skeletal
muscle cell. One
myofibril has been
extended and disrupted

to indicate its
myofilament
composition.

tics; thus, it makes sense to start the study of skeletal muscle at the cellular level.

Skeletal muscle cells (also called *muscle fibers*) are fairly large, long, cigar-shaped cells ranging from 10–100 mμ in diameter and up to 6 cm in length. However, the cells of large hard-working muscles, such as the antigravity muscles of the hip, are so coarse that they can be seen with the naked eye.

Skeletal muscle cells (Figure 5-1) are multinucleate. Many oval nuclei can be seen just beneath the cell membrane (also called the **sarcolemma** [sar″ko-lem′ah]) in these cells. The nuclei are pushed aside by long ribbonlike organelles, **myofibrils** (mi″o-fi′brils), which nearly fill the cytoplasm. Alternating light (I) and dark (A) bands along the length of the perfectly aligned myofibrils give the muscle cell, as a whole, its striped appearance. Myofibrils are made up of even smaller threadlike structures called **myofilaments.** The myofilaments are formed from two types of proteins, **actin** and **myosin,** which slide past each other during muscle activity causing the

shortening or contraction of the muscle cells. It is the precise arrangement of the myofilaments in the myofibrils that produces the banding pattern in skeletal muscle.

Muscle fibers are soft and surprisingly fragile. Thousands of muscle fibers are bundled together with connective tissue, which provides strength and support, to form the organs called skeletal muscles (Figure 5-2). Each muscle fiber is enclosed in a delicate connective tissue sheath called an **endomysium** (en″do-mis′e-um). Several sheathed muscle fibers are then wrapped by a collagen membrane called a **perimysium** to form a bundle of fibers called a **fascicle** (fas′ĭ-k′l). Many fascicles are bound together by an even tougher "overcoat" of connective tissue called an **epimysium,** or **deep fascia,** which covers the entire muscle. The epimysia blend into the strong cordlike **tendons,** which attach muscles indirectly to bones, or sheetlike **aponeuroses** (ap″o-nu-ro′sēs), which attach muscles to each other.

Tendons perform several functions: The most important are providing durability and conserving space. Since tendons are tough collagenic connective tissue, they can cross rough bony projections, which would literally rip apart the more delicate muscle tissues. Because of their relatively small size, more tendons than fleshy muscles can pass over a joint.

Many people think of muscles as always having an enlarged "belly" that tapers down to a tendon at each end. However, muscles vary considerably in the way their fibers are arranged. Many are spindle-shaped as just described; in others, the fibers are arranged in a fan-shape or a circle.

Smooth Muscle

Smooth muscle is involuntary, which means that we cannot consciously control it. Found in the walls of blood vessels and internal organs (stomach, small and large intestine, bladder, and so on), smooth muscle functions to propel substances along a definite tract, or pathway, within the body.

Smooth muscle cells are arranged in sheets or layers. Most often there are two layers, one running circularly and the other longitudinally as shown in Figure 5-3*a*. When the two layers alternately contract and relax, they change the size and shape of the organ. Movement of food through the digestive tract and emptying the bowels and bladder are typical examples of the types of "housekeeping" activities normally done by your smooth muscles.

Cardiac Muscle

Cardiac muscle is found in only one place in the body—the heart. The heart serves as a pump, propelling blood through the blood vessels to all tissues of the body. Cardiac muscle is involuntary and cannot be consciously controlled by most individuals.

The cardiac cells are cushioned by small amounts of soft connective tissue and arranged in spiral or figure 8–shaped bundles as shown in Figure 5-3*b*. When the heart contracts, its internal chambers become smaller, forcing the blood into the large arteries leaving the heart. Cardiac cells are branching cells that form tight junctions (*intercalated disks*) with each other. These two structural features and the spiral arrangement of the bundles in the heart allow heart activity to be closely coordinated.

MUSCLE ACTIVITY

Stimulation and Contraction of Muscle Cells

Skeletal muscle cells must be stimulated by nerve impulses to contract. One motor neuron may stimulate a single muscle cell or hundreds of them, depending on the particular muscle and the work it does. One neuron and all the skeletal muscle cells it stimulates is a **motor unit.**

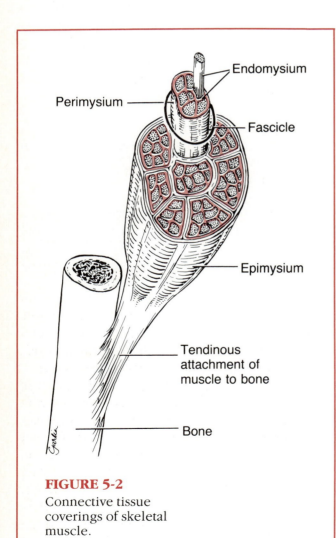

FIGURE 5-2
Connective tissue coverings of skeletal muscle.

Labels in figure: Endomysium, Perimysium, Fascicle, Epimysium, Tendinous attachment of muscle to bone, Bone

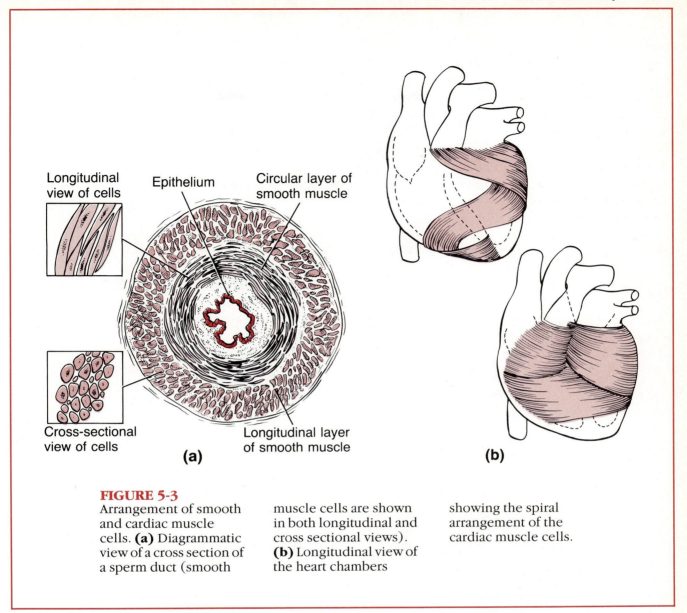

Longitudinal
view of cells

Epithelium

Circular layer of
smooth muscle

Cross-sectional
view of cells

Longitudinal layer
of smooth muscle

(a)

(b)

FIGURE 5-3
Arrangement of smooth and cardiac muscle cells. **(a)** Diagrammatic view of a cross section of a sperm duct (smooth muscle cells are shown in both longitudinal and cross sectional views). **(b)** Longitudinal view of the heart chambers showing the spiral arrangement of the cardiac muscle cells.

The junction between the motor neuron's fiber (axon), which transmits the impulse, and the muscle cell's sarcolemma is a **myoneural junction.** The end of the nerve fiber and the muscle cell's membrane are very close but never touch. The gap between them **(synaptic cleft)** is filled with tissue fluid. When the nerve impulse reaches the end of the axon, a chemical referred to as a **neurotransmitter** is released. The specific neurotransmitter that stimulates skeletal muscle cells is **acetylcholine** (as″ĕ-til-ko′lēn). Acetylcholine diffuses across the synaptic cleft and attaches to receptors (membrane proteins), which are part of the sarcolemma. If enough acetylcholine has been released, the sarcolemma at that point becomes *temporarily* permeable to sodium ions (Na^+), which go rushing into the muscle cell. This sudden inward rush of sodium ions gives the cell interior excessive positive ions, which upsets and changes the electrical conditions of the sarcolemma. This electrical upset causes an **action potential** (an electric current) to be generated. Once begun, the action potential travels over the entire surface of the sarcolemma, conducting the electrical impulse from one end of the cell to the other. This results in the contraction of the muscle cell. This series of events is explained more fully on p. 123 in the discussion of nerve physiology, but perhaps it would be helpful to compare this to some common event such as lighting a match under a small dry twig (Figure 5-4). The flame, which soon begins to

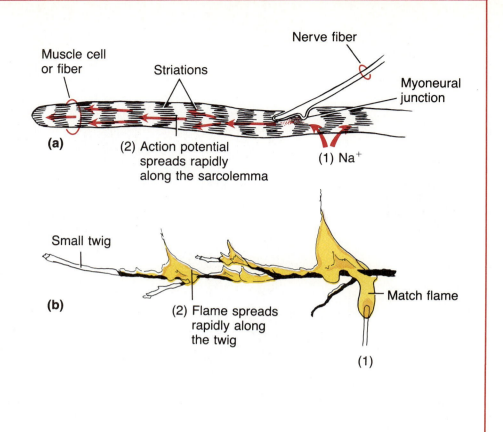

FIGURE 5-4
Comparison of the action potential to a flame consuming a dry twig. **(a)** The first event is the rapid diffusion of sodium ions (Na$^+$) into the cell when the sarcolemma permeability changes. The second event is the spreading of the electrical current along the sarcolemma when enough sodium ions have entered to upset the electrical conditions in the cell. **(b)** The first event is holding the match flame under one area of the twig. The second event is the twig bursting into flame (when heated enough) and the flame spreading to burn the entire twig.

char the twig, can be compared to the change in membrane permeability that allows sodium ions into the cell. When that part of the twig becomes hot enough (when enough sodium ions have entered the cell), the twig will suddenly burst into flame and the flame will consume the twig (the action potential will be conducted along the entire length of the sarcolemma). The events that return the cell to its resting state include (a) diffusion of potassium ions (K$^+$) out of the cell and (b) activation of the sodium-potassium pump (the active transport mechanism), which moves the sodium and potassium ions back to their initial positions.

The movement of electrical current along the sarcolemma causes calcium ions (Ca^{++}) to be released from storage areas inside the muscle cell. When calcium ions attach to actin myofilaments, the sliding of the myofilaments is triggered, and the whole cell shortens. The sliding of the myofilaments is energized by ATP. When the action potential ends, calcium ions are immediately reabsorbed back into the storage areas, and the muscle cell relaxes and settles back to its original length. This whole series of events takes just a few thousandths of a second.

It should be mentioned that while the action potential is occurring, acetylcholine, which began the process, is broken down by enzymes present on the sarcolemma. For this reason, a single nerve impulse produces only one contraction, preventing a muscle cell from contracting continuously in the absence of additional nerve impulses. The muscle cell relaxes until stimulated by the next release of acetylcholine.

Contraction of a Muscle as a Whole

GRADED RESPONSES

The "all or none" law of muscle physiology states that a muscle cell will contract to its fullest extent when it is stimulated adequately; it never partially contracts. However, skeletal muscles consist of thousands of muscle cells and react to stimuli with graded responses, or different degrees, of shortening. Thus, muscle contractions can be slight or vigorous depending on what has to be done. (The same hand that soothes can deliver a stinging slap.) How vigorously a muscle contracts depends on how many of its cells are stimulated.

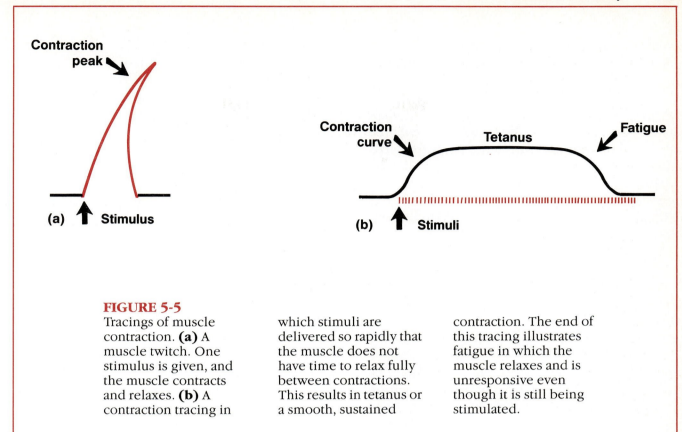

FIGURE 5-5
Tracings of muscle contraction. **(a)** A muscle twitch. One stimulus is given, and the muscle contracts and relaxes. **(b)** A contraction tracing in which stimuli are delivered so rapidly that the muscle does not have time to relax fully between contractions. This results in tetanus or a smooth, sustained contraction. The end of this tracing illustrates fatigue in which the muscle relaxes and is unresponsive even though it is still being stimulated.

When only a few cells are stimulated, the contraction of the muscle as a whole will be slight; in the strongest contractions, all the muscle cells of the muscle are being stimulated.

TETANUS

Although *muscle twitches* (single, brief, jerky contractions) sometimes occur as in certain nervous system problems, this is *not* the way our muscles normally operate. In most types of muscle activity, nerve impulses are delivered to the muscle at a very rapid rate—so rapid that the cells do not get a chance to relax completely between stimuli. As a result, the contractions are smooth and sustained, and the muscle is said to be in *tetanus* (Figure 5-5). Tetanic contractions allow muscles to work to their fullest ability and allow prolonged smooth actions.*

*Tetanic contraction is normal and desirable and is quite different from the pathologic condition of tetanus (commonly called lockjaw) caused by a toxin made by a bacterium. Lockjaw causes muscles to go into uncontrollable spasms, which finally causes respiratory arrest.

MUSCLE FATIGUE

If we exercise our muscles strenuously for a long time, *muscle fatigue* occurs. A muscle is fatigued when it is unable to contract, even though it is still being stimulated. Without rest, it begins to tire and contracts more weakly until it finally ceases reacting and stops contracting. Muscle fatigue is believed to result from the *oxygen debt* that occurs during prolonged muscle activity. A person is not able to breathe in adequate oxygen fast enough to keep the muscles supplied with all the oxygen needed when working vigorously. Obviously, then, the work that a muscle can do and how long it can work without becoming fatigued depends on how good its blood supply is. When muscles lack oxygen, lactic acid begins to accumulate in the muscles. In addition, the muscle cells' ATP supply starts to run low. The increasing acidity in the muscle and the lack of ATP cause the muscle to contract less and less effectively and finally to stop contracting altogether.

True muscle fatigue, in which the muscle quits entirely, rarely occurs in most of us because we feel fatigued long before it happens and we simply stop our activity. It *does* happen commonly in marathon runners. Many of them have literally collapsed when their muscles became fatigued and could not work any longer.

Oxygen debt, which always occurs during vigorous muscle activity, must be "paid back" whether or not fatigue occurs. During the recovery period after activity, the individual breathes rapidly and deeply. This continues until the amount of oxygen needed to get rid of the accumulated lactic acid and make ATP has been delivered to the muscles.

TYPES OF MUSCLE CONTRACTIONS—ISOTONIC/ISOMETRIC

Muscles do not always shorten when they are stimulated. For this reason, contractions can be classified as *isotonic* or *isometric*. Isotonic contractions (literally, same tone or tension) are more familiar to most of us. In isotonic contractions, the muscle shortens and movement occurs. Bending the knee, rotating the arms, and smiling are all examples of isotonic contractions.

Contractions in which the muscles do *not* shorten are called isometric contractions (literally, same measurement or length). In isometric contractions, the myofilaments are "skidding their wheels" and the tension in the muscle keeps increasing; they are trying to slide, but the muscle is pitted against some more or less immovable object. For example, muscles are contracting isometrically when you try to lift a 400 lb dresser alone. When you straighten a bent elbow the triceps muscle is contracting isotonically. But when you push against a wall with bent elbows, the wall doesn't move, and the triceps, which cannot shorten to straighten the elbows, are contracting isometrically.

MUSCLE TONE

There is one aspect of skeletal muscle activity that is not consciously controlled. Even when a muscle is voluntarily relaxed some of its fibers are contracting, first one group and then another. Their contraction is not visible, but as a result of it, the muscle remains firm, healthy, and constantly ready for action. This state of continuous partial contraction is called *muscle tone*. Muscle tone is the result of different motor units, which are scattered through the muscle, being stimulated by the nervous system in a systematic way. If the nerve

supply to a muscle is destroyed (as in an accident) the muscle is no longer stimulated in this manner, and it loses tone and becomes paralyzed. It then becomes *flaccid* (flak′sid), or soft and flabby, and begins to atrophy (waste away).

BODY MOVEMENTS AND NAMING SKELETAL MUSCLES

Types of Body Movements

Every muscle is attached to bone (or other connective tissue structures) at no less than two points. One of these points, the **origin,** is attached to the immovable or less movable bone. The **insertion** is attached to the movable bone, and when the muscle contracts, the insertion moves toward the origin.

Body movement occurs when muscles contract across joints. The type of movement depends on the mobility of the joint and on where the muscle is located in relation to the joint. The most common types of body movements are described next and shown in Figure 5-6. Attempt to demonstrate or produce each movement as you read through this material.

FLEXION.   A movement, generally in the sagittal plane, that decreases the angle of the joint and brings two bones closer together. Flexion is typical of hinge joints (bending the knee or elbow) but is also common at ball-and-socket joints (bending forward at the hip).

EXTENSION.   Extension is the opposite of flexion, so it is a movement that increases the angle, or the distance, between two bones or parts of the body (straightening the knee or elbow). If extension is greater than 180° (bending the trunk backward), it is *hyperextension*.

ABDUCTION.   Abduction is moving a limb away from the midline, or median plane, of the body

FIGURE 5-6
Movements occurring at joints of the body.

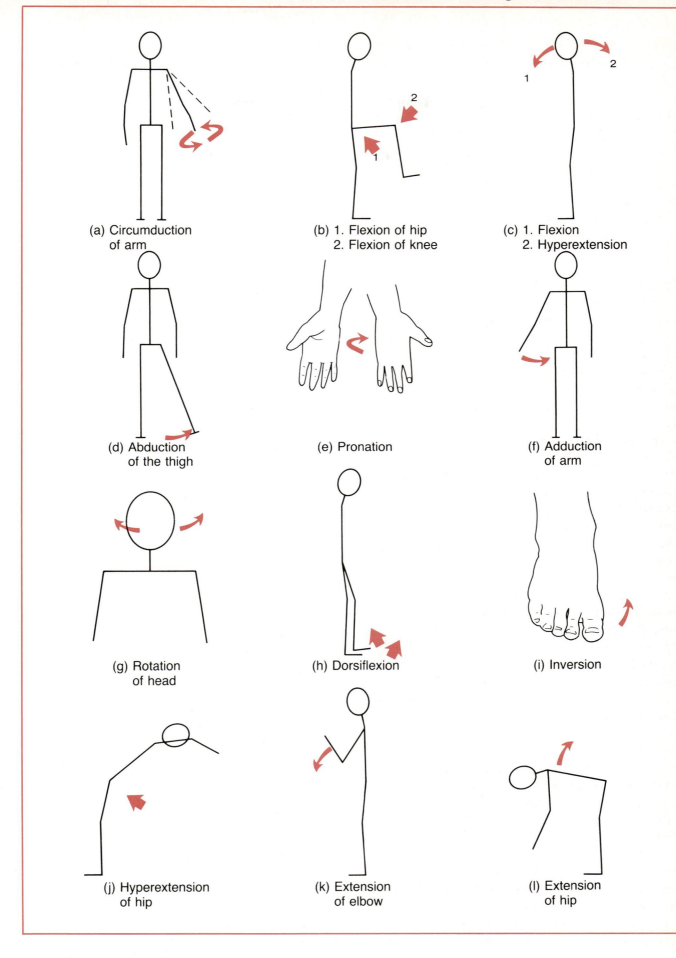

(a) Circumduction of arm

(b) 1. Flexion of hip
2. Flexion of knee

(c) 1. Flexion
2. Hyperextension

(d) Abduction of the thigh

(e) Pronation

(f) Adduction of arm

(g) Rotation of head

(h) Dorsiflexion

(i) Inversion

(j) Hyperextension of hip

(k) Extension of elbow

(l) Extension of hip

(generally on the frontal plane), or the fanning movement of the fingers or toes when they are spread apart.

ADDUCTION. Adduction is the opposite of abduction, so it is the movement of a limb toward the body midline.

ROTATION. Rotation is movement of a bone around its longitudinal axis. Rotation is a common movement of ball-and-socket joints and describes the movement of the atlas around the dens of the axis (shaking your head "no").

CIRCUMDUCTION. Circumduction is a combination of flexion, extension, abduction, and adduction commonly seen in ball-and-socket joints like the shoulder. The proximal end of the limb is stationary, and its distal end moves in a circle. The limb as a whole outlines a cone.

PRONATION. Pronation is moving the palm of the hand from an anterior, or upward-facing position, to a posterior, or downward-facing position. This action moves the distal end of the radius across the ulna.

SUPINATION. Supination is moving the palm from a posterior position to an anterior position (the anatomical position). Supination is the opposite of pronation. During supination the radius and the ulna are parallel.

The following four terms refer to movements of the foot:

INVERSION. Inversion is turning the sole of the foot medially.

EVERSION. Eversion is turning the sole of the foot laterally; eversion is the opposite of inversion.

DORSIFLEXION. Dorsiflexion is movement at the ankle that moves the instep of the foot up and dorsally, that is, standing on your heels.

PLANTARFLEXION. Plantarflexion is movement of the ankle joint in which the joint is straightened and the toes are pointed downward, or standing on your toes.

Types of Muscles

Muscles can't push, they can only pull as they contract, so most often body movements are the result of the activity of pairs or teams of muscles acting together or against each other. Muscles are arranged on the skeleton in such a way that whatever one muscle (or group or muscles) can do, another group of muscles can reverse. Because of this muscles are able to bring about an immense variety of movements.

When several muscles are contracting at the same time, the muscle that has the major responsibility for causing a particular movement is the **prime mover.** (This physiologic term has been borrowed by the business world to label a person who gets things done.) Muscles that oppose or reverse a movement are **antagonists** (an-tag'o-nists). When a prime mover is active, its antagonist is stretched and relaxed. Antagonists can be prime movers in their own right; for example, the biceps of the arm (prime mover of elbow flexion) is antagonized by the triceps (a prime mover of elbow extension).

Synergists (sin'er-jists) are muscles that help prime movers by reducing undesirable or unnecessary movement. When a muscle crosses two or more joints, its contraction will cause movement in all the joints crossed unless synergists are there to stabilize them. For example, the finger-flexor muscles cross both the wrist and the finger joints. You can make a fist without bending your wrist because synergist muscles stabilize the wrist joint and allow the prime mover to act on the finger joints.

Fixators are specialized synergists. They stabilize the origin of a prime mover so that all the tension can be used to move the insertion bone. The postural muscles that stabilize the vertebral column are fixators.

Naming Skeletal Muscles

Like bones, muscles come in many shapes and sizes to suit their particular tasks in the body. Remembering the names of skeletal muscles is a huge task, but certain clues can help. Muscles are named on the basis of several criteria:

1. **Direction of the muscle fibers.** Some muscles are named in reference to an imaginary line, usually the midline of the body or the long axis of a limb bone. When a muscle's name includes the term *rectus* (rectus means straight), its fibers run parallel to that imaginary line. For example, the rectus abdominis is the straight muscle of the abdomen, and the rectus femoris is the straight muscle of the thigh, or femur. Similarly, the term *oblique* as part of a muscle's name tells you that the muscle fibers run obliquely (slanted) to the imaginary line.

2. **Relative size of the muscle.** Such terms as *maximus* (largest), *minimus* (smallest), and *longus* (long) are often used in the names of muscles—for example, gluteus maximus.

3. **Location of the muscle.** Some muscles are named for the bone with which they are associated. For example, the temporalis and frontalis muscles overlie the temporal and frontal bones of the skull.

4. **Number of origins.** When the term *biceps, triceps,* or *quadriceps* forms part of a muscle name, one can assume that the muscle has two, three, or four origins (respectively). For example, the biceps muscle of the arm has two heads, or origins, and the triceps has three.

5. **Location of the muscle's origin and insertion.** For example, the sternocleidomastoid muscle has its origin on the sternum *(sterno)* and clavicle *(cleido)* and inserts on the *mastoid* process of the temporal bone.

6. **Shape of the muscle.** For example, the deltoid muscle is roughly triangular (deltoid means triangle).

7. **Action of the muscle.** For example, the adductor muscles of the anterior thigh all bring about its adduction, and the extensor muscles of the wrist all extend the wrist.

GROSS ANATOMY OF SKELETAL MUSCLES

It is beyond the scope of this book to describe the hundreds of skeletal muscles of the human body.

Only the more important muscles are described here. In addition, all the superficial muscles considered are summarized on pp. 106 and 108.

Head Muscles

The head muscles are an interesting group. They have many specific functions but are usually grouped into two large categories—facial muscles and chewing muscles (Figure 5-7). Facial muscles permit one to smile faintly, grin widely, frown, pout, or deliver a kiss. The chewing muscles begin the breakdown of food for the body.

FACIAL MUSCLES

FRONTALIS. The frontalis covers the frontal bone as it runs from the cranial aponeurosis to the skin of the eyebrows where it inserts. This muscle allows you to raise your eyebrows as in surprise and wrinkle your forehead.

ORBICULARIS OCULI. The orbicularis oculi (or-bik″ū-la′ris ok′u-li) has fibers that run in circles around the eyes. It allows you to close your eyes, squint, blink, and wink.

ORBICULARIS ORIS. The orbicularis oris is the circular muscle of the lips. It closes your mouth and protrudes the lips; it is often called the "kissing" muscle.

BUCCINATOR. The buccinator (buk′sĭ-na″tor) is a fleshy muscle that runs horizontally across the cheek and inserts into the orbicularis oris. It flattens the cheek (as in whistling and blowing a trumpet). It is also listed as a chewing muscle, because it compresses the cheek to hold the food between the teeth during chewing.

ZYGOMATICUS. The zygomaticus (zi″go-mat′ĭ-kus) extends from the corner of the mouth to the cheekbone. Often referred to as the "smiling" muscle, because it raises the corners of the mouth upward.

CHEWING MUSCLES

The buccinator muscle, which is a member of this group, is described with the facial muscles.

MASSETER. The masseter covers the angle of the lower jaw as it runs from the zygomatic process of

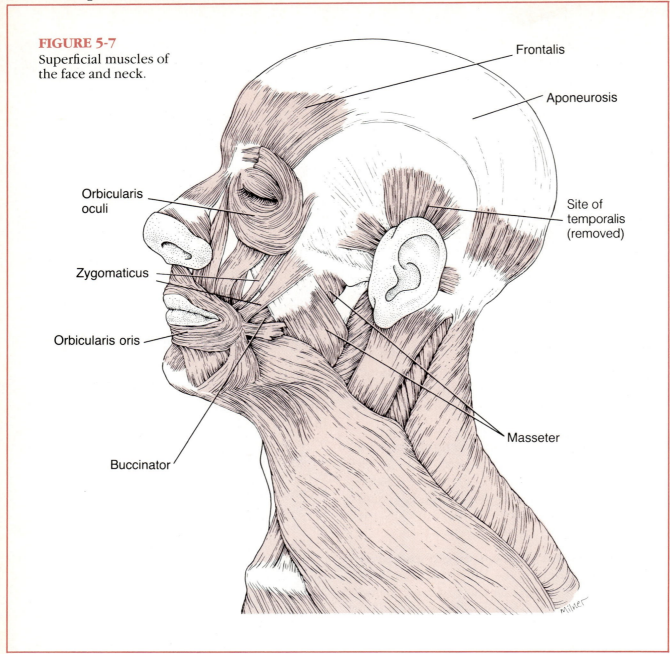

FIGURE 5-7
Superficial muscles of
the face and neck.

Frontalis

Aponeurosis

Orbicularis
oculi

Site of
temporalis
(removed)

Zygomaticus

Orbicularis oris

Masseter

Buccinator

the temporal bone to the mandible. This muscle closes the jaw by elevating the mandible.

TEMPORALIS.   The temporalis is a fan-shaped muscle overlying the temporal bone. It inserts into the mandible and acts as a synergist of the masseter in closing the jaw.

Trunk and Neck Muscles

The neck muscles, which move the head or shoulder girdle, are small, straplike muscles. The trunk muscles include those that move the vertebral column (most of which are posterior antigravity muscles); anterior thorax muscles that move the ribs, head, and arms; and muscles of the abdominal wall that help to move the vertebral column but, most importantly, form the "natural girdle" or the major part of the abdominal body wall. They also compress the abdominal contents during defecation and childbirth and are involved in forced breathing.

ANTERIOR MUSCLES (FIGURE 5-8)

STERNOCLEIDOMASTOID.   The sternocleidomastoid (ster"no-kli"do-mas'toid) muscles are a pair

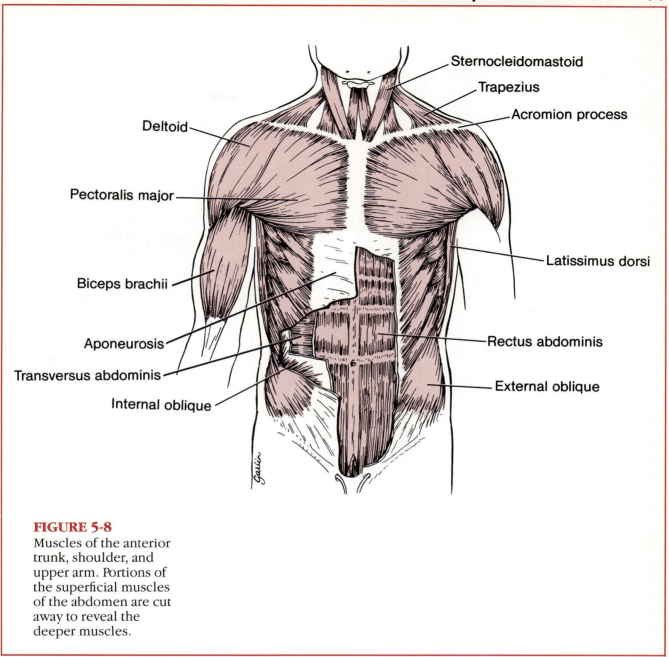

FIGURE 5-8
Muscles of the anterior trunk, shoulder, and upper arm. Portions of the superficial muscles of the abdomen are cut away to reveal the deeper muscles.

of two-headed muscles found on each side of the neck. The two heads of each muscle arise from the sternum and clavicle and insert into the mastoid process of the temporal bone. When both muscles of this muscle pair contract together, they flex your neck. If just one muscle contracts, the head is rotated toward the opposite side. (In some difficult births, one of these muscles may be injured. A baby injured in this way has *torticollis* [tor″tĭ-kol′is], or wryneck.)

PECTORALIS MAJOR. The pectoralis (pek″to-ra′-lis) major is a large fan-shaped muscle covering the upper part of the chest. Its origin is from the clavicle, sternum, and first six ribs. It inserts on the proximal end of the humerus. This muscle forms the anterior wall of the axilla and acts to adduct and flex the arm.

INTERCOSTAL MUSCLES. The intercostal muscles are deep muscles found between the ribs. (They are not shown in Figure 5-8, which only shows superficial muscles.) The *external intercostals* are important in breathing because they help to raise the rib cage for breathing air in; the *internal intercostals* depress the rib cage, which helps to move air out of the lungs when exhaling.

RECTUS ABDOMINIS.*   The paired straplike rectus abdominis muscles are the most superficial muscles of the abdomen. They run from the pubis to the sternum and fifth to seventh ribs, enclosed in an aponeurosis. Their main function is to flex the vertebral column.

EXTERNAL OBLIQUE.*   The external oblique muscles are paired superficial muscles that make up the lateral walls of the abdomen. Their fibers run downward and medially. They arise from the last eight ribs and insert into the ilium. Their functions are the same as for the rectus abdominis.

INTERNAL OBLIQUE.*   The internal oblique muscles are paired muscles deep to the external obliques. Their fibers run at right angles to those of the external obliques. They arise from the iliac crest and insert into the last three ribs. Their functions are the same as those of the rectus abdominis.

TRANSVERSUS ABDOMINIS.*   The transversus abdominis is the deepest muscle of the abdominal wall and has fibers that run horizontally across the abdomen. It arises from the iliac crests and inserts into the pubis. This muscle compresses the abdominal contents during defecation.

POSTERIOR MUSCLES (FIGURE 5-9)

TRAPEZIUS.   The trapezius (trah-pe′ze-us) muscles are the most superficial muscles of the posterior neck and upper trunk. When seen together, they form a diamond- or kite-shaped muscle mass. Their origin is very broad; each runs from the occipital bone of the skull down the vertebral column to the end of the thoracic vertebrae. They then flare laterally to insert on the scapular spine and clavicle. The trapezius muscles extend the head (thus are antagonists of the sternocleidomastoids) and adduct and stabilize the scapula.

LATISSIMUS DORSI.   The latissimus (lah-tis′ĭ-mus) dorsi is the large, flat muscle pair that covers

---

*The anterior abdominal muscles (rectus abdominis, external and internal obliques, and transversus abdominis) when taken together resemble the structure of plywood. The fibers of each muscle or muscle pair run in a different direction. Just as plywood is exceptionally strong for its thickness, the abdominal muscles form a muscular wall which is well suited for its job of containing and protecting the abdominal contents

the lower back. It originates on the lower spine and ilium and then sweeps superiorly to insert into the proximal end of the humerus. The latissimus dorsi extends and adducts the humerus. These are very important muscles when the arm must be brought down in a power stroke, as when swimming or striking a blow.

DELTOID.   The deltoids are fleshy triangle-shaped muscles that form the rounded shape of your shoulders. Because they are so bulky, they are a favorite injection site. The origin of each deltoid winds across the bones of the shoulder girdle from the spine of the scapula to the clavicle. It inserts into the proximal humerus. The deltoids are the prime movers of arm abduction.

Upper Extremity Muscles

The arm muscles fall into three groups. The first group includes muscles that arise from the shoulder girdle bones and cross the shoulder joint to insert into the humerus; these muscles move the upper arm. All of these muscles have already been considered—the pectoralis major, latissimus dorsi, and the deltoid.

The second group causes movement at the elbow joint; these muscles enclose the humerus and insert on the forearm bones. (Only the muscles of this second group will be described in this section.) The third group includes the muscles of the forearm which insert on the wrist, or hand bones, and cause movement of those body parts. The muscles of this last group are thin and spindle-shaped, and there are many of them. They will not be considered here except to reveal their general naming and function. In general the forearm muscles have names which reflect their activity—for example, the flexor carpi and flexor digitorum muscles found on the anterior aspect of the forearm cause flexion of the wrist and fingers respectively. The extensor carpi and extensor digitorum muscles found on the lateral and posterior aspect of the forearm extend the same structures. (Some of these muscles are described on p. 108 and illustrated in Figure 5-13.)

MUSCLES OF THE HUMERUS THAT ACT ON THE FOREARM

BICEPS BRACHII.   The biceps brachii (bra′ke-i) (see Figure 5-8) is the most familiar muscle of the

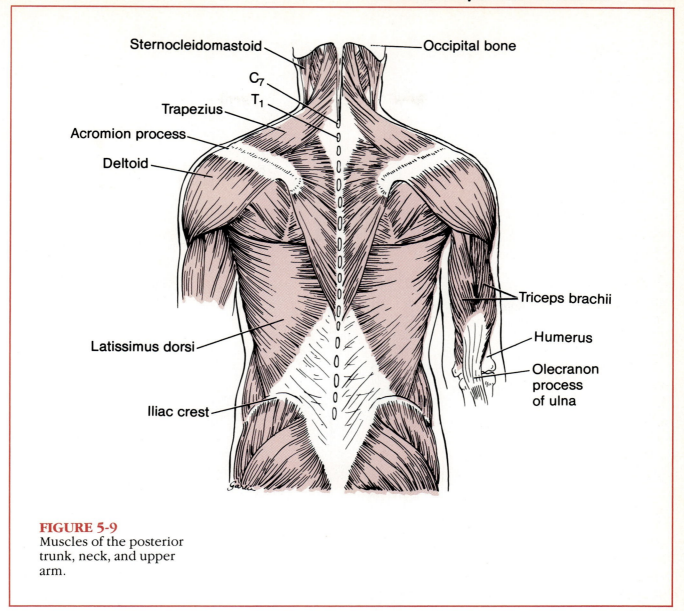

Sternocleidomastoid — Occipital bone

C7

T1

Trapezius

Acromion process

Deltoid

Triceps brachii

Humerus

Olecranon process of ulna

Latissimus dorsi

Iliac crest

FIGURE 5-9
Muscles of the posterior trunk, neck, and upper arm.

forearm, because it bulges when the elbow is flexed. It originates by two heads from the shoulder girdle and inserts into the radial tuberosity. This muscle is the powerful prime mover for flexion of the forearm and acts to supinate the forearm. The best way to remember its action is that "It turns the corkscrew and pulls the cork."

TRICEPS BRACHII.   The triceps muscle is the only muscle fleshing out the posterior humerus (see Figure 5-9). Its three heads arise from the shoulder girdle and proximal humerus, and it inserts into the olecranon process of the ulna. Being the powerful prime mover of elbow extension, it is the antagonist of the biceps brachii. This muscle is often called the "boxer's" muscle, because it delivers a straight-arm knockout punch.

Lower Extremity Muscles

Muscles that act on the lower extremity or limb cause movement at the hip, knee, and foot joints. They are among the largest, strongest muscles in the body and are highly specialized for walking and balancing the body. Because the pelvic girdle is composed of heavy, fused bones that allow little movement, no special group of muscles is necessary to stabilize it. This is very different from the shoulder girdle, which needs many fixator muscles.

Many muscles of the lower limb span two joints and can cause movement at both of them. Therefore, the terms *origin* and *insertion* are often interchangeable in referring to these muscles.

FIGURE 5-10
Pelvic, hip, and thigh
muscles of the right side
of the body. **(a)** Anterior
view of pelvis and thigh
muscles.

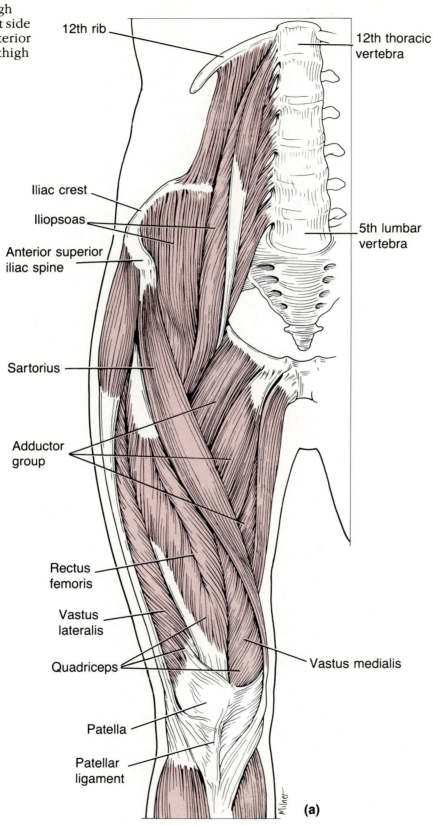

12th rib

12th thoracic
vertebra

Iliac crest

Iliopsoas

5th lumbar
vertebra

Anterior superior
iliac spine

Sartorius

Adductor
group

Rectus
femoris

Vastus
lateralis

Quadriceps

Vastus medialis

Patella

Patellar
ligament

(a)

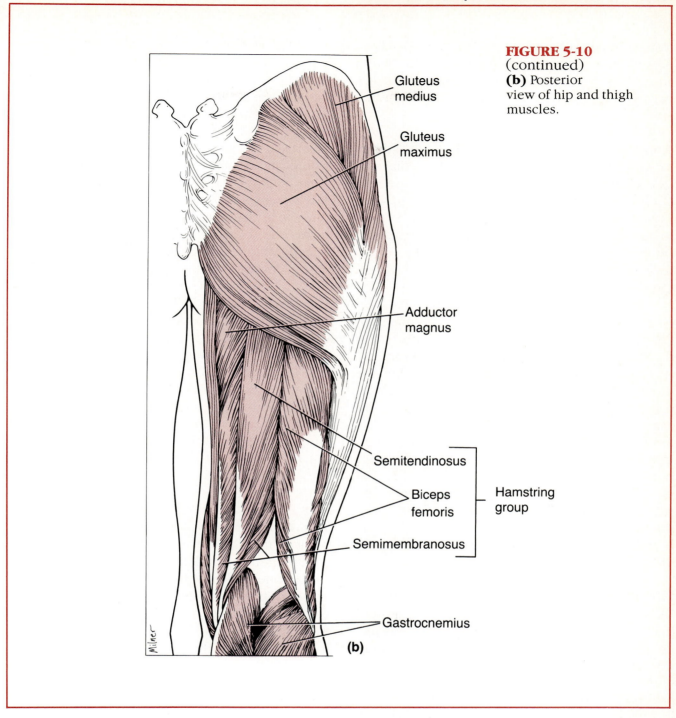

FIGURE 5-10
(continued)
(b) Posterior
view of hip and thigh
muscles.

Gluteus medius

Gluteus maximus

Adductor magnus

Semitendinosus

Biceps femoris

Hamstring group

Semimembranosus

Gastrocnemius

(b)

Muscles acting on the thigh are massive muscles that help hold the body upright against the pull of gravity and cause various movements at the hip joint. Muscles acting on the leg form the flesh of the thigh. (In common usage the term leg refers to the whole lower limb, but anatomically the term refers only to that part between the knee and the ankle.) The thigh muscles cross the knee to cause its flexion and extension. Because many of the thigh muscles also have attachments on the pelvic girdle, they can cause movement at the hip

joint as well. Muscles originating on the leg cause various movements of the ankle and foot. Only three of this group will be considered, but there are many muscles that act to extend and flex the ankle and toe joints.

MUSCLES CAUSING MOVEMENT AT THE HIP JOINT (FIGURE 5-10)

ILIOPSOAS. The iliopsoas (il″e-op-so′as) is a fused muscle composed of two other muscles. It runs from the iliac bone and lower vertebrae deep inside the pelvis to insert on the lesser trochanter of the femur. It is a prime mover of hip flexion and also acts as a postural muscle to keep the upper body from falling backwards when standing erect.

ADDUCTOR MUSCLES.   The muscles of the adductor group form the muscle mass at the medial side of each thigh. As their name indicates, they adduct or press the thighs together. However, since gravity does most of the work for them, they tend to become flabby very easily. Special exercises are usually needed to keep them toned. The adductors have their origin on the pelvis and insert on the proximal aspect of the femur.

GLUTEUS MAXIMUS. The gluteus maximus (gloo′te-us max′i-mus) is a superficial muscle of the hip that forms most of the flesh of the buttock. It is a powerful hip extensor that acts to bring the hip in a straight line with the pelvis. Although not very important in walking, it is probably the most important muscle for extending the hip when it is flexed, as in climbing stairs and jumping. It is an important site for giving intramuscular injections; however, *remember* that the lower medial part of each buttock overlies the large sacral nerve, and this area must be carefully avoided. The gluteus muscle originates from the sacrum and iliac bones and runs to insert on the gluteal tuberosity of the femur.

GLUTEUS MEDIUS.   The gluteus medius runs from the ilium to the femur and beneath the gluteus maximus for most of its length. The gluteus medius is a hip abductor. Like the gluteus maximus, it is an important injection site.

MUSCLES CAUSING MOVEMENT AT THE KNEE JOINT
(FIGURE 5-10)

SARTORIUS. The thin straplike sartorius (sar-to′rē-us) muscle is not too important, but since it is the most superficial muscle of the thigh it is rather hard to miss. It runs obliquely across the thigh from the anterior iliac crest to the medial side of the tibia. It is a weak thigh flexor, commonly referred to as the "tailor's" muscle because it acts as a synergist to bring about the cross-legged position in which old time tailors are often shown.

QUADRICEPS GROUP.   The quadriceps (kwod′rĭ-seps) group consists of four muscles (*rectus femoris* and three *vastus muscles*), which flesh out the anterior thigh. The vastus muscles originate from the femur; the rectus femoris originates on the pelvis. All four muscles insert into the tibial tuberosity by the patellar tendon. The group as a whole acts to extend the knee powerfully, as when kicking a football. Because the rectus femoris crosses two joints, the hip and the knee, it can also act alone to help flex the hip. The lateral vastus and rectus femoris are sometimes used as injection sites, particularly in infants who have poorly developed gluteus muscles.

HAMSTRING GROUP.   The muscles forming the muscle mass at the back of the thigh are the hamstrings. The group consists of three muscles (*biceps femoris, semimembranosus,* and *semitendinosus*) that originate on the ischial tuberosity and run down the thigh to insert on both sides of the proximal tibia. Their name comes from the fact that butchers use their tendons to hang "ham" (thigh and hip muscles) for smoking. These tendons can be felt at the back of the knee.

MUSCLES CAUSING MOVEMENT AT THE ANKLE AND FOOT (FIGURE 5-11)

TIBIALIS ANTERIOR.   The tibialis anterior is a superficial muscle on the anterior leg. It runs from the upper tibia paralleling the anterior crest to the tarsal bones where it inserts by a long tendon. It acts to dorsiflex and invert the foot.

PERONEUS MUSCLES   The three peroneus (per″o-ne′us) muscles of this group are found on the lateral part of the leg. They arise from the fibula and insert into the metatarsal bones of the foot. The activity of the group as a whole is to plantarflex and evert the foot.

GASTROCNEMIUS.   The gastrocnemius (gas″trok-ne′me-us) muscle is a two-bellied muscle that forms the curved calf of the posterior leg. It arises by two heads from each side of the distal femur and inserts through the large Achilles tendon into the heel of the foot. It is a prime mover for plantarflexion of the foot; for this reason it is often called the "toe dancer's" muscle. If its insertion tendon is cut, walking is very difficult; the foot drags because you cannot lift your heel.

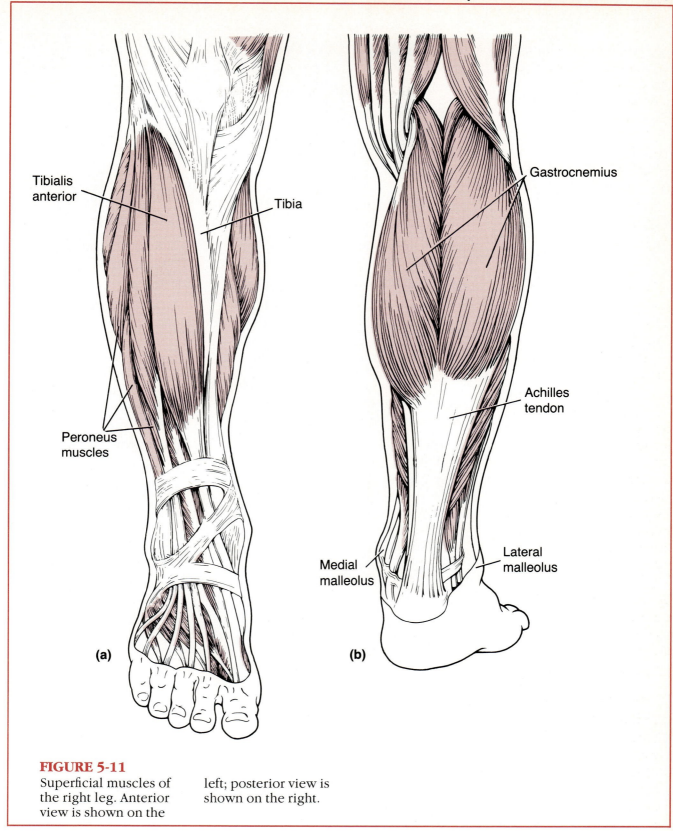

FIGURE 5-11
Superficial muscles of the right leg. Anterior view is shown on the left; posterior view is shown on the right.

Most of the superficial muscles previously described are shown in anterior and posterior views of the body as a whole (Figures 5-12 and 5-13). In addition, these muscles are summarized in Tables 5-1 and 5-2, which accompany the figures. Take the time to review these muscles again.

(Text continues on p. 110.)

TABLE 5-1   Superficial Anterior Muscles of the Body (see Figure 5-12)

Name	Origin to insertion	Action
HEAD AND NECK MUSCLES		
Frontalis	Cranial aponeurosis to eyebrows	Raises eyebrows
Orbicularis oculi	Skin around eyes	Blinks and closes eyes
Orbicularis oris	Skin around mouth	Closes and protrudes lips
Buccinator	Mandible/maxilla to skin around mouth	Flattens cheek against teeth
Temporalis	Temporal bone to mandible	Closes jaw
Zygomaticus	Zygomatic bone to corner of lips	Raises corner of mouth
Masseter	Temporal bone to mandible	Closes jaw
Sternocleidomastoid	Sternum/clavicle to temporal bone	Flexes neck, rotates head
TRUNK MUSCLES		
Pectoralis major	Sternum/clavicle/first to sixth ribs to proximal humerus	Adducts and flexes humerus
Rectus abdominis	Pubis to sternum/fifth to seventh ribs	Flexes vertebral column
External oblique	Last eight ribs to iliac crest	Flexes vertebral column
ARM AND SHOULDER MUSCLES		
Biceps brachii	Shoulder girdle to proximal radius	Flexes elbow and supinates hand
Deltoid	See Table 5-2	Abducts arm
HIP/THIGH/ LEG MUSCLES		
Iliopsoas	Ilium/lower vertebrae to femur	Flexes hip
Adductor muscles	Pelvis to proximal femur	Adducts thigh
Sartorius	Ilium to medial tibia	Flexes thigh on hip
Quadriceps group	Vasti: Femur to tibial tuberosity through patellar tendon Rectus femoris: Pelvis to tibial tuberosity through patellar tendon	All extend knee; rectus femoris also flexes hip on thigh
Tibialis anterior	Proximal tibia to tarsal bones of ankle	Dorsiflexes and inverts foot
Peroneus muscles	Fibula to metatarsals of foot	Plantarflexes and everts foot

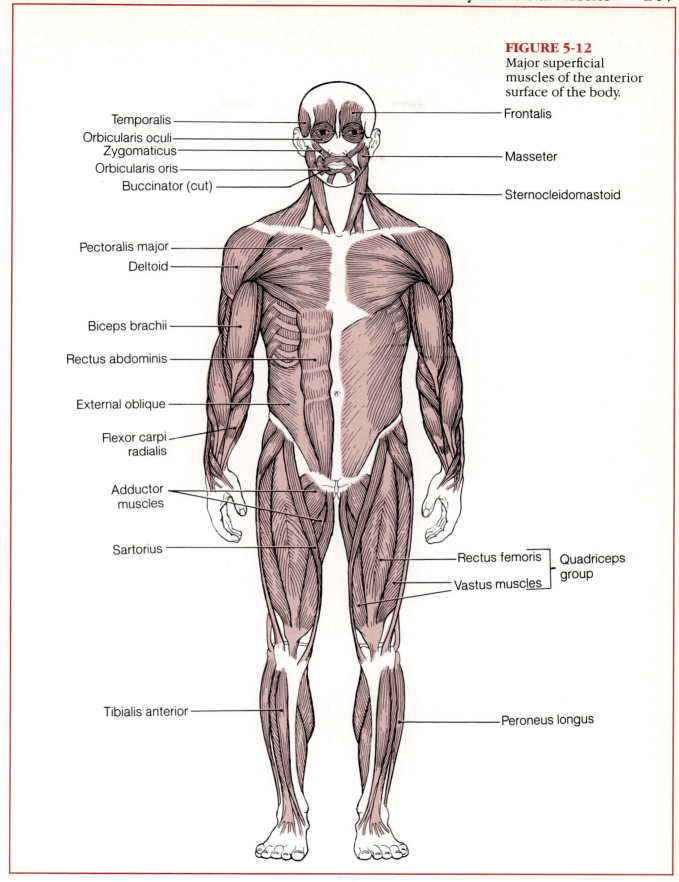

FIGURE 5-12
Major superficial
muscles of the anterior
surface of the body.

Temporalis

Orbicularis oculi

Zygomaticus

Orbicularis oris

Buccinator (cut)

Frontalis

Masseter

Sternocleidomastoid

Pectoralis major

Deltoid

Biceps brachii

Rectus abdominis

External oblique

Flexor carpi
radialis

Adductor
muscles

Sartorius

Rectus femoris

Vastus muscles

Quadriceps
group

Tibialis anterior

Peroneus longus

TABLE 5-2 Superficial Posterior Muscles of the Body (Some Forearm Muscles Also Shown) (see Figure 5-13)

Name	Origin to insertion	Action
NECK/TRUNK/ SHOULDER MUSCLES		
Trapezius	Occipital bone/vertebrae of thorax to scapular spine and clavicle	Extends neck and adducts scapula
Latissimus dorsi	Lower spine/ilium to proximal humerus	Extends and adducts humerus
Deltoid	Scapular spine/clavicle to proximal humerus	Abducts humerus
ARM/ FOREARM MUSCLES		
Triceps brachii	Shoulder girdle/proximal humerus to ulna	Extends elbow
Flexor carpi radialis	Distal humerus to second and third metacarpals	Flexes wrist and abducts hand (See Figure 5-12)
Flexor carpi ulnaris	Distal humerus/ulna to carpals of wrist	Flexes wrist and adducts hand
Flexor digitorum superficialis	Distal humerus/ulna/radius to middle phalanges of second to fifth fingers	Flexes wrists and fingers
Extensor carpi radialis	Humerus to dorsal surface of second and third metacarpals	Extends wrists and abducts hand
Extensor digitorum communis	Distal humerus to dorsum of second to fifth fingers	Extends fingers/wrist
HIP/THIGH/ LEG MUSCLES		
Gluteus maximus	Sacrum/ilium to proximal femur	Extends hip
Gluteus medius	Ilium to proximal femur	Adducts thigh
Hamstring muscles (semitendinosus, semimembranosus, biceps femoris)	Ischium to proximal tibia	Flexes knee and extends hip
Gastrocnemius	Distal femur to calcaneus (heel) through Achilles tendon	Plantarflexes ankle and flexes knee

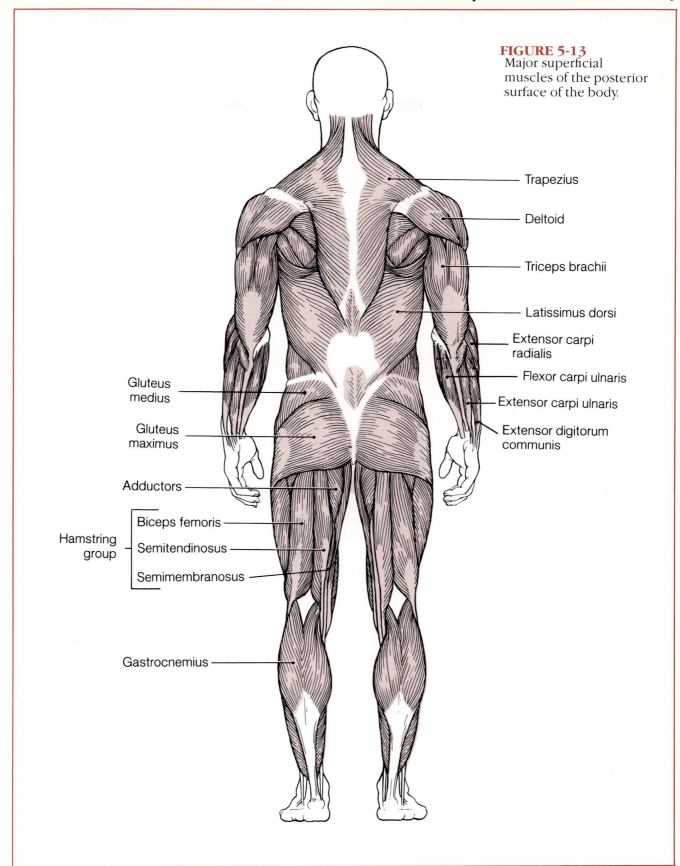

FIGURE 5-13
Major superficial muscles of the posterior surface of the body.

Trapezius

Deltoid

Triceps brachii

Latissimus dorsi

Extensor carpi radialis

Flexor carpi ulnaris

Extensor carpi ulnaris

Extensor digitorum communis

Gluteus medius

Gluteus maximus

Adductors

Hamstring group
- Biceps femoris
- Semitendinosus
- Semimembranosus

Gastrocnemius

BOX 5-1
Effects of Exercise on Muscles

Disuse of any body part leads to its gradual atrophy and general weakening. Muscles are hardly an exception to this rule and typify the saying "If you don't use it, you lose it." Exercise is absolutely vital if muscles are to remain strong and healthy. Without regular exercise, the skeletal muscles tend to stretch, become fat-laden, and generally decrease in strength.

Lack of nervous stimulation to muscles as well as their immobilization (as with complete bedrest or casting of a body part) are the major events that lead to outright muscle wasting and rapid loss of muscle mass. Conversely, regular exercise increases muscle size, strength, and endurance. However, not all types of exercise produce all of these effects—in fact, there are important differences in exercise benefits.

Largely as a result of astronauts being instructed to perform *isometric* exercises during space flights, isometric types of exercise are familiar to most people. In addition, isometrics have the distinct advantage of requiring little space to "work out in" and little special equipment. A wall can be pushed against, palms can be pressed together at chest level, and buttock muscles can be contracted even when standing in line at the grocery store or bank. There is little doubt that isometric exercises cause muscles to enlarge and become stronger, and this type of exercise is often the choice of individuals who want a rapid increase in muscle size. However, isometrics do *not* increase the efficiency or endurance of the skeletal muscles as do the *isotonic* types of exercise. Although all regularly performed isotonic exercises result

in increased nerve and capillary (blood) supply to the muscles being stressed, different benefits are also gained by varying the type of isotonic exercises.

Resistance, short-lasting types of activities such as weight lifting cause the stressed muscles to hypertrophy **(a).** On the other hand, active, endurance types of exercise such as jogging and running (often called aerobic exercise) result in increased muscle endurance and resistance to fatigue without producing dramatic increases in muscle size **(b).** Aerobic exercises additionally increase the efficiency of the cardiovascular and respiratory systems (the systems that ensure delivery of the needed nutrients and oxygen to the working muscles). Obviously, the best type of exercise program would include both types of isotonic activities.

DEVELOPMENTAL ASPECTS OF THE MUSCULAR SYSTEM

In the developing embryo, the muscular system is laid down in segments (much like the structural plan of an earthworm), and then each segment is invaded by nerves. The muscles of the thoracic and lumbar regions become very extensive since they must cover and move the bones of the limbs. Development of the muscles and their control by the nervous system occur rather early in pregnancy. The expectant mother is often astonished by the first movements (called the quickening) of the fetus, which usually occur by the sixteenth week of pregnancy.

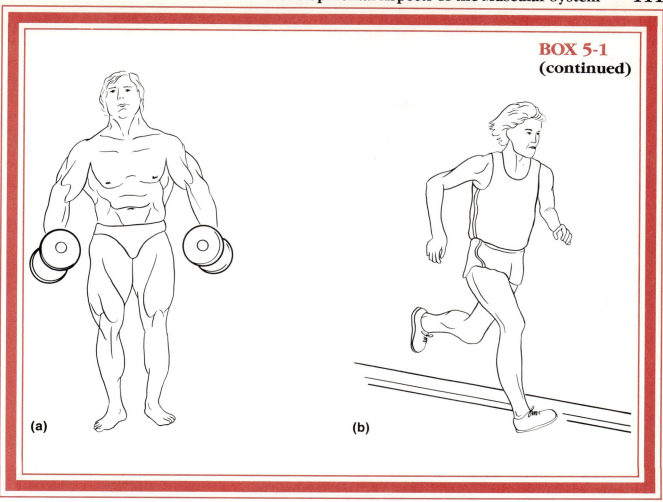

BOX 5-1
(continued)

(a)

(b)

Very few congenital muscular problems occur. The exception is *muscular dystrophy*—a group of diseases in which the muscles degenerate and become progressively weaker. Affected individuals generally do not live beyond young adulthood.

Initially after birth, a baby's movements are all gross reflex types of movements. Since the nervous system must mature before the baby can control muscles, one can usually trace the increasing efficiency of the nervous system by observing a baby's development of muscle control. Development of muscle control proceeds in a cephalic/caudal direction, and gross muscular movements precede fine ones. Babies can raise their heads before they can sit up, and sitting up comes before walking. Muscular control also proceeds in a proximal/distal direction; that is, babies can wave ''bye-bye'' and pull objects to themselves before using the pincher grasp to pick up a pin. All through childhood, the control of the skeletal muscles by the nervous system becomes more and more precise. By midadolescence, we have reached the peak level of development of this nat-

ural control and can simply accept that level or bring it to a fine edge by athletic training.

Throughout childhood and young adulthood, very few problems interfere with the functioning of the muscles. But it should be repeated that muscles, like bones, *will* atrophy, even with normal tone, if they are not used and exercised continually. Exercise not only keeps skeletal muscles healthy but also ensures healthy joints and bones and makes both respiratory and circulatory systems more efficient. The muscles of a bedridden patient atrophy right along with his or her bones. On the other hand, a lifelong program of regular exercise keeps the whole body operating in its best possible shape.

One rare disease that can affect muscles during adulthood is *myasthenia gravis* (mi″as-the′ne-ah gra′vis). The cause of this disease is unknown, but it seems to involve some sort of interference with the transmission of nerve impulses by acetylcholine. Either the acetylcholine is destroyed too quickly, or not enough of it is produced. In either case, the muscle cells are not stimulated properly and get progressively weaker. Death usually occurs as a result of the inability of the respiratory muscles to function. This is called respiratory arrest.

As we age the amount of connective tissue in the muscles increases and the amount of muscle tissue decreases, thus the muscles become stringier, or more sinewy. Since the skeletal muscles represent so much of the body weight, body weight begins to decline in the elderly person, as this natural loss in muscle mass occurs. Another result of the loss in muscle mass is a decrease in muscle strength; strength decreases by about 50% by the age of 80 years. Regular exercise can most likely help offset these effects of aging on the muscular system.

IMPORTANT TERMS*

Myofibrils (*mi″o-fi′brils*)

Actin

Myosin

Deep fascia

Tendons

Aponeuroses (*apo′ nu-ro′sēs*)

Motor unit

Myoneural junction

Synaptic cleft

Neurotransmitter

Acetylcholine (*as″e-til-ko′lēn*)

Action potential

Oxygen debt

Origin

Insertion

Flexion

Extension

Abduction

Adduction

Pronation

Supination

Dorsiflexion

Plantarflexion

Prime Mover

Antagonists (*an-tag′o-nists*)

Synergists (*sin′er-jists*)

Fixators

---

*For definitions, see Glossary.

SUMMARY

A. STRUCTURE OF MUSCLE TISSUE

1. The sole function of muscle tissue is to contract or shorten, which causes movement.

2. Skeletal muscle forms the muscles attached to the skeleton, which move the limbs and other body parts. Its cells are long, striated, and multinucleate. Connective tissue coverings (endomysium, perimysium, and epimysium) enclose and protect the muscle cells and increase the strength of skeletal muscles.

3. Smooth muscle cells are uninucleate, spindle-shaped cells arranged in opposing layers in the walls of hollow organs. When they contract, substances (food, urine, a baby) are moved along internal pathways.

4. Cardiac muscle cells are striated, branching cells that fit closely together and are arranged in spiral bundles in the heart. Their contraction pumps blood through the blood vessels of the body.

B. MUSCLE ACTIVITY

1. All skeletal muscle cells are stimulated by motor neurons. When the neuron releases a neurotransmitter (acetylcholine), the permeability of the sarcolemma is changed, allowing sodium ions to enter the muscle

cell. This changes the electrical conditions in the cell and produces an electrical current (action potential), which flows across the entire sarcolemma. The action potential results in contraction of the muscle cell.

2. Although individual muscle cells contract completely when adequately stimulated, a muscle (an organ) responds to stimuli with different degrees of shortening (graded responses), which reflects the number of muscle cells stimulated.

3. Most skeletal muscles contractions are tetanic (smooth and sustained) because rapid nerve impulses are reaching the muscle, and the muscle is not allowed to relax completely between contractions.

4. If muscle activity is strenuous and prolonged, muscle fatigue occurs. Muscle fatigue results from an accumulation of lactic acid in the muscle and a decrease in its energy (ATP) supply. After exercise the oxygen debt is repaid by rapid deep breathing.

5. Muscle contractions are isotonic (the muscle shortens and movement occurs) or isometric (the muscle does not shorten, but its tension increases).

6. Muscle tone keeps muscles healthy and ready to react. It is a result of a staggered series of nerve impulses delivered to different cells within the muscle. If the nerve supply is destroyed, the muscle loses tone, becomes paralyzed, and atrophies.

C. BODY MOVEMENTS AND NAMING SKELETAL MUSCLES

1. All muscles are attached to bones at two points. The origin is the immovable attachment; the insertion is the movable bony attachment. When contraction occurs the insertion moves toward the origin.

2. Body movements include flexion, extension, abduction, adduction, circumduction, rotation, pronation, supination, inversion, eversion, dorsiflexion, and plantarflexion.

3. On the basis of their general function in the body, muscles are classified as prime movers, antagonists, synergists, and fixators.

4. Muscles are named according to several criteria—for example, muscle size, shape, number and location of its origins, bones associated with, and the action of the muscle.

D. GROSS ANATOMY OF SKELETAL MUSCLES

1. Muscles of the head fall into two groups
   a. Muscles of facial expression include the frontalis, orbicularis oris and oculi, buccinator, and zygomaticus.
   b. Chewing muscles are the masseter, temporalis, and buccinator.

2. Muscles of the trunk and neck move the head, shoulder girdle, and trunk and form the abdominal girdle. Anterior neck and trunk muscles include the sternocleido-mastoid, pectoralis major, intercostals, rectus abdominis, external and internal obliques, and transversus abdominis. Posterior trunk and neck muscles include the trapezius, latissimus dorsi, and deltoid.

3. Muscles of the upper limb include muscles that cause movement at the shoulder joint, elbow, and wrist/fingers. Muscles causing movement at the elbow include the biceps brachii and triceps brachii.

4. Muscles of the lower extremity cause movement at the hip, knee, and ankle/foot. Lower extremity muscles include the iliopsoas, gluteus maximus and medius, adductors, quadriceps and hamstring groups, gastrocnemius, anterior tibialis, and peroneus muscles.

E. DEVELOPMENTAL ASPECTS OF THE MUSCULAR SYSTEM

1. Increasing muscular control reflects the maturation of the nervous system. Muscle control is achieved in a cephalocaudal and proximal-distal direction.

2. To remain heatlhy, muscles must be regularly exercised. Without exercise they atrophy; with extremely vigorous exercise they hypertrophy.

3. As we age muscle mass decreases and the muscles become more sinewy.

REVIEW QUESTIONS

1. What is the major function of muscle?

2. Compare skeletal, smooth, and cardiac muscles as to their microscopic anatomy, location and arrangement in body organs, and function in the body.

3. Specifically, what is responsible for the banding pattern seen in skeletal muscle cells?

4. Why are the connective tissue wrappings of skeletal muscles important? Name these connective tissue coverings beginning with the finest and ending with the most coarse.

5. What is the function of tendons? How does a tendon differ from an aponeurosis? How is it the same?

6. Define myoneural junction, motor unit, tetanus, graded response, muscle fatigue, and neurotransmitter.

7. If blue litmus paper is pressed against the cut surface of a fatigued muscle, it changes color to red, indicating a low pH or acid conditions. No such change happens when blue litmus paper is pressed to the cut surface of a rested or nonfatigued muscle. Explain.

8. Describe the events that occur from the time a motor neuron releases acetylcholine at the myoneural junction until muscle cell contraction occurs.

9. Explain how isotonic and isometric contractions differ.

10. Muscle tone keeps muscles healthy. What is muscle tone, and what causes it? What happens to a muscle that loses its tone?

11. A skeletal muscle is attached to bones at two points. Name these attachment points and indicate which is the movable end and which is the nonmovable end.

12. List the 12 body movements studied and demonstrate each.

13. How is a prime mover different from a synergist muscle? How can a prime mover also be an antagonist?

14. Name at least four criteria that are used as a basis for naming skeletal muscles and give an example to illustrate each.

15. Name the prime mover for chewing. Name three other muscles of the face and give the location and function of each.

16. The sternocleidomastoid muscles help to flex the neck. What posterior neck muscles are their antagonists?

17. Name two muscles that reverse the movement of the deltoid muscle.

18. Name the prime mover of elbow flexion. Name its antagonist.

19. Other than acting to flex the spine and compress the abdominal contents, the abdominal muscles are extremely important in protecting and containing the abdominal viscera. What is it about the arrangement of these muscles that makes them so well suited for their job?

20. The hamstring and quadriceps muscle groups are antagonists of each other, and each group is a prime mover in its own right. What action does each muscle group perform?

21. What two-bellied muscle makes up the calf region of the leg? What is its function?

22. Name three muscles or muscle groups used as the site for intramuscular injections. Which is most often used in babies?

23. What happens to muscles when they are exercised regularly? Exercised vigorously as in weight lifting? Not used?

24. What is the effect of aging on skeletal muscles?

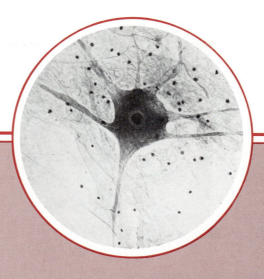

CHAPTER 6

Nervous System

Chapter Contents

Functions of the nervous system • to maintain homeostasis with electrical signals (a fast-acting system); to provide for sensation, higher mental functioning, and emotional response; and to activate the muscles and glands

After completing this chapter, you should be able to:

- List the general functions of the nervous system.

- Explain the anatomical and functional classifications of the nervous system.

- Define central nervous system and peripheral nervous system and list the major parts of each.

- State the function of neurons and neuroglia.

- Describe the general structure of a neuron and name its important anatomical regions.

- List the two major functional properties of neurons.

- Classify neurons according to function.

- List the types of general sensory receptors and describe the function of each type.

- Describe the events that lead to the conduction of a nerve impulse and its transmission from one neuron to another.

- Define reflex arc and list its elements.

- Identify or locate the major regions of the cerebral hemispheres, brain stem, and cerebellum on a human brain model or appropriate diagram and state their functions.

- Locate the major functional areas of the human cerebrum.

- Describe the composition of gray and white matter.

- Name the three meningeal layers and state their function.

- Discuss the formation and function of cerebrospinal fluid.

- List two major functions of the spinal cord.

- Describe the structure of the spinal cord.

- Discuss the function of the fiber tracts in the spinal cord white matter.

- Describe the general structure of a nerve.

- Identify the cranial nerves by number and name and list the major functions of each.

- Describe the origin and fiber composition of (a) ventral and dorsal roots, (b) the spinal nerve proper, and (c) ventral and dorsal rami.

- Discuss the distribution of the dorsal and ventral rami of the spinal nerves.

- Name the four major nerve plexuses with the major nerves of each and state their distribution.

- Identify the site of origin and explain the function of the sympathetic and parasympathetic divisions of the autonomic nervous system.

- State the difference in effect of the parasympathetic and sympathetic divisions on the following organs: heart, lungs, digestive system, and blood vessels.

You are driving down the freeway, and a horn blares to your right—you swerve to your left. Charles leaves a note on the table, "See you later—have the stuff ready at 6." You know that the "stuff" is chili with taco chips. You are dozing, and your infant son makes a soft cry. Your eyes flash open even though a fire engine could normally pass the house at night without waking you. What do all these examples have in common? They are all routine examples of the workings of your nervous system, which effectively has your

body cells humming with activities, some vital and others just fun, all the time.

The nervous system is the master controlling and communicating system of the body. Every thought, action, and sensation reflects its activity. Much like a sentry, the nervous system gathers information about what is occurring both inside and outside the body. It then processes this information and makes decisions about what should be done at each moment. Its signaling device, or means of communicating with body cells, is electrical impulses, which are rapid, specific, and cause almost immediate responses.

We are what our brain has experienced. If all past sensory input could suddenly be erased, we would be unable to walk, talk, or communicate in any manner. We would feel no pain and no pleasure. Spontaneous movements would occur (as in a fetus), but no type of voluntary activity would be possible. Clearly we would cease to be the same people.

The nervous system does not work alone to regulate and maintain body homeostasis. The endocrine system is a second important regulating system. While the nervous system controls with rapid electrical nerve impulses, the endocrine system organs produce hormones, or chemicals, that direct the body's cells to promote the well-being of the body as a whole.

DIVISIONS OF THE NERVOUS SYSTEM

Because of the complexity of the nervous system, it is difficult to "swallow it whole." So we divide it up into smaller areas when we want to discuss it in terms of its structures (anatomic classification) or its activities (functional classification).

Anatomical Classification

The anatomical classification includes all the nervous system organs. The nervous system is divided on the basis of its structures, and there are two subdivisions—the central nervous system and the peripheral nervous system.

The central nervous system (CNS) consists of the brain and spinal cord, which occupy a *central,* or medial, position in the body. The brain and cord primarily interpret incoming sensory information and issue instructions based on past experience.

The peripheral (pĕ-rif′er-al) nervous system (PNS) consists of nervous system structures (nerves) *outside* the CNS, which carry impulses to

and from the brain (cranial nerves) and to and from the cord (spinal nerves). The nerves serve as communication lines. They link all parts of the body by carrying impulses from the sensory receptors to the CNS and from the CNS to the appropriate glands or muscles.

The CNS and PNS are discussed in length later in the chapter.

Functional Classification

The structures of the nervous system also may be divided functionally in terms of motor activities or actions that the nerves of the PNS generate. This classification also has two subdivisions—the somatic and the autonomic divisions.

The somatic nervous system allows us to consciously, or voluntarily, control our skeletal muscles.

The autonomic nervous system regulates activities that are automatic, or involuntary, such as the activation of smooth and cardiac muscles and glands.

Although simpler to study the nervous system in terms of its subdivisions, you should recognize that these subdivisions are made for the sake of convenience only. Remember, the nervous system acts as a coordinated unit both structurally and functionally.

NERVOUS TISSUE

Even though it is complex, nervous tissue is made up of just two principal types of cells—neuroglia and neurons.

Neuroglia

The **neuroglia** (nu-rog′le-ah) (literally, nerve glue) include many types of cells that generally support and protect the neurons in the CNS (Figure 6-1). In addition, each of the different types of glial cells has special functions. **Oligodendroglia** (ol″ĭ-go-den-drog′le-ah) deposit a fatty insulating substance around the processes, or extensions, of the neurons. **Microglia** (mi-krog′le-ah) are phagocytes, which dispose of debris (dead

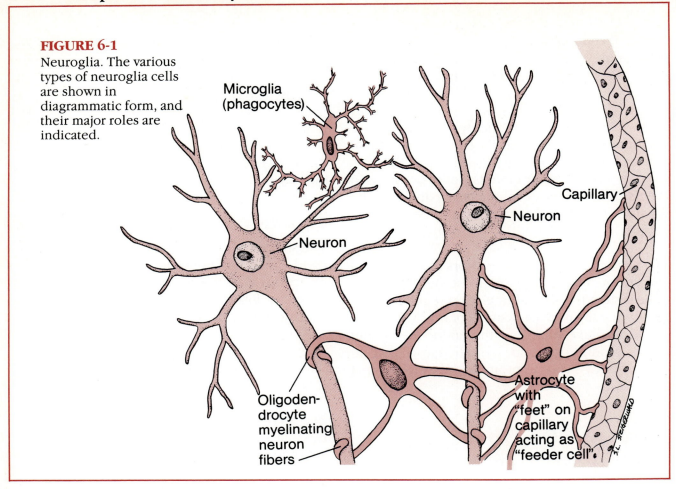

FIGURE 6-1
Neuroglia. The various types of neuroglia cells are shown in diagrammatic form, and their major roles are indicated.

Microglia (phagocytes)

Capillary

Neuron

Neuron

Oligodendrocyte myelinating neuron fibers

Astrocyte with "feet" on capillary acting as "feeder cell"

brain cells, bacteria, and the like). **Astrocytes** (as'tro-sīts) seem to act as "feeder cells" for the delicate neurons. Forming a living barrier between the capillary blood supply and the neurons, astrocytes make the exchanges between neurons and blood. In this way, they protect the neurons from harmful substances that might be in the blood.

Although they resemble neurons structurally, neuroglia are not able to transmit nerve impulses, a function that is highly developed in neurons. Another important difference is that neuroglia never lose their ability to divide whereas neurons do. For this reason, most brain tumors are *gliomas*, or tumors formed by glial cells.

Neurons

ANATOMY

The delicate **neurons** are highly specialized to conduct, or transmit, messages (nerve impulses) from one part of the body to another. Although neurons differ structurally, they have many common features (Figure 6-2). All have (a) a **cell body,** which contains the nucleus, and (b) one or more slender *processes,* or fibers, extending from the cell body. The processes vary in length from microscopic in size to a length of 3–4 feet. The longest ones in humans reach from the inferior part of the spinal cord to the big toe. Neuron processes that conduct impulses *toward* the cell body are **dendrites** (den'drīts), whereas those that conduct impulses *away* from the cell body are **axons** (ak'sons). Usually neurons have only one axon but may have many dendrites, depending on the neuron type.

Typically a neuron is excited by other neurons when their axons release neurotransmitters (chemicals) close to its dendrites or nerve cell body. The impulse travels across the nerve cell body and down the axon. The axonal endings store the neurotransmitter chemical in tiny vesicles, or sacs. An enlarged view of the axonal ending in Figure 6–2 shows that it is separated from

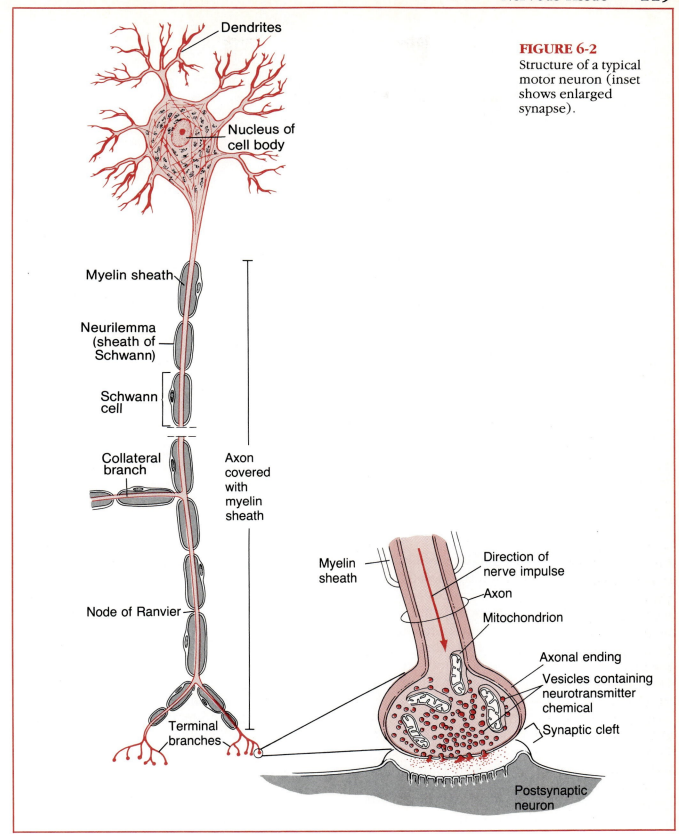

Dendrites

Nucleus of cell body

Myelin sheath

Neurilemma (sheath of Schwann)

Schwann cell

Collateral branch

Axon covered with myelin sheath

Node of Ranvier

Terminal branches

Myelin sheath

Direction of nerve impulse

Axon

Mitochondrion

Axonal ending

Vesicles containing neurotransmitter chemical

Synaptic cleft

Postsynaptic neuron

FIGURE 6-2
Structure of a typical motor neuron (inset shows enlarged synapse).

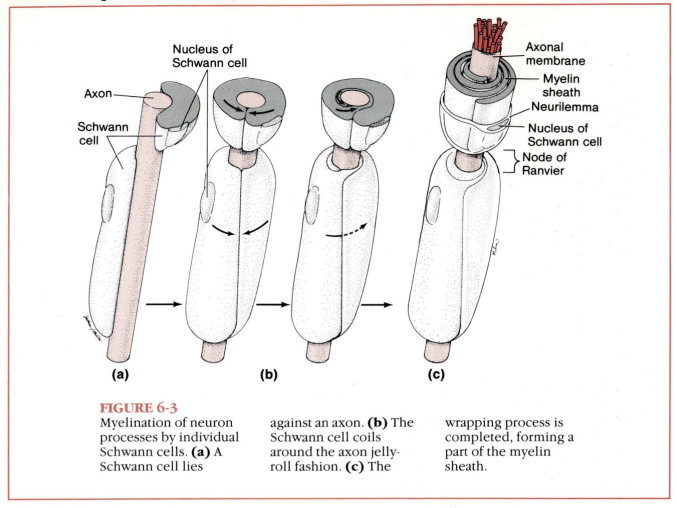

FIGURE 6-3

Myelination of neuron processes by individual Schwann cells. **(a)** A Schwann cell lies against an axon. **(b)** The Schwann cell coils around the axon jelly-roll fashion. **(c)** The wrapping process is completed, forming a part of the myelin sheath.

the next neuron by a tiny gap, the **synaptic** (sĭ-nap'tik) **cleft.** Although they are close, neurons never actually touch other neurons. When an impulse reaches the axonal ending, the vesicles break and release the neurotransmitter into the **synapse**. The neurotransmitter then diffuses across the synapse and binds to the cell membrane of the next neuron, triggering the nerve impulse.

Most long nerve fibers (both axons and dendrites) are covered with a whitish fatty material, **myelin** (mi'e-lin). Myelinated fibers have a waxy appearance. Myelin protects and insulates the fibers and increases the rate of transmission of the nerve impulses. Axons *leaving* the CNS are typically heavily myelinated by **Schwann cells**, which wrap themselves tightly around the axon jelly-roll fashion (Figure 6–3). When the wrapping process is done, a tight core of cell membrane material encloses the axon; this wrapping is the **myelin sheath**. Most of the Schwann cell cy-

toplasm ends up just beneath the outermost part of its cell membrane. The exposed part of the Schwann cell membrane is the **neurilemma** (nu"rĭ-lem'mah). Since the myelin sheath is formed by many individual Schwann cells, it has gaps or indentations called **nodes of Ranvier** (rahn-ve-āz') (see Figure 6-2). Neuron fibers that have myelin sheaths conduct nerve impulses *much* faster than unmyelinated fibers, because the nerve impulse literally jumps, or leaps, from node to node along the length of the fiber. This rapid type of impulse transmission is called *saltatory* (sal'tah-to"re) *conduction*.

The importance of the myelin insulation to nerve transmission is best illustrated by observing what happens when it is not there. In people with *multiple sclerosis,* the myelin coating around the fibers gradually disappears. As this happens, the affected person begins to lose the ability to control his or her muscles and becomes increasingly disabled.

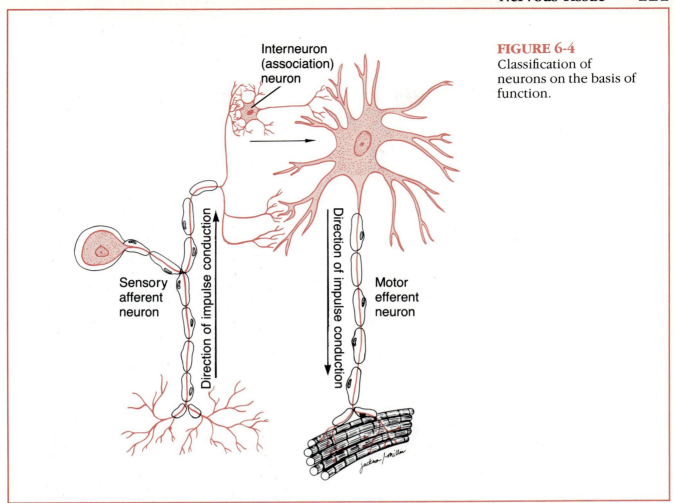

Interneuron
(association)
neuron

Sensory
afferent
neuron

Direction of impulse conduction

Motor
efferent
neuron

Direction of impulse conduction

Jackson/Miller

FIGURE 6-4
Classification of
neurons on the basis of
function.

Nerve cell bodies, which are found only in the CNS or in **ganglia** (gang′le-ah), make up the gray matter of the nervous system. Ganglia are collections of nerve cell bodies *outside* the CNS. Clusters of nerve cell bodies inside the CNS are **nuclei** (nu′kle-i). The neuron processes running through the CNS form **tracts** of white matter; in the PNS they form the **nerves**, which are bundles of fibers extending from the CNS to the body periphery (body muscles and glands).

CLASSIFICATION

Neurons may be classified, or grouped, according to their structure or how they function in the body. We will emphasize the functional classification scheme here.

The functional classification that groups neurons according to the direction the nerve impulse is traveling includes afferent, efferent, and association neurons (Figure 6-4). Neurons carrying impulses from sensory receptors (in the internal or-

gans or the skin) to the CNS are **afferent**, or **sensory, neurons**. The activity of sensory neurons keeps us informed about what is happening both inside and outside the body. The dendrite endings of the sensory neurons usually carry specialized receptors that are activated by specific changes occurring in their vicinity.

The receptors of the special sense organs (vision, hearing, equilibrium, taste, and smell) are quite complex and are covered separately in Chapter 7. The simpler types of sensory receptors seen in muscles and tendons (**proprioceptors** [pro″pre-o-sep′tors]) and in the skin (**cutaneous sense organs**) are shown in Figure 6-5. The pain receptors (actually bare dendrite endings) are the least specialized of the cutaneous receptors and the most numerous, because pain warns one that some type of damage is occurring in the body. Vigorous stimulation of any of the cutaneous receptors (for example, heat, cold, or pressure receptors) is also interpreted as pain. The proprioceptors detect the

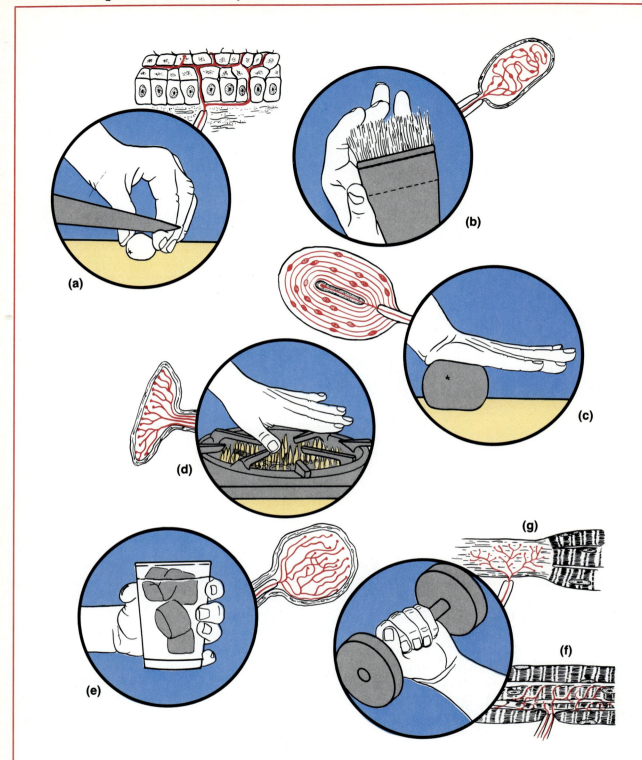

FIGURE 6-5

Types of sensory receptors. **(a)** Naked nerve endings (pain receptors).
(b) Meissner's corpuscle (touch receptor).
(c) Pacinian corpuscle (deep pressure receptor). **(d)** Ruffini's corpuscle (heat receptor). **(e)** Krause's end bulb (cold receptor). **(f)** Muscle spindle (proprioceptor). **(g)** Golgi tendon organ (proprioceptor). (Note: The proprioceptors respond to the degree of stretch, or tension, in the muscles and tendons.)

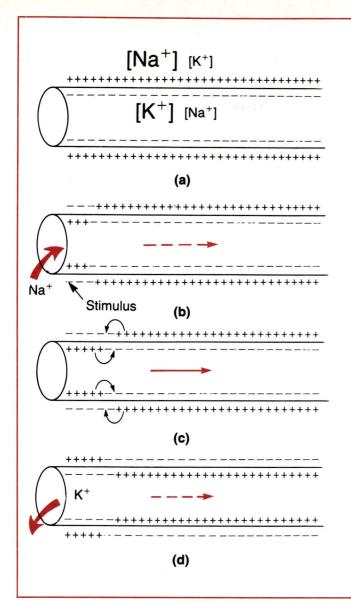

$[Na^+]$ $[K^+]$

$[K^+]$ $[Na^+]$

(a)

Na^+ Stimulus

(b)

(c)

K^+

(d)

FIGURE 6-6
The nerve impulse. **(a)** Electrical conditions of a resting (polarized) membrane. The predominant extracellular ion is sodium (Na^+) whereas the predominant ion inside the cells is potassium (K^+). The membrane is relatively impermeable to both ions. **(b)** Depolarization. The resting potential is reversed. A stimulus changes the membrane permeability and sodium ions diffuse rapidly into the cell. **(c)** Generation of the action potential or nerve impulse. If the stimulus is strong enough, the action potential is initiated and spreads rapidly along the entire length of the membrane. **(d)** Repolarization. The resting potential is restored. Potassium ions diffuse out of the cell as the membrane permeability changes once again. This restores the negative charge on the inside of the membrane and the positive charge on its external surface. Repolarization occurs in the same direction as depolarization. The original ionic concentrations of the resting state are restored later by activation of the sodium–potassium pump.

amount of stretch, or tension, in the skeletal muscles and their tendons. They send this information to the brain so that the proper adjustments can be made to maintain balance and normal posture. The nerve cell bodies of sensory neurons are always found in a ganglion outside the CNS.

Neurons carrying impulses from the CNS to the viscera and/or body muscles and glands are **efferent,** or **motor, neurons**. The cell bodies of motor neurons are always located in the CNS.

The third category of neurons are the **association neurons**, or **interneurons**. They connect the motor and sensory neurons in neural pathways. Like the motor neurons, their nerve cell bodies are always located in the CNS.

PHYSIOLOGY

NERVE IMPULSES   Neurons have two major functional properties: *irritability*, the ability to respond to a stimulus and convert it into a nerve impulse, and *conductivity*, the ability to transmit the impulse to other neurons, muscles, or glands.

The cell membrane of a resting, or inactive, neuron is **polarized**, which means that there are fewer positive ions inside the neuron than there are in the tissue fluid that surrounds it (Figure 6-6). The major positive ions inside the cell are potassium (K^+), whereas the major positive ions outside the cell are sodium (Na^+). As long as the inside remains negative and the outside remains positive, the neuron will stay inactive.

Many different types of stimuli can excite neurons so that they become active and generate an impulse. For example, light excites the eye receptors, sound excites some of the ear receptors, and pressure excites some cutaneous receptors of the skin. However, most neurons in the body are excited by neurotransmitters, released by other neurons. Regardless of what the stimulus is, the result is always the same—that is, the permeability properties of the cell membrane change for a very brief period. *Normally,* sodium ions cannot diffuse through the cell membrane. But when the neuron is stimulated, the "sodium gates" in the membrane are opened. Because sodium is in much higher concentration outside the cell, it will diffuse quickly into the neuron whenever the sodium gates are opened. (Remember the laws of diffusion?) This inward rush of sodium ions changes the polarity of the neuron's membrane; the inside is now more positive and the outside is less positive. This event is **depolarization**, and it activates the neuron to transmit the impulse, or **action potential**. When the action potential reaches the axonal endings, the axonal endings release neurotransmitter, which diffuses across the synapse and probably causes the next neuron to become activated. Notice that the transmission of an impulse is an *electrochemical event;* transmission down the length of the neuron's membrane is basically electrical, but the next neuron is stimulated by a neurotransmitter, which is a chemical.

Almost immediately after the sodium ions have rushed into the neuron, the membrane permeability changes again. Once again it becomes impermeable to sodium ions, but now potassium ions are allowed to diffuse out of the neuron into the tissue fluid, and they do so very rapidly. This outflow of positive ions from the cell restores the electrical conditions at the membrane to the polarized, or resting, state; this event is **repolarization**. *Until repolarization occurs, a neuron cannot conduct another impulse.* After repolarization has occurred, the initial concentrations of the sodium and potassium ions inside and outside the neuron are restored by the activation of the sodium–potassium pump. This pump uses ATP (cellular energy) to pump excess sodium ions out of the cell and bring potassium ions back into the cell. This whole sequence of events, which takes just a few thousandths of a second, is responsible for the major part of body control and coordination.

REFLEX ARC.   There are many types of communication between neurons, but much of what the body *must* do every day is programmed as **reflexes**. Reflexes are rapid, predictable, and involuntary responses to stimuli. They are much like one-way streets—once a reflex begins, it always goes in the same direction. Reflexes occur over neural pathways called **reflex arcs**.

The types of reflexes that occur in the body are classed either as autonomic or somatic reflexes. **Autonomic** (aw"to-nom'ik) **reflexes** are not voluntarily, or consciously, controlled. These reflexes result in the activation of smooth muscles, the heart, and glands. For example, the production of saliva (salivary reflex) and changes in the size of the eye pupils (pupillary reflex) are two such reflexes. The autonomic reflexes regulate such body functions as digestion, elimination, blood pressure, and perspiration. **Somatic** (so-mat'ik) **reflexes** include all reflexes that stimulate the skeletal muscles. When you quickly pull your hand away from a hot object, a somatic reflex is working.

All reflex arcs have a sensory receptor (which reacts to a stimulus), an effector organ (the muscle or gland eventually stimulated), and efferent and afferent neurons to connect the two. The simple *patellar* (pah-tel'ar), or *knee-jerk, reflex* shown in Figure 6-7a is an example of a two-neuron reflex arc. The patellar reflex is familiar to most of us. This is one of the tests a physician or nurse generally does during a physical exam to determine the general health of the motor portion of the nervous system. Most reflexes are much more complex than the two-neuron reflex, and they involve one or more interneurons in the pathway. A three-neuron reflex arc, the *flexor reflex,* is diagrammed in Figure 6-7b. The three-neuron reflex arc consists of five elements—receptor, afferent neuron, interneuron, efferent neuron, and effector. Since there is always a delay at synapses, the more synapses there are in a reflex pathway, the longer the reflex takes to happen.

Many spinal reflexes involve only spinal cord neurons and occur without brain involvement. As long as the spinal cord is functional, spinal reflexes such as the flexor reflex will work. On the other hand, some reflexes require that the brain become involved, because many different types of information have to be evaluated to arrive at

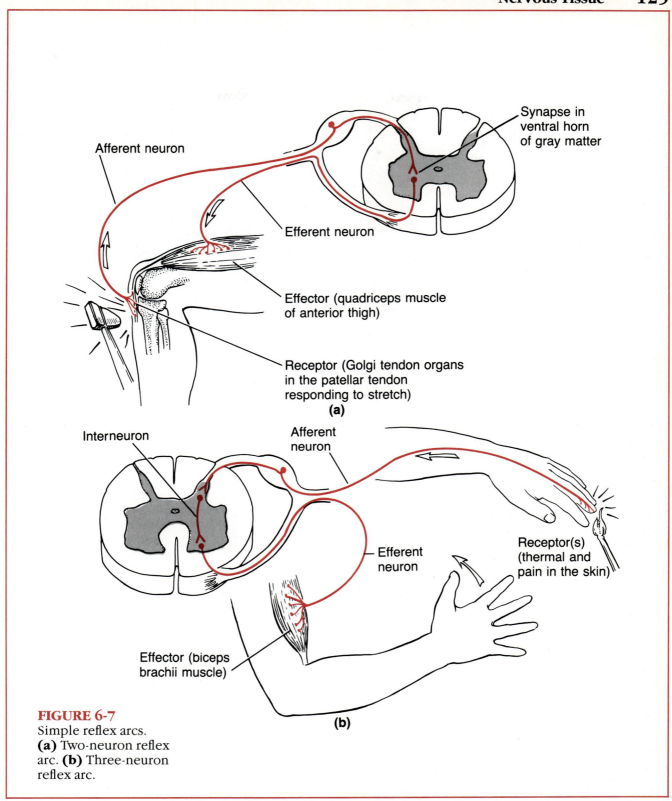

FIGURE 6-7
Simple reflex arcs.
(a) Two-neuron reflex
arc. **(b)** Three-neuron
reflex arc.

the "right" response. The response of the pupils of the eyes to light is a reflex of this type.

As noted earlier, reflex testing is an important tool in evaluating the condition of the nervous system. When reflexes are exaggerated, distorted, or absent, nervous system disorders are indicated. Reflex changes often occur before the pathologic condition has become obvious in other ways.

FIGURE 6-8
Left lateral view of the human brain. **(a)** Major structural areas and **(b)** major functional areas of the cerebral hemispheres.

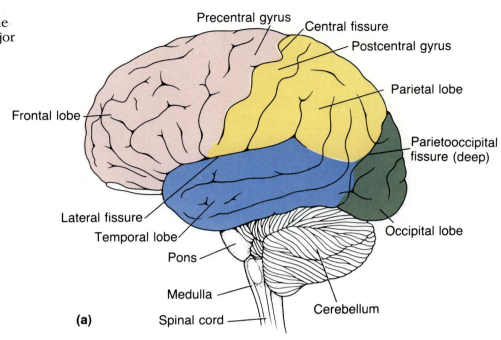

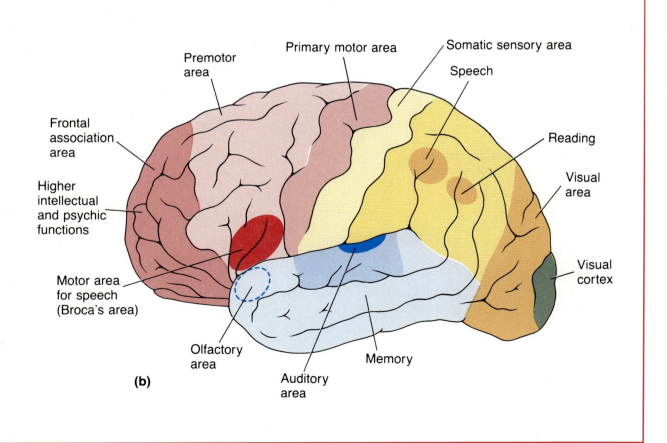

CENTRAL NERVOUS SYSTEM

During embryonic development the CNS first appears as a simple tube, the **neural tube**, which extends down the dorsal median plane. By the fourth week, the anterior end of the neural tube begins to expand, and brain formation begins. The rest of the neural tube posterior to the forming brain becomes the spinal cord. The central canal of the neural tube, which is continuous between the brain and spinal cord, becomes enlarged in four regions of the brain to form chambers called **ventricles** (see p. 133).

Brain

The brain's unimpressive appearance gives no hints as to its remarkable abilities. It is a gray, wrinkled organ, weighing about 3 pounds. Because the brain is the largest and most complex mass of nervous tissue in the body, it is generally discussed in terms of three major regions—cerebral hemispheres, brain stem, and cerebellum.

CEREBRAL HEMISPHERES

The paired **cerebral** (ser'ĕ -bral) **hemispheres** are the most superior part of the brain, and together are a good deal larger than the other two brain regions combined (Figure 6-8a). In fact, the cerebral hemispheres enclose and obscure most of the brain stem so that many brain-stem structures cannot be seen unless a sagittal section is made. Picture how a mushroom cap covers the top of its stalk: This is a fairly good representation of how the cerebral hemispheres cover the superior part of the brain stem.

The entire surface of the cerebral hemispheres is thrown into elevated ridges of tissue called **gyri** (jī'ri), which are separated by grooves called **fissures**. Many of these fissures and gyri are important anatomical landmarks. The cerebral hemispheres are separated by a single deep fissure, the *longitudinal fissure*. Other fissures divide each cerebral hemisphere into a number of **lobes,** which are named for the cranial bones that lie over them.

Many of the functional areas of the cerebral hemispheres have been located (Figure 6-8b). The **somatic sensory area** is located posterior to the *central fissure* in the **parietal lobe**. Impulses traveling from the body's sensory receptors are localized and interpreted in this area of the brain. The somatic sensory area allows you to recognize pain, coldness, or a light touch. As illustrated in Figure 6-9, the body is represented in an upside down manner in the sensory area. Body regions with the most sensory receptors, the face and mouth region, send impulses to neurons which account for a large part of the sensory cortex. Also, the sensory pathways are crossed pathways—meaning that the left side of the sensory cortex receives impulses from the right side of the body, and the right cortex receives impulses from the left side of the body.

Impulses from the special sense organs are interpreted in other cortical areas. For example, the visual area is located in the posterior part of the **occipital lobe**, the auditory area is in the **temporal lobe** bordering the *lateral fissure,* and the olfactory area is found deep inside the **temporal lobe.**

The **primary motor area** that allows us to consciously move our skeletal muscles is anterior to the central fissure in the **frontal lobe.** The axons of these motor neurons form the major voluntary motor tract, which descends to the cord—the **pyramidal**, or **corticospinal** (kor"tĭ-ko-spi'nal) **tract.** Like the somatic sensory cortex, the body is represented in an upside down fashion, and the pathways are crossed pathways. Most of the neurons in this primary motor area control body areas having the finest motor control, that is, the face, mouth, and hands (Figure 6-9).

A specialized motor speech area, **Broca's** (bro'-kahz) **area**, is found at the base of the precentral gyrus (the gyrus anterior to the central fissure). Damage to this area, which is located only in one cerebral hemisphere (usually the left), results in an inability to say words properly.

Areas involved in higher intellectual reasoning are believed to be in the anterior part of the frontal lobes; complex memories appear to be stored in the temporal and frontal lobes. The *speech area* is located at the junction of the temporal, parietal, and occipital lobes. The speech area allows one to understand words, whether they are spoken or read, and respond vocally to them. Since this area (like Broca's area) is usually in one cerebral

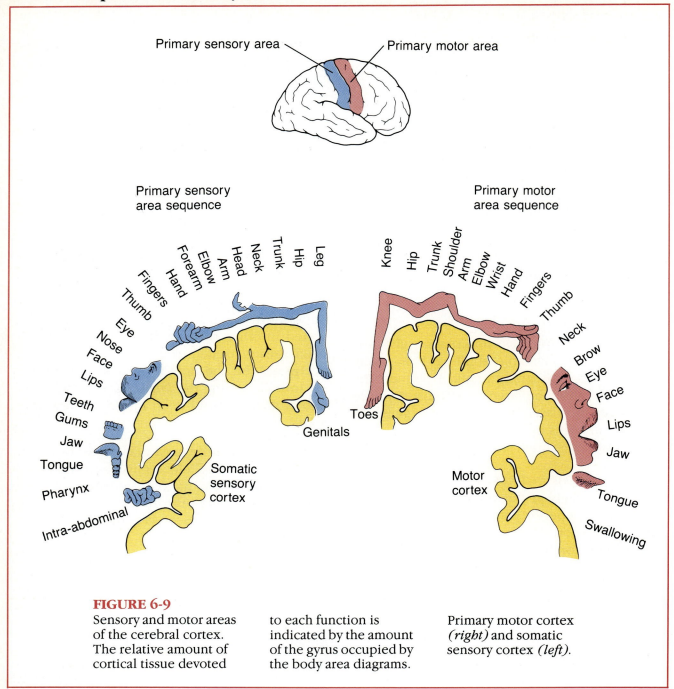

Primary sensory
area sequence

Primary motor
area sequence

Fingers
Thumb
Eye
Nose
Face
Lips
Teeth
Gums
Jaw
Tongue
Pharynx
Intra-abdominal

Hand
Forearm
Elbow
Arm
Head
Neck
Trunk
Hip
Leg

Somatic
sensory
cortex

Toes

Genitals

Knee
Hip
Trunk
Shoulder
Arm
Elbow
Wrist
Hand
Fingers
Thumb
Neck
Brow
Eye
Face
Lips
Jaw
Tongue
Swallowing

Motor
cortex

FIGURE 6-9

Sensory and motor areas
of the cerebral cortex.
The relative amount of
cortical tissue devoted
to each function is
indicated by the amount
of the gyrus occupied by
the body area diagrams.
Primary motor cortex
(right) and somatic
sensory cortex *(left).*

hemisphere, *aphasia* (ah-fa′ze-ah) is a common result of damage to the left cerebral hemisphere. There are many types of aphasias, which can involve a loss of ability to speak, write, or understand written or spoken language. It is a frequent result of a *cerebrovascular* (ser″ĕ-bro-vas′ku-lar) *accident* (CVA), or stroke.*

Speech, memory, logical and emotional response, as well as consciousness, interpretation of sensation, and voluntary movement are all functions of the cerebral cortex neurons. Because the functional areas of the cerebral hemispheres are fairly well known, it is often possible after a CVA to determine the area of damage to the brain by observing the symptoms of the patient. For example, if the patient has left-sided paralysis, the right motor cortex of the frontal lobe is most likely involved. Conversely, if the person has aphasia or

*A CVA is a situation in which there is death of brain neurons most often caused by an interruption in the blood supply to them.

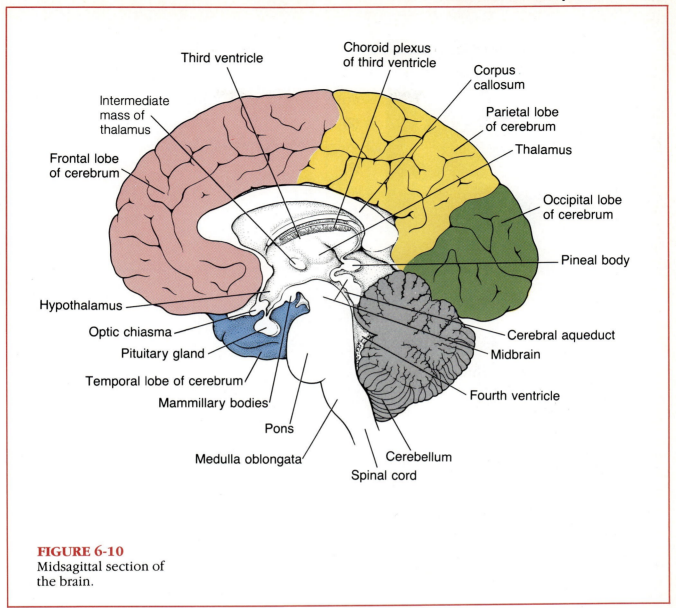

FIGURE 6-10
Midsagittal section of
the brain.

motor speech problems, it is not usually a problem with his or her intellect; most often the speech initiation area or Broca's area have been damaged in such cases. There is no question that brain lesions can cause marked changes in a person's disposition (for example, a change from a sunny to a foul personality). In such cases, a tumor as well as a CVA might be suspected.

The cell bodies of neurons involved in the cerebral hemisphere functions are found only in the outermost **gray matter** of the cerebrum, the **cerebral cortex**. As noted earlier, the cortical region is highly ridged and convoluted, providing more room for the thousands of neurons found there.

Most of the rest of the cerebral hemisphere tissue—the deeper **white matter**, or **cerebral medulla**—is composed of fiber tracts (bundles of nerve fibers) carrying impulses to or from the cortex. One very large fiber tract, the **corpus callosum** (kah-lo′sum) connects the cerebral hemispheres (Figure 6-10). The corpus callosum arches above the structures of the brain stem and allows the cerebral hemispheres to communicate with one another. (This is important because, as already discussed, some of the cortical functional areas are only in one hemisphere.)

Although most of the gray matter is in the cerebral cortex, there are several "islands" of gray matter, the **basal nuclei**, buried deep within the white

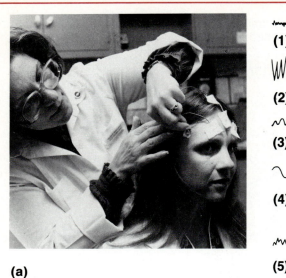

(a)

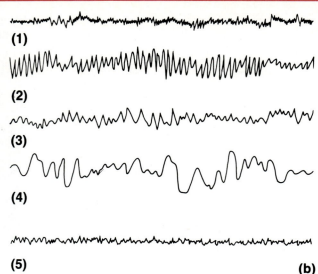

(1)

(2)

(3)

(4)

(5)

(b)

BOX 6-1
Electroencephalography

Normal brain function involves the continuous transmission of electrical impulses by neurons. A record of their activity can be made by placing electrodes at various points on the scalp and connecting these to a recording device **(a).** Such a recording is an **electroencephalogram** (e-lek″tro-en-sef′ah-lo-gram″) (EEG), and the patterns of electrical activity of the neurons are called brain waves. Five of the most common brain waves are (1) awake and alert, (2) awake and relaxed, (3) drowsy, (4) deep sleep, and (5) asleep and dreaming **(b).** As

might be expected, brain-wave patterns typical of the alert wide-awake state are different from those that occur during relaxation or deep sleep.

Some details about brain waves are known. They vary in frequency in different brain areas; and they change with age, sensory stimuli, brain disease, and the chemical state of the body. Brain waves are always present, even during unconsciousness, which indicates that the brain is constantly active. Absence of brain waves (a flat EEG) is evidence of clinical death. The

EEG is used to aid in the diagnosis and localization of many different types of brain lesions (such as epileptic lesions, tumors, abscesses, and infections). Interference with the function of the cerebral cortex is suggested by brain waves that are too fast or too slow, and unconsciousness occurs at both extremes. Sleep and coma result in brain-wave patterns that are slower than normal, whereas fright, epileptic seizures, and some kinds of drug overdose cause abnormally fast brain waves.

matter of the cerebral hemispheres. The basal nuclei help regulate voluntary motor activities by modifying instructions sent to the skeletal muscles by the primary motor cortex. Individuals who have problems with their basal nuclei are often

unable to walk normally or carry out other voluntary movements. *Huntington's chorea* (ko-re′ah), in which the individual is unable to control muscles and waves his or her arms uncontrollably, is one example of basal nuclei problems; *Parkin-*

son's disease is another. People with Parkinson's disease have trouble initiating movement or getting their muscles going. They also have a persistent hand tremor in which their thumb and index finger make continuous circles with one another in what is called the "pill-rolling" movement.

BRAIN STEM

The **brain stem** is about the size of a thumb in diameter and 4–5 inches long. Its structures are the diencephalon, midbrain, pons, and medulla oblongata. In addition to providing a pathway for tracts to ascend and descend through, the brain stem has many small gray matter areas. These collections of neurons, or nuclei, form the cranial nerves and are involved in the control of vital activities such as breathing and the control of blood pressure. Extending for the entire length of the brain stem is a diffuse mass of gray matter, the **reticular formation**. The neurons of the reticular formation are involved in the arousal response, consciousness, and the awake/sleep cycles. Damage to this area results in permanent unconsciousness. Identify the brain-stem areas in Figure 6-10 as you read through their descriptions next.

DIENCEPHALON.   The **diencephalon** (di″en-sef′-ah-lon), or **interbrain**, is the most superior part of the brain stem and is enclosed by the cerebral hemispheres. The major structures of the diencephalon are the thalamus, hypothalamus, and epithalmus. The **thalamus** consists of two large lobes of gray matter that enclose the shallow *third ventricle* of the brain. The thalamus is a relay station for sensory impulses passing upward to the sensory cortex. As impulses pass through the thalamus, we have a crude recognition of whether the sensation we are about to have is pleasant or unpleasant. The actual localization and interpretation of the sensation is done by the neurons of the sensory cortex.

The **hypothalamus** (literally, under the thalamus) makes up the floor of the diencephalon. It is an important autonomic nervous system center since it plays a role in the regulation of body temperature, water balance, and metabolism. The hypothalmus is also the center for many drives. For example, thirst, appetite, sex, pain, and pleasure centers are in the hypothalamus. The **pituitary gland** hangs from the anterior floor of the hypothalamus by a slender stalk. (Its function is dis-

cussed in Chapter 8.) The **mammillary bodies**, reflex centers involved in swallowing, bulge from the floor of the hypothalamus posterior to the pituitary gland.

The **epithalamus** (ep″y-thal′ah-mus) forms the roof of the third ventricle. Important parts of the epithalamus are the **pineal body** (part of the endocrine system) and the **choroid** (ko′roid) **plexus** of the third ventricle. The choroid plexuses, knots of capillaries within each ventricle, form the cerebrospinal fluid.

MIDBRAIN.   The **midbrain** is a relatively small part of the brain stem. It extends from the mammillary bodies to the rounded pons inferiorly. The **cerebral aqueduct** is a tiny canal that travels through the midbrain and connects the third ventricle of the diencephalon to the fourth ventricle below. The midbrain is composed primarily of ascending and descending fiber tracts.

PONS.   The **pons** (ponz) is the rounded structure that protrudes just below the midbrain. Pons means bridge, and this area of the brain stem is mostly fiber tracts; however, it does have important nuclei involved in the control of breathing.

MEDULLA OBLONGATA.   The **medulla oblongata** (mě-dul′ah ob″long-ga′tah) is the most inferior part of the brain stem. It continues into the spinal cord below without any obvious change in structure. Like the pons, the medulla is an important fiber tract area. However, the medulla has many nuclei involved in the regulation of vital visceral activities. It contains centers that control heart rate, blood pressure, breathing, swallowing, and vomiting, as well as others. The *fourth ventricle* is posterior to the pons and medulla and anterior to the cerebellum.

CEREBELLUM

The large cauliflowerlike **cerebellum** (ser″ě-bel′um) projects dorsally from under the occipital lobe of the cerebrum. Like the cerebrum, it has two hemispheres and a convoluted surface. The cerebellum also has an outer cortex made up of gray matter and an inner region of white matter.

The cerebellum coordinates skeletal muscle activity and controls balance and equilibrium. Fibers reach the cerebellum from the equilibrium apparatus of the inner ear, the eye, the proprioceptors of the skeletal muscles and tendons, and many other areas. Thus, the cerebellum constantly monitors body position and the amount of tension in various body parts and initiates the appropriate responses. If the cerebellum is damaged (for example, by a blow to the head, a tumor, or a CVA) the person cannot keep his or her balance and may appear to be drunk because of the loss of muscle coordination.

Protection of the Central Nervous System

Nervous tissue is very soft and delicate. It has the consistency of slightly dried-out oatmeal, and the fragile neurons are injured by even the slightest pressure. Nature has tried to protect the brain and spinal cord by enclosing them within bone (the skull and vertebral column), membranes (the meninges), and a watery cushion (cerebrospinal fluid).

MENINGES

The three connective tissue membranes covering and protecting the CNS structures are **meninges** (mě-nin′jēz) (Figure 6-11). The outermost layer, the leathery **dura mater** (du′rah ma′ter) (literally, hard mother) is a double-layered membrane. One of its layers is attached to the inner surface of the skull, forming the periosteum; the other forms the outermost brain covering and continues as the dura mater of the spinal cord. The dural layers are fused together except in three areas where the inner membrane extends inward to form a fold that attaches the brain to the cranial cavity.

The middle meningeal layer is the weblike **arachnoid** (ah-rak′noid) **mater**. Arachnida means spider, and some think that the arachnoid membrane looks like a cobweb. Its threadlike extensions span the subarachnoid space to attach it to the innermost membrane, the **pia** (pi′ah) **mater**. The delicate pia mater (which means gentle mother) clings tightly to the surface of the brain (and cord), following every fold.

The subarachnoid space is filled with cerebrospinal fluid. Specialized projections of the arachnoid membrane, **arachnoid villi**, protrude through the dura mater. The arachnoid villi drain the cerebrospinal fluid into the venous blood in the dural sinuses.

Meningitis (men″in-ji′tis), an inflammation of the meninges, is a serious threat to the brain because of the closeness of the brain and meninges. Meningitis is usually diagnosed by taking a sample of cerebrospinal fluid from the subarachnoid space.

CEREBROSPINAL FLUID

Cerebrospinal (ser″ĕ-bro-spi′nal) **fluid** (CSF), which is similar to plasma, is continually formed by the choroid plexuses. Choroid plexuses are clusters of capillaries hanging from the roof in each of the brain's ventricles. The CSF in and around the brain and cord forms a watery cushion, which protects the delicate nervous tissue from blows and other trauma.

Inside the brain, CSF is continually moving. It circulates from the two lateral ventricles (in the cerebral hemispheres) into the third ventricle in the diencephalon and then through the cerebral aqueduct of the midbrain into the fourth ventricle dorsal to the pons and medulla oblongata (see Figure 6-11). Some of the fluid reaching the fourth ventricle continues down the central canal of the spinal cord, but most of it circulates into the subarachnoid space through three openings in the walls of the fourth ventricle. The fluid returns to the blood in the dural sinuses through the arachnoid villi.

Ordinarily, CSF forms and drains at a constant rate. However, if something obstructs its drainage (for example, a tumor) the CSF begins to accumulate and exert pressure on the brain. This condition is *hydrocephalus* (hi-dro-sef′ah-lus) (literally, water on the brain). Hydrocephalus in a newborn baby causes the head to enlarge as the brain increases in size. This is possible because the skull bones have not yet fused. However, in adults this condition is likely to result in brain damage because the skull is hard, and the accumulating fluid crushes the soft nervous tissue. Today hydrocephalus is treated surgically by putting in a shunt to drain off the excess fluid into a vein in the neck region.

Spinal Cord

The cylindrical **spinal cord**, which is approximately 17–18 inches (42–43 cm) long depending on one's height, is a continuation of the brain stem. The spinal cord provides a two-way conduction pathway to and from the brain, and it is a

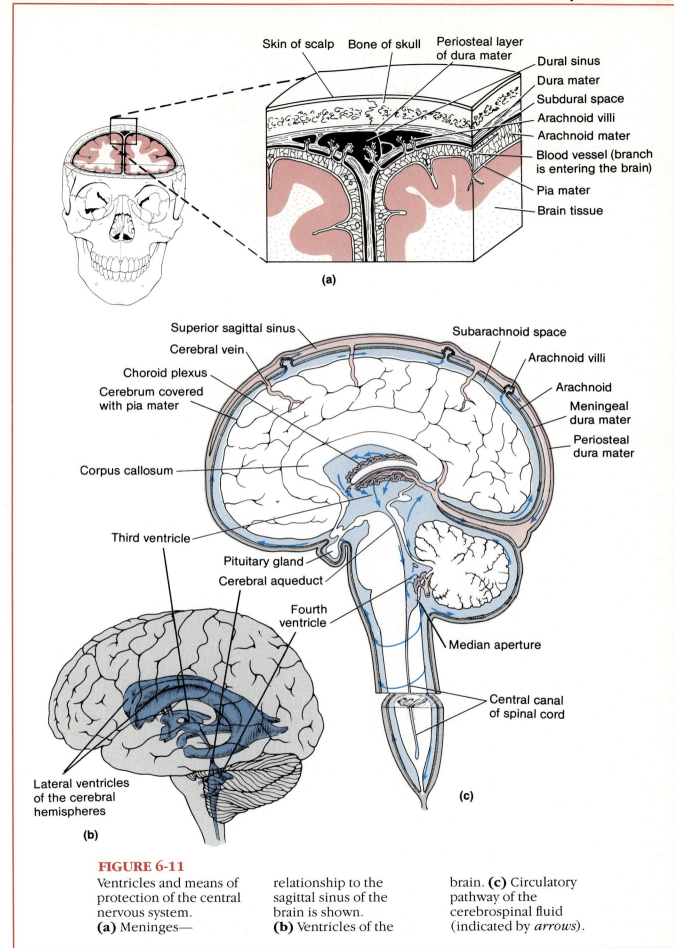

Skin of scalp   Bone of skull   Periosteal layer of dura mater
Dural sinus
Dura mater
Subdural space
Arachnoid villi
Arachnoid mater
Blood vessel (branch is entering the brain)
Pia mater
Brain tissue

(a)

Superior sagittal sinus
Cerebral vein
Choroid plexus
Cerebrum covered with pia mater
Corpus callosum
Third ventricle
Pituitary gland
Cerebral aqueduct
Fourth ventricle
Lateral ventricles of the cerebral hemispheres

(b)

Subarachnoid space
Arachnoid villi
Arachnoid
Meningeal dura mater
Periosteal dura mater
Median aperture
Central canal of spinal cord

(c)

FIGURE 6-11
Ventricles and means of protection of the central nervous system.
(a) Meninges— relationship to the sagittal sinus of the brain is shown.
(b) Ventricles of the brain. **(c)** Circulatory pathway of the cerebrospinal fluid (indicated by *arrows*).

FIGURE 6-12
Anatomy of the spinal cord, dorsal view.

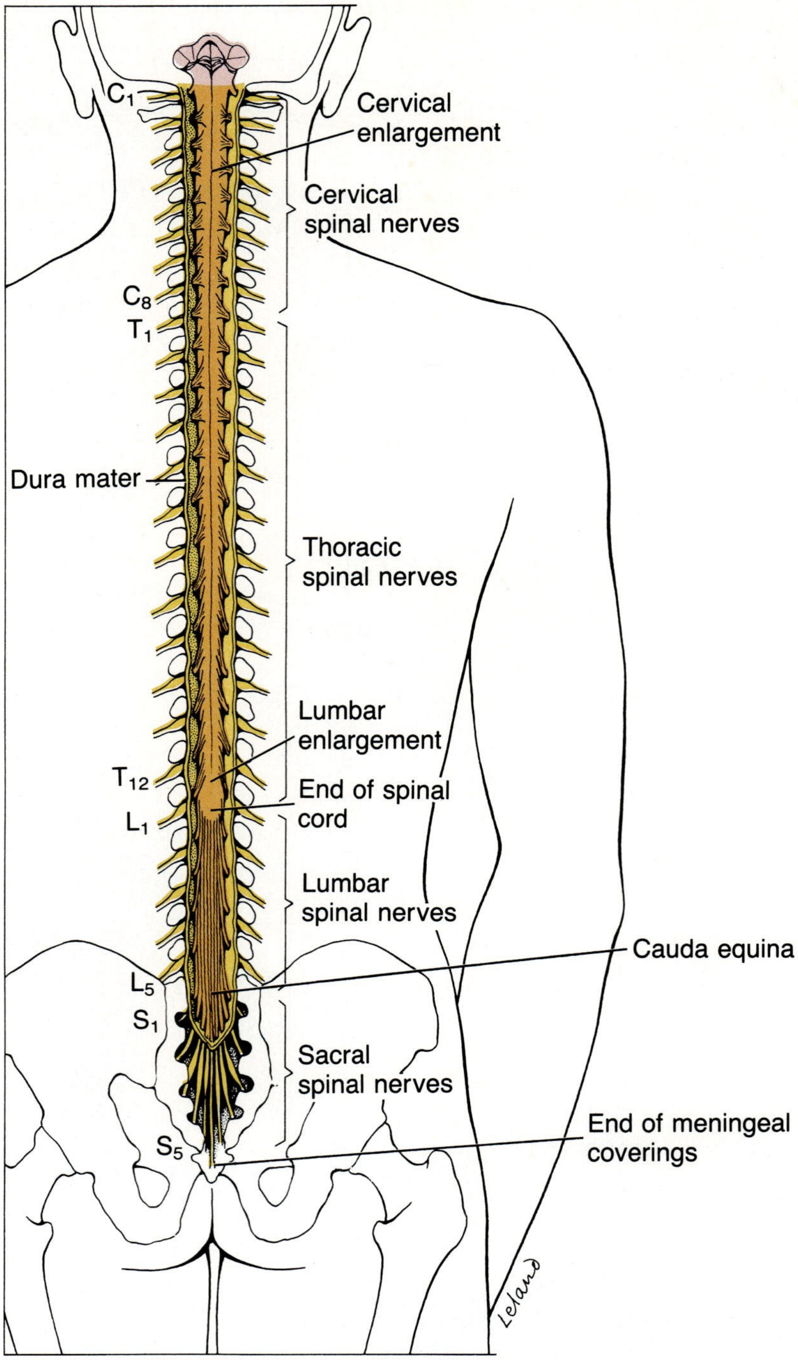

C₁

Cervical enlargement

Cervical spinal nerves

C₈
T₁

Dura mater

Thoracic spinal nerves

Lumbar enlargement

End of spinal cord

T₁₂
L₁

Lumbar spinal nerves

Cauda equina

L₅
S₁

Sacral spinal nerves

End of meningeal coverings

S₅

Leland

major reflex center (the spinal reflexes are completed at this level). Enclosed within the vertebral column, the spinal cord extends from the foramen magnum of the skull to the first or second lumbar vertebra, where it ends just below the ribs (Figure 6-12). Like the brain, the spinal cord is cushioned and protected by meninges. Meningeal coverings do not end at L₂, but instead extend well beyond the end of the spinal cord in the vertebral canal. Since there is no possibility of damaging the cord beyond L₄, the meningeal sac inferior to that point provides a nearly ideal spot for removing CSF for testing. (This procedure is a *lumbar puncture.*)

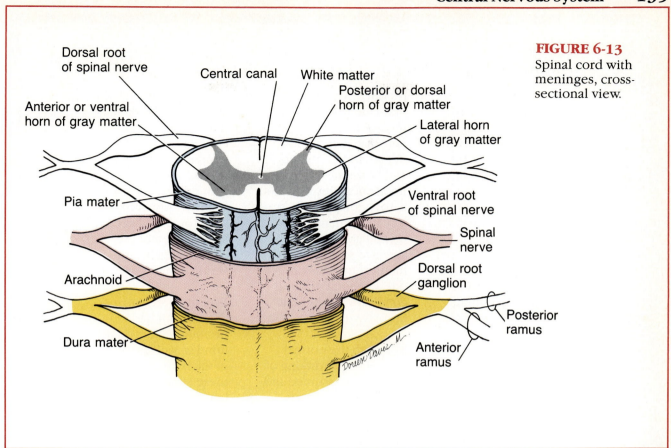

FIGURE 6-13
Spinal cord with
meninges, cross-
sectional view.

In humans 31 pairs of spinal nerves arise from the cord and exit from the vertebral column to serve the body area close by. The spinal cord is about the size of a thumb for most of its length, but it is obviously enlarged in the cervical and lumbar regions where the nerves serving the upper and lower limbs leave the cord. Because the cord does not reach the end of the vertebral column, the spinal nerves leaving the inferior end of the cord must travel through the vertebral canal for some distance before exiting. This collection of spinal nerves at the inferior end of the vertebral canal is the **cauda equina** (kaw'dah e-ki'nah), because it looks so much like a horse's tail (the literal translation of cauda equina).

GRAY MATTER OF THE SPINAL CORD AND SPINAL ROOTS

The gray matter of the cord looks like a butterfly or the letter H in cross section (Figure 6-13). The two posterior projections are the **dorsal** (posterior) **horns**; the two anterior projections are the **ventral** (anterior) **horns**. The gray matter surrounds the **central canal** of the cord, which contains CSF.

Neurons with specific functions can be located in the gray matter. The dorsal horns contain association neurons, or interneurons. The nerve cell bodies of the sensory neurons, whose fibers enter the cord by the **dorsal root**, are found in the enlarged area of the dorsal root called the **dorsal root ganglion**. If the dorsal root (or its ganglion) is damaged, sensation from the body area served will be lost. The ventral horns of the gray matter contain nerve cell bodies of motor neurons of the somatic nervous system (voluntary system), which send their axons out the **ventral root** of the cord. Damage to the ventral root results in a *flaccid paralysis* of the muscles served. In flaccid paralysis nerve impulses do not reach the muscles affected; thus, no voluntary movement of those muscles is possible. The muscles begin to atrophy, or waste away, because they are no longer stimulated. The dorsal and ventral roots fuse to form the spinal nerves.

WHITE MATTER OF THE SPINAL CORD

White matter of the cord is composed of myelinated fibers—some running to higher centers, some

BOX 6-2
The Shrinking Brain

It is commonly and mistakenly assumed that everyone is destined for senility in old age. Although the brain does shrink with age (as does every other body tissue) and some slight cognitive deficits may appear, true degenerative changes accompanied by lesions (such as plaques and tangled intracellular fibrils) and leading to the true senile state are not inevitable. In fact, less than 5% of people over age 65 years illustrate true senility (senile dementia), which is characterized by forgetfulness, irritability, difficulty in concentrating or thinking clearly, and confusion.

Research reveals that many elderly individuals assumed to have degenerative brain disease in fact show cognitive changes or seem senile for other reasons that often go undiagnosed. For example, "reversible" senility can be produced by inadequate brain circulation, impacted bowels, poor nutrition, depression, dehydration, infections, and hormone imbalance. Because this is so, perhaps the best way to maintain one's mental abilities in old age is to seek continual medical updates on one's general physical condition throughout life.

Although the normal age-related shrinking of the brain is assumed and anticipated, it seems that some individuals accelerate the process long before aging plays its part. Professional boxers and chronic alcoholics provide such examples.

Whether a boxer wins the match or not, the likelihood of brain damage and atrophy increases with each fight as the brain bounces and rebounds within the skull with each blow. The delicate, exceedingly complex brain is vulnerable to such repeated physical abuse, and the damage that occurs is cumulative. The expression "punch drunk" applies to many former boxers because of the nearly universal symptoms of slurred speech, tremors, abnormal gait, and senile dementia.

Everyone recognizes that alcohol has a profound effect on the mind as well as the rest of the body. However, these effects may not be temporary, especially in chronic alcoholics. Computerized axial tomography (CAT) scans of chronic alcoholics reveal reductions in the size and density of their brains (particularly, the left cerebral hemisphere) and document that these changes occur at a fairly early age. Like boxers, chronic alcoholics tend to be mental function dropouts and exhibit signs of senile dementia unrelated to the aging process.

---

traveling from the brain to the cord, and some conducting impulses from one side of the cord to the other.

Because of the irregular shape of gray matter, white matter on each side of the cord is divided into three regions—the **posterior, lateral,** and **anterior columns**. Each of the columns contains a number of fiber tracts made up of axons with the same destination and function. Tracts conducting sensory impulses to the brain are called *sensory,* or *afferent, tracts;* those carrying impulses from the brain to the skeletal muscles are *motor,* or *efferent, tracts.*

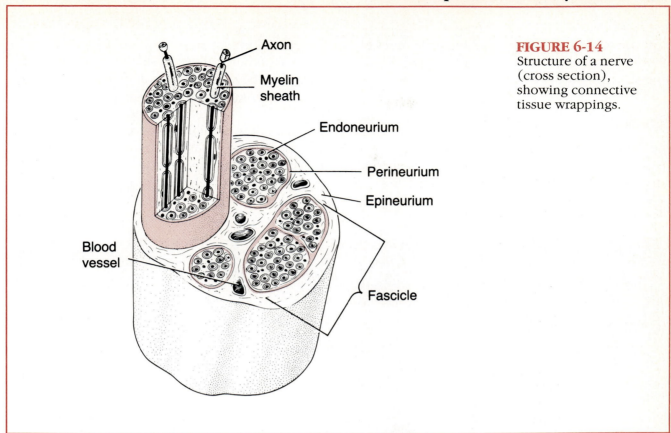

FIGURE 6-14
Structure of a nerve (cross section), showing connective tissue wrappings.

Labels: Axon, Myelin sheath, Endoneurium, Perineurium, Epineurium, Blood vessel, Fascicle

If the cord is transected or crushed, *spastic paralysis* results. In spastic paralysis, the affected muscles stay healthy because they are still stimulated by spinal reflex arcs, and movement of those muscles *does* occur. However, the movements are nonvoluntary and not controllable, and this can be as much of a problem as complete lack of mobility. In addition, since the cord carries both sensory and motor impulses, a loss of feeling or sensory input occurs in the body areas below the point of cord destruction. Physicians often use a pin to see if a person can feel pain after spinal cord injury as a way of finding out if regeneration is occurring. Pain is a hopeful sign in such cases. If the cord injury occurs high in the cord so that all four limbs are affected, the individual is a *quadraplegic* (kwod″rah-plej′ik). If only the legs are paralyzed, the individual is a *paraplegic* (par″ah-plej′ik).

PERIPHERAL NERVOUS SYSTEM

The PNS consists of nerves and scattered groups of nerve cell bodies (ganglia) found outside the CNS. One type of ganglion has already been considered—the dorsal root ganglion of the spinal cord. Others will be covered in the discussion of the autonomic nervous system. Here, we will only concern ourselves with nerves.

Structure of a Nerve

A **nerve** is a bundle of neuron fibers found outside the CNS. Within a nerve, neuron fibers, or processes, are wrapped in protective connective tissue coverings. Each fiber is surrounded by a delicate connective tissue sheath, an **endoneurium** (en″do-nu′re-um). Groups of fibers are bound by a coarser connective tissue wrapping, the **perineurium** (per″i-nu′re-um), to form bundles of fibers called **fascicles.** Finally, all the fascicles are bound together by a tough fibrous sheath, the **epineurium,** to form the cordlike nerve (Figure 6-14).

Like neurons, nerves are classified according to the direction in which they transmit impulses. Nerves carrying both sensory and motor fibers are called **mixed nerves;** all spinal nerves are mixed nerves. Nerves that carry impulses toward the CNS only are called **afferent,** or **sensory nerves,** whereas those that carry only motor fibers are **efferent,** or **motor, nerves.**

Cranial Nerves

The 12 pairs of **cranial nerves** primarily serve the head and neck. Only 1 pair (the vagus nerves) extends to the thoracic and abdominal cavities.

The cranial nerves are numbered in order, and in most cases their names reveal the most important structures they control. The cranial nerves are described by name, number, origin, course, and function in Table 6-1. The last column of the table describes how cranial nerves are tested, which is an important part of any neurologic examination. You do not need to memorize these tests, but this information may help you to understand cranial

TABLE 6-1 The Cranial Nerves

Name and number	Origin and course	Function	How tested
I. Olfactory	Fibers arise from the nasal olfactory epithelium and synapse with the olfactory bulbs, which transmit impulses to the temporal lobe of the cerebral cortex	Purely sensory Carries impulses for sense of smell	Person is asked to sniff aromatic substances, such as oil of cloves and vanilla, and identify them
II. Optic	Fibers arise from the retina of the eye to form the optic nerve, which passes through the sphenoid bone; the two optic nerves then form the optic chiasma (with partial crossover of fibers) and continue on to the occipital cortex as the optic tracts	Purely sensory Carries impulses for vision	Vision and visual field are tested with an eye chart and by testing the point at which the person first sees an object (finger) moving into the visual field; the inside of the eye is viewed with ophthalmoscope to observe blood vessels of the eye interior
III. Oculomotor	Fibers emerge from midbrain and exit from skull to run to eye	Contains motor fibers to the superior, inferior, and medial rectus muscles that direct the eyeball, to the muscles of the eyelid, and to the iris and smooth muscle controlling lens shape; contains	Pupils are examined for size, shape, and equality; pupillary reflex is tested with a penlight (pupils should constrict when illuminated); convergence for near vision is tested, as is the ability to follow moving objects

TABLE 6-1 The Cranial Nerves (continued)

Name and number	Origin and course	Function	How tested
III. Oculomotor (continued)		proprioceptor fibers from the external eye muscles to the brain.	
IV. Trochlear	Fibers emerge from the midbrain and exit from skull to run to eye	Proprioceptor and motor fibers for superior oblique muscle of the eye (external eye muscle)	Tested in common with cranial nerve III relative to the ability to follow moving objects
V. Trigeminal	Fibers emerge from pons and form three divisions that exit from skull and run to face	Both motor and sensory for face; conducts sensory impulses from mouth, nose, and surface of eyes; also contains motor fibers that stimulate chewing muscles	Sensations of pain, touch, and temperature are tested with safety pin and hot and cold objects; corneal reflex tested with a wisp of cotton; motor branch tested by asking the subject to clench teeth, open mouth against resistance, and move jaw from side to side
VI. Abducens	Fibers leave inferior pons and exit from skull to run to eye	Contains motor fibers to lateral rectus muscle and proprioceptor fibers from same muscle to brain	Tested in common with cranial nerve III relative to the ability to move each eye laterally
VII. Facial	Fibers leave pons and travel through temporal bone to reach face	Mixed: (a) Supplies motor fibers to muscles of facial expression and to lacrimal and salivary glands, and (b) carries sensory fibers from taste buds of anterior part of tongue	Anterior two-thirds of tongue is tested for ability to taste sweet (sugar), salty, sour (vinegar), and bitter (quinine) substances; symmetry of face is checked; subject is asked to close eyes, smile, whistle, and so on; tearing is tested with ammonia fumes
VIII. Vestibulocochlear (acoustic)	Fibers run from inner ear hearing and equilibrium receptors in temporal bone to	Purely sensory Vestibular branch transmits impulses for sense of equilibrium;	Hearing is checked by air and bone conduction using a tuning fork

(Continued.)

TABLE 6-1   The Cranial Nerves (continued)

Name and number	Origin and course	Function	How tested
VIII. Vestibulocochlear (continued)	enter brain stem just below pons	cochlear branch transmits impulses for sense of hearing	
IX. Glossopharyngeal	Fibers emerge from medulla and leave skull to run to throat	Mixed: (a) Motor fibers serve pharynx (throat) and salivary glands, and (b) sensory fibers carry impulses from pharynx, posterior tongue (taste buds), and pressure receptors of the carotid artery	Gag and swallowing reflexes are checked; subject is asked to speak and cough; posterior one-third of tongue may be tested for taste
X. Vagus	Fibers emerge from medulla, pass through skull, and descend through neck region into thorax and abdominal region	Fibers carry sensory and motor impulses for pharynx and larynx; a large part of this nerve is parasympathetic motor fibers, which supply the heart and smooth muscles of abdominal organs; transmits sensory impulses from the viscera	Same as for cranial nerve IX (IX and X are tested in common since they both serve muscles of the throat)
XI. Spinal accessory	Fibers arise from medulla and superior spinal cord and travel to the muscles of neck and back	Provides sensory and motor fibers for sternocleidomastoid and trapezius muscles and muscles of soft palate, pharynx, and larynx	Sternocleidomastoid and trapezius muscles are checked for strength by asking subject to rotate head and shrug shoulders against resistance
XII. Hypoglossal	Fibers arise from medulla and exit from skull to travel to tongue	Carries motor fibers to muscles of tongue and sensory impulses from tongue to brain	Subject is asked to stick out tongue, and any position abnormalities are noted

nerve function. As you read through the table, also look at Figure 6-15 which shows the location of the cranial nerves on the brain's anterior surface. Most cranial nerves are mixed nerves; however, three pairs (optic, olfactory, and vestibulocochlear [ves-tib″u-lo-kok′le-ar]) are purely sensory in their function. (The older name for the vestibulocochlear nerve is acoustic nerve.)

For many years students have used this catchy little saying to help them remember the names of the cranial nerves in order; perhaps it will help

FIGURE 6-15
Cranial nerves.

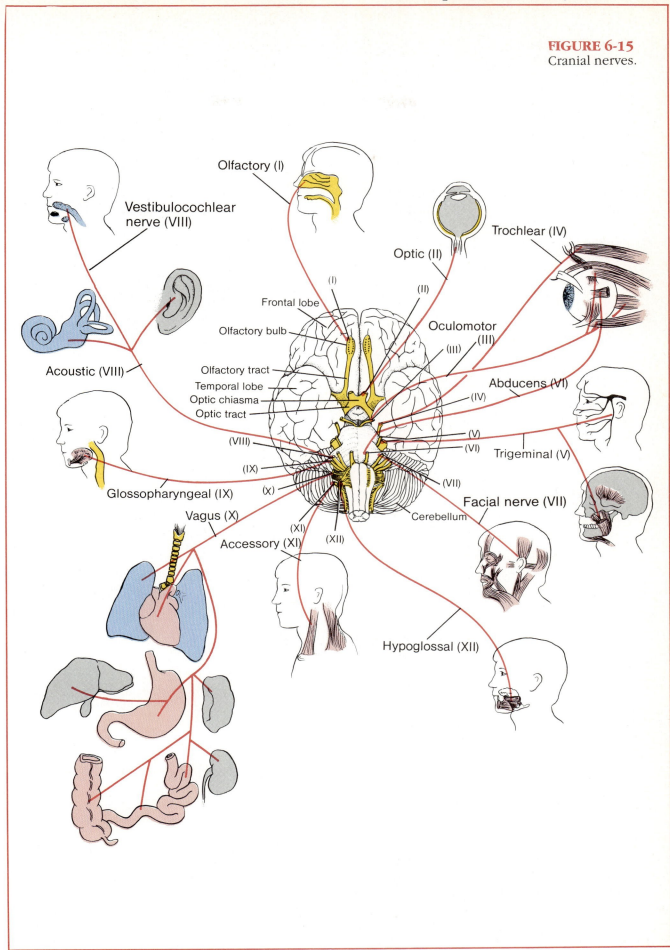

Olfactory (I)

Vestibulocochlear
nerve (VIII)

Optic (II)

Trochlear (IV)

Oculomotor
(III)

(I)

(II)

Frontal lobe

Olfactory bulb

Olfactory tract

Temporal lobe

Optic chiasma

Optic tract

Acoustic (VIII)

Abducens (VI)

(III)

(IV)

(V)

(VI)

(VIII)

(IX)

(VII)

Trigeminal (V)

Glossopharyngeal (IX)

(X)

Facial nerve (VII)

Vagus (X)

Cerebellum

Accessory (XI)

(XI)

(XII)

Hypoglossal (XII)

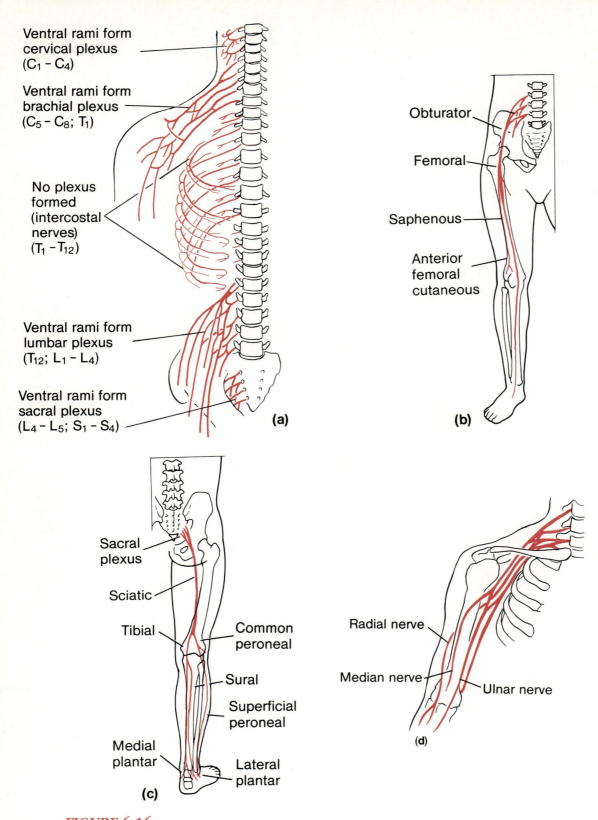

Ventral rami form
cervical plexus
$(C_1 - C_4)$

Ventral rami form
brachial plexus
$(C_5 - C_8; T_1)$

No plexus
formed
(intercostal
nerves)
$(T_1 - T_{12})$

Ventral rami form
lumbar plexus
$(T_{12}; L_1 - L_4)$

Ventral rami form
sacral plexus
$(L_4 - L_5; S_1 - S_4)$

(a)

Obturator

Femoral

Saphenous

Anterior
femoral
cutaneous

(b)

Sacral
plexus

Sciatic

Tibial

Common
peroneal

Sural

Superficial
peroneal

Medial
plantar

Lateral
plantar

(c)

Radial nerve

Median nerve

Ulnar nerve

(d)

FIGURE 6-16
Spinal nerves.
(a) Relationship of
spinal nerves to the
vertebral column. Areas
of plexuses formed by
the anterior rami are
indicated for the right
side of the body.
(b) Major nerves of the
lumbar plexus.
(c) Major nerves of the
sacral plexus.
(d) Major nerves of the
brachial plexus.

TABLE 6-2 Spinal Nerve Plexuses

Plexus	Origin (from ventral rami)	Important nerves	Body areas served	Result of damage to Plexus or its nerves
Cervical	C_1–C_4	Phrenic	Diaphragm and muscles of shoulder and neck	Respiratory paralysis (and death if not treated quickly)
Brachial	C_5–C_8 and T_1	Radial	Triceps and other extensor muscles of arm	Wristdrop—inability to extend hand at wrists
		Median	Flexor muscles of forearm and some muscles of the hand	Decreased ability to flex and abduct hand and flex and abduct thumb and index finger—therefore, unable to pick up small objects
		Musculocutaneous	Flexor muscles of the upper arm	Decreased ability to flex forearm on upper arm
		Ulnar	Wrist and many hand muscles	Clawhand—inability to spread fingers apart
Lumbar	T_{12} and L_1–L_4	Femoral	Lower abdomen, buttocks, anterior thighs, and skin of anteromedial leg and thigh	Inability to extend leg and flex hip
Sacral	L_4–L_5 and S_1–S_4	Sciatic (largest nerve in body)	Lower trunk and posterior surface of thigh and leg	Inability to extend hip and flex knee; sciatica
		Peroneal	Foot and lateral aspect of leg	Footdrop—inability to dorsiflex foot

you too. The first letter of each word in the saying is the first letter of the cranial nerve to be remembered. "On Old Olympus Towering Tops A Finn And German Viewed Some Hops."

Spinal Nerves and Nerve Plexuses

The 31 pairs of human **spinal nerves** are formed by the combination of the ventral and dorsal roots of the spinal cord. Although each of the cranial nerves is specifically named, the spinal nerves are named for the region of the cord from which they arise. Figure 6-16 shows how the nerves are named according to this scheme.

Almost immediately after being formed, each spinal nerve divides into **dorsal** and **ventral rami** (ray′mi), making each spinal nerve only about ½-inch long. The rami, like the spinal nerves, contain both motor and sensory fibers. Thus, damage to a spinal nerve or either of its rami results in both loss of sensation and flaccid paralysis of the area of the body served. The smaller dorsal rami serve the skin and muscles of the posterior body trunk. The ventral rami of spinal nerves T_1 through T_{12} form the *intercostal nerves,* which supply the muscles between the ribs and the skin and muscles of the anterior and lateral trunk. The ventral rami of all other spinal nerves form complex networks of nerves called **plexuses**, which serve the motor and sensory needs of the limbs. The four nerve plexuses and their major peripheral nerves are shown in Figure 6-16 and described in Table 6-2.

Autonomic Nervous System

The **autonomic nervous system** is the motor subdivision of the PNS that controls body activities automatically, that is, we do not consciously control them. It is composed of a special group of neurons that conduct impulses to cardiac muscle (the heart), smooth muscles (found in the walls of the visceral organs and blood vessels), and glands. Because this system functions without our conscious control or will, it is often referred to as the *involuntary nervous system.*

There is an important difference between the motor pathways of the autonomic nervous system and those of the somatic nervous system, which stimulates the skeletal muscles. In the somatic division, nerve cell bodies of the motor neurons are inside the CNS, and their axons (in spinal nerves) extend all the way to the skeletal muscles they serve. On the other hand, the autonomic nervous system has a chain of *two* motor neurons. The first motor neuron of each pair is in the brain or cord. Its axon, the **preganglionic axon** (literally, the axon before the ganglion), leaves the CNS to synapse with the second motor neuron in a ganglion outside the CNS. The axon of this neuron, the **postganglionic axon,** then extends to the organ it serves. These differences are summarized in Figure 6-17.

The autonomic nervous system has two major subdivisions, the sympathetic and parasympathetic. Both serve the same organs but cause opposite effects (Figure 6-18, p. 146).

SYMPATHETIC DIVISION
The **sympathetic division** is also called the *thoracolumbar* (tho″rah-ko-lum′bar) *division,* because its first neurons are in the gray matter of the spinal cord from T_1 through L_3. The preganglionic axons leave the cord in the ventral root, enter the spinal nerve, and then pass through a small branch to enter a sympathetic chain ganglion (Figure 6-19, p. 147). The **sympathetic chain,** or **trunk,** lies alongside the vertebral column on each side. After it reaches the ganglion, the axon

may synapse with the second neuron in the sympathetic chain (the postganglionic axon then reenters the spinal nerve to travel to the skin), or the axon may pass through the ganglion without synapsing and form part of the *splanchnic* (splank′nik) *nerves.* The splanchnic nerves travel to the viscera to synapse with the second neuron found in a **collateral ganglion**. (The major collateral ganglia—the celiac and superior and inferior mesenteric ganglia—supply the abdominal and pelvic organs.) The postganglionic axon then leaves the collateral ganglion and travels to a nearby visceral organ, which it serves.

PARASYMPATHETIC DIVISION
The first neurons of the **parasympathetic division** are located in brain nuclei of several cranial nerves (the vagus being the most important of these) and in the S_1 through S_3 (or S_4) level of the spinal cord (see Figure 6-18). The neurons of the cranial region send their axons out in cranial nerves to the head and neck organs, which they serve. There they synapse with the second motor neuron in a **terminal ganglion**. From the terminal ganglion, the postganglionic axon extends a short distance to the organ it serves. In the sacral region, the preganglionic axons leave the spinal cord and form the *pelvic nerve,* which travels to the pelvic cavity. In the pelvic cavity, the preganglionic axons synapse with the second motor neurons in terminal ganglia on or close to the organs served.

AUTONOMIC FUNCTIONING
Body organs served by the autonomic nervous system receive fibers from both divisions. The only exceptions are the structures of the skin, some glands, and the adrenal medulla, all of which receive only sympathetic fibers. When both divisions serve the same organ they cause opposite effects, because their postganglionic axons release different neurotransmitters. The parasympathetic fibers, called *cholinergic* (ko″lin-er′jik) *fibers,* release acetylcholine; the sympathetic postganglionic fibers, called *adrenergic* (ad″ren-er′jik) *fibers,* release norepinephrine (nor″ep-ĭ-nef′rin). (The preganglionic axons of both divisions release acetylcholine.)

The parasympathetic division is often referred to as the ''housekeeping system,'' because it keeps the visceral organs working normally and main-

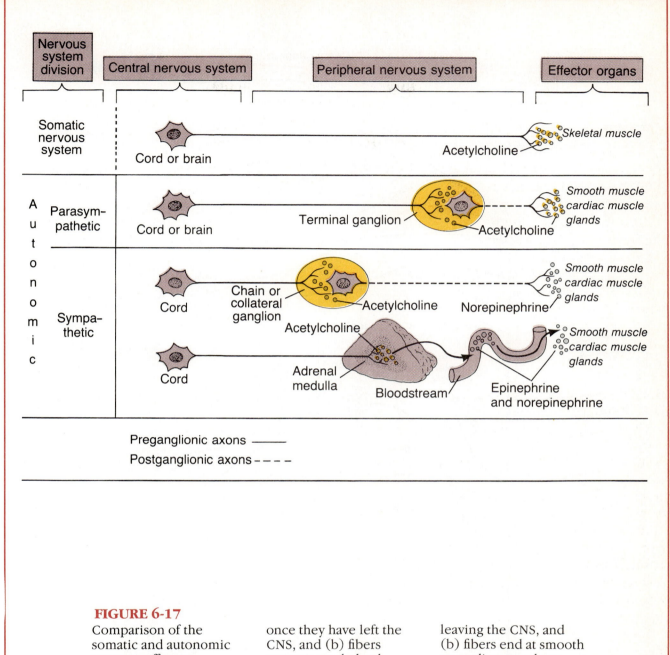

FIGURE 6-17

Comparison of the somatic and autonomic systems, efferent division. The differences between the somatic and autonomic nervous systems are as follows: Somatic–(a) Fibers do not synapse once they have left the CNS, and (b) fibers synapse on skeletal muscle cells. Their effect is always stimulatory (assuming a threshold stimulus). Autonomic–(a) Fibers synapse in ganglia after leaving the CNS, and (b) fibers end at smooth or cardiac muscle or glands. This can lead to excitation *or* inhibition of the effector, depending on which division is in control.

tains homeostasis when the body is not threatened. This division promotes normal digestion and elimination and acts to decrease demands on the heart and circulatory system.

The sympathetic division is referred to as the "fight or flight" system, because it prepares the body to deal with stressful situations that may interfere with homeostasis. Under such conditions, the sympathetic nervous system causes an increase in heart rate, blood pressure, and blood glucose levels; dilates the bronchioles of the lungs; and brings about many other effects that help the individual cope with stress. Dilation of

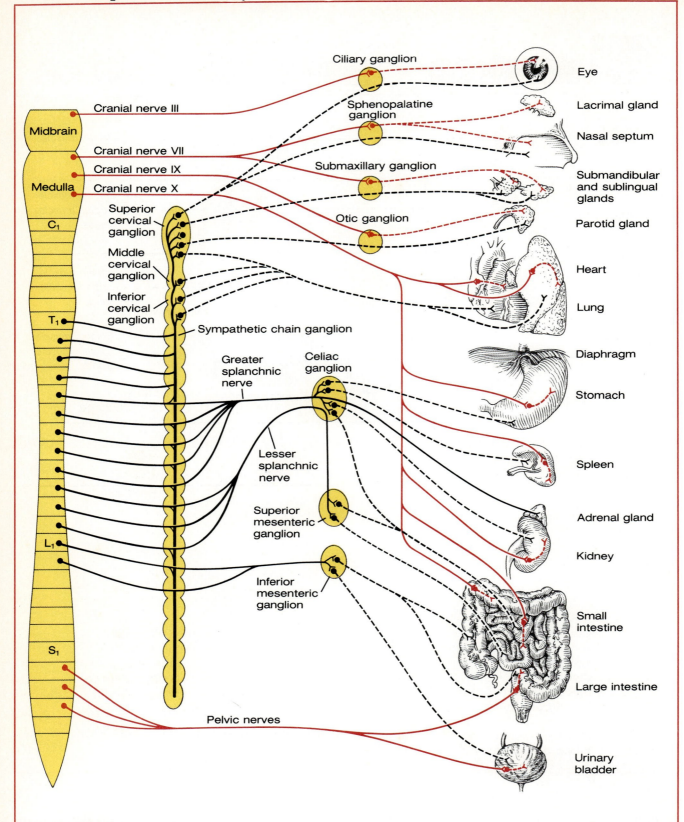

FIGURE 6-18

The autonomic nervous system. Parasympathetic fibers are shown in red while sympathetic fibers are indicated by black lines. Solid lines represent preganglionic fibers; dashed lines are postganglionic fibers.

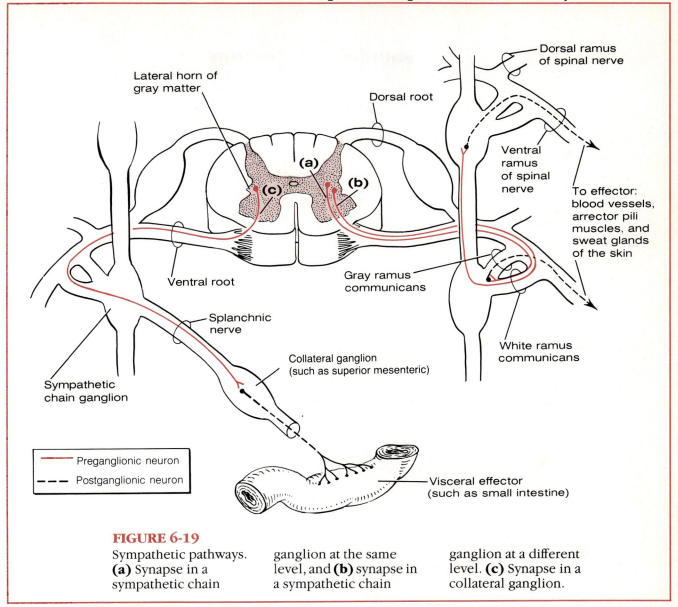

FIGURE 6-19
Sympathetic pathways.
(a) Synapse in a
sympathetic chain

ganglion at the same
level, and **(b)** synapse in
a sympathetic chain

ganglion at a different
level. **(c)** Synapse in a
collateral ganglion.

the blood vessels in skeletal muscles (so that one can run faster or fight better) and withdrawal of blood from the digestive organs (so that the bulk of the blood can be used to serve the heart, brain, and skeletal muscles) are other examples. The sympathetic nervous system is working at full speed any time you are angry, frightened, excited, or experiencing any other type of emotional upset. It is also operating when you are physically stressed. For example, if you had just had surgery or just run a marathon, your sympathetic nervous system would be pumping out norepinephrine. Some illnesses or diseases are believed to be at least aggravated, if not caused, by excessive sympathetic nervous system stimulation. Some individuals, called Type A people, are always running at breakneck speed and pushing themselves con-

tinually. These are the people who are likely to have heart disease, high blood pressure, and ulcers, all of which can result from extended sympathetic nervous system activity or the rebound from it. The major effects of the sympathetic and parasympathetic divisions on the body organ systems are summarized in Table 6-3.

DEVELOPMENTAL ASPECTS OF THE NERVOUS SYSTEM

Since the nervous system is formed early in embryonic development, any maternal disease during the first part of pregnancy can have extremely

TABLE 6-3 Comparison of the Effects of the Sympathetic and Parasympathetic Divisions

Target organ/system	Parasympathetic effects	Sympathetic effects
Digestive system	Increases mobility (peristalsis) and amount of secretion by the digestive system glands; relaxes sphincters	Decreases activity of the digestive system and constricts digestive system sphincters (for example, anal sphincter)
Liver	No effect	Causes glucose to be released to the blood
Lungs	Constricts bronchioles	Dilates bronchioles
Bladder/urethra	Relaxes sphincters (allows voiding)	Constricts sphincters (prevents voiding)
Heart	Decreases rate; slows and steadies	Increases rate and force of heartbeat
Blood vessels	No effect	Constricts blood vessels in viscera and skin (dilates those in skeletal muscle and the heart); increases blood pressure
Salivary glands	Stimulates; increases production of saliva	Inhibits—result is a dry mouth
Eye (iris)	Stimulates constrictor muscles; constricts pupils	Stimulates dilator muscles; dilates pupils
Eye (ciliary muscle)	Stimulates to increase bulging of lens for close vision	Inhibits; decreases bulging of lens; prepares for distant vision
Adrenal medulla	No effect	Stimulates medulla cells to secrete epinephrine and norepinephrine
Sweat glands of skin	No effect	Stimulates to produce perspiration
Smooth muscles attached to hair follicles	No effect	Stimulates; produces "goosebumps"

harmful effects on the fetal nervous system. For example, maternal measles (rubella) often causes deafness and other types of CNS damage. Also, since nervous tissue has the highest metabolic rate in the body, lack of oxygen for even a few minutes leads to death of neurons. (Because smoking decreases the amount of oxygen that can be carried in the blood, a smoking mother may be sentencing her infant to possible brain damage.) In difficult deliveries, the temporary lack of oxygen often leads to *cerebral palsy* (pawl'ze), but this is only one of the suspected causes. Cerebral palsy is a neuromuscular disability in which the voluntary muscles are poorly controlled due to

brain damage. It is the largest single cause of crippling in children.

One of the last areas of the CNS to mature is the hypothalamus, which contains centers for regulating body temperature. For this reason, premature babies usually have problems in controlling their loss of body heat and must be carefully monitored. No more neurons are formed after birth (because neurons are amitotic), but growth of the nervous system continues all through childhood as a result of myelination that goes on during this period. A good indication of the degree of myelination of particular neural pathways is the level of neuromuscular control in that body area. For example, fine hand control occurs before the baby is able to walk well, which indicates that myelination occurs in a superior/inferior direction in the body. Because infants can wave their arms before they can pick up pennies, we know that myelination occurs in a proximal/distal direction in the nerves.

All through life, neurons are damaged and die and, since they cannot reproduce themselves, our storehouse of neurons continually decreases. However, there is an unlimited number of neural pathways that are always available and ready to be developed. We never run out of "recording tape" and can continue to learn throughout our lives.

As we grow older, the sympathetic nervous system gradually becomes less and less efficient, particularly in vasoconstriction of blood vessels. When elderly people stand up quickly after sitting or lying down, they often become lightheaded or faint. This is because the sympathetic nervous system is not able to react quickly enough to counteract the pull of gravity by activating the vasoconstrictor fibers, and blood pools in the feet. This condition, *orthostatic* (or"tho-stat'ik) *hypotension,* is a type of low blood pressure resulting from changes in body position as described. Orthostatic hypotension can be prevented to some degree if *slow* changes in position are made. This gives the sympathetic nervous system a little more time to adjust and react.

The normal cause of nervous system deterioration is circulatory system problems. For example, arteriosclerosis (ar-te"re-o-sklĕ-ro'sis) and high blood pressure result in a decreasing supply of oxygen to the brain neurons. A gradual lack of oxygen finally leads to *senility* (changes in intellect, memory, and other nervous system functions due to the aging process). A *rapid* loss of blood and oxygen delivery to the brain resulting from a clot or rupture of blood vessels in the brain, results in a CVA. The effect of a CVA depends on the extent and area of brain damage.

IMPORTANT TERMS*

Neuroglia *(nu-rog'le-ah)*

Neurons

Dendrites *(den'drĭts)*

Synapse

Myelin *(mi'e-lin)*

Nodes of Ranvier *(rahn-ve-āz')*

Ganglion

Tracts

Afferent neurons

Efferent neurons

Interneurons

Polarization

Depolarization

Repolarization

Reflex

Central nervous system

Cerebral *(ser'ĕ-bral)* **hemispheres**

Gray matter

White matter

Brain stem

Cerebellum *(ser"ĕ-bel'um)*

Meninges *(mĕ-nin'jez)*

Cerebrospinal *(ser"ĕ-bro-spi'nal)* **fluid**

Peripheral *(pe-rif'er-al)* **nervous system**

Plexus

Autonomic *(aw"to-nom'ik)* **nervous system**

Receptor

Nerve

Somatic *(so-mat'ik)* **nervous system**

*For definitions, see Glossary.

BOX 6-3
Post-Polio Syndrome

It is now some thirty plus years after the great polio epidemics of the late 1940s and 1950s, but it appears that many of its "recovered" survivors are not yet done with the viral disease that destroys spinal motor neurons, rendering associated spinal nerves useless. Vital people, today in their late thirties to mid-forties, who fought their way back to health and active vigorous lives have begun to notice a new, disturbing set of symptoms. Examples include extreme lethargy, sensitivity to cold, sharp burning pains in their muscles and joints, and weakness and gradual loss of muscle mass in the muscles that for so many years after their initial illness served them well.

Initially, these symptoms were summarily dismissed as the "flu" or psychosomatic illness. In many cases where the baffling symptoms were not dismissed, they were often misdiagnosed as indicative of multiple sclerosis, arthritis, lupus, or Lou Gehrig's disease, a devastating neurological disease that kills its victims swiftly. However, as the number of similar complaints has burgeoned, increasing attention has been given to this group of symptoms, which is now named *post-poliomyelitis muscular atrophy (PPMA)* or *post-polio syndrome*.

According to a 1983 study conducted by the Mayo Clinic in Minnesota, 25% of polio survivors are affected. It now appears that this estimate is extremely low. Certain survivors seem to be particularly vulnerable: those who contracted the disease after the age of 10, were severely affected, needed ventilatory support, and/or had all four limbs affected.

The cause of PPMA is not known. The earlier belief that the polio virus was reactivated has now been discarded. A more likely explanation is that polio survivors, like all of us, continue to lose neurons throughout life. Whereas unimpaired nervous systems can enlist nearby neurons to compensate for the losses, polio survivors, having lost many neurons during the early disease assault, have already drawn on that "pool" and may have few neurons left to take over. What is particularly ironic is that PPMA victims seem to be overachievers, those who worked the hardest to overcome their disease. The grueling hours of exercise that they assumed would make them strong may be the factor that has ultimately caused their overworked motor neurons to "burn out."

Post-polio syndrome progresses slowly and it is not life-threatening. Currently its victims are advised to conserve their energy by resting more, and by using devices such as canes, walkers, and braces to support their wasting muscles. A major concern is determining how much muscle relief is required, so that the muscle atrophy caused by PPMA will not be accentuated by disuse atrophy.

SUMMARY

A. DIVISIONS OF THE NERVOUS SYSTEM

1. Anatomical: All nervous system structures are classified as part of the CNS (brain and cord) or PNS (nerves and ganglia).

2. Functional: Motor nerves of the PNS are classified on the basis of whether they stimulate skeletal muscle (somatic division) or smooth/cardiac muscle and glands (autonomic division).

B. NERVOUS TISSUE

1. Supportive connective tissue cells
   a. Neuroglia protect neurons in the CNS. Specific glial cells are part of the blood–brain barrier; others myelinate neuron processes in the CNS or are phagocytes.
   b. Schwann cells myelinate neuron processes in the PNS.

2. Neurons
   a. Anatomy: All neurons have a nerve cell body containing the nucleus and processes of two types: (a) axons (one per cell) conduct impulses away from the nerve cell body and release a neurotransmitter, and (b) dendrites (one to many per cell) conduct impulses toward the nerve cell body. Axons are myelinated; myelin increases the rate of nervous transmission.
   b. Classification: On the basis of function (direction of impulse transmission) there are sensory (afferent), motor (efferent), and association (interneurons) neurons. The dendrite endings of sensory neurons are bare (pain receptor) or modified with special cells to form sensory receptors.
   c. Physiology
      (1) A nerve impulse is an electrochemical event initiated by various stimuli that causes a change in neuron cell membrane permeability allowing sodium ions (Na^+) to enter the cell (depolarization). Once begun, the impulse continues over the entire surface of the cell. Electrical conditions of the resting cell are restored by the diffusion of potassium ions (K^+) out of the cell (repolarization). Ion concentrations of the resting state are restored by the sodium–potassium pump.
      (2) A reflex arc is a rapid predictable response to stimulus. There are two types—autonomic and somatic. The minimum number of components of a reflex arc is four: receptor, effector, sensory, and motor neurons (most, however, have one or more interneurons). Normal reflexes indicate normal nervous system function.

C. CENTRAL NERVOUS SYSTEM

1. The brain is located within the cranial cavity of the skull and consists of the cerebral hemispheres, brainstem structures, and cerebellum.
   a. The two cerebral hemispheres are the largest part of the brain. Their surface or cortex is gray matter and their interior (medulla) is white matter. The cortex is convoluted and has gyri and fissures. The cerebral hemispheres are involved in logical reasoning, moral conduct, emotional responses, sensory interpretation, and the initiation of voluntary muscle activity. The functional areas of the cerebral lobes have been identified (see p. 126). The basal nuclei, regions of gray matter deep within the white matter of the cerebral hemispheres, modify voluntary motor activity.
   b. Brain stem structures
      (1) The diencephalon (interbrain) is the most superior part of the brain stem and is enclosed by the cerebral hemispheres. The major structures include:
         (a) The thalamus encloses the third ventricle and is the relay station for sensory impulses passing to the sensory cortex for interpretation.
         (b) The hypothalamus makes up the floor of third ventricle and is the most important autonomic nervous system regulatory center (regulates water balance, metabolism, thirst, temperature, and the like).
         (c) The epithalamus includes the pineal body (an endocrine gland) and the choroid plexus of the third ventricle.

    (2) The midbrain is inferior to the hypothalamus and is primarily fiber tracts.

    (3) The pons is inferior to the midbrain and has fiber tracts and nuclei involved in respiration.

    (4) The medulla oblongata is the most inferior part of brainstem. In addition to fiber tracts, it contains many autonomic nuclei involved in the regulation of vital life activities (for example, breathing, heart rate, and blood pressure).

  c. The cerebellum is a large cauliflowerlike part of the brain dorsal to the fourth ventricle. It coordinates muscle activity and body balance.

2. Protection of the CNS
  a. Bone—the skull and verebral column.
  b. Meninges are three connective tissue membranes—dura mater (tough outermost), arachnoid mater (middle weblike), and pia mater (innermost delicate). The meninges extend beyond the end of the spinal cord.
  c. Cerebrospinal fluid (CSF) provides a watery cushion around the brain and cord. CSF is formed by the choroid plexuses of the brain. It is found in the subarachnoid space, ventricles, and central canal. CSF is continually formed and drained.

3. The spinal cord is a reflex center and conduction pathway. Found within the vertebral canal, the cord extends from the foramen magnum to L_1 or L_2. The cord has a central bat-shaped area of gray matter surrounded by columns of white matter, which carry motor and sensory tracts from and to the brain.

D. PERIPHERAL NERVOUS SYSTEM

1. Structure of a nerve: A nerve is a bundle of neuron processes wrapped in connective tissue coverings (endoneurium, perineurium, and epineurium).

2. Cranial nerves: Twelve pairs of nerves that extend from the brain to serve the head and neck region. The exception is the vagus nerves, which extend to the thorax and abdomen.

3. Spinal nerves: Thirty-one pairs of nerves that are formed by the union of the dorsal and ventral roots of the spinal cord on each side. The spinal nerve proper is very short and splits into dorsal and ventral rami. Dorsal rami serve the posterior body trunk; ventral rami (except T_1 through T_{12}) form plexuses (cervical, brachial, lumbar, and sacral) that serve the limbs.

4. Autonomic nervous system: The autonomic nervous system is part of the PNS and is composed of neurons that stimulate smooth and cardiac muscle and glands. This system differs from the somatic nervous system in that there is a chain of two neurons from the CNS to the effector. There are two subdivisions that serve the same organs with different effects.

  a. The parasympathetic division is the "housekeeping system" and is in control most of the time. This division maintains homeostasis by seeing that normal digestion, elimination, and the like occur. The first motor neurons are in the brain or the sacral region of the cord and the second motor neurons are in the terminal ganglia close to the organ served. Postganglionic axons secrete acetylcholine.

  b. The sympathetic division is the "fight or flight" subdivision, which prepares the body to cope with stress. Activation of the sympathetic nervous system results in an increased heart rate and blood pressure. The first motor neurons are in the gray matter of the cord and the postganglionic neurons are in sympathetic chain or in collateral ganglia. Postganglionic axons secrete norepinephrine.

REVIEW QUESTIONS

1. What are the two great controlling systems of the body?

2. Explain both the anatomical and the functional classifications of the nervous system.

Include in your explanation the subdivisions of each.

3. List the structures of the CNS and PNS.

4. Two major cell groups make up the nervous system—neurons and connective tissue cells such as neuroglia and Schwann cells. Which are "nervous cells"? Why? What are the major functions of the other cell groups?

5. Give the basis for the functional classification of neurons.

6. Briefly explain why one-way conduction at synapses always happens.

7. Name four types of cutaneous sensory receptors. Which of the cutaneous receptor types is most numerous? Why?

8. What is a reflex arc? Name its minimum number of components.

9. Name at least five functions performed by the cerebral hemispheres.

10. What is meant when we say that most cerebral hemisphere pathways are *crossed pathways?*

11. Other than serving as a conduction pathway, what is a major function of the pons? Why is the medulla the most vital part of the brain?

12. What is the function of the thalamus? The hypothalamus? The cerebellum?

13. What is gray matter? White matter? How does the arrangement of gray and white matter differ in the cerebral hemispheres and the spinal cord?

14. What are two functions of the spinal cord?

15. How many pairs of cranial nerves are there? Which are purely sensory?

16. With the exception of the vagus nerves, what general area of the body do the cranial nerves serve?

17. How many pairs of spinal nerves are there? How do they arise?

18. What region of the body is served by the dorsal rami? The ventral rami?

19. Name the four major nerve plexuses formed by the ventral rami and the body region served by each.

20. How does the autonomic nervous system differ from the somatic nervous system?

21. What is the difference in function of the parasympathetic and sympathetic divisions of the autonomic nervous system?

22. Since the sympathetic and parasympathetic fibers serve the same organs, how can their opposing effects be explained?

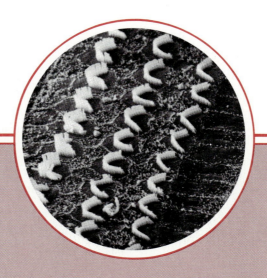

CHAPTER 7

Special Senses

Chapter Contents

Function of the special senses • to respond to different types of energetic stimuli; involved in vision, hearing, balance, taste, and smell

The above electron micrograph from *Tissues and Organs: A Text-Atlas of Scanning Electron Microscopy* by Richard G. Kessel and Randy H. Kardon. W. H. Freeman and Co. Copyright © 1979.

After completing this chapter, you should be able to:

- Identify the structures of the eye and the ear when provided with a model or diagram and list the functions of each.

- Explain the difference in rod and cone function.

- Describe image formation on the retina.

- Trace the pathway of light through the eye to the retina.

- Discuss the importance of an ophthalmoscopic examination.

- Define the following terms: accommodation, astigmatism, blind spot, cataract, emmetropia, glaucoma, hyperopia, myopia, and refraction.

- Trace the visual pathway to the optic cortex.

- Discuss the importance of the pupillary and convergence reflexes.

- Explain the function of the organ of Corti in hearing.

- Describe how the equilibrium organs help maintain balance.

- Define sensineural and conductive deafness and list possible causes of each.

- Explain how one is able to localize the source of sounds.

- Describe the location, structure, and function of the olfactory and taste receptors.

- Name the four basic taste sensations and factors that modify the sense of taste.

- Define photoreceptor, mechanoreceptor, and chemoreceptor and give one example of each.

- Discuss briefly what visual maturation means.

- Describe changes that occur in the special sense organs with aging.

There is no question that all people are irritable creatures. Hold a sizzling steak before us and our mouths water. Stroke our arms gently and we smile. These "irritants" (the steak and the soft touch) and many others are stimuli that continually greet us and must be interpreted by our nervous system.

Usually we are told that we have five senses, but that list includes only some of the senses—taste, touch, smell, sight, and hearing—which relay to the brain what is going on in the external world.

Actually we have sense organs that react to stimuli or changes occurring inside the body as well as in the external environment. The tiny sensory receptors, or "pick-up units," of the general senses react to touch, pressure, pain, heat, cold, stretch, and changes in pull on your muscles. These receptors are scattered throughout the body in the skin, the viscera, and muscles. Without these additional receptors, you would not know when you

had eaten too much, you could burn or freeze your fingers without realizing it, and you would have no recognition of what was happening with your arms or legs. These general sensory receptors have already been considered in Chapter 6.

In contrast to the widely distributed general receptors, the special senses have large, complex sensory organs or small but localized groups of receptors. The special senses include sight, hearing, equilibrium, smell and taste (four of the five senses normally referred to). The ear, which provides hearing, also houses the equilibrium receptors that help maintain balance. Touch, as already noted, is handled by the general cutaneous receptors of the skin.

EYE AND VISION

How we see has captured the curiosity of researchers, and vision is the sense that has been most studied. Of all the sensory receptors in the

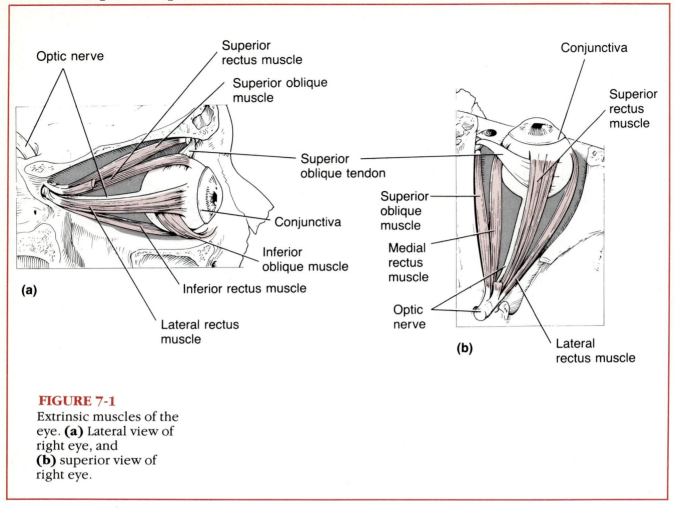

FIGURE 7-1
Extrinsic muscles of the
eye. **(a)** Lateral view of
right eye, and
(b) superior view of
right eye.

body, 70% are in the eyes. The optic tracts that carry information from the eyes to the brain are massive bundles, which contain over a million nerve fibers. Vision is the sense that requires the most "learning," and the eye appears to delight in being fooled. The old expression "You see what you expect to see" is often very true.

Anatomy of the Eye

EXTERNAL AND ACCESSORY STRUCTURES

The adult eye is a sphere which measures about 1 inch (2.5 cm) in diameter. Only about one-sixth of the eye's anterior surface can normally be seen. The rest of it is enclosed and protected by a cushion of fat and the walls of the bony orbit.

Six **external,** or **extrinsic, eye muscles** are attached to the outer surface of each eye. These

muscles allow gross eye movements and make it possible for the eyes to follow a moving object. The names and locations of the extrinsic muscles are shown in Figure 7-1 and their action is given in Table 7-1.

Anteriorly the eyes are protected by the **eyelids,** which meet at the medial and lateral corners of the eye (the medial and lateral canthus respectively) (Figure 7-2). Projecting from the border of each eyelid are the **eyelashes. Meibomian** (mibo′me-an) **glands,** a type of sebaceous gland, are associated with the eyelash hair follicles. These glands produce an oily secretion that lubricates the eye. An inflammation of one of these glands is called a *sty.* A mucous membrane, the **conjunctiva** (kon-junk″tǐ′vah), lines the inner surface of the eyelids and covers the anterior surface of the eyeball. The conjunctiva secretes mucus, which also helps to lubricate the eyeball and keep it moist.

TABLE 7-1 Extrinsic Eye Muscles—Nerve Supply and Action

Name	Controlling cranial nerve	Action
Lateral rectus	VI (Abducens)	Moves eye laterally
Medial rectus	III (Occulomotor)	Moves eye medially
Superior rectus	III	Elevates eye or rolls it superiorly
Inferior rectus	III	Depresses eye or rolls it inferiorly
Inferior Oblique	III	Elevates eye and rolls it laterally
Superior Oblique	IV (Trochlear)	Depresses eye and turns it laterally

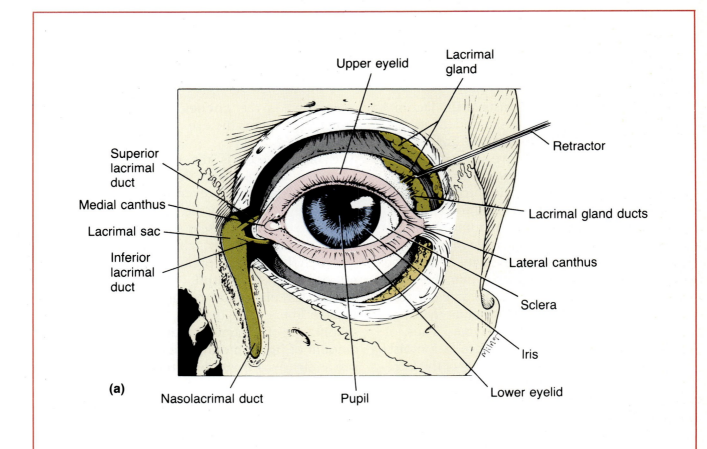

(a)

FIGURE 7-2
External anatomy of the eye and accessory structures. **(a)** Anterior view. Lacrinal gland is shown pulled superiorly by a retractor to expose its ducts.

(*Continued.*)

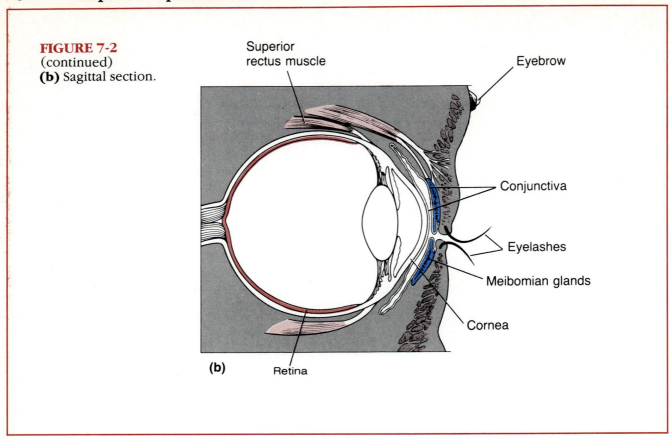

FIGURE 7-2
(continued)
(b) Sagittal section.

Superior rectus muscle

Eyebrow

Conjunctiva

Eyelashes

Meibomian glands

Cornea

(b) Retina

The **lacrimal apparatus** consists of the lacrimal gland, lacrimal ducts, lacrimal sac, and the nasolacrimal duct. The **lacrimal glands** are located above the lateral end of each eye. They continually release a dilute salt solution (tears) onto the anterior surface of the eyeball through several small ducts. The tears flush across the eyeball into the **lacrimal ducts** medially, then into the **lacrimal sac,** and finally into the **nasolacrimal duct,** which empties into the nasal cavity. When you have a good cry, the tears empty into the nasal cavities so rapidly that you soon become congested. Lacrimal secretion also contains *lysozyme* (li'so-zim), an antibacterial enzyme. Thus, it cleanses and protects the eye surface as it moistens and lubricates it. Since the activity of the lacrimal glands decreases as we age, the eyes tend to become dry and more vulnerable to bacterial invasion and irritation during the later years of life.

INTERNAL STRUCTURES
As illustrated in Figure 7-3, the wall of the eye is formed by three tunics or coats. The outermost coat, the protective **sclera** (skle'rah), is thick, white connective tissue. The sclera is seen as the "white of the eye." Its central anterior portion is modified so that it is transparent. This transparent

"window" is the **cornea** (kor'ne-ah) through which light enters the eye.

The middle coat is the **choroid** (co'roid), a bloodrich nutritive layer that contains a dark pigment. The pigment prevents light scattering inside the eye. Anteriorly the choroid is modified to form two smooth muscle structures, the **ciliary** (sil'e-er-e) **body,** to which the **lens** is attached, and then the **iris.** The pigmented iris has a rounded opening, the **pupil,** through which light passes. Circularly and radially arranged smooth muscle fibers form the iris, which acts like the diaphragm of a camera; that is, it regulates the amount of light entering the eye so that one can see as clearly as possible in the available light. In close vision and bright light the circular muscles contract, and the pupil constricts. In distant vision and dim light the radial fibers contract to enlarge (dilate) the pupil, which allows more light to enter the eye.

The innermost sensory tunic of the eye is the delicate white **retina** (ret'i-nah) which extends anteriorly only to the ciliary body. The retina contains millions of receptor cells, the **rods** and **cones.** Rods and cones are called *photoreceptors;* because

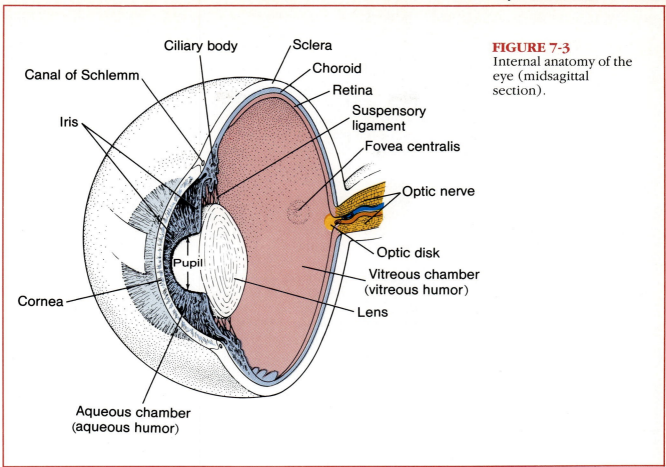

FIGURE 7-3
Internal anatomy of the eye (midsagittal section).

they convert light energy into nerve impulses. These impulses are transmitted to the optic cortex of the brain, and the result is vision. The photoreceptor cells are distributed over the entire retina, except where the optic nerve leaves the eyeball; this site is called the **optic disk,** or **blind spot.** When light from an object is focused on the optic disk, it disappears from our view and we cannot see it. To illustrate this, perform the blind spot demonstration. Hold Figure 7-4 about 18 inches from your eyes. Close your left eye and stare at the **X** with your right eye. Move the figure slowly toward your face, keeping your right eye focused on the **X**. When the dot focuses on your blind spot, which lacks photoreceptors, the dot will disappear. Repeat the test for your left eye. This time close your right eye and stare at the dot with your left eye. Move the page as before until the **X** disappears.

Photoreceptors are not evenly distributed in the retina. The rods are most dense at the periphery, or edge, of the retina and decrease in number as the center of the retina is approached. The rods allow us to see in gray tones and under conditions

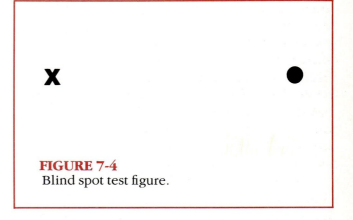

FIGURE 7-4
Blind spot test figure.

of dim light; they provide for our peripheral vision. On the other hand, the cones are color receptors that allow us to see the world in technicolor under bright light conditions. They are densest in the center of the retina and decrease in number toward the retinal edge. Lateral to each blind spot is the **fovea centralis** (fo've-ah sentra'lis), a tiny pit which contains only cones (see Figure 7-3). This is the area of greatest visual acuity, or point of sharpest vision. Anything we wish to view critically is focused on the fovea centralis.

BOX 7-1
Visual Pigments—the Actual Photoreceptors

The minute photoreceptor cells of the retina have names that reflect their general shapes. As shown in **a**, rods are skinny elongated neurons, whereas the fatter cones taper to their more pointed tips. In each, there is an outer segment (corresponding to a dendrite) attached to the nerve cell body in which the visual pigments are stacked like a row of pennies.

The behavior of the visual pigments is dramatic. When light strikes them, they lose their color, or are "bleached," and then shortly after they are regenerated once again. The absorption of light and pigment bleaching excites the photoreceptor cells and causes nerve impulses to be transmitted to the brain for visual interpretation. The regeneration of the pigment that follows ensures that one is not blinded and unable to see in bright sunlight.

Much is known about the structure and function of *rhodopsin,* the purple pigment found in rods **(b).** It is formed from the union of a protein portion *(opsin)* and a modified vitamin A product *(retinal).* When combined as rhodopsin, the retinal has a kinked shape that allows it to bind to opsin. But when light strikes rhodopsin, retinal straightens out and releases the protein. Once straightened out, the retinal continues its conversion until it is once again vitamin A. As these changes occur, the purple color of rhodopsin changes to the yellow of retinal and then finally becomes colorless as the change to vitamin A occurs. Thus the term "bleaching of the pigment" accurately describes the color changes that occur when light hits the pigment. Regeneration of rhodopsin occurs as vitamin A is again converted to the kinked form of retinal and recombined with opsin. The cone pigments, while similar to rhodopsin, differ in the specific kinds of proteins and modified vitamin A molecules involved.

There are three varieties, or types, of cones. Each type is most sensitive to one of the *primary* colors of light. One type responds most vigorously to blue light, another to red light, and the third to green light. Impulses received at the same time from more than one type of cone by the visual cortex are interpreted as *intermediate* colors. For example, simultaneous impulses from blue and red color receptors are seen as purple or violet tones. Also, if someone shines red light into one of your eyes and green light into the other, you will see yellow indicating that the "mixing" and interpretation of colors occurs in the brain, not in the retina. Lack of all cone types results in total color blindness, whereas lack of one cone type leads to partial color blindness. Most common is the lack of red or green receptors, which leads to two different varieties of red-green color blindness. Since the genes regulating color vision are on the X (female) sex chromosome, color blindness is a sex-linked condition. It occurs almost exclusively in males.

Light entering the eye is focused on the retina by the lens, a flexible biconvex crystallike structure. The **lens** is held upright in the eye by suspensory ligaments that are attached to the ciliary body (see Figure 7-3). In youth the lens is transparent and has the consistency of hardened jelly, but in the elderly person it becomes increasingly hard and opaque. *Cataracts,* which result from this process, cause vision to become hazy, and eventually blindness of the affected eye occurs. Current treatment of cataracts is surgical removal of the

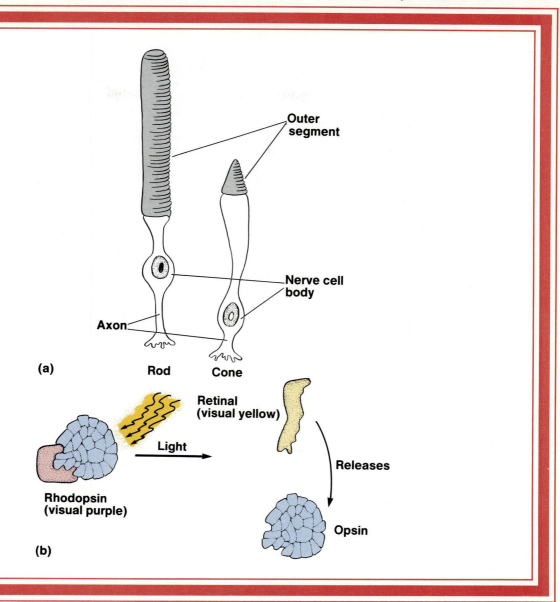

(a)

Outer segment

Nerve cell body

Axon

Rod   Cone

(b)

Rhodopsin (visual purple)

Light

Retinal (visual yellow)

Releases

Opsin

lens and its replacement by a lens implant or cataract glasses.

The lens divides the eye into two chambers. The *aqueous chamber,* anterior to the lens, contains a clear watery fluid called **aqueous humor.** The *vitreous chamber,* which is posterior to the lens, is filled with a gel-like substance, **vitreous** (vit're-us) **humor,** or **vitreous body** (see Figure 7-3). Vitreous humor helps prevent the eyeball from collapsing inward by reinforcing it internally. Aqueous humor is similar to blood plasma and is continually secreted by a special area of the choroid. It helps maintain the pressure inside the eye and provides nutrients for the lens and cornea, which lack a blood supply. Aqueous humor is reabsorbed into the venous blood (at the junction

of the sclera and cornea) into the **canal of Schlemm** (shlem) (see Figure 7-3). If drainage is blocked, pressure within the eye increases dramatically and begins to compress the delicate retina and optic nerve. This condition, *glaucoma* (glaw-ko'mah), results in pain and possible blindness. Glaucoma is a common cause of blindness in the elderly person. Since it progresses slowly and there are relatively few symptoms at first, it tends to occur without obvious signs. Later signs include seeing halos around lights, headaches, and blurred vision. A simple instrument called a *tonometer* (to-nom'ĕ-ter) is used to measure the intraocular (in"trah-ok'u-lar) (within the eye) pressure. This examination should be performed yearly in people over 40 years of age. When diagnosed, glaucoma is treated with eye-

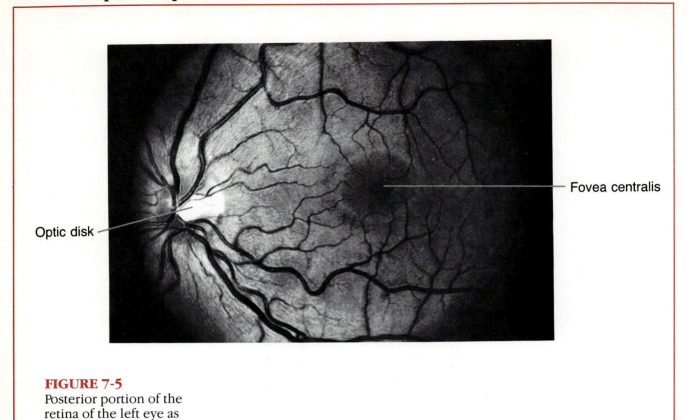

Optic disk

Fovea centralis

FIGURE 7-5
Posterior portion of the
retina of the left eye as
seen with an
ophthalmoscope.

drops (miotics), which increase the rate of aqueous humor drainage.

The *ophthalmoscope* (of-thal′mo-skōp) is an instrument that illuminates the interior of the eyeball, allowing observation of the retina, optic disk, and internal blood vessels. Certain pathological conditions such as diabetes, arteriosclerosis, and degeneration of the optic nerve and retina can be detected by such an examination. When the ophthalmoscope is correctly set, the posterior wall **(fundus)** of the healthy eye should appear as shown in Figure 7-5.

Pathway of Light Through the Eye and Refraction

When light passes from one substance to another substance with a different density, its speed changes, and its rays are bent, or *refracted*. Thus, the light rays are bent as they encounter the cor-

nea, aqueous humor, lens, and vitreous humor of the eye.

The refractive, or bending, power of the cornea and humors is constant. However, that of the lens can be changed by changing its shape—that is, by making it more or less convex, so that light can be properly focused on the retina. The greater the lens convexity, or bulge, the more the light bends. On the other hand, the flatter the lens is, the less it bends the light. In general, light from a distant source (over 20 feet away) approaches the eye as parallel rays (Figure 7-6), and no change in lens shape is necessary for it to be focused properly on the retina. However, light from a close object tends to diverge, or spread out, and scatter, and the lens must bulge more to make close vision possible. To achieve this the ciliary body contracts, allowing the lens to become more convex. This ability of the eye to focus specifically for close objects (those less than 20 feet away) is called *accommodation*. The image formed on the retina as a result of the light-bending activity of

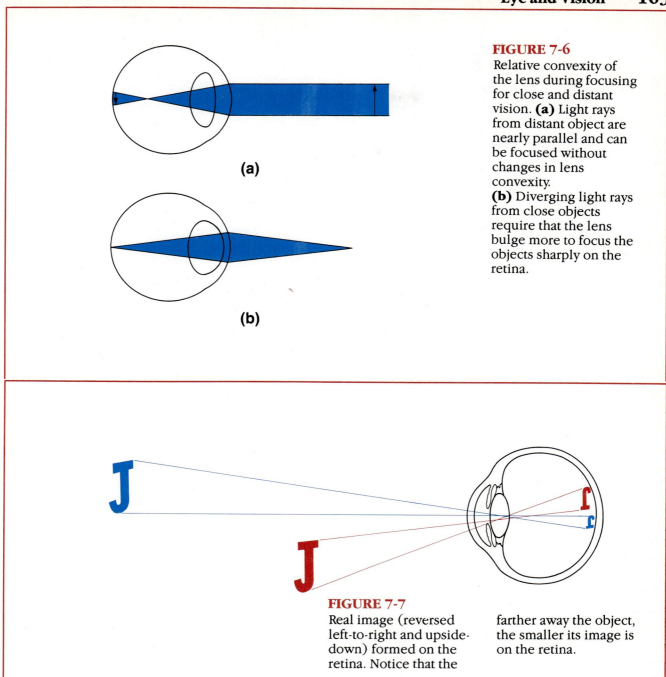

FIGURE 7-6
Relative convexity of the lens during focusing for close and distant vision. **(a)** Light rays from distant object are nearly parallel and can be focused without changes in lens convexity.
(b) Diverging light rays from close objects require that the lens bulge more to focus the objects sharply on the retina.

(a)

(b)

FIGURE 7-7
Real image (reversed left-to-right and upside-down) formed on the retina. Notice that the farther away the object, the smaller its image is on the retina.

the lens is a *real image*—that is, it is reversed from left-to-right, upside-down (inverted), and smaller than the object (Figure 7-7).

The normal, or *emmetropic* (em"ĕ-trop'ik), eye is able to accomodate properly. However, vision problems occur when a lens is too strong or too weak (over converging and under converging respectively) or from structural problems of the eyeball.

When the image normally focuses in front of the retina, the individual has *myopia* (mi-o'pe-ah), or is *nearsighted*. Such people can see close objects without any problem, but distant objects are blurred. Myopia results from a lens that is too strong or a long eyeball. Correction requires a concave lens, which diverges the light before it enters the eye (Figure 7-8).

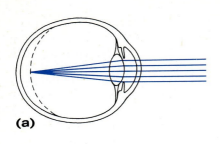

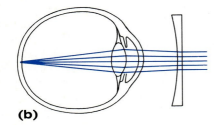

FIGURE 7-8
Myopia. **(a)** In myopia, light rays from distant objects focus in front of the retina. **(b)** Myopia may be corrected by a concave lens that diverges light rays before they enter the eye.

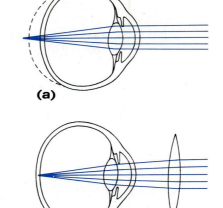

FIGURE 7-9
Hyperopia. **(a)** In hyperopia, light from distant objects focuses behind the retina. **(b)** Hyperopia may be corrected by a convex lens that converges light rays as they enter the eye.

If the image focuses behind the retina, the individual is said to have *hyperopia* (hi″per-o′pe-ah), or *farsightedness*. These people have no problems with their distant vision, but they need glasses with convex lenses to converge the light more strongly for close vision (Figure 7-9). Hyperopia results from a "lazy" lens or an eyeball that is too short.

Unequal curvatures of the lens or cornea lead to a blurred vision problem called *astigmatism* (ah-stig′mah-tizm). Specially ground lenses that correct the imbalance are prescribed.

Visual acuity, or sharpness of vision, is generally tested with a Snellen eye chart, which consists of letters of different sizes printed on a white card. This test is based on the fact that letters of a certain size can be clearly seen by eyes with normal vision at a specific distance. The distance at which the normal, or emmetropic, eye can read a line of letters is printed at the end of that line. The emmetropic eye is normally referred to as having 20/20 vision. When the vision is reported with a larger second number (for example, 20/40), vision is poorer than normal. On the other hand, if the second number is smaller (20/10), the persons's vision is sharper than normal.

Visual Pathways to the Brain

Axons carrying impulses from the retina are bundled together at the posterior aspect of the eyeball and leave the eye as the optic nerve. At the **optic chiasma** (ki-as′mah) the fibers from the medial side of each eye cross over to the opposite side. The fiber tracts that result are the **optic tracts.** Each optic tract contains fibers from the lateral side of the eye on the same side and the medial side of the opposite eye. The optic tract fibers synapse with neurons in the thalamus, whose axons form the **optic radiation,** which runs to the occipital lobe of the brain. Here they synapse with the cortical cells and visual interpretation, or seeing, occurs. The visual pathway from the eye to the brain is shown in Figure 7-10.

Each side of the brain receives visual imput from the lateral field of vision of the eye on its own side

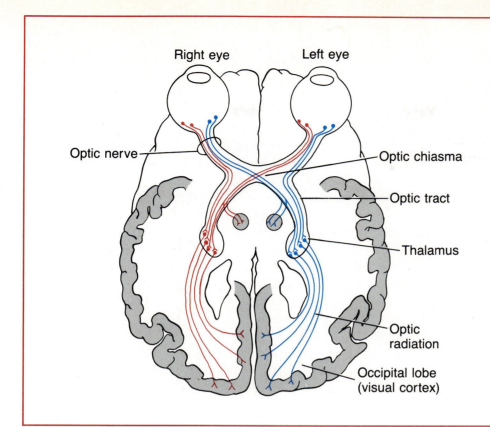

Right eye   Left eye

Optic nerve

Optic chiasma

Optic tract

Thalamus

Optic radiation

Occipital lobe (visual cortex)

FIGURE 7-10
Visual pathway to the brain. (Notice that fibers from the lateral portion of each retinal field do not cross over at the optic chiasma.)

and the medial field of the other eye. If there is damage to the visual cortex on one side only (as occurs in some CVAs), the result is *hemianopia* (hem″e-ah-no′pe-ah). Hemianopia is the loss of the same side of the visual field of both eyes; thus, the person would not be able to see things past the middle of his or her visual field on either the right or left side, depending on the site of the CVA. Such individuals should be carefully attended and warned of objects in the nonfunctional (or nonseeing side) of the visual field. Their food and personal objects should always be placed on their functional side, or they might miss them.

Eye Reflexes

Both the internal and the external eye muscles are necessary for proper eye function. The internal muscles are controlled by the autonomic nervous system. These muscles include the ciliary body (which alters lens curvature) and the radial and circular muscles of the iris (which control pupil size). The external muscles are the rectus and oblique muscles, which are attached to the eyeball exterior. The external muscles control eye movements and make it possible to follow moving objects. They are also responsible for *conver-*

gence, which is the reflex movement of the eyes medially when we view close objects. When convergence occurs, both eyes are aimed toward the near object being viewed. The extrinsic muscles are controlled by somatic fibers of cranial nerves III, IV and VI as indicated in Table 7-1.

When the eyes are exposed to bright light suddenly, the pupil constricts rapidly; this is the *photopupillary reflex.* This protective reflex prevents damage to the delicate photoreceptors by the excessive light. The pupils also constrict reflexly when we view close objects. This is the *accommodation pupillary reflex,* which provides for more acute vision.

Reading requires almost continuous work by both sets of muscles. The ciliary body must make the lens bulge, and the circular (or constrictor) muscles of the iris are called upon to produce the accomodation pupillary reflex. In addition, the extrinsic muscles must converge the eyes as well as move them to follow the printed lines. This is why long periods of reading tire the eyes and often result in what is commonly called *eyestrain.* When reading for an extended time, it is helpful to look up from time to time and stare into the distance. This relaxes all the eye muscles temporarily.

EAR: HEARING AND BALANCE

The machinery for hearing and balance senses in the ear is a product of evolution and, at first glance, appears very crude—fluids must be stirred to stimulate the ear receptors. Sound vibrations stimulate the hearing receptors, whereas gross movements of the head and body disturb the fluids surrounding the balance organs. Receptors that respond to such physical forces are *mechanoreceptors* (mek"ah-no-re-sep'tors). However, for all the simplicity of the hearing apparatus, it allows us to hear an extraordinary range of sound. Similarly, our highly sensitive equilibrium receptors keep our nervous system continually up to date, letting it know whether or not the head is uppermost and if the body is rotating, still, or moving in a straight line. Without this information, balance would be difficult, if not impossible, to maintain.

Anatomy of the Ear

Anatomically the ear is divided into three major areas: the outer, or external ear; the middle ear; and the inner, or internal ear (Figure 7-11). The outer and middle ear structures are involved with hearing *only;* the inner ear functions both in equilibrium and hearing.

EXTERNAL EAR
The external ear is composed of the **pinna** (pin'-nah), or **auricle** (aw'rě-k'l), and the **external auditory canal.** The pinna is a shell-shaped structure surrounding the auditory canal opening. In many animals, it collects and directs sound waves into the auditory canal, but in humans this functions is largely lost.

The auditory canal is a short, narrow (about 1-inch long by ¼-inch wide) chamber carved into the temporal bone of the skull. In its skin-lined walls are the **ceruminous** (sě-roo'mĭ-nus) **glands,** which secrete a waxy substance. Sound waves entering the external auditory canal eventually hit the **tympanic** (tim-pan'ik) **membrane,** or **eardrum,** and cause it to vibrate. The canal

ends at the eardrum, which separates the outer from the middle ear.

MIDDLE EAR
The middle ear is a small chamber, the **tympanic cavity,** within the temporal bone. The cavity is spanned by three tiny bones, the **ossicles** (os'si-k'ls), that transmit the vibratory motion of the eardrum to the fluids of the inner ear. When the eardrum moves, the **hammer** moves with it and transfers the vibration to the **anvil.** The anvil in turn passes it on to the third ossicle, the **stirrup,** which presses against a small membrane, the **oval window** of the inner ear. The movement of the oval window sets the fluids of the inner ear into motion.

The **eustachian** (u-sta'ke-an) **tube** connects the middle ear chamber with the throat. Normally the eustachian tube is flattened and closed, but swallowing or yawning can open it briefly to equalize the pressure in the middle ear cavity with the external, or atmospheric, pressure. This is an important function because the ear drum does not vibrate freely unless the pressure on both of its surfaces is the same. When the pressures are not equal, there is hearing difficulty (voices may sound far away) and sometimes earaches. This sensation is familiar to anyone who has flown in an airplane.

Because the mucosa of the middle ear is continuous with the mucosa lining the throat (through the eustachian tube), inflammation of the middle ear, *otitis media* (o-ti'tis me'de-ah), is fairly common, especially in children who have frequent sore throats. In otitis media, the eardrum bulges and often becomes inflamed and red. When large amounts of fluid or pus accumulate in the middle ear, an emergency *myringotomy* (mir"in-got'o-me), or lancing, of the eardrum may need to be done to relieve the pressure.

INTERNAL EAR
The inner ear is a system of mazelike bony chambers called the **osseous,** or **bony, labyrinth** (lab'i-rinth) that is filled with a fluid called **perilymph** (per'i-limf). The three subdivisions of the bony labyrinth are the **cochlea** (kok'le-ah), the **vestibule** (ves'tĭ-bul), and the **semicircular canals.** The vestibule is situated between the semicircular canals and the cochlea. Suspended in the

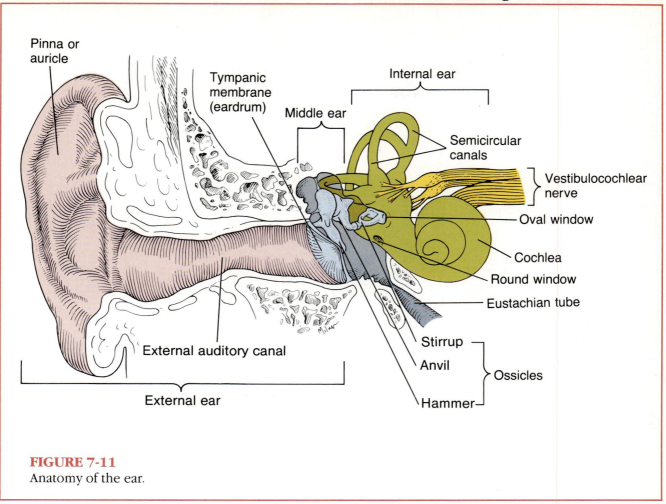

Pinna or auricle

Tympanic membrane (eardrum)

Middle ear

Internal ear

Semicircular canals

Vestibulocochlear nerve

Oval window

Cochlea

Round window

Eustachian tube

Stirrup

Anvil

Hammer

Ossicles

External auditory canal

External ear

FIGURE 7-11
Anatomy of the ear.

perilymph is a **membranous labyrinth,** a system of membranes that (more or less) follows the shape of the bony labyrinth. The membranous labyrinth is filled with a thicker fluid called **endolymph** (en'do-limf).

Within the membranes of the snail-like cochlea is the **organ of Corti** (kor'te), which contains the hearing receptors or hair cells (Figure 7-12). Sound waves that reach the cochlea through vibrations of the eardrum, ossicles, and the oval window set the cochlear fluids into motion. The receptor cells, positioned on the basilar membrane, in the organ of Corti, are stimulated when their ''hairs'' are bent or pulled by the movement of the gel-like **tectorial** (tek-to're-al) **membrane** that lies over them. In general, high-pitch sounds disturb receptor cells close to the oval window whereas low-pitch sounds stimulate specific hair cells further along the cochlea. Once stimulated, the hair cells transmit impulses along the **cochle-**

ar nerve (a division of the eighth cranial nerve—the vestibulocochlear nerve) to the auditory cortex in the temporal lobe where interpretation of the sound, or hearing, occurs. Since sound usually reaches the two ears at different times, we could say that we hear in ''stereo.'' Functionally this allows us to determine where sounds are coming from in our environment.

When the same sounds, or tones, keep reaching the ears, the auditory receptors tend to *adapt* or stop responding to those sounds, and we are no longer aware of them. This is why the drone of a continually running motor does not demand our attention after the first few seconds.

The semicircular canals and vestibule contain the balance, or equilibrium, receptors. The **dynamic equilibrium receptors,** found in the semicircular canals, respond to changes in angular motion rather than to motion itself. The semicircular ca-

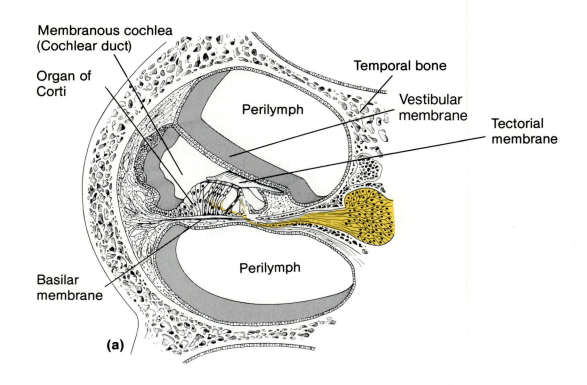

Membranous cochlea
(Cochlear duct)

Organ of
Corti

Temporal bone

Vestibular
membrane

Perilymph

Tectorial
membrane

Basilar
membrane

Perilymph

(a)

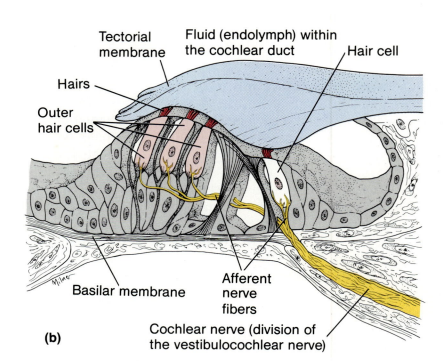

Tectorial
membrane

Fluid (endolymph) within
the cochlear duct

Hair cell

Hairs

Outer
hair cells

Basilar membrane

Afferent
nerve
fibers

Cochlear nerve (division of
the vestibulocochlear nerve)

(b)

FIGURE 7-12
Organ of hearing in the
inner ear. **(a)** Cross
section through the
cochlea. **(b)** Enlarged
view of the organ of
Corti.

nals (each about ½-inch around) are oriented in the three planes of space. Thus, regardless of which plane one moves in, there will be receptors to detect the movement.

Within each membranous semicircular canal is a receptor region called a **crista ampullaris** (kris′tă am″pu-lar′is) that consists of a tuft of hair cells covered with a gelatinous cap, or **cupula** (ku′po-lah) (Figure 7-13). When your head moves in an angular direction or rotates (as when twirling on the rides at an amusement park, or when taking a rough boat ride), the endolymph in the canal lags behind and moves in the *opposite* direction, pushing the cupula—like a swinging

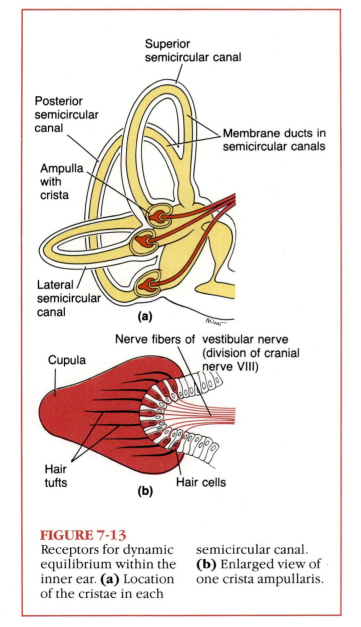

FIGURE 7-13
Receptors for dynamic equilibrium within the inner ear. **(a)** Location of the cristae in each semicircular canal. **(b)** Enlarged view of one crista ampullaris.

door—in a direction opposite to that of your angular motion. This stimulates the hair cells and impulses are transmitted up the **vestibular nerve** (division of the eighth cranial nerve) to the cerebellum of the brain. When the angular motion stops suddenly, the endolymph again flows in the opposite direction and reverses the cupula's movement, which stimulates the hair cells again. It is this backflow that causes the reversed motion sensation you feel when you stop suddenly after twirling. When you are moving at a constant rate, the receptors gradually stop transmitting impulses, and you no longer have the sensation of motion until your speed or direction of movement changes.

The vestibule contains receptors, **maculae** (mak′u-le), within its membrane sacs that are essential to our sense of *static equilibrium* (Figure 7-14). The maculae respond to the pull of gravity and provide information on which way is up or down. A gel-like material containing small "stones" of calcium salts (**otoliths** [o′to-liths]) lies over the hair cells in each macula. As the head moves, the otoliths move in the gel in response to changes in the pull of gravity. This pulls on the gel, which in turn pulls on the hair cells in the various maculae. This triggers nerve impulses in the hair cells that are transmitted to the brain.

Although the receptors of the semicircular canals and vestibule are responsible for dynamic and static equilibrium (respectively), they usually act together. It is also important to recognize that these equilibrium senses do not act alone in providing information to the cerebellum needed to control balance. Sight and the proprioceptors of the muscles and tendons are also very important.

Hearing and Equilibrium Deficits

Children with ear problems or hearing deficits often pull on their ears or fail to respond when spoken to. Under such conditions, tuning fork tests are done to try to diagnose the problem. *Deafness* is defined as hearing loss, and the degree may range from a slight loss to a total inability to hear sound. Generally speaking, there are two kinds of deafness, conduction and sensineural. *Conduction deafness* results when something interferes with the conduction of sound vibrations to the

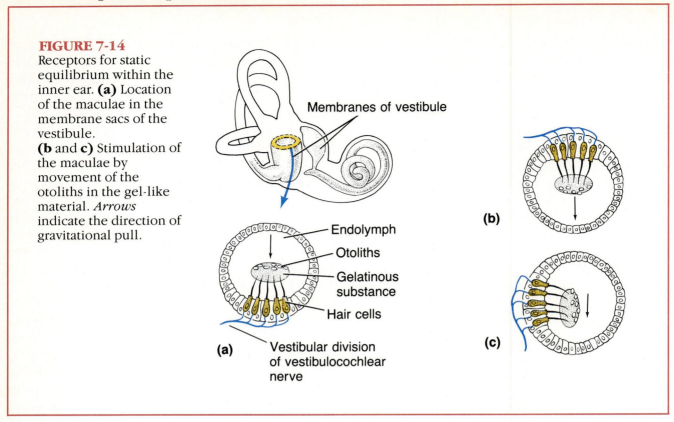

FIGURE 7-14
Receptors for static equilibrium within the inner ear. **(a)** Location of the maculae in the membrane sacs of the vestibule.
(b and **c)** Stimulation of the maculae by movement of the otoliths in the gel-like material. *Arrows* indicate the direction of gravitational pull.

Membranes of vestibule

Endolymph
Otoliths
Gelatinous substance
Hair cells

(a) Vestibular division of vestibulocochlear nerve

(b)

(c)

fluids of the inner ear. Something as simple as a build-up of earwax may be the cause. Other causes of conduction deafness include fusion of the ossicles (*otosclerosis* [o″to-skle-ro′sis]), a ruptured eardrum, and otitis media.

Sensineural deafness occurs when there is degeneration or damage to the receptor cells in the organ of Corti, to the cochlear nerve, or to neurons of the auditory cortex. (This often results from extended listening to excessively loud sounds.) Thus, whereas conduction deafness results from mechanical factors, sensineural deafness is a problem of nervous system structures.

A person who has a hearing loss due to conduction deafness will still be able to hear by bone conduction even though his or her ability to hear air-conducted sounds (the normal conduction route) is decreased or lost. On the other hand, individuals with sensineural deafness cannot hear better by *either* conduction route. Hearing aids, which use skull bones to conduct sound vibrations to the inner ear, are generally very successful in helping those with nonreversible conduction deafness to hear. They are less successful with sensineural deafness.

Equilibrium problems are usually obvious. Nausea, dizziness, and problems in maintaining balance are common symptoms. Also, there may be strange (jerky or rolling) eye movements. Simple types of tests (for example, observing a person after he or she has been rotated, or watching a person stand on one leg) are used to screen for possible problems with the equilibrium apparatus. If the results of these tests seem to indicate that there are such problems, then much more sophisticated tests are performed to try to track down the exact cause.

A serious pathology of the inner ear is *Menière's* (men″e-ārz′) *syndrome*. The exact cause of this condition is not fully known, but suspected causes are arteriosclerosis, degeneration of the eighth cranial nerve, and an increase in the pressure of the inner ear fluids. In Menière's syndrome, progressive deafness occurs. Affected individuals become nauseated and often have *vertigo* (a sensation of spinning), which is so severe that they cannot stand up without extreme discomfort. Antimotion sickness drugs are often prescribed to decrease the discomfort.

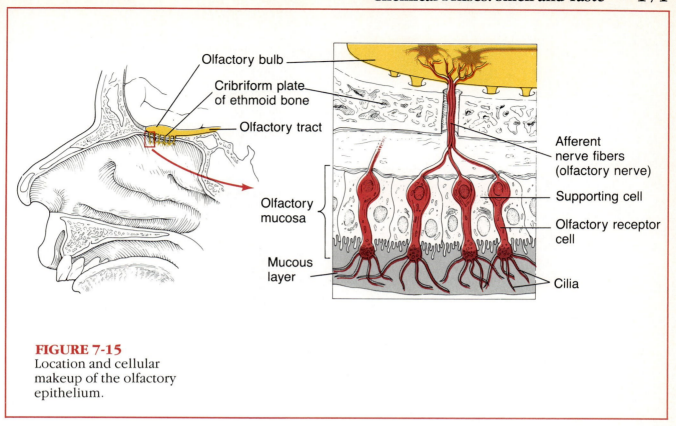

FIGURE 7-15
Location and cellular
makeup of the olfactory
epithelium.

CHEMICAL SENSES: SMELL AND TASTE

The receptors for taste and olfaction are classified as *chemoreceptors* (ke″mo-re-sep′tors), because they respond to chemicals in solution. Although four types of taste receptors have been identified, the olfactory receptors are believed to be sensitive to a much wider range of chemicals. The sense of smell is the least understood of the special senses.

Olfactory Receptors and the Sense of Smell

The thousands of receptors for the sense of smell occupy a postage stamp–sized area in the roof of each nasal cavity (Figure 7-15). Since air that enters the nasal cavities must make a hairpin turn to enter the respiratory passageway below, the nasal epithelium is in a very poor position for doing its job. This is why sniffing, which causes more air to flow across the olfactory receptors, intensifies the sense of smell. Even though our sense of smell is far less acute than that of other animals, the human nose is still much keener in picking up small differences in odors than any manmade machine. Some people capitalize on this ability when they become tea and coffee blenders or work in the wine making industry.

The olfactory receptor cells are neurons equipped with olfactory hairs, or cilia, that protrude outward from the nasal epithelium. When the receptors are stimulated by chemicals in solution, they transmit impulses along the first cranial nerve (olfactory nerve) to the olfactory cortex of the brain where interpretation of the odor occurs, and something similar to an "odor snapshot" is made. The olfactory pathways are closely tied into the limbic system (emotional-visceral part of the brain). Thus, olfactory impressions are long lasting and very much a part of our memories and emotions. For example, the smell of freshly baked bread may remind you of your grandmother, and the smell of a particular brand of pipe tobacco may make you think of your father. There are hospital smells, school smells, mother smells, baby smells, travel smells; the list can be continued without end. Our reactions to odors are rarely neutral; we tend to either like or dislike certain odors, and we change, avoid, or add odors according to our preferences.

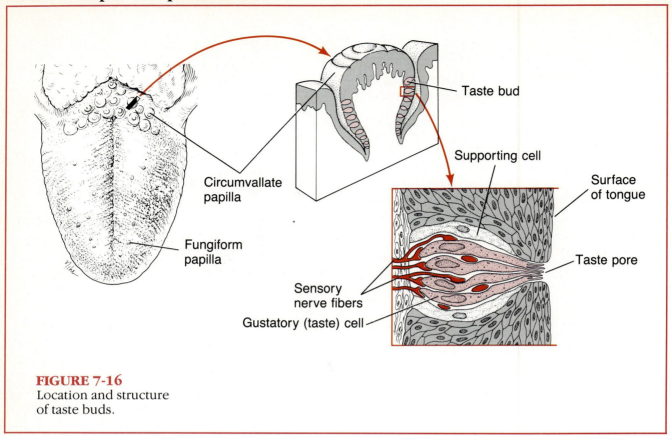

FIGURE 7-16
Location and structure
of taste buds.

Like the auditory receptors, the olfactory neurons tend to adapt rather quickly when they are continually exposed to the same stimulus, in this case, an odor. This is why a woman stops smelling her own perfume after a while but will quickly pick up the scent of perfume on someone else.

Taste Buds and the Sense of Taste

The **taste buds,** or specific receptors for the sense of taste, are widely scattered in the oral cavity. Of the 10,000 or so taste buds that we have, most are located on the tongue. A few are found on the soft palate and inner surface of the cheeks.

The dorsal tongue surface is covered with small projections, or **papillae** (pah-pil'e), of three types: sharp filiform (fil'ĭ-form) papillae and the rounded fungiform (fun'ji-form) and circumvallate (ser″kum-val'āt) papillae. The taste buds are found on the sides of the circumvallate papillae and on the more numerous fungiform papillae (Figure 7-16). The specific cells that respond to chemicals in solution in the saliva are epithelial

cells called *gustatory cells,* which are surrounded by supporting cells in the taste bud. Their microvilli protrude through the *taste pore.* When they are stimulated they depolarize, and impulses are transmitted to the brain. Three cranial nerves—the seventh, ninth, and tenth—carry taste impulses from the various taste buds. The facial nerve (VII) serves the anterior part of the tongue, whereas the other two nerves serve the other taste bud–containing areas.

There are four basic taste sensations; each corresponds to the stimulation of one of the four major types of taste buds. The *sweet receptors* respond to many substances such as sugars, saccharine, and some amino acids. Some believe that the common factor is the hydroxyl (OH^-) group. *Sour receptors* respond to hydrogen ions (H^+), or the acidity of the solution; *bitter receptors* to alkaloids; and *salty receptors* to metal ions in solution. A "taste map" showing the major location of each receptor type on the tongue is shown in Figure 7-17. It has been suggested that the bitter receptors are protective taste receptors since many poisonous substances are alkaloids. With this in mind, the position of the bitter receptors on the most posterior part of the tongue seems a bit strange since by the time we get to taste the bitter substance, some of it has already been swallowed.

The word "taste" is believed to come from the Latin word *taxare* which means to touch, estimate, or judge. When we taste things, we are, in fact, testing or judging our environment in an intimate way, and the sense of taste is considered by many to be the most pleasurable of our special senses. There is no question that what does not taste good to us will usually not be allowed to enter the body.

Taste is affected by many other factors, and what is commonly referred to as our sense of taste depends heavily on stimulation of our olfactory receptors. Think of how bland, almost tasteless, food is when your nasal passages are congested by a cold. In addition to the sense of smell, the temperature and texture of foods eaten either enhances or spoils its taste for us. For example, some people will not eat foods that have a pasty texture (avocados) or that are gritty (pears), and almost everyone considers a cold greasy hamburger unfit to eat.

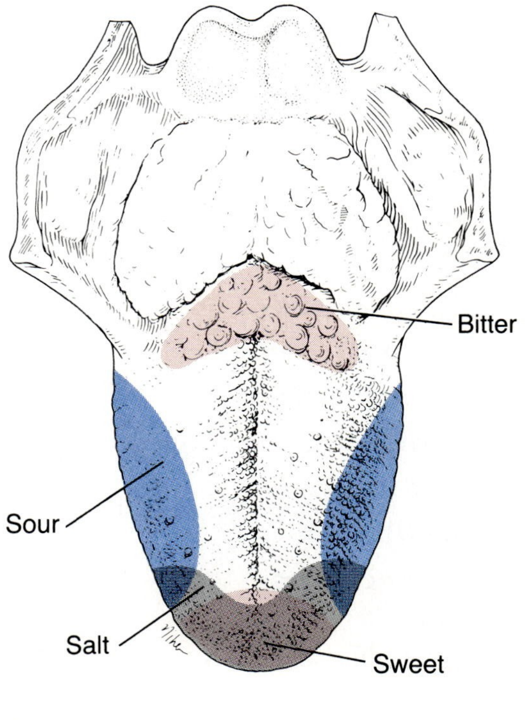

DEVELOPMENTAL ASPECTS OF THE SPECIAL SENSES

The special sense organs, essentially part of the nervous system, are formed very early in embryonic development. For example, the eyes, which are literally outgrowths of the brain, are developing by the fourth week of the embryonic period.

Congenital eye problems are relatively uncommon, but some examples can be given. *Strabissmus* (strah-biz′mus), commonly called "crossed-eyes," results from unequal pull by the external eye muscles so that the baby can't coordinate the movement of the two eyes. Exercises and/or surgery are always used to correct the condition since *ambylopia* (am″ble-o′pe-ah), or "lazy eye," may occur in the deviating eye. This is a condition in which the brain stops recognizing signals from the deviating eye, and that eye becomes functionally blind.

Maternal infections, particularly rubella (measles) which occur during early pregnancy, may lead to congenital blindness or cataracts. If the mother has a type of venereal disease, called *gonorrhea* (gon″o-re′ah), the baby's eyes will be infected by the bacteria during delivery and an inflammation of the conjunctiva or *conjunctivitis* (kon-junk″tǐ-vi′tis) occurs. In this condition, specifically called *opthalmia neonatorum* (of-thal′me-ah ne″o-na-to′rum), the baby's eyelids become red and swollen, and pus is produced. All states now legally require that all newborn babies' eyes be routinely treated with silver nitrate or antibiotics shortly after birth.

Generally speaking, the sense of vision is the only one of the special senses that is not fully functional when the baby is born, and many years of "learning" are needed before the eyes are fully mature. The eyeballs continue to enlarge until the age of 8 or 9 years, but the lens continues to grow throughout life. At birth the eyeballs are foreshortened, and all babies are hyperopic. As the

FIGURE 7-17

A taste "map." Areas of the tongue most sensitive to the four basic taste sensations.

eyes grow, this condition is usually corrected. The newborn infant sees only in gray tones, eye movements are uncoordinated, and often only one eye at a time is used. Because the lacrimal glands are not completely developed until about 2 weeks after birth, the baby is tearless for this period, even though he or she may cry lustily.

By 5 months the infant is able to focus on objects within easy reach and follows moving objects, but visual acuity is still poor (20/200). By the time the child is 5 years old, color vision is well developed, visual acuity has improved to about 20/30, and depth perception is present providing a readiness to begin reading. By school age the earlier hyperopia has usually been replaced by emmetropia. This condition continues until about age 40 when *presbyopia* (pres"be-o'pe-ah) begins to set in. Presbyopia (literally, old vision) results from a decreasing lens elasticity that occurs with aging. This condition makes it difficult to focus for close vision, and it is basically farsightedness. The person who holds the newspaper at arms length to read it provides the most familiar example of this developmental change in vision.

As aging occurs, the lens also loses its crystal clarity and becomes yellow or discolored. The radial (dilator) muscles of the iris become less efficient; thus, the pupils are always somewhat constricted. These two conditions work together to decrease by half the amount of light reaching the retina so that visual acuity is dramatically lower by the 70s. In addition to these changes, elderly persons are susceptible to certain conditions that may result in blindness, such as glaucoma and cataracts. Other common types of aging-related problems, such as arteriosclerosis and diabetes, may lead to the death of the delicate photoreceptors because of an increasing lack of oxygen and nutrients.

Congenital abnormalities of the ears are fairly common. Examples include partly or completely missing pinnas or closed or absent external ear canals. Maternal infections can have a devastating effect on ear development, and maternal rubella during the early weeks of pregnancy results in sensineural deafness.

A newborn infant can hear after his first cry, but early responses to sound are mostly reflex in nature (for example, crying and clenching the eye-lids). By age 3 or 4 months, the infant is able to localize sounds and will turn to the voice of close family members. Critical listening begins to occur in the toddler as he or she begins to imitate sounds, and good language skills are very closely tied to the ability to hear well.

Except for ear inflammations (otitis) resulting from bacterial infections or allergies, few problems affect the ears during childhood and adult life. By the 60s, however, a gradual deterioration

---

IMPORTANT TERMS*

Sclera *(skle'rah)*

Cornea *(kor'ne-ah)*

Choroid *(co'roid)*

Ciliary *(sil'e-er-e)* **body**

Lens

Iris

Retina *(ret'i-nah)*

Blind spot

Accommodation

Tympanic *(tim-pan'ik)* **cavity**

Ossicles *(Os'si-k'ls)*

Labyrinth *(lab'ĭ-rinth)*

Cochlea *(kok'le-ah)*

Vestibule *(ves'tĭ-bul)*

Semicircular canals

Organ of Corti *(kor'te)*

Dynamic equilibrium

Crista ampullaris *(kris'tă am"pu-lar'is)*

Maculae *(mak'u-le)*

Static equilibrium

Olfactory receptors

Taste buds

Refraction

Chemoreceptors *(ke"mo-re-sep'tors)*

Photoreceptors

*For definitions, see Glossary.

and atrophy of the organ of Corti begins and leads to a loss in the ability to hear high tones and speech sounds. This condition, *presbycusis* (pres″bĕ-ku′sis), is a type of sensineural deafness. Because many elderly people refuse to accept their hearing loss and resist using hearing aids, they begin to rely more and more on their vision for clues as to what is going on around them and may be accused of ignoring people. Although presbycusis is considered to be a disability of old age, it is becoming much more common in younger people as our world grows noisier day by day. Noise pollution has become a major health problem, and the damage caused by excessively loud sounds is progressive and cumulative. Each insult causes a bit more damage. Rock music played and listened to at deafening levels is definitely a contributing factor to the deterioration of the hearing receptors.

The chemical senses, taste and smell, are sharpest at birth. This explains why infants seem to relish food adults consider to be bland or tasteless. Some researchers claim that the sense of smell is just as important as the sense of touch in guiding newborn babies to their mother's breast.

There appear to be few interferences with the chemical senses throughout childhood and young adulthood; however, a decrease in chemical sensitivity may result from certain nutritional deficiencies such as lack of zinc. Beginning in the mid-40s our ability to taste and smell diminishes, which reflects the gradual decrease in the number of these receptor cells. This helps to explain why older adults often prefer highly seasoned (although not necessarily spicy) foods, and why many elderly people do not seem to notice many odors that younger people find very disagreeable.

SUMMARY

A. EYE AND VISION

1. External/accessory structures of the eye
   a. External eye muscles aim the eyes for following moving objects and convergence.
   b. Lacrimal apparatus includes a series of ducts and the lacrimal glands that produce a saline secretion, which washes and lubricates the eyeball.
   c. Eyelids protect the eyes. The meibomian glands associated with the eyelashes produce an oily secretion, which helps keep the eye lubricated.
   d. Conjunctiva is a mucous membrane that covers the anterior eyeball and lines the eyelids. It produces a lubricating mucus.

2. Three tunics form the eyeball
   a. The sclera is the outer tough protective tunic. Its anterior portion is the cornea, which is transparent to allow light to enter the eye.
   b. The choroid is the middle coat that provides nutrition to the internal eye structures and prevents light scattering in the eye. Anterior modifications include two smooth muscle structures, the ciliary body, and the iris (which controls the size of the pupil).
   c. The retina is the innermost (sensory) coat that contains the photoreceptors. Rods are dim light receptors. Cones are receptors that provide for color vision and high visual acuity. The fovea centralis, on which acute focusing occurs, contains only cones.

3. The blind spot is the point where the optic nerve leaves the back of the eyeball.

4. The lens is the major light-bending (refractory) structure of the eye. Its convexity is increased by the ciliary body for close focus. Anterior to the lens is the aqueous humor; posterior to the lens is the vitreous humor. Both humors reinforce the eye internally. The aqueous humor also provides nutrients to the avascular lens and cornea.

5. Errors of refraction include myopia, hyperopia, and astigmatism. All are correctable with specially ground lenses.

6. The pathway of light through the eye is the cornea → aqueous humor → (through the pupil) → aqueous humor → lens → vitreous humor → retina.

7. The pathway of nerve impulses from the retina of the eye is the optic nerve → optic chiasma → optic tract → thalamus → optic radiation → visual cortex in the occipital lobe of the brain.

8. Eye reflexes include the photopupillary, accommodation pupillary, and convergence reflexes.

B. EAR: HEARING AND BALANCE

1. The ear is divided into three major areas
   a. External ear structures are the pinna, external auditory canal, and tympanic membrane. Sound entering the external auditory canal sets the eardrum into vibration. These structures are involved with sound transmission only.
   b. Middle ear structures are the ossicles and eustachean tube within the tympanic cavity. Ossicles transmit the vibratory motion from the eardrum to the oval window. The eustachean tube allows pressure to be equalized on both sides of the eardrum. These structures are involved with sound transmission only.
   c. Internal ear or bony labyrinth consists of bony chambers (cochlea, vestibule, and semicircular canals) in the temporal bone. The bony labyrinth contains perilymph and membranes filled with endolymph. Within the membranes of the vestibule and semicircular canals are equilibrium receptors; hearing receptors are found within the membranes of the cochlea.

2. Receptors of the semicircular canals (cristae) are dynamic equilibrium receptors that respond to angular or rotational movements of the body. Receptors of the vestibule (maculae) are static equilibrium receptors that respond to the pull of gravity and report on head position. Visual and proprioceptor input are also necessary for normal balance.

3. The organ of Corti within the cochlea is stimulated by sound vibrations.

4. Deafness is hearing loss in any degree. Conduction deafness results when the transmission of sound vibrations through the external and middle ears is interfered with in any way. Sensineural deafness occurs when there is damage to the nervous system structures involved in hearing.

5. Symptoms of equilibrium apparatus problems include involuntary rolling of the eyes, nausea, vertigo, and an inability to stand erect.

C. CHEMICAL SENSES: SMELL AND TASTE

1. Chemical substances must be dissolved in water to excite these receptors.

2. The olfactory receptors are located at the superior aspect of each nasal cavity; sniffing helps to bring more air (containing odors) over the olfactory mucosa.

3. Olfactory pathways are closely linked to the limbic system; therefore, odors recall memories and arouse emotional responses.

4. Gustatory cells are located in the taste buds primarily on the tongue. There are four major taste sensations: sweet, salt, sour, and bitter.

5. Taste and appreciation of foods is influenced by the sense of smell and the temperature and texture of foods.

D. DEVELOPMENTAL ASPECTS OF THE SPECIAL SENSES

1. Special sense organs are formed early in embryonic development. Maternal infections during the first 5–6 weeks of pregnancy may cause visual abnormalities as well as sensineural deafness in the developing child. An important congenital eye problem is strabissmus; the most important congenital ear problem is lack of the external auditory canal.

2. Vision requires the most learning. The infant has poor visual acuity (is farsighted) and lacks color vision and depth perception at birth. The eye continues to grow and mature until the eighth or ninth year of life.

3. Problems of aging associated with vision include presbyopia, glaucoma, cataracts, and arteriosclerosis of the eye blood vessels.

4. The newborn infant can hear sounds, but initial responses are reflexive. By the toddler stage, the child is listening critically and beginning to imitate sounds as language development begins.

5. Sensineural deafness (presbycusis) is a normal consequence of aging.

6. Taste and smell are most acute at birth and decrease in sensitivity after the age of 40 years as the number of olfactory and gustatory receptors decreases.

REVIEW QUESTIONS

1. Name three accessory eye structures that help to lubricate the eyeball and name the secretory product of each.

2. Why do you often have to blow your nose after crying?

3. Diagram and label the internal structures of the eye and then give the major function for each structure.

4. Name the external eye muscles that allow you to direct your eyes.

5. Locate and describe the function of the two humors of the eye.

6. What is the blind spot, and why is it called this?

7. What is the difference in the function of the rods and cones?

8. What is the fovea centralis, and why is it important?

9. Trace the pathway of light from the time it hits the cornea until it excites the rods and cones.

10. Trace the pathway of nerve impulses from the photoreceptors in the retina to the visual cortex of the brain.

11. How is the right optic *tract* anatomically different from the right optic *nerve*.

12. Define refraction and name the refractory structures or substances of the eye.

13. Define hyperopia, myopia, and emmetropia.

14. Why do most people develop presbyopia as they age? Which of the conditions in Question 13 does it most resemble?

15. Since there are only three types of cones, how can you explain the fact that we see many more colors?

16. Why is the ophthalmoscopic examination an important test?

17. Many students struggling through mountains of reading assignments are told that they need glasses for eyestrain. Why is it more a strain on the extrinsic and intrinsic eye muscles to look at close objects than far objects?

18. When a light is shone into one eye, the pupils constrict. Why is this an important protective reflex?

19. Name the structures of the outer, middle, and inner ears and give the general function of each structure and each group of structures.

20. Sound waves hitting the eardrum set it into motion. Trace the pathway of vibrations from the eardrum to the organ of Corti where the hair cells are stimulated.

21. Clearly explain the difference between sensineural deafness and conductive deafness and then list two causes of each type of deafness.

22. Distinguish between static and dynamic equilibrium.

23. Normal balance depends on information transmitted from a number of sensory receptor types. Name at least three of these receptors.

24. What name is given to the taste receptors, and where are they found?

25. Name the four primary taste sensations.

26. Where are the olfactory receptors located, and why is that site poorly suited for their job?

27. What common name is used to describe both the taste and the olfactory receptors and why?

28. If you just had your *special* sensory receptors, what type of sensory information would you miss out on?

29. Describe the effect or results of aging on the special sense organs.

30. You sometimes hear the expression "The eye must mature." What does that mean?

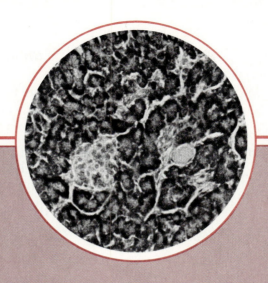

CHAPTER 8

Endocrine System

Chapter Contents

Functions of the endocrine system • to maintain homeostasis by releasing chemicals called hormones (a slow-acting system), and to control prolonged or continuous processes such as growth and development, reproduction, and metabolism

After completing this chapter, you should be able to:

- Describe the difference between endocrine and exocrine glands.

- Identify the major endocrine glands and tissues when given an appropriate diagram.

- List hormones produced by the endocrine glands and discuss their general functions.

- Discuss ways in which hormones promote body homeostasis by giving examples of hormonal actions.

- State how various endocrine glands are stimulated to release their hormonal products.

- Explain negative feedback and its role in regulating blood levels of the various hormones.

- Describe the functional relationship between the hypothalamus and the pituitary gland.

- Describe major pathologic consequences of hypersecretion and hyposecretion of the hormones considered.

- Describe the effect of aging on the endocrine system and body homeostasis.

The endocrine system is the second major controlling and communicating system of the body. Along with the nervous system, it helps to coordinate and direct the activity of the body's cells. However, the speed of control in these two major regulating systems is very different. One could say that the nervous system is built for speed. It uses nerve impulses to prod the body organs into immediate action so that rapid adjustments can be made in response to changes occurring both inside and outside the body. On the other hand, the more slowly acting endocrine system uses chemical messengers, or **hormones,** which are released into the blood to be transported throughout the body. The endocrine system regulates continuing processes that go on for relatively long periods of time.

THE ENDOCRINE SYSTEM AND HORMONE FUNCTION: AN OVERVIEW

Compared to other organ systems of the body, which are made up of fairly large organs, the organs (or glands) of the endocrine system seem insignificantly small and unimportant. This impression is reinforced when you realize that all

together the endocrine glands weigh less than ½ pound. In addition, the endocrine organs have no anatomic continuity. Instead bits and pieces of endocrine tissue appear to be tucked away in widely separated regions of the body (Figure 8–1). However, functionally the endocrine organs are very impressive and when their role in maintaining body homeostasis is considered, they are true giants.

The key to the incredible power of the endocrine glands is the hormones they produce and secrete. The term hormone comes from a Greek word meaning "to arouse." The body's hormones, in fact, do just that. They arouse the body's cells by stimulating changes in their metabolism. These changes lead to growth and development and functional homeostasis of many body systems. Normal functioning on the part of the endocrine glands is critical since even a slight change in hormonal balance can throw the whole body out of balance.

Although the blood-borne hormones circulate to all the organs of the body, a given hormone affects the activity of only a specific organ or organs. Organs that respond to a particular hormone are the *target organs* of that hormone. The ability of the cells of the target tissue to respond to the hormone seems to depend on the ability of

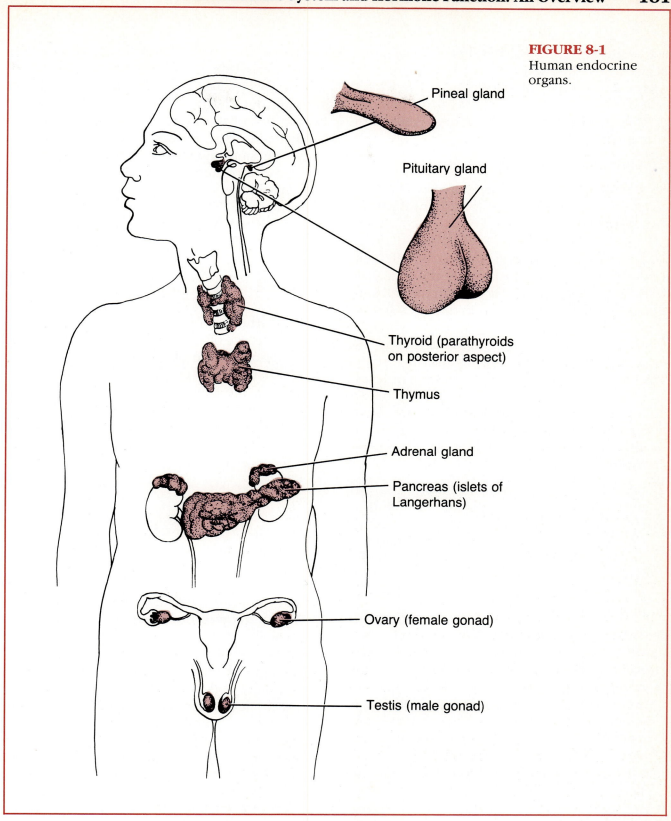

FIGURE 8-1
Human endocrine organs.

Pineal gland

Pituitary gland

Thyroid (parathyroids on posterior aspect)

Thymus

Adrenal gland

Pancreas (islets of Langerhans)

Ovary (female gonad)

Testis (male gonad)

the hormone to attach to specific receptors (proteins) on the cell membranes or inside the cells. Only when this binding occurs can the hormone influence the workings of the cell.

Hormones affect the body cells primarily by *altering* a cellular activity—that is, by increasing or decreasing a certain normal process rather than by stimulating a new one. Precisely how hormones act is still being researched, but it appears that changes that follow hormonal binding include one or more of the following:

1. Changes in the selective permeability of the cell membrane

2. Activation or inactivation of enzymes

3. Stimulation of the genetic material itself to produce the instructions for making a particular enzyme

In cases where a hormone binds to the cell membrane, but does not actually enter the cell, a second messenger (cyclic AMP) within the cell stimulates the changes that occur in the cell's activity.

There are many different endocrine glands, and each of them has a unique gross and microscopic anatomy. However, there are some similarities.

1. All endocrine glands have a very rich blood supply. This is easy to remember once you realize that their product is released into the blood.

2. Although many different hormones are produced by the various endocrine glands, all of them are either protein/peptide or steroid (based on cholesterol) molecules. Steroid hormones include the sex hormones made by the gonads (ovaries and testes) and the hormones produced by the adrenal cortex. All others are proteins or peptide in nature.

3. There are three categories of stimuli that activate the endocrine organs to make and release their hormones. Some are stimulated by other hormones; this is the most common stimulus and is called a *hormonal stimulus.* Others are stimulated by the nervous system (*neural stimulus*) or by the concentration of chemical substances other than hormones (such as calcium ions and glucose) in the blood (*humoral stimulus*).

Although the function of some hormone-producing glands (the anterior pituitary, thyroid, adrenals, and parathyroids) is purely endocrine, the function of others (pancreas and gonads) is mixed—both endocrine and exocrine. Both types of glands are formed from epithelial tissue, but the endocrine glands are ductless glands that pro-

duce hormones for release to the blood. Conversely, exocrine glands release their products at the body's surface or into body cavities through ducts. The formation, differences, and similarities of these two types of glands has already been discussed in Chapter 2. In addition, there are large numbers of hormone-producing cells within organs, such as the intestine, stomach, and kidneys, whose functions are primarily nonendocrine. These will be discussed in later chapters when the organs they are part of are studied.

PITUITARY GLAND

The **pituitary** (pĭ-tu'ĭ-tār"e) **gland** is approximately the size of a grape. It hangs by a stalk from the inferior surface of the hypothalamus of the brain where it is snugly surrounded by the "turk's saddle" of the sphenoid bone. It has two functional areas, the anterior pituitary (glandular tissue) and the posterior pituitary (nervous tissue).

Hormones of the Anterior Pituitary

There are several anterior pituitary hormones that affect many body organs. These hormones are described next, and their relationships to the various body organs are shown in Figure 8-2.

Somatotropic (so"mah-to-trop'ik) **hormone (STH),** or **growth hormone,** is a general metabolic hormone, but its major effects are directed to the growth of skeletal muscles and long bones of the body. It plays an important role in determining final body size. STH is a protein-sparing and anabolic hormone that causes amino acids to be built into proteins. At the same time, it causes fats to be broken down for energy. Hyposecretion of STH during childhood leads to a type of dwarfism called *pituitary dwarfism.* Body proportions are fairly normal for the affected person's age, but the person as a whole is a living miniature. Hypersecretion during childhood results in *gigantism.* The individual becomes extremely tall, 8–9 feet is not uncommon. Again, body proportions are fairly normal. If hypersecretion occurs after long bone growth has ended, *acromegaly* (ak"ro-meg'ah-le) results. The facial bones, particularly the lower jaw and the bony ridges underlying the eye-

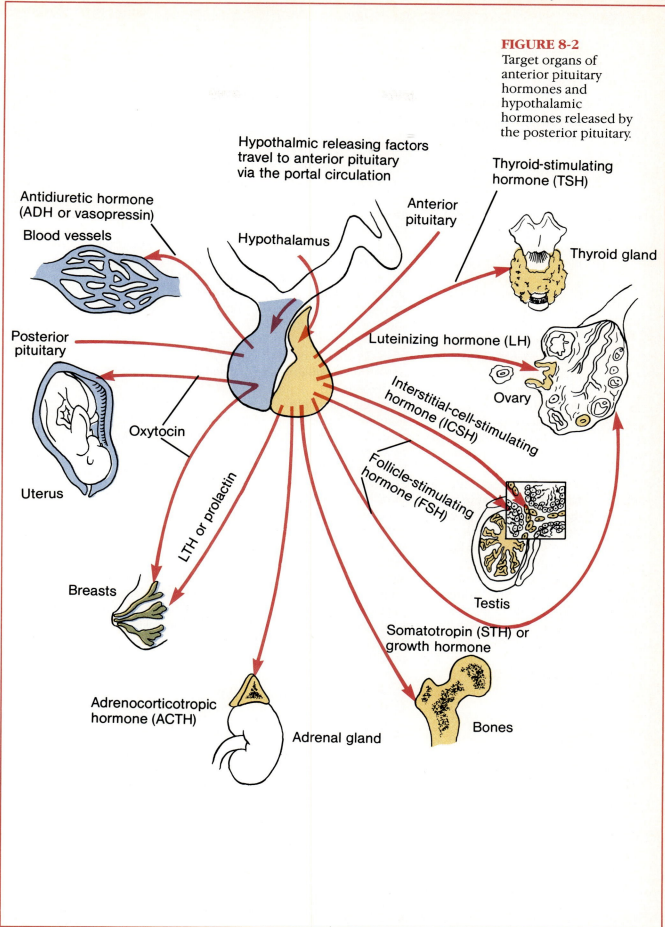

FIGURE 8-2
Target organs of
anterior pituitary
hormones and
hypothalamic
hormones released by
the posterior pituitary.

Hypothalmic releasing factors
travel to anterior pituitary
via the portal circulation

Anterior
pituitary

Thyroid-stimulating
hormone (TSH)

Antidiuretic hormone
(ADH or vasopressin)

Blood vessels

Hypothalamus

Thyroid gland

Posterior
pituitary

Luteinizing hormone (LH)

Interstitial-cell-stimulating
hormone (ICSH)

Ovary

Oxytocin

Follicle-stimulating
hormone (FSH)

Uterus

LTH or prolactin

Breasts

Testis

Somatotropin (STH) or
growth hormone

Adrenocorticotropic
hormone (ACTH)

Adrenal gland

Bones

BOX 8-1
Prostaglandins— Remarkable Local Hormones

Prostaglandins, often referred to as local hormones, are among the most shortlived, yet potent, chemicals ever discovered in the body. First reported in 1930 as a component of secretions produced by the male prostate gland (thus, their naming), they have since been found to be made by practically every tissue of the body. Nobel prize-winning researchers have found these fatty acid molecules to be part and parcel of the cell membrane surrounding each cell and that local disturbances near a cell can result in the release of a stream of these fatty acids from the membrane (much like wasps swarming from a jolted nest). Once the fatty acids are released from the cell membrane, extracellular enzymes then convert them into their active prostaglandin forms, which cause profound effects on various body functions including digestion, respiration, circulation, and reproduction.

At present, the 16 known prostaglandins are divided into four groups—E, A, F, and B—based on slight variations in their structures. Many prostaglandins act on smooth muscle structures. For example, some are powerful vasoconstrictors whereas others are powerful vasodilators. In the lungs, prostaglandins help regulate the size of the bronchiole passageways by acting on the smooth muscle in their walls. Still others, like oxytocin, are effective as activators of uterine muscle as during the labor phase of the birth process. On the other side of the reproductive coin, it is likely that many prostaglandins have played a role (unknown until recently) in birth control. There is now evidence that IUDs (intrauterine devices) inserted in the uterus to prevent pregnancy in fact provoke the release of prostaglandins, which increase uterine contractions and prevent successful implantation of the embryo.

Some prostaglandins are known to be pain-provoking substances in diseases such as arthritis, and many antiprostaglandin drugs have recently been marketed to inhibit this distressing prostaglandin function. (However, it should be noted that aspirin is the oldest and most widely used of antiprostaglandin medications.) In addition to acting hormonelike in their own right, prostaglandins have a modulating and mediating effect on the activity of hormones on their target cells— acting to enhance or diminish their effects. In this role, they mimic or inhibit circulating hormones and regulate most of the basic processes of life.

It appears that the prostaglandins' arena of activity is wider than that of any biochemical substance known. They are indeed remarkable, and the study of these substances is currently one of the most explosive areas of biomedical research.

brows, enlarge tremendously as do the feet and hands. Most cases of hypersecretion by endocrine organs (the pituitary as well as the other endocrine organs) result from tumors of the affected gland. The tumor cells act much in the same way as the normal glandular cells do, that is, they produce the hormones normally made by that gland.

The **gonadotropic** (gon″ah-do-trōp′ik) **hormones** regulate the hormonal activity of the gonads (ovaries and testes). In females **follicle stimulating hormone (FSH)** stimulates graafian (graf′e-an) follicle development in the ovaries. As the follicles mature they produce estrogen, and eggs are readied for ovulation. In males, FSH stimulates sperm development by the testes. **Luteinizing** (lu′te-niz-ing) **hormone (LH)** triggers ovulation of an egg from the female ovary and causes the ruptured follicle to be converted to a **corpus luteum** (lu-te′um). It then stimulates the corpus luteum to produce progesterone and some estrogen. In men, LH is also referred to as **interstitial cell-stimulating hormone (ICSH)** because it stimulates testosterone production by the interstitial cells of the testes. The third gonadotropic hormone is **lactogenic hormone (LTH)** or **prolactin**. The main target organ of LTH is the breast. After childbirth, it stimulates and maintains milk production by the mother's breasts. Its function in the male is not known.

Hyposecretion of FSH or LH leads to sterility in both males and females. In general, hypersecretion does not appear to cause any problems. However, some drugs used to promote fertility appear to stimulate the release of the gonadotropic hormones. Multiple births (indicating multiple ovulations at the same time rather than the usual single ovulation each month) are fairly common after their use.

Adrenocorticotropic (ad-re″no-kor″-te-ko-trop′ik) **hormone (ACTH)** regulates the endocrine activity of the cortex portion of the adrenal gland. **Thyroid-stimulating hormone (TSH)** influences the growth and activity of the thyroid gland.

All of the anterior pituitary hormones just described, except STH, are **tropic hormones.** Tropic hormones stimulate their target organs, also endocrine glands, to secrete their hormones, which in turn exert their effects on other body organs and tissues. The interaction between the anterior pituitary hormones and the endocrine glands they stimulate into action can be compared to the operation of the heating system in your home. As the temperature in the house begins to drop, it is sensed by the thermostat, which then triggers the circuit to fire the furnace into activity. As a result the temperature begins to rise and when it is high enough (as determined by the thermostat setting), the thermostat kicks off, and the furnace is shut down temporarily. Such a system is called a **negative feedback system.** When a tropic hormone is released by the anterior pituitary, it stimulates its target endocrine gland to produce its hormones. For example, TSH causes the thyroid gland to release thyroxine into the blood. When thyroxine blood levels have reached a certain peak, this "feeds back" to the anterior pituitary and blocks or turns off its production of TSH. Then when the thyroid gland is no longer prodded by TSH, it stops releasing thyroxine. When thyroxine blood levels have declined to a certain point, this stimulates the anterior pituitary to again begin releasing TSH, and the whole cycle of increasing and decreasing TSH and thyroxine blood levels is repeated again and again. Negative feedback mechanisms are the major means of regulating blood levels of nearly all hormones.

Pituitary–Hypothalamus Relationship

Despite its insignificant size, the anterior pituitary controls the activity of so many other endocrine glands that it has often been called the master endocrine gland. Its removal or destruction has a dramatic effect on body metabolism. The gonads and adrenal and thyroid glands atrophy, and results of hyposecretion by those glands become obvious. However, the anterior pituitary is not as all powerful in its control as it might appear since the release of each of its hormones is controlled by a *releasing factor* produced by the hypothalamus. The hypothalamus liberates its releasing factors into the blood of the portal circulation, which serves the anterior pituitary. This relationship is shown in Figure 8-2.

In addition, the hypothalamus itself can be called an endocrine organ because it makes two hormones, antidiuretic hormone and oxytocin.

These hormones are transported along the axons of the hypothalmic neurosecretory cells to the posterior pituitary for storage. They are released into the blood in response to nerve impulses from the hypothalamus.

Hormones of the Posterior Pituitary

The posterior pituitary is not really an endocrine gland in a strict sense because it *does not make* the hormones it releases. Instead, as mentioned previously, it simply acts as a reservoir, or storage area, for hormones made by hypothalmic neurons (see Figure 8-2).

Oxytocin (ok"se-to'sin) stimulates powerful contractions of the uterine muscle during birth, sexual relations, and when a woman breast feeds her baby. It also causes milk ejection (the letdown reflex) in a nursing woman. Because oxytocin is so effective in producing uterine contractions, it is often administered during the end stages of labor to hasten the delivery of a baby.

The second hormone released by the posterior pituitary is **antidiuretic** (an"ti-di"u-ret'ik) **hormone (ADH).** (Diuresis means secretion of urine, and anti means against.) ADH causes the kidneys to decrease the amount of urine produced, thus conserving body water. It also increases blood pressure by causing constriction of the arterioles of the body. (For this reason, it is sometimes referred to as **vasopressin** [vas"o-pres'in]). Hyposecretion of ADH leads to dehydration because of excessive urine output, a condition called *diabetes insipidus* (di"ah-be'tēz in-sip'i-dus). Individuals with this problem are continually thirsty and drink huge amounts of water.

THYROID GLAND

The **thyroid gland** is a hormone-producing gland that is familiar to most people primarily because many obese individuals blame their overweight condition on their "glands" (meaning the thyroid). Actually the effect of thyroid hormones on body weight is not as great as many believe it to be.

The thyroid gland is located at the base of the throat, just inferior to the Adam's apple where it is easily palpated during a physical examination. It is a fairly large gland consisting of two lobes joined by a central mass, or **isthmus** (is'mus). The thyroid gland makes two hormones, thyroxine and thyrocalcitonin.

Thyroxine (thi-rok'sin) is often called the body's metabolic hormone, because it controls the rate at which glucose is "burned," or oxidized, and converted to body heat and chemical energy. Since all body cells depend on a continuous supply of chemical energy to power their activities, every cell in the body is thyroxine's target. Thyroxine is important for normal tissue growth and development, especially in the reproductive and nervous systems. Thyroxine is a small peptide molecule bound to iodine. Without iodine, functional hormones cannot be made. The source of iodine is our diet, and the foods richest in iodine are seafoods. Years ago many people who lived in areas of the Midwest, which had iodine-deficient soil and was far from the seashore (and a supply of fresh seafood), developed *goiters* (goi'ters). Thus, that region of the country came to be known as the "goiter belt." A goiter is an enlargement of the thyroid gland, which results when the diet is deficient in iodine. TSH keeps "calling" for thyroxine, and the thyroid gland continues to enlarge so that it can put out more of the thyroxine. However, since there is no iodine only the peptide part of the molecule is produced, which is nonfunctional. The simple goiter is uncommon today because most salt purchased is iodized, and modern transportation systems make seafood available throughout the country.

Hyposecretion of thyroxine may also occur in situations where iodine deficiency is not a problem, such as lack of TSH stimulation. If it occurs in early childhood the result is *cretinism* (kre'tin-izm). Cretinism results in a dwarf. In this type of dwarfism adult body proportions are not normal. Instead the individual has childlike proportions in which the head and trunk are about 1½ times the length of the legs rather than being approximately the same length as in normal adults. Untreated cretins are mentally retarded. Their hair is scanty, and their skin is very dry. If the hyposecretion problem is discovered early, hormone replacement will prevent mental retardation and other signs and symptoms of the deficiency. Hypothyroidism occurring in adults results in *myxedema* (mik"sĕ-de'mah) in which there is both physical and mental sluggishness (however, mental retardation does not occur). Other signs are a puffiness of the face, fatigue, poor muscle tone, low body temperature (the individual is always cold), obesity, and dry skin. Oral thyroxine is prescribed to treat this condition.

Hyperthyroidism generally results from a tumor of the thyroid gland. Extreme overproduction of thyroxine results in a high basal metabolic rate, intolerance of heat, rapid heart rate, weight loss, nervous agitated behavior, and an general inability to relax or slow down. *Grave's disease,* or *exophthalmos* (ek"sof-thal'mos), is one form of hyperthyroidism. In addition to the symptoms of hyperthyroidism given earlier, the thyroid gland enlarges and the eyes begin to bulge, or protrude, anteriorly. There are several avenues of treatment for hyperthyroidism. It may be treated surgically by removal of part of the thyroid (and/or a tumor if present) or chemically by administration of thyroid-blocking drugs or radioactive iodine, which destroys some of the thyroid cells.

The second thyroid gland hormone, **thyrocalcitonin** (thi"ro-kal"sĭ-to'nin) (or simply **calcitonin**), decreases blood calcium levels by causing calcium to be deposited in the bones. It acts antagonistically to parathyroid hormone, the hormone produced by the parathyroid glands. Whereas thyroxine is stored in chambers called follicles before it is released to the blood, calcitonin is made by cells found between the follicles. It is released directly to the blood in response to increasing levels of blood calcium. There are few effects of hypo- or hypersecretion of calcitonin known, but is believed that calcitonin production ceases in elderly adults. This may help to explain (at least in part) the increasing decalcification of bones that occurs in the elderly.

PARATHYROID GLANDS

The **parathyroid glands** are tiny masses of glandular tissue found on the posterior surface of the thyroid gland. Typically, there are two glands on each lobe of the thyroid, that is, a total of four glands. The parathyroid glands secrete **parathormone** (par"ah-thor'mōn) or **parathyroid hormone (PTH),** which is the most important regulator of calcium-ion (Ca^{++}) homeostasis of the blood. When blood calcium levels fall below a certain critical level, the parathyroids release PTH, which stimulates bone destruction cells (osteoclasts) to break down bone matrix and release calcium into the blood. Thus, PTH is a *hypercalcemic* hormone (that is, it acts to increase blood levels of calcium) whereas calcitonin is a *hypocalcemic* hormone.

If blood calcium levels fall too low, neurons become extremely irritable and overactive. They deliver impulses to the muscles at such a rapid rate that the muscles go into spasms (*tetany*), which may be fatal. Before surgeons knew the importance of these tiny glands on the backside of the thyroid, they would remove a hyperthyroid patient's thyroid gland entirely. Many times this resulted in death. Once it was revealed that the parathyroids are functionally very different from the thyroid gland, surgeons began to leave at least some parathyroid-containing thyroid tissue (if at all possible) to take care of blood calcium homeostasis.

Severe hyperparathyroidism causes massive bone destruction to the extent that an X-ray examination of the bones shows large punched out holes in the bony matrix. The bones become very fragile, and spontaneous fractures begin to occur.

ADRENAL GLANDS

As illustrated in Figure 8-1, the two beanshaped **adrenal glands** curve over the top of the kidneys. Although the adrenal gland looks like a single organ, it is actually much like the pituitary gland, because it has glandular (cortex) and neural tissue (medulla) parts. The central medulla region is enclosed by the adrenal cortex, which contains three separate layers of cells.

Hormones of the Cortex

The **adrenal cortex** produces three major groups of steroid hormones collectively called **cortico-**

steroids (kor″tĭ-ko-ste′roids)—the mineralocorticoids, the glucocorticoids, and the sex hormones.

The **mineralocorticoids,** mainly **aldosterone** (al″do-ster′ōn), are produced by the outermost adrenal cortex cell layer. As their name suggests, the mineralocorticoids are important in regulating the mineral (or salt) content of the blood. Their target is the kidney tubules that selectively resorb the minerals or allow them to flush out of the body in the urine. When blood levels of aldosterone rise, the kidney tubule cells resorb increasing amounts of sodium ions and allow potassium ions to go out in the urine. When sodium is resorbed, water follows. Thus, the mineralocorticoids help regulate both water and electrolyte balance in body fluids. The release of aldosterone is stimulated by humoral factors such as fewer sodium ions or increased numbers of potassium ions in the blood (and by ACTH to a lesser degree). A hormone (**renin**) released by the kidneys when blood pressure drops also causes the release of aldosterone.

The middle cortical layer produces the **glucocorticoids,** which include **cortisone, hydrocortisone,** and **cortisol.** Glucocorticoids promote normal cell metabolism and help the body to resist *long-term* stress primarily by increasing blood glucose levels. When blood levels of glucocorticoids are high, fats and even proteins are broken down by body cells and converted to glucose, which is released to the blood. For this reason, glucocorticoids are said to be *hyperglycemic hormones.* Glucocorticoids also seem to control the more unpleasant effects of inflammation by decreasing edema, and they decrease pain by inhibiting some pain-causing molecules called *prostaglandins* (pros″tah-glan′dins). Because of their anti-inflammatory properties, glucocorticoids are often prescribed as drugs for patients with rheumatoid arthritis to suppress inflammation. Glucocorticoids are released from the adrenal cortex in response to increased blood levels of ACTH.

The **sex hormones,** produced by the innermost cortex layer, are chiefly **androgens** (male sex hormones), but some **estrogens** (female sex hormones) are also formed. The sex hormones are produced by the adrenal cortex throughout life in relatively small amounts.

A generalized hyposecretion of all the adrenal cortex hormones leads to *Addison's disease.* A major sign of Addison's disease is a peculiar bronze tone of the skin. Because aldosterone levels are low, sodium and water are lost from the body, which leads to problems with electrolyte and water balance. This, in turn, causes the muscles to become weak, and shock is a possibility. Other signs and symptoms of Addison's disease include those resulting from a lack of glucocorticoids, such as hypoglycemia, a lessened ability to cope with stress (burnout), and an increased susceptibility to infection. Depression of glucocorticoids is incompatible with life.

Hypersecretion problems are generally the result of tumors and the resulting condition depends on the cortex area involved. Hyperactivity of the outermost cortical area results in *hyperaldosteronism* (hi″per-al″do-ster′ōn-izm). Excessive water and sodium are retained leading to high blood pressure and edema, and potassium is lost to such an extent that the activity of the heart and nervous system may be interfered with. When the tumor is in the middle cortical area, *Cushing's syndrome* occurs. Excessive output of glucocorticoids results in a "moon face," and the appearance of a "buffalo hump" of fat on the upper back. Other common (and undesirable) effects include high blood pressure, hyperglycemia (and possible diabetes), weakening of the bones (as protein is withdrawn to be converted to glucose), and a severe depression of the immune system. Hypersecretion of the sex hormones leads to *masculinization* (a condition of increased body hairiness), regardless of sex.

Hormones of the Medulla

The **adrenal medulla,** like the posterior pituitary, develops from nervous tissue. When the medulla is stimulated by sympathetic nervous system neurons, its cells release two similar hormones, **epinephrine** (ep″ĭ-nef′rin) (also called adrenalin) and **norepinephrine** (noradrenalin), into the blood stream. Since the nerve endings of the sympathetic neurons of the sympathetic nervous system release norepinephrine as a neurotransmitter, the adrenal medulla is often thought of as a "misplaced sympathetic nervous system ganglion."

When you are (or feel) threatened physically or emotionally, your sympathetic nervous system brings about the "fight or flight" response to help you cope with the stressful situation. One of the organs it stimulates is the adrenal medulla, which literally pumps its hormones into the bloodstream to enhance and prolong the effects of the neurotransmitters of the sympathetic nervous system. Basically, the effect of the catecholamines (kat"ĕ-kol-am'ins) (a term used to describe epinephrine and norepinephrine) is to increase heart rate, blood pressure, and blood glucose levels, and to dilate the small passageways of the lungs. These events result in more oxygen and glucose in the blood and a faster circulation of blood to the body organs (most importantly, the brain, muscles, and heart). Thus, the body is better able to deal with a short-term stressor whether the job at hand is to run, begin the inflammatory process, or make you more alert so that you can think more clearly.

Although the catecholamines of the adrenal medulla are most important in preparing the body to cope with a brief or short-term stressful situation and cause the so-called *alarm stage* of the stress response, glucocorticoids produced by the adrenal cortex are more important in helping the body to cope with longer-term or continuing stress such as dealing with the death of a family member or having a major operation. Thus, glucocorticoids operate primarily during the *resistance stage* of the stress response. If they are successful in their protection of the body, the problem will eventually be resolved successfully without lasting damage to the body. In cases in which the stress continues on and on, the adrenal cortex may simply "burn out," which is usually fatal. The relationship of catecholamines and glucocorticoids in the stress response is shown in Figure 8-3.

Damage or destruction of the adrenal medulla has no major effects as long as the sympathetic nervous system neurons continue to function normally. However, hypersecretion of catecholamines leads to symptoms typical of excessive sympathetic nervous system activity—that is, a

rapidly beating heart, high blood pressure, and a tendency to perspire and be very irritable. Surgical removal of the adrenal gland easily corrects this condition.

ISLETS OF LANGERHANS OF THE PANCREAS

The **pancreas,** located close to the stomach in the abdominal cavity, is a mixed gland. Probably the best hidden endocrine glands in the body are the **islets of Langerhans** (lahng'er-hanz). These little masses of hormone-producing tissue are scattered among the enzyme-producing tissue of the pancreas. The exocrine (enzyme-producing) part of the pancreas, which acts as part of the digestive system, will be discussed later. Only the islets of Langerhans will be considered here.

Although there are more than a million islets, separated by exocrine cells, each of these tiny clumps of cells goes busily about its business manufacturing its hormones and working like an organ within an organ. Two different hormones are produced by the islet cells, **insulin** and **glucagon** (gloo'kah-gon). Both help to regulate the amount of sugar (glucose) in the blood, but in exactly opposite ways.

High blood glucose levels stimulate the release of insulin from the **beta** (be'tah) **cells** of the islets. Insulin acts on just about all body cells and increases their ability to transport glucose across their cell membranes. Once inside the cells, glucose is oxidized for energy or converted to glycogen or fat for storage. These activities are also speeded up by insulin. Since insulin sweeps the glucose out of the blood, its effect is said to be *hypoglycemic*. Many hormones have hyperglycemic effects (glucagon, glucocorticoids, and epinephrine to name a few), but insulin is the only hormone which decreases blood glucose levels. Insulin is absolutely necessary for the use of glucose by the body cells. Without it, essentially no glucose can get into the cells to be used.

Without insulin, blood levels of glucose (which normally range between 80–120 mg/100 ml of blood) increase to dramatically high levels (for example, 600 mg/100 ml of blood). In such in-

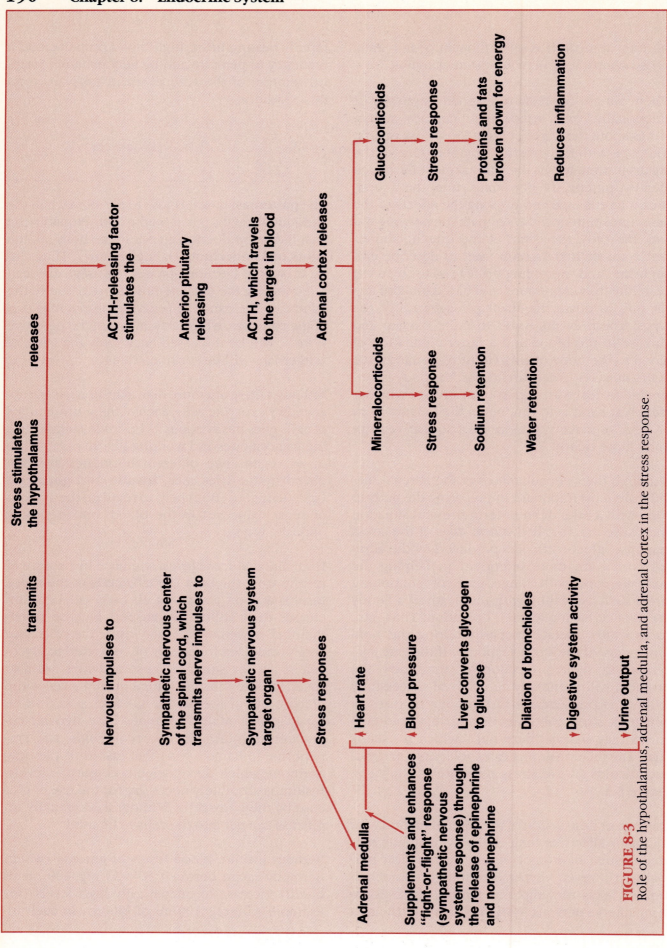

FIGURE 8-3
Role of the hypothalamus, adrenal medulla, and adrenal cortex in the stress response.

stances, glucose begins to spill into the urine because the kidney tubule cells cannot resorb it fast enough. As glucose flushes from the body, water follows leading to dehydration. The clinical name for this condition is *diabetes mellitus* (di"ah-be'tēz mĕ-li'tus), which literally means that something sweet (mel = honey) is passing through or siphoning (diabetes = Greek "siphon") from the body. Since cells cannot use glucose for their energy needs, fats and even proteins start to be broken down and used to meet the energy requirements of the body. As a result, weight loss begins. Loss of body proteins leads to a decreased ability to fight infections. For this reason, diabetics must be careful with their hygiene and in caring for even small cuts and bruises. When large amounts of fats (instead of sugars) are used for energy, the blood becomes very acidic (*acidosis* [as"ĭ-do'sis]) as ketones (intermediate products of fat breakdown) appear in the urine. Unless corrected, coma and death result. The cardinal signs of diabetes mellitus are *polyuria* (pol"e-u're-ah)—excessive urination to flush out the glucose and ketones—*polydipsia* (pol"e-dip'se-ah)—excessive thirst resulting from water loss—and *polyphagia* (pol"e-fa'je-ah)—hunger due to inability to use sugars and the loss of fat and proteins from the body. Mild cases of diabetes mellitus (many cases of adult-onset diabetes) are treated with special diets or oral medications which prod the sluggish islets into action. In severe cases (juvenile or brittle diabetics), injections of insulin must be given periodically through the day to regulate blood glucose levels.

As explained earlier, glucagon acts as an antagonist of insulin. Its release by the **alpha** (al'fah) **cells** of the islets is stimulated by low blood levels of glucose. Its action is basically hyperglycemic. Its primary target organ is the liver, which it stimulates to break down stored glycogen to glucose and release the glucose into the blood. No important disorders resulting from the hypo- or hypersecretion of glucagon are known.

GONADS

Hormones of the Ovaries

The female **gonads** (go'nads), or **ovaries,** are paired, almond-sized organs located in the pelvic cavity. Besides producing female sex cells (ova, or eggs), ovaries produce two steroid hormone groups, estrogens and progesterone. Ovaries do not really begin to function until puberty when the anterior pituitary gonadotropic hormones stimulate them into activity. This results in the rhythmic ovarian cycles in which ova develop and hormonal levels rise and fall.

Estrogens, primarily **estrone** (es'trōn) and **estradiol** (es"trah-di'ol), produced by the **graafian follicles** of the ovaries stimulate the development of the secondary sex characteristics in females (primarily growth and maturation of the reproductive organs and the appearance of hair in the pubic and axillary regions). Estrogens also help maintain pregnancy and prepare the breasts to produce milk (lactation). However, the placenta and not the ovary is the source of the estrogens at this time. In addition, the estrogens work with progesterone to prepare the uterus to receive a fertilized egg. This results in cyclic changes in the uterine lining, which is called the *menstrual cycle.*

Progesterone (pro-jes'te-ron), as already described, acts with estrogen to bring about the menstrual cycle. During pregnancy, it quiets the muscles of the uterus so that an implanted embryo will not be aborted and helps prepare the breast tissue for lactation. Progesterone is produced by another glandular structure of the ovary, the **corpus luteum.** The corpus luteum produces both estrogen and progesterone, but progesterone is secreted in larger amounts.

Ovaries are stimulated to release their estrogens and progesterone in a cyclic way by the anterior pituitary gonadotropic hormones. More detail on this feedback cycle and on the structure and function of the ovary is given in Chapter 15, but it should be obvious at this time that hyposecretion of the ovarian hormones severely hampers the ability of a woman to conceive and bear children.

Hormones of the Testes

The paired oval **testes** of the male are suspended in a sac, the **scrotum,** outside the pelvic cavity. In addition to male sex cells, or **sperm,** the testes also produce the male sex hormone, **testosterone** (tes-tos'tĕ-ron). Testosterone, made by the **interstitial cells** of the testes, causes the development of the adult male sex characteristics. It promotes the growth and maturation of the reproductive system organs to prepare the young man for reproduction. It also causes development of the male's secondary sex characteristics (growth of the beard, development of heavy bones and muscles, and lowering of the voice) as well as stimulating the male sex drive.

In adulthood, testosterone is necessary for continuous production of sperm. In cases of hyposecretion, the man becomes sterile; such cases are usually treated by testosterone injections. Both the endocrine and exocrine functions of the testes begin at puberty under the influence of the anterior pituitary gonadotropic hormones. Testosterone production is specifically stimulated by LH. Chapter 15, which deals with the reproductive system, contains more information on the structure and exocrine function of the testes.

OTHER HORMONE-PRODUCING TISSUES AND ORGANS

Pineal Gland

The **pineal** (pin'e-al) **gland** is a small cone-shaped gland that is found in the roof of the third ventricle of the brain. The endocrine function of this tiny gland is still somewhat of a mystery. Although many chemical substances have been identified in the pineal gland, only the hormone **melatonin** (mel"ah-to'nin) is believed to be secreted in substantial amounts. The levels of melatonin rise and fall during the course of the day and night, and many believe that melatonin plays an important role in establishing the body's day–night cycle. Melatonin also helps regulate mating behavior (and rhythms) in lower animals; and in humans, it is believed to inhibit the reproductive system (especially the ovaries of females) so that sexual maturation is prevented from occurring during childhood, before adult body size has been reached.

Thymus Gland

The **thymus gland** is located in the upper thorax, posterior to the sternum, and has a limited life. It is large in infants and during childhood. By adult-

hood it is no longer functional and has been replaced by fat. Researchers believe that the thymus produces a hormone called **thymosin** (thi'mo-sin) and that during childhood the thymus acts as an "incubator" for the maturation of a special group of white blood cells, *T-lymphocytes*. T-lymphocytes are important in the immune response. They reject foreign grafts or tumors and appear to activate another group of lymphocytes that makes antibodies. Antibodies are important in inactivating bacteria or viruses that enter the body. (The immune system is described in Chapter 11.)

Placenta

The **placenta** (plah-sen'tah) is a remarkable, but temporary, organ that is formed only in pregnant women. In addition to its roles as the respiratory, excretory, and nutrition-delivery systems for the fetus, it also produces hormones from about the third month of pregnancy until the delivery of the baby.

During early pregnancy, a hormone called **human chorionic** (ko"re-on'ik) **gonadotropin (HCG)** is produced by the fetal part of the placenta. HCG is similar to LH. It stimulates the corpus luteum of the ovary to *keep* producing estrogen and progesterone so that the lining of the uterus is not sloughed off in menses. In the third month, the placenta assumes the job of producing estrogen and progesterone, and the ovaries become inactive for the rest of the pregnancy. The high estrogen and progesterone blood levels maintain the lining of the uterus (thus, the pregnancy), and prepare the breasts for producing milk. The placenta is also believed to produce a placental growth hormone and a thyroid-stimulating hormone, but less is known about these hormones.

DEVELOPMENTAL ASPECTS OF THE ENDOCRINE SYSTEM

The embryonic development of the endocrine glands varies. The pituitary gland is an epithelial (epithelium from the oral cavity or mouth) and neural (inferior projection of the hypothalamus) structure. The pineal gland is entirely composed of neural tissue. Most of the strictly epithelial glands appear to develop as little saclike out-pocketings of the epithelial lining of the digestive tract. These would include the thyroid, thymus, and pancreas. The formation of the gonads and the adrenal and parathyroid glands is much more complex; thus, the formation of these glands will not be considered here.

Barring outright malfunctions of the endocrine glands (hypoactivity and hyperactivity), which have already been discussed, most endocrine organs seem to operate smoothly throughout life until old age. In late middle age, the efficiency of the ovaries begins to decline. This causes the onset of *menopause* (commonly called the change of life). During this period, a woman's reproductive organs begin to atrophy, and the ability to bear children ends. Problems commonly associated with estrogen deficiency are expected at this time—such as arteriosclerosis, osteoporosis, decreased skin elasticity, increased fat deposits, and changes in the operation of the sympathetic nervous system that result in "hot flashes," chills, and tingling sensations. In addition, mood changes such as depression, nervousness, and fatigue are common. Although hormone replacement is sometimes prescribed, there is a great deal of controversey over this, and it is never suggested for women with a history of cancer, high blood pressure, or diabetes mellitus. No such dramatic changes seem to happen in men. In fact, many men remain fertile throughout their lifespan indicating that testosterone is still being produced in adequate amounts.

While striking changes occur in aging women, owing to the decreasing levels of female hormones, there are many other much less noticeable hormone-related changes that occur in both sexes. The efficiency of the endocrine system as a whole gradually declines in old age. There is no

question that hormone output by the anterior pituitary decreases. Since the anterior pituitary affects so many other endocrine glands through its tropic hormones, it is assumed that its target organs also become less productive; for example, older people are often mildly hypothyroid. It is common knowledge that elderly persons are less able to resist stress and infection. This may result from overproduction as well as underproduction of the defensive hormones, since both "derail" the stress defense equilibrium and alter general body metabolism as a whole. All older people have some decline in insulin production, and adult-onset diabetes is most common in this age group.

IMPORTANT TERMS*

Target organ

Pituitary *(pĭ-tu′ĭ-tar″e)* **gland**

Releasing factors

Negative feedback

Tropic hormones

Thyroid gland

Parathyroid glands

Adrenal cortex

Adrenal medulla

Islets of Langerhans *(lahng′er-hanz)*

Gonads *(go′nads)*

Thymus gland

Pineal *(pin′e-al)* **gland**

Placenta

---

*For definitions, see Glossary.

SUMMARY

A. THE ENDOCRINE SYSTEM AND HORMONE FUNCTION, AN OVERVIEW

1. The endocrine system is a major controlling system of the body. Through hormones it stimulates such long-term processes as growth and development, metabolism, reproduction, and body defense.

2. Endocrine organs are small and widely separated in the body. Some are mixed glands (both endocrine and exocrine in

function); others are purely hormone-producing.

3. Endocrine organs are activated to release their hormones into the blood by hormonal, humoral, or neural stimuli. Negative feedback is important in regulating hormone levels in the blood.

4. Blood-borne hormones alter the metabolic activities of their target organs. The ability of a target organ to respond to a hormone depends on the presence of receptors in or on its cells to which the hormone binds or attaches.

5. All hormones are protein or steroid in chemical nature.

B. PITUITARY GLAND

1. The pituitary gland hangs from the base of the brain by a stalk and is enclosed by bone. It consists of a glandular portion (anterior pituitary) and a neural portion (posterior pituitary).

2. Hormones of the anterior pituitary are all tropic hormones (except STH). They include:
   a. Somatotropic hormone (STH): Also called growth hormone, STH is an anabolic and protein-conserving hormone that promotes total body growth. Its most important effect is on skeletal muscles and bones. Hyposecretion during childhood results in pituitary dwarfism; hypersecretion produces gigantism (during childhood) and acromegaly (during adulthood).
   b. Follicle-stimulating hormone (FSH): Beginning at puberty, FSH stimulates follicle development and estrogen production by the female ovaries. In the male, it promotes sperm production.
   c. Luteinizing hormone (LH): Beginning at puberty, LH stimulates ovulation, converts the ruptured ovarian follicle to a corpus luteum, and causes the corpus luteum to produce progesterone. LH stimulates the male's testes to produce testosterone.
   d. Lactogenic hormone (LTH), or prolactin: LTH stimulates milk production by the breasts.
   e. Adrenocorticotropic hormone (ACTH): ACTH stimulates the adrenal cortex to release its hormones.
   f. Thyroid-stimulating hormone (TSH): TSH

stimulates the thyroid gland to release thyroxine.

3. Releasing factors made by the hypothalamus regulate the release of hormones made by the anterior pituitary. The hypothalamus also makes two hormones that are transported to the posterior pituitary for storage and later release.

4. The posterior pituitary stores and releases hyothalamic hormones on command. The hormones released are:
   a. Oxytocin: Oxytocin stimulates powerful uterine contractions and causes milk ejection in the nursing woman.
   b. Antidiuretic hormone (ADH): ADH causes the kidney tubule cells to resorb and conserve body water and increases blood pressure by constricting the arterioles. Hyposecretion leads to diabetes insipidus.

C. THYROID GLAND

1. The thyroid gland is located in the anterior throat.

2. Thyroxine is released from the thyroid follicles when blood levels of TSH rise. Thyroxine is the body's metabolic hormone. It increases the rate at which the cells oxidize glucose and it is necessary for normal growth and development. Lack of iodine leads to an enlargement of the thyroid gland called a goiter. Hyposecretion of thyroxine results in cretinism in children and myxedema in adults. Hypersecretion results from Grave's disease or other forms of hyperthyroidism.

3. Thyrocalcitonin is released by the cells surrounding the thyroid follicles in response to high blood levels of calcium. It causes calcium to be deposited in the bones.

D. PARATHYROID GLANDS

1. The parathyroid glands are four small glands located on the posterior aspect of the thyroid gland.

2. Low blood levels of calcium stimulate the parathyroid glands to release parathyroid hormone (PTH). PTH causes bone calcium to be liberated to the blood. Hyposecretion

of PTH results in tetany; hypersecretion leads to extreme bone wasting and fractures.

E. ADRENAL GLANDS

1. The adrenal glands are paired glands perched on the kidneys. Each gland has two functional endocrine portions, the cortex and the medulla.

2. Three groups of hormones are produced by the cortex.
   a. Mineralocorticoids (primarily aldosterone) regulate sodium ion ($Na+$) and potassium ion ($K+$) resorption by the kidneys. Their release is stimulated primarily by low $Na+$/ high $K+$ levels in the blood.
   b. Glucocorticoids enable the body to resist long-term stress by increasing blood glucose levels and by decreasing the inflammatory response.
   c. Sex hormones (mainly male sex hormones) are produced in small amounts throughout life.

3. Generalized hypoactivity of the adrenal cortex results in Addison's disease. Hypersecretion can result in hyperaldosteronism, Cushing's disease, and/or masculinization.

4. The adrenal medulla produces catecholamines (epinephrine and norepinephrine) in response to sympathetic nervous system stimulation. Its catecholamines enhance and prolong the effects of the fight or flight (sympathetic nervous system) response to short-term stress. Hypersecretion leads to symptoms typical of sympathetic nervous system overactivity.

F. ISLETS OF LANGERHANS OF THE PANCREAS

1. The pancreas, located in the abdomen close to the stomach, is both an exocrine and an endocrine gland. The endocrine portion (islets) releases insulin and glucagon to the blood.

2. Insulin is released when blood levels of glucose are high. It increases the rate of glucose uptake and metabolism by the body cells. Hyposecretion of insulin results in diabetes mellitus, which severely disturbs body metabolism. Cardinal signs

include polyuria, polydipsia, and polyphagia.

3. Glucagon is released when blood levels of glucose are low. It stimulates the liver to release glucose to the blood, thus increasing blood glucose levels.

G. GONADS

1. The ovaries of the female, located in the pelvic cavity, release two hormones.
   a. Estrogen: The release of estrogen by the ovarian follicles begins at puberty under the influence of FSH. Estrogen stimulates the maturation of the female reproductive system and the development of the secondary sex characteristics of the female. In cooperation with progesterone, it causes the menstrual cycle.
   b. Progesterone: Progesterone is released from the corpus luteum of the ovary in response to high blood levels of LH. It works with estrogen in establishing the menstrual cycle.

2. The testes of the male begin to produce testosterone at puberty in response to LH stimulation. Testosterone promotes the maturation of the male reproductive organs, the secondary sex characteristics, and the production of sperm by the testes.

3. Hyposecretion of gonadal hormones results in sterility in both females and males.

H. OTHER HORMONE-PRODUCING TISSUES AND ORGANS

1. The pineal gland, located in the third ventricle of the brain, releases melatonin, which affects biological rhythms and reproductive behavior.

2. The thymus gland, located in the upper thorax, functions until the end of adolescence. Its hormone, thymosin, is believed to cause the maturation of T lymphocytes, which are important in body defense.

3. The placenta is a temporary organ formed in pregnant women. Its endocrine function, established by the third month of pregnancy, is to produce estrogen and

progesterone that maintain the pregnancy and ready the breasts for lactation.

I. DEVELOPMENTAL ASPECTS OF THE ENDOCRINE SYSTEM

1. Excluding pathologic excesses and lack of hormones, the efficiency of the endocrine system remains high until old age.

2. Decreasing function of the female ovaries at menopause leads to osteoporosis, an increased chance of heart disease, and possible mood changes.

3. The efficiency of all endocrine glands seems to gradually decrease as aging occurs. This leads to a generalized increase in the incidence of diabetes mellitus, immune system depression, and a lower metabolic rate in the elderly.

REVIEW QUESTIONS

1. The two major controlling systems of the body are the nervous and endocrine systems. Explain how these two systems differ in: (a) the rate of their control, (b) the way in which they communicate with body cells, and (c) the types of body processes they control.

2. There are two general types of glands in the body. Explain how the endocrine and exocrine glands differ in their products and in the way their products reach their final destination.

3. Which endocrine organs are actually mixed (endocrine and exocrine) glands? Which are purely endocrine?

4. Define hormone.

5. Define negative feedback and explain how it regulates blood levels of the various hormones.

6. Define target organ and then explain why all organs are not target organs for all hormones.

7. Name three ways in which endocrine glands are stimulated to release their hormones and give one hormone example for each way.

8. Describe the body location for each of the following endocrine organs: anterior pituitary, pineal gland, thymus, pancreas, ovaries, and testes. Then, for each organ name its hormones and the effect(s) they have on body processes. Finally, for each hormone list the important results of its hypersecretion or hyposecretion.

9. Name two endocrine-producing glands (or regions) that are important in the stress response and explain *why* they are important.

10. The anterior pituitary is often referred to as the master endocrine gland, but it too has a "master." What controls the release of hormones by the anterior pituitary?

11. What are tropic hormones?

12. The posterior pituitary is not really an endocrine gland. Why not? What is it?

13. What is the most common cause of hypersecretion by endocrine organs?

14. Name three hormone antagonists of insulin and one of PTH.

15. Salt (or electrolyte) balance and water balance are very important. Two hormones are closely involved in the regulation of the fluid and electrolyte balance of the body. Name them and then explain their effects on their common target organ.

16. A simple goiter is not really the result of a malfunction of the thyroid gland. What does cause it?

17. In general, the endocrine system becomes less efficient as we age. List some examples of problems that elderly individuals have as a result of decreasing hormone production.

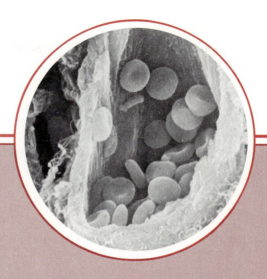

CHAPTER 9

Blood

Chapter Contents

Functions of blood • to serve as a vehicle for transporting nutrients, respiratory gases, and other substances throughout the body, and to distribute body heat

After completing this chapter, you should be able to:

- Describe the composition of whole blood.

- Describe the composition of plasma and discuss its importance in the body.

- List the cell types composing the formed elements and describe their major functions.

- Explain what type of information is revealed by the following blood tests: hematocrit, hemoglobin determination, and differential white blood cell count.

- Describe briefly the blood-clotting process.

- Explain the basis of a transfusion reaction.

- Define anemia, polycythemia, leukopenia, and leukocytosis.

Blood is the life-sustaining "river of life" that flows within our bodies. It transports everything that must be carried from one place to another within the body—nutrients, wastes (headed for final elimination from the body), and body heat—through blood vessels. For centuries, long before modern medicine, man recognized that blood was vital (some believed "magical"), and its loss was always considered to be a possible cause of death. In this chapter the composition and function of blood is considered. The means by which it is propelled throughout the body is discussed in Chapter 10.

BLOOD: COMPOSITION AND FUNCTION

Blood is a thick, or viscous, substance composed of a fluid portion (plasma) in which living cells (formed elements) are suspended. Its color varies from bright scarlet to a dull brick red, depending on the amount of oxygen it is carrying. The cardiovascular system of the average adult contains about 5.5 L, or approximately 6 quarts, of blood.

Plasma

Plasma (plaz'mah), which is approximately 90% water, is the liquid part of the blood. Over 100 different substances are dissolved in the plasma; examples are nutrients (fats, amino acids, and glucose), gases (carbon dioxide, oxygen, and nitrogen), hormones, various wastes and products of cell activity (lactic acid, urea, and uric acid), proteins (albumin, antibodies, and clotting proteins), and salts. Composition of plasma varies continuously as cells remove or add substances to the blood. However, assuming a normal healthy diet, the composition of plasma is kept relatively constant by various homeostatic mechanisms of the body. For example, when blood protein levels drop to undesirable levels, the liver is stimulated to make more blood proteins; when the blood starts to become too acid *(acidosis)* or too basic *(alkalosis),* both the respiratory system and kidneys are called into action to restore it to its normal slightly alkaline pH range of 7.35–7.45. There are literally dozens of adjustments that various body organs make, day-in and day-out, to maintain the many plasma solutes at life-sustaining levels. In addition to transporting various substances around the body, plasma has another important role—it helps distribute body heat evenly through the body tissues.

Formed Elements

Three types of formed elements, or blood **corpuscles,** (kor'pus'ls) are present in the blood. The most numerous are erythrocytes (ĕ-rith'ro-sīts), which are oxygen-carriers. Leukocytes (lu'ko-sīts), are part of the body's immune system. Thrombocytes (throm'bo-sīts), or platelets (plāt'lets) help repair leaks in the blood vessels. The various types of formed elements are shown in Figure 9-1, and their characteristics are summarized in Table 9-1 (p. 200–201). Formed elements make up 40%–45% of whole blood whereas plasma accounts for the remaining 55%–60%.

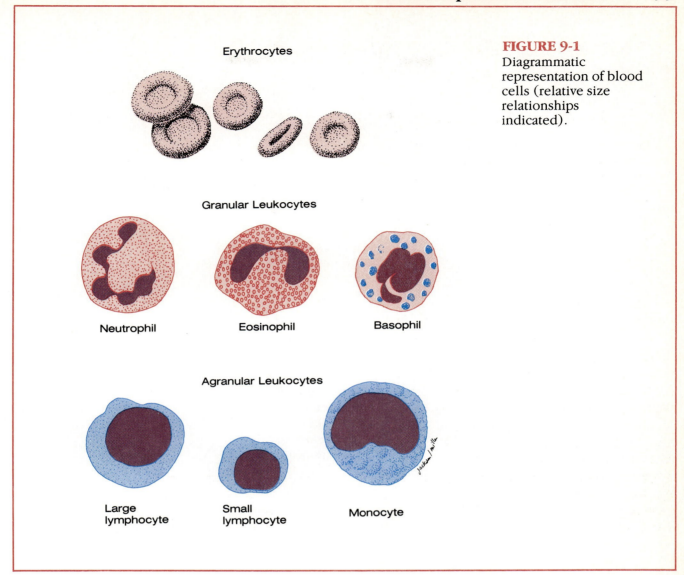

Erythrocytes

Granular Leukocytes

Neutrophil   Eosinophil   Basophil

Agranular Leukocytes

Large lymphocyte   Small lymphocyte   Monocyte

FIGURE 9-1
Diagrammatic representation of blood cells (relative size relationships indicated).

ERYTHROCYTES

Erythrocytes, or red blood cells (RBCs), are small cells with a distinct biconcave disk shape. RBCs differ from other blood cells because they are anucleate (a-nu′kle-āt); that is, they lack a nucleus. When mature and circulating in the blood, RBCs are literally sacs of hemoglobin molecules. **Hemoglobin** (he″mo-glo′bin) (Hb) transports the bulk of the oxygen that is carried in the blood. (It also binds with a small amount of carbon dioxide.) Because they are anucleate, RBCs are unable to divide and have a limited life span of 100–120 days. After that time, they begin to fragment, or fall apart, and their remains are eliminated by the spleen, liver, and other body tissues. Lost cells are replaced more-or-less continuously by the division of cells called **hemocytoblasts** (he″mo-

si′to-blasts), located in the red bone marrow (Figure 9-2, p. 201). The rate of RBC production, *hemopoiesis* (he″mo-poi-e′sis), is controlled by a hormone, *erythropoietin* (ĕ-rith″ro-poi′ĕ-tin), which is released by the kidneys in response to decreasing oxygen levels in the blood.

RBCs outnumber white blood cells by about 1000 to 1 and are the major factor contributing to blood viscosity. Although the numbers of RBCs in the circulation do vary, there are normally 4.5–5.5 million cells per cubic millimeter of blood. (A cubic millimeter [mm^3] is a very tiny drop of blood, almost not enough to be seen.) When the number of RBC/mm^3 increases, blood viscosity increases. Similarly, as the number of RBCs decreases, blood thins and flows more rapidly.

TABLE 9-1 Characteristics of Formed Elements of the Bood

Cell type	Occurrence in blood (per mm³)	Cell anatomy	Function
Erythrocytes (RBCs)	4.5–5.5 million	Biconcave disks; anucleate; literally, a sac of hemoglobin molecules; most organelles have been ejected	Transport oxygen bound to hemoglobin molecules; also transport a small amount carbon dioxide
Leukocytes (WBCs) Granulocytes	5000–10,000		
Neutrophils	3000–7000 (55%–65% of WBCs)	Cytoplasm stains pink to blue with Wright's stain and contains fine granules, which are difficult to see; nucleus consists of three to seven lobes connected by thin strands of nucleoplasm and stains deep purple	Active phagocytes; their number increases rapidly during short-term or acute infections
Eosinophils	100–400 (1%–3% of WBCs)	Large cytoplasmic granules, which stain red with Wright's stain; figure-8 or bilobed deep staining nucleus	Exact function not known; increase during allergy attacks; might phagocytize antigen–antibody complexes
Basophils	0–50 (0.5% of WBCs)	Cytoplasm has a few large granules, which stain deep purple with Wright's stain, large U- or S-shaped nucleus with constrictions, which stains dark blue	Granules believed to contain histamine (a vasodilator chemical) and heparin (an anticoagulant), which are discharged on exposure to foreign substances (bacteria, viruses, and toxins)
Agranulocytes			
Lymphocytes	1000–3000 (20%–35% of WBCs)	Cytoplasm stains pale blue and appears as a thin rim around the nucleus; dark purple nucleus that is spherical or slightly indented	Concerned with body immunity; one group (B lymphocytes) produces antibodies; the other group (T lymphocytes) is involved in graft rejection and in fighting tumors and viruses; activates B lymphocytes

TABLE 9-1   Characteristics of Formed Elements of the Bood (continued)

Cell type	Occurrence in blood (per mm³)	Cell anatomy	Function
Agranulocytes *(continued)*			
Monocytes	100–500 (3%–8% of WBCs)	Abundant cytoplasm, which stains gray-blue with Wright's stain; dark blue-purple nucleus is often kidney-shaped	Active phagocytes—the long-term "clean-up team"; increase in number during long-term or chronic infections such as tuberculosis
Thrombocytes (platelets)	250,000– 400,000	Essentially cell fragments, therefore are irregularly shaped bodies; stain deep purple	Needed for normal clotting of the blood; initiate the clotting cascade by clinging to the broken area; help to control blood loss from broken blood vessels

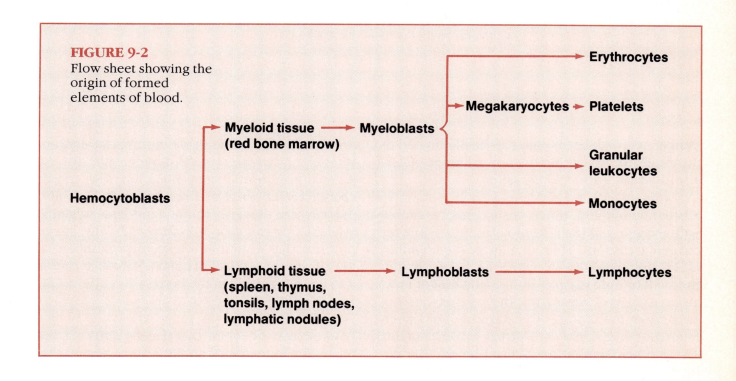

FIGURE 9-2
Flow sheet showing the origin of formed elements of blood.

The **hematocrit** (he-mat′o-krit) (Hct), or packed cell volume, is a blood test routinely done to determine if an individual has the normal amount of RBCs. Spinning whole blood in a centrifuge causes the formed elements to collect at the bottom of the tube, with plasma forming the top layer. Since the bulk of formed elements are RBCs, the Hct is usually considered to be equal to the red blood cell volume. Normal Hct values range between 42%–47% of the volume of blood in the

TABLE 9-2   Types of Anemias

Direct cause	Resulting from	Leading to
Decrease in RBC number	Sudden hemorrhage	Hemorrhagic anemia
	Lysis of RBCs—due to bacterial infections	Hemolytic (he″mo-lit′ik) anemia
	Lack of Vitamin B_{12}	Pernicious (per-nish′us) anemia
	Depression/destruction of bone marrow by cancer, radiation, or certain medications	Aplastic anemia
Decrease in hemoglobin content or abnormal hemoglobin in RBCs	Lack of iron in diet or slow/prolonged bleeding (such as results from heavy menstrual flow or a bleeding ulcer), which depletes the iron reserves needed to make hemoglobin; RBCs are small and pale because they lack hemoglobin	Iron-deficiency anemia
	Genetic defect leads to an abnormal hemoglobin, which becomes sharp and sickle-shaped under conditions of increased oxygen use by the body; occurs mainly in members of the black race.	Sickle cell anemia

sample, with values for males at the higher end of the range. An Hct within the normal range generally indicates a normal RBC value.

Since hemoglobin binds easily and reversibly with oxygen, it is the RBC protein responsible for oxygen transport. Thus, the more hemoglobin molecules the RBCs contain, the more oxygen they will be able to carry. Perhaps the best or most accurate way of measuring the oxygen-carrying capacity of the blood is to determine how much hemoglobin it contains. Normal blood contains 12–18 g hemoglobin per 100 ml blood. The hemoglobin content is slightly higher in men (14–18 g) than in women (12–14 g). A decrease in the oxygen-carrying ability of the blood (whatever the reason) is *anemia* (ah-ne′me-ah). Anemia may be the result of: (a) a lower than normal *number* of RBCs or, (b) a lower than normal hemoglobin content in the RBCs. Several types of anemia are classified and described briefly in Table 9-2.

An abnormal increase in the number of erythrocytes is *polycythemia* (pol″e-si-the′me-ah). Polycythemia may result from bone marrow cancer *(polycythemia vera)* or from living at high altitudes where the air is thinner and less oxygen is available *(secondary polycythemia)*. The major problem that results from excessive numbers of RBCs is increased viscosity of blood, which causes it to flow sluggishly in the body.

LEUKOCYTES

Leukocytes, or white blood cells (WBCs), are much less numerous than RBCs and average from 5000–10,000 cells mm^3. WBCs are more like most body cells than RBCs are—that is, they contain nuclei (see Figure 9-1).

Basically WBCs are the blood's protective "movable army," which helps protect the body from damage by bacteria, viruses, and tumor cells. WBCs are carried to all parts of the body in the blood (or lymph), and some WBCs are able to slip into and out of the blood vessels—a process called *diapedesis* (di"ah-pĕ-de'sis). WBCs have the ability to locate areas of tissue damage and infection in the body by responding to certain chemicals that diffuse from the damaged cells. Once they have "caught the scent," the WBCs then move through the tissue spaces by *ameboid* (ah-me'boid) *motion.* By following the diffusion gradient they are able to pinpoint areas of tissue damage and rally round in large numbers to destroy foreign substances or dead cells. Whenever WBCs mobilize for action, the body speeds up their production, and as many as double the normal number of WBCs may appear in the blood within a few hours. A total WBC count over 10,000 cells/mm³ is referred to as *leukocytosis* (lu"ko-si-to'sis). Leukocytosis generally indicates that bacterial or viral infection is stewing in the body. *Leukemia* (lu-ke'me-ah) (literally, white blood) is a situation in which the bone marrow becomes cancerous, and huge numbers of WBCs are turned out rapidly. While this might not appear to present a problem, the WBCs are immature and, thus, incapable of carrying out their normal protective functions. Generally when a total WBC count is done, a *differential WBC count* is also ordered at the same time. The differential count provides information on the *relative* numbers of each type of WBC. This information may be helpful in diagnosing the disease. For example, an excessively high number of monocytes may indicate mononucleosis or some long-term infection in the body.

The WBCs are classified into two major groups, depending on whether or not they contain visible granules in their cytoplasm.

Granulocytes (gran'u-lo-sīts") are granule-containing WBCs. They are formed in the red bone marrow from the same stem cell (hemocytoblast) as the RBCs. They have lobed nuclei, which often consist of rounded nuclear areas connected by thin strands of nuclear material. The granules in their cytoplasm stain specifically with Wright's stain. The granulocytes include the **neutrophils** (nu'tro-fils), **eosinophils** (e"o-sin'o-fils), and **basophils** (ba'so-fils).

The second group, **agranulocytes,** lack visible cytoplasmic granules. They also arise from hemocytoblasts but migrate to lymphatic tissues. Their nuclei are closer to the norm—that is, they are spherical, oval, or kidney-shaped. The agranulocytes include **lymphocytes** (lim'fo-sīts) and **monocytes** (mon'o-sīts). The specific characteristics of the leukocytes are listed in Table 9-1.

THROMBOCYTES

Thrombocytes, or **platelets,** are believed to be fragments of large multinucleate cells called **megakaryocytes** (meg"ah-kar'e-o-sīts), which are formed in the bone marrow. They are darkly staining, irregularly shaped bodies scattered among the other blood cells. The normal platelet count in blood is 300,000 mm³. Platelets are needed for the clotting process that occurs in plasma when blood vessels are ruptured or broken.

BLOOD CLOTTING

Normally, blood flows smoothly past the intact endothelium of the blood vessel walls, but if the blood vessel wall is damaged, a whole series of reactions called the *blood clotting cascade* is set into motion (Figure 9-3). Blood clotting, or coagulation, reduces blood loss when blood vessels are ruptured. This process involves many substances that are normally present in the plasma (clotting factors) as well as some that are released by platelets and injured tissues.

Basically clotting occurs as follows:

1. The platelets cling to the damaged site and pile up to form a small mass of platelets called a white thrombus, or platelet plug.

2. Once anchored, the platelets then release **serotonin** (ser"o-to'nin), which causes that blood vessel to go into spasms. The spasms narrow the blood vessel at that point; this decreases blood loss until clotting can occur.

3. At the same time, the injured tissues and the platelets are releasing **thromboplastin** (throm"bo-plas'tin), which starts or triggers the clotting cascade.

4. Thromboplastin interacts with other blood protein clotting factors (for example, Christmas factor V) and calcium ions (Ca^{++}) to convert **prothrombin** (pro-throm'bin) (present in the plasma) to **thrombin.**

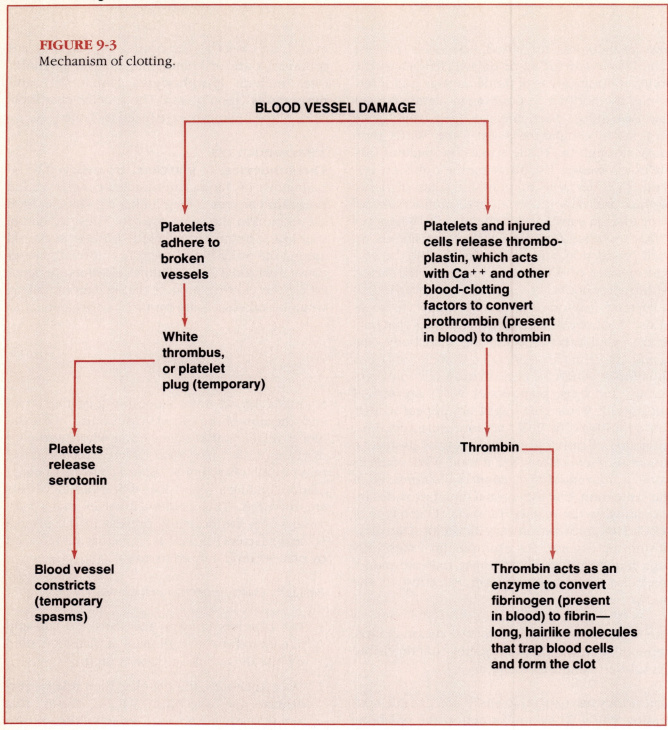

FIGURE 9-3
Mechanism of clotting.

BLOOD VESSEL DAMAGE

Platelets adhere to broken vessels

White thrombus, or platelet plug (temporary)

Platelets release serotonin

Blood vessel constricts (temporary spasms)

Platelets and injured cells release thromboplastin, which acts with Ca^{++} and other blood-clotting factors to convert prothrombin (present in blood) to thrombin

Thrombin

Thrombin acts as an enzyme to convert fibrinogen (present in blood) to fibrin— long, hairlike molecules that trap blood cells and form the clot

5. Thrombin then joins soluble **fibrinogen** (fi-brin'o-jen) proteins into long hairlike molecules of insoluble **fibrin,** which forms a meshwork of strands that traps the RBCs and forms the basis of the clot (Figure 9-4).

Normally blood clots within 2–6 minutes. As a rule, once the clotting cascade has started, the triggering factors are rapidly inactivated to prevent widespread clotting ("solid blood").

A lack of *any* of the factors needed for clotting can lead to a decreased ability to form clots, a condition called *hemophilia* (he"mo-fil'e-ah), or bleeder's disease. When a bleeding problem occurs, people with hemophilia are given blood transfu-

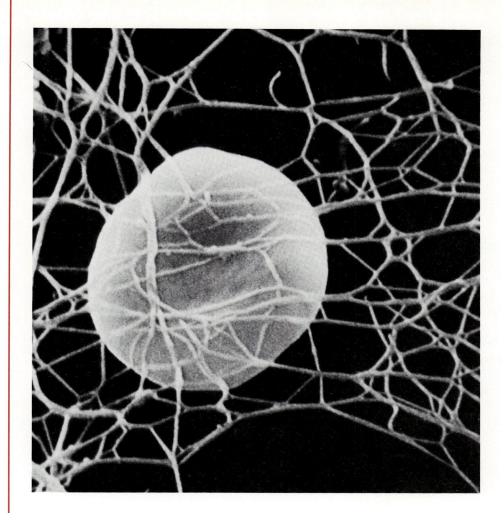

FIGURE 9-4
Photomicrograph of a RBC trapped in a fibrin mesh.

sions (which would contain all the factors) or receive injections of the specific factors they lack.

In some individuals, abnormal clotting in intact, or unbroken, blood vessels is a problem, particularly in the legs. This may result when the interior of the blood vessel is roughened by trauma or an accumulation of fatty material in the vessel. Any change in the smoothness (or electrical nature) of the blood vessel encourages platelets to cling to the area, forming a **thrombus** (throm′bus), or clot. Undesirable thrombus formation is itself a problem because it may block the blood vessel. But if the thrombus detaches and becomes a free-floating clot (**embolus** [em′bo-lus]), the situation can rapidly become more serious particularly if the clot lodges in the blood vessels of the lungs or heart.

BLOOD GROUPS AND TRANSFUSIONS

The different blood groups are classified on the basis of specific proteins that are present on the outer surface of the RBC plasma membrane. These proteins—**antigens,** or **agglutinogens** (ag″loo-tin′o-jens)—are genetically determined. In many cases, these antigens are accompanied by other proteins found in the plasma—**antibodies,** or **agglutinins** (ah-gloo′ti-nins). During blood transfusions, these antibodies attach to RBCs bearing antigens different from those on the patient's RBCs. Binding of the antibodies causes the RBCs to become clumped and eventually ruptured *(hemolyzed)*. Because of this reaction, a person's blood must be carefully typed before he or

BOX 9-1
Inflammation: The Body's Second Line of Defense

Even though the body is continuously assaulted by many potentially damaging agents, it has a remarkable ability to defend itself. Its primary line of defense is the mechanical barrier that covers the exterior (the skin) and lines cavities open to the exterior (the mucous membranes). As long as these barriers are intact, they are usually able to guard against the entry of bacteria (and other agents). But when harmful agents do succeed in invading the body, other defense mechanisms—the inflammatory response and the immune response—are activated in an attempt to prevent them from seriously damaging the body.

The inflammatory response is a nonspecific response that is triggered whenever body tissues are injured. It occurs in response to trauma, intense heat, and irritating chemicals as well as to infection by viruses and bacteria. The major symptoms of an inflammation are *redness, heat, swelling,* and *pain.* It is easy to understand why these symptoms occur if the events of the inflammatory response are understood. When cells are injured, they release various chemical substances including histamine and kinins that (a) cause blood vessels in the involved area to dilate and capillaries to become leaky, (b) activate pain receptors, and (c) attract phagocytes and white blood cells to the area. Dilation of the blood vessels increases the blood flow to the area, accounting for the redness and heat observed. Increased permeability of the capillaries allows plasma to leak from the bloodstream into the tissue spaces, causing local edema (swelling) that also activates pain receptors in the area.

Within an hour or so after the inflammation process has begun, many neutrophils squeeze through the capillary walls to enter the area and begin the cleanup detail by engulfing damaged/ dead tissue cells and/or pathogens. The neutrophil invasion is followed several hours later by the appearance of monocytes in the area. They continue to wage the battle, replacing the shortlived neutrophils in the battlefield. Besides phagocytosis, other protective events are also occurring at the inflammed site. Clotting proteins, leaked into the area from the blood, are activated and begin to wall off the damaged area with a meshwork of fibrin to prevent the invasion of pathogens or injurious agents to neighboring tissues.

If the area contains pathogens that have previously invaded the body, the third line of defense, the immune response mediated by lymphocytes, also occurs. Both antibodies (formed by B lymphocytes) and T lymphocytes (which directly act against the pathogen) invade the area to act in a specific manner against the damaging agents.

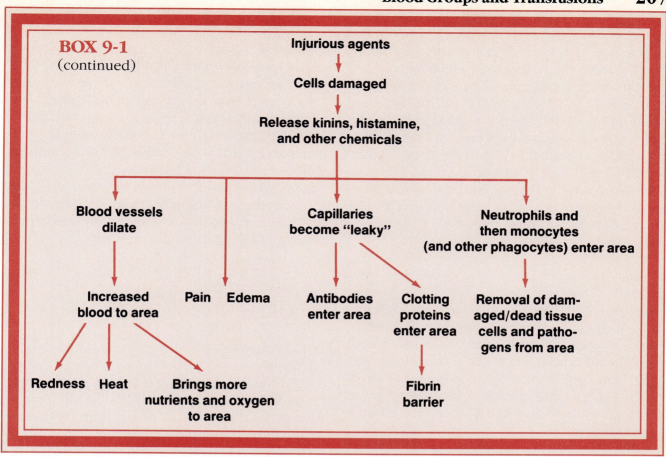

BOX 9-1
(continued)

Injurious agents
↓
Cells damaged
↓
Release kinins, histamine, and other chemicals

Blood vessels dilate → Increased blood to area → Redness / Heat / Brings more nutrients and oxygen to area

Pain   Edema

Capillaries become "leaky" → Antibodies enter area / Clotting proteins enter area → Fibrin barrier

Neutrophils and then monocytes (and other phagocytes) enter area → Removal of damaged/dead tissue cells and pathogens from area

ABO blood type	Antigens present on RBC membranes	Antibodies present in plasma	Blood that can be received
A	A	Anti-B	A and O
B	B	Anti-A	B and O
AB (universal recipient)	A and B	None	A, B, AB, and O
O (universal donor)	Neither	Anti-A and anti-B	O

FIGURE 9-5
Basis of ABO blood types and blood that can be received by each.

she is given a whole blood or packed cell transfusion.

Blood Typing

There are several blood-typing systems based on the many possible antigens, but the factors most often typed for are the antigens of the ABO and Rh blood groups, which are usually involved in transfusion reactions. Other blood factors, such as Kell, Lewis, and N, are not routinely typed for unless the individual is expected to need many transfusions.

The basis of the ABO typing is shown in Figure 9-5. The two antigens of this system are called A and

B. Absence of both antigens results in type O blood, presence of both antigens leads to type AB, and the possession of either A or B antigen results in type A or B blood respectively. In the ABO blood group, antibodies are formed early in life against the ABO antigens *not* present on your own RBCs. Nearly one-half (45%) of the United States population is type O. The second most common ABO blood type is type A, which is seen in approximately 40%; the least common is type AB, which is present in only about 3% of the populace.

In situations in which a person receives mismatched or incompatible blood, his or her plasma antibodies will vigorously attack the foreign RBCs, leading to a *transfusion reaction*. The initial event, clumping of RBCs, may clog small blood vessels, while the lysis of RBCs that follows releases hemoglobin into the blood stream. If large amounts of hemoglobin molecules are free in the blood, they may block the kidney tubules and cause kidney failure.

The Rh blood–typing system is so named because the Rh antigen (agglutinogen D) was originally identified in *Rh*esus monkeys; later it was discovered that human beings carry the same antigen. Most Americans are Rh$^+$, meaning that their RBCs carry the Rh antigen. However, unlike what is seen in the ABO system, anti-Rh antibodies are *not* automatically formed in the blood of Rh$^-$ individuals. However, if an Rh$^-$ person receives mismatched blood (that is, Rh$^+$), his or her immune system becomes sensitized and begins producing antibodies (anti-Rh$^+$ antibodies) against the foreign blood type shortly after the transfusion. Hemolysis does not occur with the first transfusion because it takes time for the body to react and start making antibodies. But the second time and every time thereafter, a typical transfusion reaction occurs in which the patient's antibodies attack and rupture the donor's RBCs.

An important Rh-related problem occurs in pregnant Rh$^-$ women whose children are fathered by Rh$^+$ men and who are carrying Rh$^+$ babies. The first such pregnancy usually presents no major problems and results in the delivery of a healthy baby. But since the mother is sensitized by Rh$^+$ antigens that passed through the placenta into her bloodstream, she will form anti-Rh$^+$ antibodies unless treated with RhoGAM (which prevents this sensitization and prevents her immune response) shortly after the birth of her baby. If she is not treated and becomes pregnant again with a Rh$^+$ baby, her antibodies will cross through the placenta and destroy the baby's RBCs, a condition called *erythroblastosis fetalis* (e-rith″-ro-blas-to′sis fe-ta′lis). The baby is anemic and becomes hypoxic. Brain damage and even death may result unless fetal transfusions are done *before* birth to provide more RBCs for oxygen transport.

DEVELOPMENTAL ASPECTS OF BLOOD

In the young embryo, development of the entire circulatory system occurs early. Generally, embryonic blood cells are being circulated in the newly formed blood vessels by day 28 of development. Fetal hemoglobin (HbF) differs from the hemoglobin formed after birth. It has a greater ability to pick up oxygen, a characteristic that is highly desirable since fetal blood is less oxygen rich than that of the mother. After birth fetal blood cells are gradually replaced by RBCs that contain the more typical hemoglobin A. In situations in which the fetal RBCs are destroyed at such a rapid rate that the immature infant liver cannot keep pace with the need to rid the body of hemoglobin breakdown products in the bile, the infant becomes *jaundiced* (jawn′dis'd). This type of jaundice is generally not serious and is referred to as physiologic jaundice to distinguish it from various disease conditions that result in jaundiced or yellowed tissues.

Various congenital diseases result from genetic factors (such as hemophilia and sickle cell anemia) and from interactions with maternal blood factors (such as erythroblastosis fetalis). Since all of these conditions have already been discussed, they will not be reconsidered here.

As indicated earlier, dietary factors can lead to abnormalities in blood cell formation as well as hemoglobin production. Women are particularly at risk for iron-deficiency anemia because of their monthly blood loss during menses. Increased production of the protective leukocytes typically follows bacterial or viral attacks on the body and is considered to be a normal homeostatic response. Conversely, excessive production of abnormal leukocytes, as seen in the various leukemias, is distinctly pathologic. Young and elderly individuals are particularly at risk for leukemia.

With increasing age, the tendency to form thrombi becomes a problem. This increasing incidence of undesirable clot formation reflects the gradual and progressive blood vessel damage that accompanies the onset of arteriosclerosis.

IMPORTANT TERMS*

Plasma *(plaz′ mah)*

Formed elements

Erythrocytes *(ĕ-rith′ ro-sīts)*

Hemoglobin *(he″mo-glo′ bin)*

Leukocytes *(lu′ ko-sīts)*

Neutrophils *(nu′ tro-fils)*

Eosinophils *(e″o-sin′ o-fils)*

Basophils *(ba′ so-fils)*

Lymphocytes *(lim′ fo-sīts)*

Monocytes *(mon′ o-sīts)*

Platelets *(plāt′ lets)*

Coagulation

Prothrombin *(pro-throm′ bin)*

Fibrin

Agglutinogens *(ag″loo-tin′ o-jens)*

Agglutinins *(ah-gloo′ ti-nins)*

---

*For definitions, see Glossary.

SUMMARY

A. COMPOSITION AND FUNCTION

1. Blood is composed of a nonliving fluid matrix (plasma) and formed elements. It is scarlet to dull brick red, depending on the amount of oxygen it is carrying.

2. Dissolved in plasma, which is primarily water, are nutrients, gases, hormones, wastes, proteins, salts, and so on. Its composition changes as body cells remove or add substances to it, but homeostatic mechanisms act to keep it relatively constant. Plasma makes up 55% of whole blood.

3. Formed elements are the living blood cells and make up 40%–45% of whole blood. They include:
   a. Erythrocytes, or RBCs, are disklike anucleate cells that transport oxygen bound to their hemoglobin molecules. Their life span is 100–120 days.
   b. Leukocytes, or WBCs, are ameboid cells involved in the protection of the body.
   c. Thrombocytes, or platelets, are cell fragments that act in blood clotting.

4. A decrease in the oxygen-carrying ability of the blood is anemia. It may result from a decrease in the number of functional RBCs or a decrease in the amount of hemoglobin they contain.

5. Leukocytes are nucleated cells that are classed into two groups.
   a. The granulocytes include neutrophils, eosinophils, and basophils.
   b. The agranulocytes include monocytes and lymphocytes.

6. When bacteria, viruses, or other foreign substances invade the body, WBCs mobilize to fight them in various ways.

WBCs are attracted to various sites of inflammation by chemicals that diffuse from injured cells.

B. BLOOD CLOTTING

1. Blood clotting is started by a tear or roughening of the blood vessel lining. Platelets adhere to the damaged site and initiate the clotting cascade. Platelets and injured tissue cells release chemicals that cause vasoconstriction and the formation of fibrin threads. Fibrin traps RBCs as they flow past, forming the clot. An attached clot is a thrombus; a clot traveling in the blood stream is an embolus.

C. BLOOD GROUPS AND TRANSFUSIONS

1. Blood groups are classified on the basis of proteins (antigens) on the RBC membranes. Complementary antibodies may (or may not) be present in the blood. Antibodies act to lyse foreign RBCs.

2. The blood group most commonly typed for is the ABO grouping. Type O is the most common; the least common is type AB. The ABO antigens are accompanied by preformed antibodies in the plasma.

3. The Rh factor is found in most Americans. Rh$^-$ people do not have preformed antibodies to Rh$^+$ RBCs, but form them once exposed to the foreign blood.

D. DEVELOPMENTAL ASPECTS OF BLOOD

1. Congenital blood defects include various types of hemolytic anemias and bleeder's disease. An incompatibility between maternal and fetal blood can result in fetal cyanosis, resulting from the destruction of fetal blood cells.

2. Fetal hemoglobin bonds more readily with oxygen than does HbA.

3. Lack of important dietary factors (such as iron and B vitamins) may cause anemia.

4. Leukocytosis is a normal protective response to infection. Excessive leukocytosis may be indicative of malignancy of the blood-forming organs.

REVIEW QUESTIONS

1. What is the blood volume of an average-size adult?

2. What determines whether blood is bright red or a dull brick red in color?

3. Name as many different categories of substances carried in blood as you can.

4. Define formed elements.

5. What is the average life span of a RBC? How does the fact that it has no nucleus affect its life span?

6. Name the granular and agranular WBCs. Give the major function *each* type has in the body.

7. If you had a severe infection, would you expect your total WBC count to be closest to 5000, 10,000, or 15,000/mm^3? Why? What is this condition called?

8. What is anemia? Give three possible causes of anemia.

9. If you had a high hematocrit, would you expect your hemoglobin determination to be high or low? Why?

10. Describe the process of blood clotting. Indicate *what* starts the process.

11. What is the *basis* of blood groups? What are agglutinins?

12. Name the four ABO blood groups. Which is most common? Which is least common?

13. What is a transfusion reaction, and why does it happen?

14. Explain why an Rh$^-$ person does not have a transfusion reaction on the *first* exposure to Rh$^+$ blood. Why is there a transfusion reaction the second time he receives the Rh$^+$ blood?

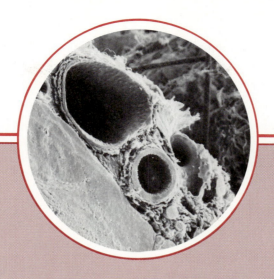

Circulatory System

Chapter Contents

Function of the heart • to pump blood

Function of the blood vessels • to provide the conduits within which blood circulates to all body tissues

Function of the lymphatic system • to return leaked plasma to the blood vessels after removing bacteria and other foreign matter from it

The above electron micrograph from *Tissues and Organs: A Text-Atlas of Scanning Electron Microscopy* by Richard G. Kessel and Randy H. Kardon. W. H. Freeman and Co. Copyright © 1979.

After completing this chapter, you should be able to:

- Describe the location of the heart in the body and identify its major anatomical areas when given an appropriate model or diagram.

- Trace the pathway of blood through the heart.

- Compare the pulmonary and systemic circuits.

- Name the functional blood supply of the heart.

- Define systole, diastole, cardiac cycle.

- Explain the operation of the semilunar and atrioventricular valves during the cardiac cycle.

- Define murmur.

- Name the elements of the Purkinje system of the heart and describe the pathway of impulses through this system; define heart block.

- Describe what information can be gained from an electrocardiogram.

- Describe the effect of the following on heart rate: stimulation by the vagus nerve, atropine, epinephrine, and various ions.

- Compare the structure and function of arteries, veins, and capillaries.

- List and/or identify the major arteries and veins and name the body region supplied by each.

- Discuss the unique features of special circulations (arterial circulation of the brain, hepatic portal circulation, and fetal circulation) of the body.

- Define blood pressure and pulse and name several pulse points.

- List factors affecting and/or determining blood pressure.

- Define hypertension, thrombus, embolus, and arteriosclerosis.

- Name the two major types of structures composing the lymphatic system and explain how the lymphatic system is functionally related to the cardiovascular system.

- Describe the formation and composition of lymph and explain how it is transported through the lymphatic vessels.

- Describe the function(s) of lymph nodes, tonsils, thymus, and spleen.

- Explain how regular exercise and a diet low in fats and cholesterol may help maintain cardiovascular health.

When most people hear the term circulatory system, they immediately think of the heart. We have all felt our own heart "pound" from time to time and it tends to make us a bit nervous when this happens. The importance of the heart has been recognized for a long time, and both serious and comical songs referring to it have been written. However, the circulatory system is much more than just the heart, and from a scientific and medical standpoint, it is important to understand *why* the circulatory system is so vital to life.

The almost continuous traffic into and out of a busy factory occurs at a snail's pace when compared to the endless activity going on within our bodies. Night and day, minute after minute, needed materials are taken up by or loaded into our trillions of cells; wastes, as well as useful substances, are moved out of the cells. Although the pace of these exchanges slows during sleep, the exchanges must go on continuously because when they stop, we die. Like the bustling factory, the body must have a transportation system to carry its various "cargos" back and forth. But instead of roads, railway tracks, and airways, the body's delivery routes are its hollow blood vessels.

Most simply stated, the major function of the circulatory system is transportation. Using blood as the transport vehicle, the system carries oxygen, nutrients, cell wastes, and many other substances vital for body homeostasis to and from the cells. The force to move the blood around the body is provided by the beating heart.

The circulatory system has two major subdivisions—the cardiovascular (kar″de-o-vas′ku-lar) system and the lymphatic (lim-fat′ik) system. The cardiovascular system can be compared to a muscular pump equipped with one-way valves and a closed system of large and small plumbing tubes within which the blood travels. Blood (the substance transported) is discussed in Chapter 9. Here we will consider the heart (the pump) and the blood vessels (the network of tubes). The lymphatic system, a pumpless system of vessels and lymph nodes that aids the cardiovascular system in its function, will be the final topic of this chapter.

HEART

Anatomy

EXTERNAL STRUCTURE

The **heart** is a cone-shaped organ about the size of your fist; it weighs less than a pound. The heart is located within the bony thorax and is flanked on each side by the lungs. Its more pointed **apex** extends slightly to the left and rests on the diaphragm, approximately at the level of the fifth intercostal space. (This is exactly where one would place a stethoscope to count the heart rate for an *apical pulse*.) Its broader **base,** from which the great vessels of the body emerge, is more superior; it lies beneath the second rib.

Figure 10-1 shows two views of the heart—an external anterior view and a frontal section. As the anatomical areas of the heart are described next, keep referring to the figure to locate each of the heart structures or heart regions.

The heart is enclosed by a double sac of serous membrane, the **pericardium** (per″ĭ-kar′de-um). The thin **visceral pericardium,** or **epicardium,** tightly hugs the external surface of the heart. The outer, loosely applied **parietal pericardium** (not indicated in the figure) is attached at the heart apex to the diaphragm. A slippery lubricat-

ing fluid (serous fluid) is produced by the pericardial membranes. This fluid allows the heart to beat easily in a relatively frictionless environment as the pericardial layers slide smoothly across each other. Inflammation of the pericardium, *pericarditis* (per″ĭ-kar-di′tis), often results in a decrease in the amount of serous fluid. This causes the pericardial layers to bind and stick to each other forming painful *adhesions.* These adhesions interfere with heart movements.

The heart walls are composed of thick cardiac muscle, the **myocardium** (mi″o-kar′de-um), which is twisted and whorled into ringlike arrangements. The myocardium is reinforced internally by a dense fibrous connective tissue network. This network is called the "skeleton of the heart."

HEART CHAMBERS

The heart is divided into four hollow chambers—two **atria** (a′tre-ah) and two **ventricles** (ven′trĭ-k′ls). Each of these chambers is lined with a thin serous lining, the **endocardium** (en″do-kar′de-um), which helps blood flow smoothly through the heart. The superior atria are primarily *receiving chambers;* they are not important in the pumping activity of the heart. Blood flows into the atria under low pressure from the veins of the body and then continues on to fill the ventricles below. The inferior thick-walled ventricles are the *discharging chambers,* since they have the major responsibility for forcing blood out of the heart into the large arteries that rise upward from its base. As illustrated in Figure 10-1, the heart is somewhat twisted and the right ventricle forms most of its anterior surface whereas the left ventricle forms its apex. A septum divides the heart longitudinally. It is referred to as the **interventricular,** or **interatrial, septum** depending on which chambers it divides and separates.

PULMONARY AND SYSTEMIC CIRCULATIONS

The heart functions as a double pump. The right side works as the pulmonary circuit pump. It receives oxygen-poor blood from the veins of the body through the large **superior** and **inferior vena cavae** (ka′ve) and pumps it out through the **pulmonary trunk.** The pulmonary trunk splits into the right and left **pulmonary arteries,** which carry blood to the lungs where oxygen is

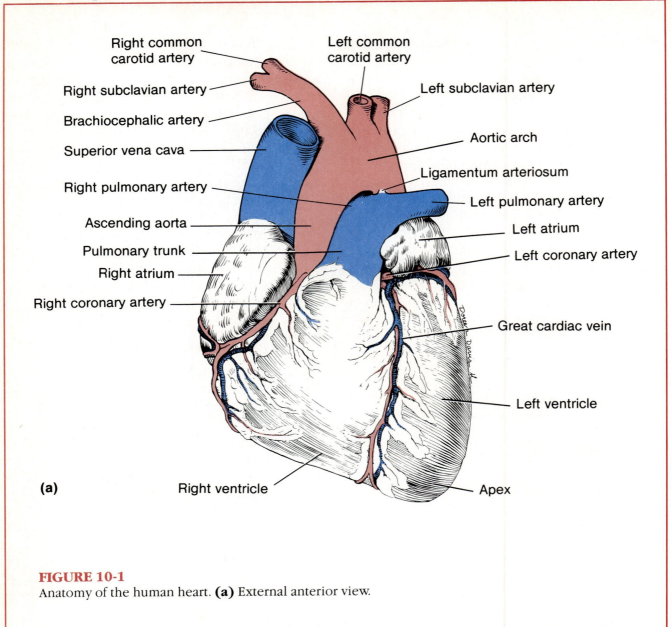

Right common carotid artery

Right subclavian artery

Brachiocephalic artery

Superior vena cava

Right pulmonary artery

Ascending aorta

Pulmonary trunk

Right atrium

Right coronary artery

Left common carotid artery

Left subclavian artery

Aortic arch

Ligamentum arteriosum

Left pulmonary artery

Left atrium

Left coronary artery

Great cardiac vein

Left ventricle

(a)

Right ventricle

Apex

FIGURE 10-1
Anatomy of the human heart. **(a)** External anterior view.

picked up and carbon dioxide is unloaded. Oxygen-rich blood drains from the lungs and is returned to the left side of the heart through the four **pulmonary veins.** The circulation just described, from the right side of the heart to the lungs and back to the left side of the heart, is called the *pulmonary circulation.* Its only function is to carry blood to the lungs for gas exchange and then return it to the heart.

Blood returned to the left side of the heart is pumped out of the heart into the **aorta** (a-or′tah) from which the systemic arteries branch to supply

the capillary beds of body tissues. Oxygen-poor blood is returned from the tissues through systemic veins, which return the blood to the right atrium through the superior and inferior vena cavae as mentioned earlier. This second circuit, from the left heart through the body tissues and back to the right heart, is called the *systemic circulation.* It provides the oxygen-rich, nutrient blood supply to all body organs. Because the left ventricle is the systemic pump that must pump blood over a much longer pathway through the body, its walls are substantially thicker than those of the right ventricle.

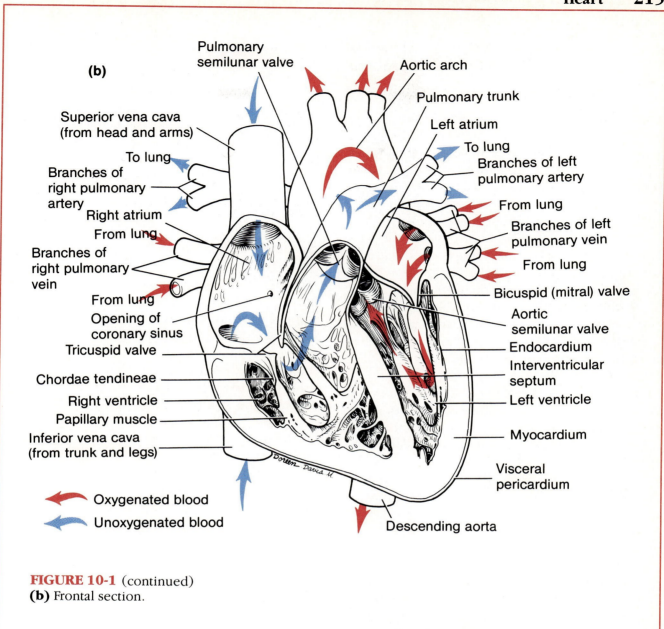

(b)

Pulmonary
semilunar valve

Aortic arch

Pulmonary trunk

Left atrium

Superior vena cava
(from head and arms)

To lung

Branches of
right pulmonary
artery

Right atrium

From lung

Branches of
right pulmonary
vein

From lung

Opening of
coronary sinus

Tricuspid valve

Chordae tendineae

Right ventricle

Papillary muscle

Inferior vena cava
(from trunk and legs)

To lung
Branches of left
pulmonary artery

From lung
Branches of left
pulmonary vein

From lung

Bicuspid (mitral) valve

Aortic
semilunar valve

Endocardium

Interventricular
septum

Left ventricle

Myocardium

Visceral
pericardium

Descending aorta

Oxygenated blood

Unoxygenated blood

FIGURE 10-1 (continued)
(b) Frontal section.

VALVES

The heart is equipped with four valves that allow blood to flow in *only* one direction through the heart chambers. The **atrioventricular valves** (a″tre-o-ven-trik′u-lar), or AV valves, are located between the atrial and ventricular chambers on each side. The AV valves prevent backflow into the atria when the ventricles begin to contract. The left AV valve—the **bicuspid,** or **mitral** (mi′ tral) **valve**—consists of two cusps, or flaps, of endocardium. The right AV valve, the **tricuspid valve,** has three cusps (see Figure 10-1). Tiny white cords, the **chordae tendineae** ([kor′de ten-

din′e] literally, heart strings), anchor the cusps to the walls of the ventricles. When the heart is relaxed and blood is passively filling its chambers, the AV-valve flaps hang downward into the ventricles. As the ventricles contract, they compress the blood in their chambers, and the intraventricular pressure (pressure inside the ventricles) begins to rise. This causes the AV-valve flaps to be forced upward, and the valve to be closed; at this point the chordae tendineae are working to anchor the flaps in a closed position. If the flaps were unanchored, they would blow upward into the atria (like an umbrella being turned inside out by a strong wind). In this manner, backflow into the atria is prevented when the ventricles are contracting.

The second set of valves, the **semilunar** (sem"e-lu'nar) **valves,** guard the bases of the two large arteries leaving the ventricular chambers. Thus, they are known as the **pulmonary** and **aortic semilunar valves.** Each semilunar valve has three cusps that fit tightly together when the valves are closed. But when the ventricles are contracting and forcing blood out through the large arteries, the cusps are forced open and flattened against the walls of the arteries by the tremendous force of the blood rushing out of the heart. Then, when the ventricles relax, the blood begins to flow backward toward the heart and the cusps fill with blood, closing the valves. This prevents arterial blood from reentering the heart.

Each set of valves operates at different times. The AV valves are open during heart relaxation and closed when the ventricles are contracting. The semilunar valves are closed during heart relaxation and are forced open when the ventricles contract. As they open and close in response to pressure changes in the heart, the valves assure that the blood continually moves forward in its journey through the heart.

CARDIAC CIRCULATION

Although the heart chambers are filled with blood almost continuously, this contained blood does not nourish the myocardium. The functional blood supply of the heart is provided by the right and left **coronary arteries.** The coronary arteries branch from the base of the aorta and encircle the heart in the groove at the junction of the atria and ventricles. The coronary arteries and their branches are compressed when the ventricles are contracting and fill when the heart is relaxed. When the heart beats at a very rapid rate, the myocardium may not receive an adequate blood supply because the relaxation periods (when the blood is able to flow to the heart tissue) are shortened. Situations in which the myocardium is deprived of oxygen often result in a crushing chest pain called *angina pectoris* (an-ji'nah pek'tor-is). Anginal pain is a warning that should *never* be ignored. The myocardium is drained by the **coronary veins,** which empty into the **coronary sinus,** which in turn empties into the right atrium.

Physiology

As the heart beats or contracts, the blood makes continuous round trips—in and out of the heart, through the rest of the body, and then back to the heart, only to be sent out again. The amount of work that a heart does is almost too incredible to believe. In one day it pushes the body's supply of 5 quarts or so of blood through the blood vessels over a 1000 times, meaning that it actually pumps about 5000 quarts of blood in a single day.

CARDIAC CYCLE AND HEART SOUNDS

In a healthy heart, the atria contract simultaneously and as they begin to relax, contraction of the ventricles begins. **Systole** (sis'to-le) and **diastole** (di-as'to-le) mean contraction and relaxation, respectively. Since most of the pumping work is done by the ventricles, the terms are almost always used to refer to the contraction and relaxation of the ventricles unless otherwise stated. The term *cardiac cycle* is used to refer to the events of one complete heart beat, during which both atria and ventricles contract and then relax. Since the average heart beats approximately 72 beats/minute, the length of the cardiac cycle is normally about 0.8 seconds.

The discussion of the cardiac cycle starts with the heart in complete relaxation. At this point, the pressure in the heart is low, and blood is flowing passively into and through the atria into the ventricles from the pulmonary and systemic circulations. The semilunar valves are closed and the AV valves are open. Then the atria contract and force the blood remaining in their chambers into the ventricles. Shortly after, ventricular contraction (systole) begins and the pressure within the ventricles increases rapidly, closing the AV valves. When the intraventricular pressure in the ventricles is higher than the pressure in the large arteries leaving the heart, the semilunar valves are forced open and blood rushes through them out of the ventricles. During ventricular systole, the atria are relaxed, and their chambers are again filling with blood. At the end of systole, the ventricles relax, the semilunar valves snap shut (preventing backflow), and, for a moment, the ventricles are completely closed chambers. During ventricular diastole, the intraventricular pressure drops. When it drops below the pressure in the atria (which has been increasing as blood has been filling their chambers), the AV valves are forced open and the ventricles again begin to refill rapidly with blood, completing the cycle.

When using a stethoscope, you can hear two distinct sounds during each cardiac cycle. These heart sounds are commonly described by the two syllables "lup" and "dup," and the sequence is lup-dup, pause, lup-dup, pause, and so on. The first heart sound (lup) is caused by the closing of the AV valves. The second heart sound (dup) occurs when the semilunar valves close at the end of systole. The first heart sound is longer and louder (or more booming) than the second heart sound, which tends to be short and sharp.

Abnormal heart sounds are referred to as *murmurs* and often indicate problems with the heart valves. In valves that do not close tightly, their closure is followed by a swishing sound caused by the backflow of blood. Distinct sounds also can be heard when blood flows turbulently through *stenosed* (narrowed) valves. Like a leaky pump, a heart with faulty valves can still operate unless the defect is very serious. In such cases, valve replacements are done surgically.

CONDUCTION SYSTEM OF THE HEART

Unlike skeletal muscle cells that must be stimulated before they will contract, cardiac muscle cells can and do contract spontaneously and independently, even if all nervous connections are severed. Moreover, these spontaneous contractions occur in a regular and continuous way. Although cardiac muscle *can* beat independently, the muscle cells in different areas of the heart have different rhythms. For example, the atrial cells beat about 60 times/minute whereas the ventricular cells contract much more slowly (40/minute). Therefore, without external controls, the activity of the heart would be uncoordinated and inefficient as a pump.

Two types of controlling systems act to regulate heart activity. One of these involves the nerves of the autonomic nervous system that act like "spurs" and "reins" to increase or decrease the heart rate depending on which division is activated (see p. 148). The second system is the **Purkinje** (pur-kin′jē), or **nodal, system** that is built into the heart tissue (Figure 10-2). The Purkinje system is composed of a special tissue found nowhere else in the body; it is much like a cross between muscle and nervous tissue. The Purkinje system causes heart muscle depolarization in one direction and one direction only—that is, from the atria to the ventricles. In addition, it enforces a contraction rate of approximately 72 beats/minute on the heart; thus, the heart beats as a coordinated unit.

One of the most important parts of the Purkinje system is a tiny node of tissue called the **sinoatrial** (si″no-a′tre-al) **(SA) node,** which is located in the right atrium. Other components include the **atrioventricular (AV) node** at the junction of the atria and ventricles, the **bundle of His,** and the right and left **bundle branches** located in the interventricular septum, and finally the **Purkinje fibers,** which spread within the muscle of the ventricle walls.

The SA node, which has the highest rate of depolarization in the whole system, starts each heart beat. Because it sets the pace for the whole heart, the SA node is often called the *pacemaker.* From the SA node, the impulse spreads through the atria to the AV node and then the atria contract. At the AV node, the impulse is delayed briefly to give the atria time to finish contracting. It then passes rapidly through the bundle of His, the bundle branches, and the Purkinje fibers, resulting in the contraction of the ventricles.

Because the atria and ventricles are separated from one another by a region of "insulating" connective tissue, the depolarization wave can reach the ventricles only through the AV node. Thus, any damage to the AV node can partially or totally release the ventricles from the control of the SA node. When this occurs, the ventricles (thus the heart) begin to beat at their own rate, which is much slower, some or all of the time. This condition is *heart block.* There are other conditions that can interfere with the regular conduction of impulses across the heart—for example, damage to the SA node results in a slower heart rate. When this is a problem, artificial pacemakers are usually installed surgically. *Ischemia* (is-ke′me-ah), or lack of an adequate blood supply to the heart muscle, may lead to *fibrillation*—a rapid uncoordinated shuddering of the heart muscle (it looks like a bag of worms). Fibrillation makes the heart

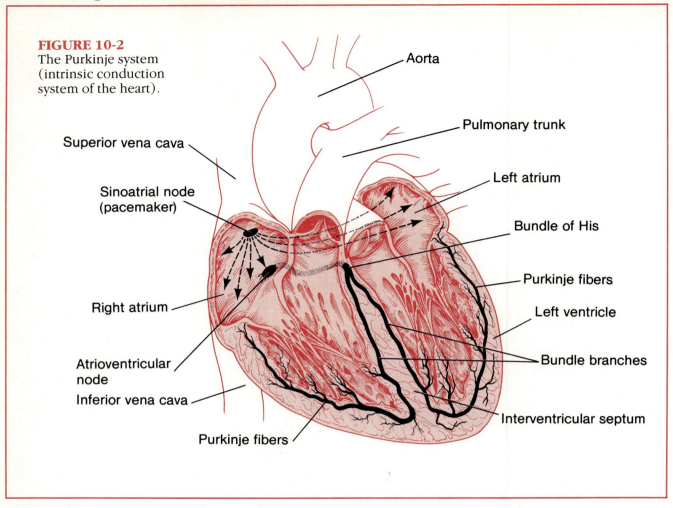

FIGURE 10-2
The Purkinje system (intrinsic conduction system of the heart).

Aorta

Superior vena cava

Pulmonary trunk

Sinoatrial node (pacemaker)

Left atrium

Bundle of His

Purkinje fibers

Left ventricle

Right atrium

Atrioventricular node

Bundle branches

Inferior vena cava

Interventricular septum

Purkinje fibers

totally useless as a pump and is a major cause of death owing to heart attacks in adults.

Tachycardia (tak″e-kar′de-ah) is a rapid heart rate (over 100 beats/minute). *Bradycardia* (brad″e-kar′de-ah) is a heart rate which is substantially slower than normal (less than 60 beats/minute). Although neither condition is pathologic, prolonged tachycardia may progress to fibrillation.

FACTORS INFLUENCING HEART RATE

As already explained, the heart's efficiency as a pump depends on both intrinsic (within the heart) and extrinsic (external to the heart) controls. The Purkinje system, in which the pacemaker sets the rate of the rest of the heart, is the major intrinsic factor.

Although heart contraction does not depend on external nervous impulses, its rate *can* be changed temporarily by the autonomic nerves. In addition, heart activity is modified by various chemicals, hormones, and ions.

During times of physical or emotional stress, the nerves of the sympathetic division of the autonomic nervous system stimulate the SA node, and the heart beats more rapidly. This is a familiar phenomenon to anyone who has ever been frightened or has had to run to catch a bus. As fast as the heart pumps under ordinary conditions, its speed is dramatically increased when special demands are placed on it. During strenuous exercise, the heart may pump up to 16 gallons of blood per minute. Since a faster heart rate increases the rate at which fresh blood is reaching the body cells, more oxygen and glucose are made available to them during periods of stress. When demand lessens, the heart adjusts. Parasympathetic nerves, primarily the vagus nerves, slow and steady the heart, giving it more time to rest during noncrisis times. In patients with *congestive heart failure,*

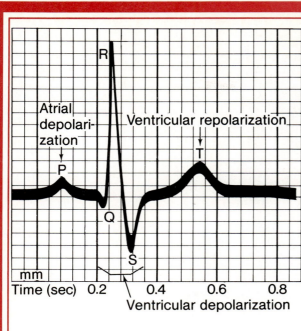

BOX 10-1
Electrocardiogram

When impulses pass through the heart, electrical currents are generated that spread throughout the body. These impulses can be detected on the body surface and recorded with an *electrocardiograph.* The recording that is made, the electrocardiogram (ECG), traces the flow of current through the heart. A normal ECG tracing is shown.

The typical ECG has three recognizable waves. The first wave, the *P wave,* is a small wave that follows the depolarization of the atria immediately before they contract. The large *QRS complex,* which results from the depolarization of the ventricles, has a complicated shape; it precedes the contraction of the ventricles. The *T wave* results from currents flowing during the repolarization of the ventricles. (The repolarization of the atria is generally hidden by the large QRS complex, which is being recorded at the same time.)

Abnormalities in the shape of the waves and changes in their timing send signals that something may be wrong with the Purkinje system or may indicate a *myocardial infarct* (present or past). A myocardial infarct is an area of heart tissue in which the cardiac cells have died; it is generally a result of *ischemia.* During *fibrillation,* the normal pattern of the ECG is totally lost and the heart ceases to act as a functioning pump.

their heart is nearly "worn out" owing to age or hypertensive heart disease, and it pumps weakly. For those patients, digitalis is the drug that is routinely prescribed. It acts much in the same way as the vagus nerves; that is, it slows and steadies the heart, which results in a stronger heart beat.

Various chemicals and ions can have a dramatic effect on heart activity. Epinephrine, which mimics the effect of the sympathetic nerves, and atropine, which inhibits the parasympathetic nerves, both cause increased heart rate. Excesses or a lack of needed blood ions like sodium, potassium, and calcium also modify heart activity. For example, excesses of potassium ions in the blood decreases the ability of the heart to contract and may stop it entirely. This is why the chemical balance of the blood is so vital.

BLOOD VESSELS

Blood circulates inside the blood vessels, which form a closed transport system. The idea of circulation, or blood that "makes rounds," through the body is only about 300 years old. Originally, as a Greek physician proposed, it was thought that blood moved through the body like a tide, first moving out from the heart and then back to it in the same vessels to get rid of its impurities in the

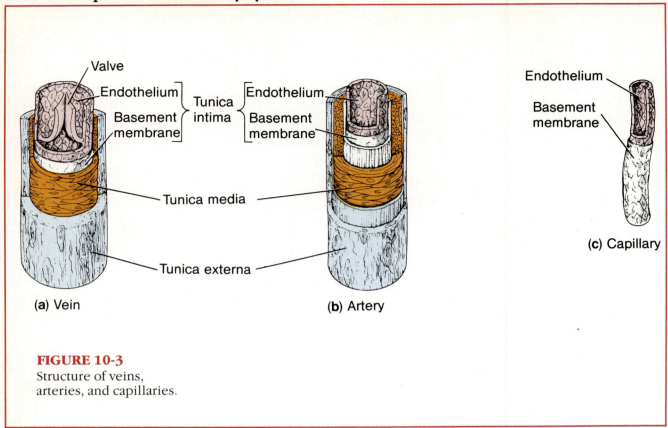

FIGURE 10-3
Structure of veins,
arteries, and capillaries.

lungs. It was not until the seventeenth century that William Harvey, an English scientist, proved that blood did, in fact, move in circles.

Like any other system of roads, the circulatory system has its freeways, secondary roads, and alleys. As the heart beats, blood is propelled into the large **arteries** leaving the heart. It then moves into successively smaller and smaller arteries and then into the **arterioles** (ar-te′re-ōls), which feed the **capillary** (kap′ĭ-lar″e) **beds** in the tissues. Capillary beds are drained by **venules** (ven′ūls), which in turn empty into **veins** that finally empty into the great veins entering the heart. Thus arteries, which carry blood away from the heart, and the veins, which drain the tissues and return the blood to the heart, are simply conducting vessels—the freeways and secondary roads. Only the tiny hairlike capillaries, which ramify through the tissues and connect the smallest arteries (arterioles) to the smallest veins (venules), directly serve the needs of the body cells. The capillaries are the alleys that intimately intertwine among the body cells. It is only through their walls that exchanges between the tissue cells and the blood can occur. Respiratory gases, nutrients, and wastes move along diffusion gradients; thus, oxygen and nutrients diffuse from the blood to the tissue cells, and carbon dioxide and metabolic wastes move from the cells to the blood.

Microscopic Anatomy

Except for the microscopic capillaries, the walls of blood vessels have three coats, or tunics (Figure 10-3). The **tunica intima** (tu′nĭ-kah in′tĭ-mah), which lines the lumen or interior of the vessels, is a single thin layer of endothelium. Its cells fit closely together and form an extremely smooth lining that decreases friction as blood flows through the vessels.

The **tunica media** (me′de-ah) is a more bulky middle coat and is primarily smooth muscle and elastic tissue. The smooth muscle, which is controlled by the sympathetic nervous system, is active in changing the diameter of the vessels, which in turn increases or decreases the blood pressure.

The **tunica externa** (eks′tern-ah) is the outermost tunic; it is composed of connective tissue.

Its function is basically to support and protect the vessels.

The walls of arteries are usually much thicker than the walls of veins. The tunica media, in particular, tends to be much heavier. This structural difference is related to a difference in the functioning of these two types of vessels. Arteries, which are closer to the pumping action of the heart, must be able to expand as blood is forced into them and then recoil passively as the blood flows off into the circulation during diastole. Their walls must be strong and stretchy enough to take these continuous changes in pressure.

On the other hand, veins are far away from the heart in the circulatory pathway, and the pressure in them tends to be low all the time. Thus veins have thinner walls. However, since the blood pressure in veins is usually too low to force the blood back to the heart and blood returning to the heart is often flowing against gravity, veins are modified to assure that the amount of blood returning to the heart (*venous return*) equals the amount being pumped out of the heart (*cardiac output*) at any time. The lumens of veins tend to be much larger than those of corresponding arteries, and the larger veins have valves that prevent backflow of blood. Skeletal muscle activity also helps venous return. As the muscles surrounding the veins contract and relax, the blood is "milked" through the veins toward the heart. Finally, pressure changes that occur in the thorax during breathing also help to return blood to the heart.

To demonstrate the efficiency of the venous valves in preventing backflow of blood, perform the following simple experiment. Allow one hand to hang by your side until the blood vessels on its dorsal aspect become distended or swollen with blood. Place two fingertips against one of the distended veins. Then pressing firmly, move your proximal finger along the vein toward your heart. Now release that finger. As you can see, the vein remains flattened and collapsed in spite of gravity. Now remove your distal finger and watch the vein fill rapidly.

The transparent walls of the tiny capillaries are only one-cell layer thick, just the endothelium or tunica intima. Because of this exceptional thinness, exchanges are easily made between the blood and the tissue cells.

Gross Anatomy

MAJOR ARTERIES OF THE SYSTEMIC CIRCULATION

The **aorta** is the largest artery of the body; it is a truly splendid vessel, about the same size as your thumb. The aorta curves upward from the left ventricle of the heart as the **ascending aorta,** arches as the **aortic arch** to the left, and then it plunges downward through the thorax following the spine **(descending aorta)** to finally pass through the diaphragm into the abdominopelvic cavity (see Figure 10-1).

The major branches of the aorta and the organs they serve are listed next in sequence from the heart. Figure 10-4 shows the course of the aorta and its major branches. As you locate the arteries on the figure, be aware of ways in which you can make your learning easier. In many cases the name of the artery tells you the body region or organs served (iliac artery, brachial artery, and coronary artery) or the bone followed (femoral artery and ulnar artery).

Arterial branches of the aorta are as follows:

- The **coronary** (right [R.] and left [L.]) **arteries** serve the heart.

- The **brachiocephalic** (brak″e-o-sĕ-fal′ik) **artery** splits into the **R. common carotid** (kah-rot′id) **artery** and **R. subclavian** (sub-kla′ve-an) **artery.** (See same named vessels for left side of body for organs served.)

- The **left common carotid artery** divides, forming the **L. internal carotid** that serves the brain, and the **L. external carotid** that serves the skin and muscles of the head and neck.

- The **left subclavian artery** gives off an important branch—the **vertebral artery,** which serves part of the brain. In the axillary region, the subclavian artery becomes the **axillary artery** and then continues into the arm as the **brachial artery,** which serves the upper arm. At the elbow, the brachial artery splits to form the **radial** and **ulnar arteries,** which serve the forearm.

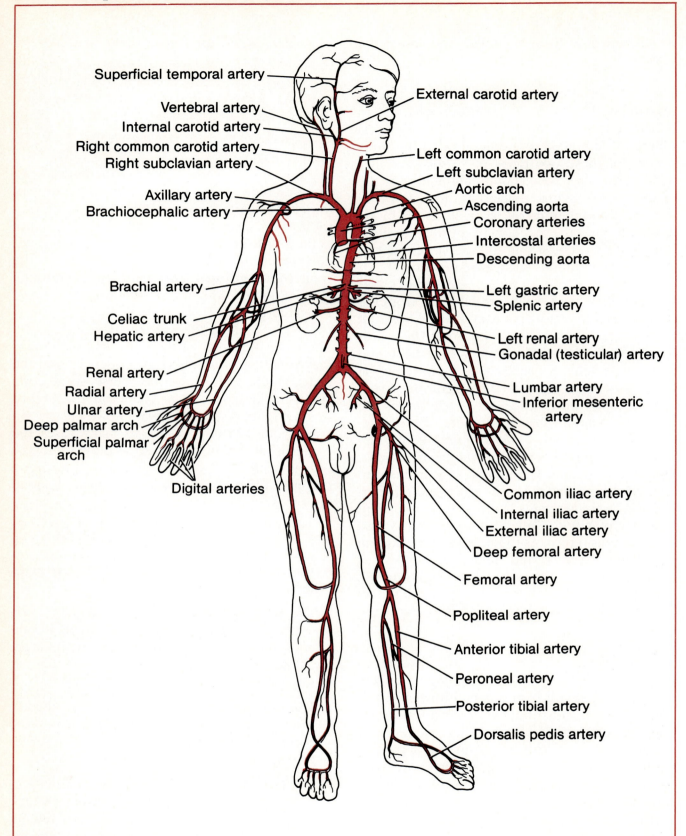

Superficial temporal artery
Vertebral artery
Internal carotid artery
Right common carotid artery
Right subclavian artery
Axillary artery
Brachiocephalic artery
Brachial artery
Celiac trunk
Hepatic artery
Renal artery
Radial artery
Ulnar artery
Deep palmar arch
Superficial palmar arch
Digital arteries

External carotid artery
Left common carotid artery
Left subclavian artery
Aortic arch
Ascending aorta
Coronary arteries
Intercostal arteries
Descending aorta
Left gastric artery
Splenic artery
Left renal artery
Gonadal (testicular) artery
Lumbar artery
Inferior mesenteric artery
Common iliac artery
Internal iliac artery
External iliac artery
Deep femoral artery
Femoral artery
Popliteal artery
Anterior tibial artery
Peroneal artery
Posterior tibial artery
Dorsalis pedis artery

FIGURE 10-4
Major systemic arteries
of the body.

- The **intercostal arteries** (ten pairs) supply the muscles of the thorax wall.

- The **celiac trunk** is a single vessel that has three branches: (a) the **gastric artery** supplies the stomach, (b) the **splenic artery** supplies the spleen, and (c) the **hepatic artery** supplies the liver.

- The **superior mesenteric** (mes"en-ter'ik) **artery** supplies most of the small intestine and the first half of the large intestine or colon.

- The **renal** (R. and L.) **arteries** serve the kidneys.

- The **gonadal** (R. and L.) **arteries** serve the gonads. They are called the **ovarian arteries** in females (serve the ovaries) and the **testicular arteries** in males (serve the testes).

- The **lumbar arteries** are several pairs of arteries serving the heavy muscles of the abdomen and trunk walls.

- The **inferior mesenteric artery** is a small artery supplying the last half of the large intestine.

- The **common iliac** (R. and L.) **arteries** are the final branches of the descending aorta. Each divides into an **internal iliac artery**, which supplies the pelvic organs (bladder, rectum, and so on), and an **external iliac artery**, which enters the thigh where it becomes the **femoral artery.** The femoral artery and its branch, the **deep femoral artery,** serve the thigh. At the knee, the femoral artery becomes the **popliteal artery** which then splits into the **anterior** and **posterior tibial arteries,** which supply the lower leg, ankle, and foot. The anterior tibial artery terminates in the **dorsalis pedis artery,** which supplies the dorsum of the foot. (The dorsalis pedis is often palpated in patients with circulatory problems of the legs to determine if the distal part of the leg has an adequate circulation.)

Major Veins of the Systemic Circulation

Although arteries are generally located in deep, well-protected body areas, many veins are more superficial and are often easily seen and palpated on the body surface. Most deep veins follow the course of the major arteries; in many cases, the naming of the veins and arteries is identical except for designating the vessels as veins. While major systemic arteries branch off the aorta, the veins converge on the vena cavae, which enter the right atrium of the heart. Veins draining the head and arms empty into the **superior vena cava** and those draining the lower body empty into the **inferior vena cava.** These veins are described next and shown in Figure 10-5. As before, locate the veins on the figure as you read through their descriptions.

Veins Draining into the Superior Vena Cava. Veins draining into the superior vena cava are named in a distal/proximal direction; that is, in the same direction as the blood flow into the superior vena cava.

- The **radial** and **ulnar veins** are deep veins draining the forearm. They unite to form the deep **brachial vein,** which drains the upper arm and empties into the **axillary vein** in the axillary region.

- The **cephalic** (sĕ-fal'ik) **vein** provides for the superficial drainage of the lateral aspect of the arm and empties into the axillary vein.

- The **basilic** (bah-sil'ik) **vein** is a superficial vein that drains the medial aspect of the arm and empties into the brachial vein proximally. The basilic and cephalic veins are joined anteriorly at the anterior aspect of the elbow by the **median cubital vein.** (The median cubital vein is often chosen for removing blood for testing purposes.)

- The **subclavian vein** receives venous blood from the arm through the axillary vein and from the skin and muscles of the head through the small **external jugular vein.**

- The **vertebral vein** drains the posterior part of the head.

- The **internal jugular vein** drains the dural sinuses of the brain.

- The **brachiocephalic** (R. and L.) **veins** are large veins that receive venous drainage from the subclavian, vertebral, and internal jugular veins on their respective sides. The brachiocephalic veins join to form the **superior vena cava,** which enters the heart.

- The **azygos** (az'ĭ-gos) **vein** is a single vein that drains the thorax and enters the superior vena cava just before it joins the heart.

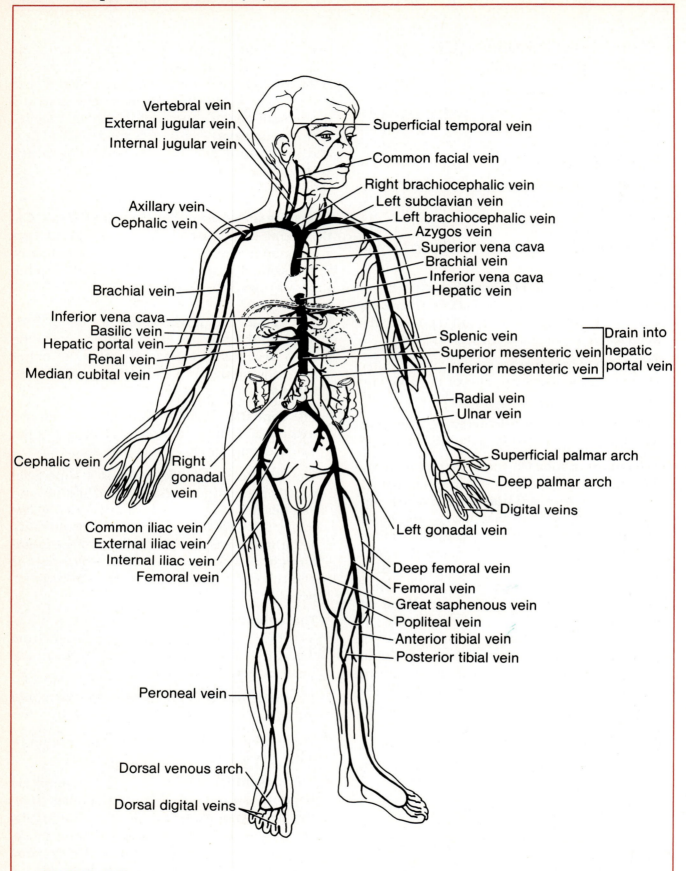

Vertebral vein
External jugular vein
Internal jugular vein
Superficial temporal vein
Common facial vein
Right brachiocephalic vein
Left subclavian vein
Left brachiocephalic vein
Azygos vein
Superior vena cava
Brachial vein
Inferior vena cava
Hepatic vein
Axillary vein
Cephalic vein
Brachial vein
Inferior vena cava
Basilic vein
Hepatic portal vein
Renal vein
Median cubital vein
Splenic vein
Superior mesenteric vein
Inferior mesenteric vein
Drain into hepatic portal vein
Radial vein
Ulnar vein
Superficial palmar arch
Deep palmar arch
Digital veins
Cephalic vein
Right gonadal vein
Left gonadal vein
Common iliac vein
External iliac vein
Internal iliac vein
Femoral vein
Deep femoral vein
Femoral vein
Great saphenous vein
Popliteal vein
Anterior tibial vein
Posterior tibial vein
Peroneal vein
Dorsal venous arch
Dorsal digital veins

FIGURE 10-5
Major systemic veins of
the body.

VEINS DRAINING INTO THE INFERIOR VENA CAVA. The inferior vena cava, which is much longer than the superior vena cava, returns blood to the heart from all body regions below the diaphragm. As before, we will trace the venous drainage in a distal/proximal direction.

- The **anterior and posterior tibial veins** drain the leg (calf and foot). The posterial tibial vein becomes the **popliteal vein** at the knee and then the **femoral vein** in the thigh. The femoral vein becomes the **external iliac vein** as it enters the pelvis.

- The **great saphenous** (sah-fe′nus) **veins** are the longest veins in the body. They receive the superficial drainage of the leg. They begin at the **dorsal venous arch** in the foot and travel up the medial aspect of the leg to empty into the femoral vein in the thigh.

- Each **common iliac** (R. and L.) **vein** is formed by the union of the **external iliac vein** and the **internal iliac vein** (which drains the pelvis) on its own side. The common iliac veins join to form the inferior vena cava, which then ascends superiorly in the abdominal cavity.

- The **R. gonadal vein** drains the right ovary in females and the right testicle in males. (The **left gonadal veins** empty into the **left renal veins.**)

- The **renal** (R. and L.) **veins** drain the kidneys.

- The **hepatic portal vein** is a single vein that drains the digestive tract organs and carries this blood through the liver before it enters the systemic circulation. (The hepatic portal system is discussed in the next column.)

- The **hepatic** (R. and L.) **veins** drain the liver.

SPECIAL CIRCULATIONS

ARTERIAL SUPPLY OF THE BRAIN AND THE CIRCLE OF WILLIS. A continuous blood supply to the brain is crucial, since a lack of blood for even a few minutes causes the delicate brain cells to die. The brain is supplied by two pairs of arteries, the internal carotid arteries and the vertebral arteries (Figure 10-6).

The **internal carotid arteries,** branches of the common carotid arteries, run through the neck and enter the skull through the temporal bone.

Once inside the cranium, each divides into the **anterior** and **middle cerebral arteries,** which supply most of the cerebrum.

The paired **vertebral arteries** pass upward from the subclavian arteries at the base of the neck. Within the skull, the vertebral arteries join to form the single **basilar artery,** which serves the brain stem and cerebellum as it travels upward. At the base of the cerebrum, the basilar artery divides to form the **posterior cerebral arteries,** which supply the posterior part of the cerebrum.

The anterior and posterior blood supplies of the brain are joined by small communicating arterial branches. The result is a complete circle of connecting blood vessels called the **circle of Willis,** which surrounds the base of the brain. The circle of Willis protects the brain, because it provides more than one route for blood to reach brain tissue in case of a clot or impaired blood flow anywhere in the system.

HEPATIC PORTAL CIRCULATION. The veins of the hepatic portal circulation drain the digestive organs, spleen, and pancreas, and deliver this blood to the liver through the **hepatic portal vein.** After you have just eaten, the hepatic portal blood contains large amounts of nutrients. Since the liver is a key body organ involved in maintaining the proper glucose, fat, and protein concentrations in the blood, this system "takes a detour" to ensure that the liver processes these substances before they enter the systemic circulation. As blood flows slowly through the liver, some of the nutrients are removed to be stored or processed in various ways for later release to the blood. The liver is drained by the hepatic veins that enter the inferior vena cava. The hepatic portal circulation is a unique and unusual circulation. Normally, arteries feed capillary beds, which in turn drain into veins; here we see *veins* feeding into the liver circulation.

The **inferior mesenteric vein,** draining the terminal part of the large intestine, drains into the **splenic vein,** which itself drains the spleen, pancreas, and the left side of the stomach. The splenic vein and **superior mesenteric vein** (which drains the small intestine and the first part of the colon) join to form the hepatic portal vein. The **gastric vein,** which drains the right side of the

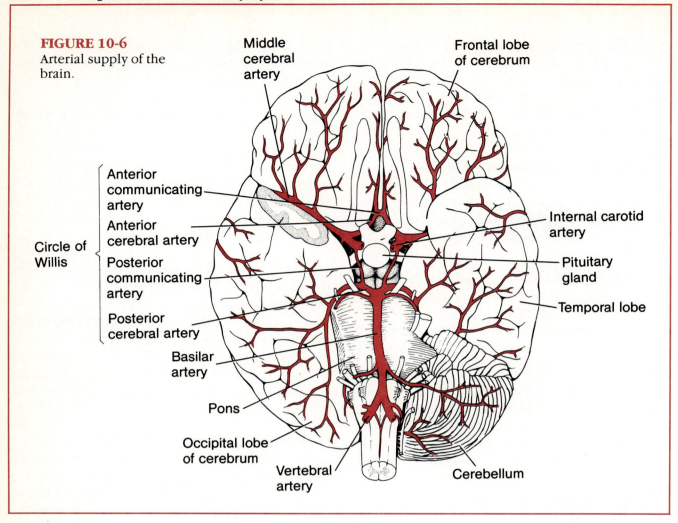

FIGURE 10-6
Arterial supply of the brain.

Middle cerebral artery

Frontal lobe of cerebrum

Anterior communicating artery

Anterior cerebral artery

Circle of Willis

Posterior communicating artery

Posterior cerebral artery

Basilar artery

Pons

Occipital lobe of cerebrum

Vertebral artery

Internal carotid artery

Pituitary gland

Temporal lobe

Cerebellum

---

stomach, drains directly into the hepatic portal vein (Figure 10-7).

FETAL CIRCULATION.   Since the lungs and digestive system are not yet functioning in a fetus, all nutrient, excretory, and gas exchanges occur through the placenta. Thus, nutrients and oxygen move from the mother's blood into the fetal blood, and fetal wastes move in the opposite direction. As shown in Figure 10-8 (p. 228), the umbilical cord contains three blood vessels: two smaller **umbilical arteries** and one large **umbilical vein.** The umbilical vein carries blood rich in nutrients and oxygen to the fetus; the umbilical arteries carry carbon dioxide and debris-laden blood from the fetus to the placenta. As blood flows superiorly toward the heart of the fetus, most of it bypasses the immature liver through the **ductus venosus** (duk′tus ve-nos′us) and enters the inferior vena cava, which carries the blood to the right atrium of the heart.

Since fetal lungs are nonfunctional and collapsed, two shunts see to it that they are almost entirely bypassed. Some of the blood entering the right atrium is shunted directly into the left atrium through a flaplike opening in the interatrial septum, the **foramen ovale** (fo-ra′men o-val′e). Blood that does manage to enter the right ventricle is pumped out the pulmonary trunk where it meets second shunt, the **ductus arteriosus** (ar-ter″i-o′sus), a short vessel that connects the aorta and the pulmonary trunk. Because the lungs are collapsed, blood tends to enter the systemic circulation through the ductus arteriosus. The aorta carries blood to the tissues of the fetal body and ultimately back to the placenta through the umbilical arteries.

At birth, or shortly after, the foramen ovale closes and the ductus arteriosus collapses and is converted to fibrous **ligamentum arteriosum** (lig″ah-men′tum ar-ter″i-o′sum) (see Figure 10-1). As

FIGURE 10-7
The hepatic portal
circulation.

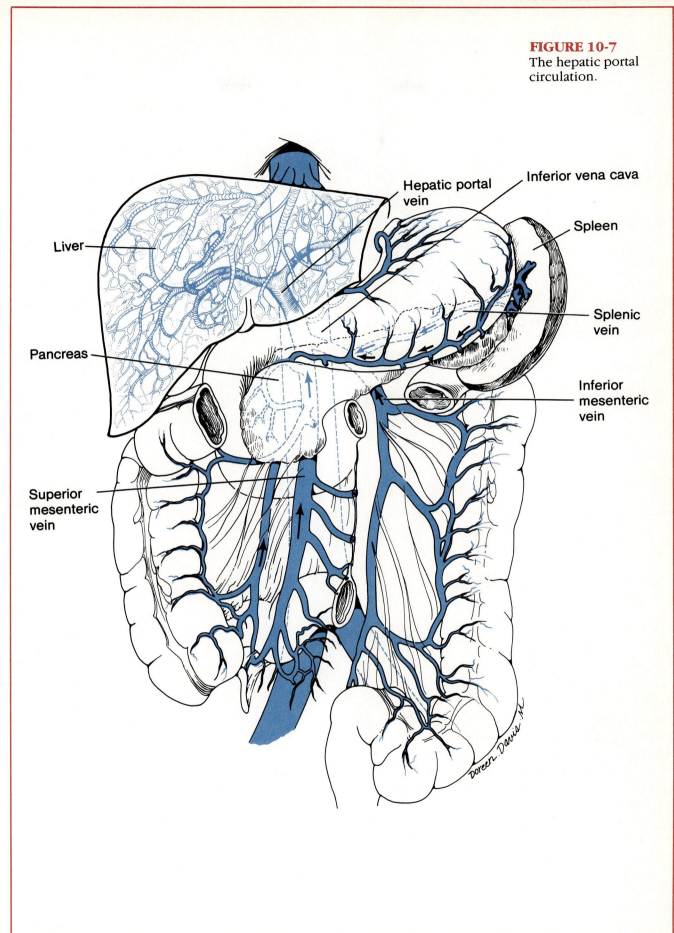

Inferior vena cava

Hepatic portal
vein

Spleen

Liver

Splenic
vein

Pancreas

Inferior
mesenteric
vein

Superior
mesenteric
vein

FIGURE 10-8
The fetal circulation.

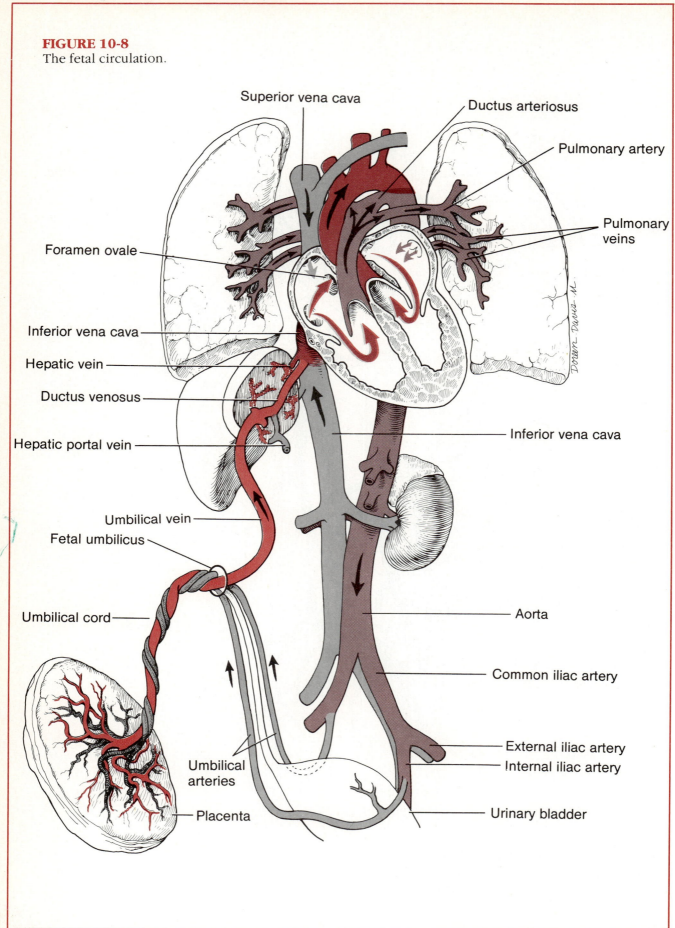

Superior vena cava

Ductus arteriosus

Pulmonary artery

Pulmonary veins

Foramen ovale

Inferior vena cava

Hepatic vein

Ductus venosus

Hepatic portal vein

Inferior vena cava

Umbilical vein

Fetal umbilicus

Umbilical cord

Aorta

Common iliac artery

External iliac artery

Internal iliac artery

Umbilical arteries

Urinary bladder

Placenta

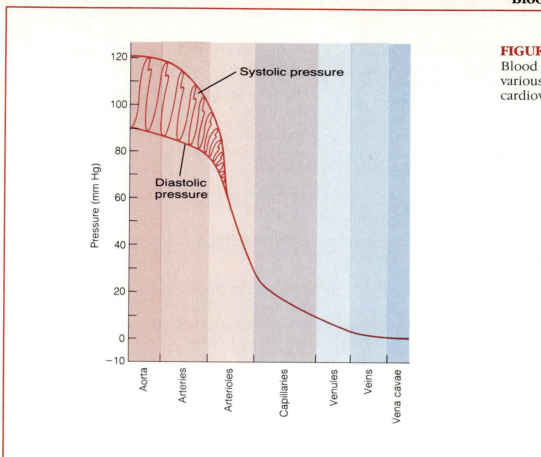

FIGURE 10-9
Blood pressure in
various areas of the
cardiovascular system.

blood stops flowing through the umbilical vessels, they become obliterated, and the circulatory pattern becomes that of an adult.

Physiology

BLOOD PRESSURE

Any system equipped with a pump that forces fluid through a network of closed tubes operates under pressure. The closer you get to the pump, the higher the pressure. Blood pressure is the pressure the blood exerts against the inner walls of the blood vessels, and it is the force that keeps the blood continuously circulating even between heart beats.

When the ventricles contract, they force blood into large thick-walled elastic arteries which *expand* as the blood is pushed into them. The high pressure in these arteries forces the blood to continually move into areas where the pressure is lower. The pressure is highest in the large arteries and continues to drop throughout the pathway,

reaching zero or negative pressure at the venae cavae (Figure 10-9). Thus, the blood flows into the smaller arteries, then arterioles, capillaries, venules, veins, and finally back to the large vena cavae entering the right heart. Blood then flows continually along a pressure gradiant from high-to-low pressure as it makes its circuits day-in and day-out. If venous return depended entirely on a high blood pressure throughout the system, blood would probably never be able to complete its circuit back to the heart. This is why the valves in the larger veins, the milking activity of the skeletal muscles, and pressure changes in the thorax are so important.

Continual flow of blood absolutely depends on the stretchiness of the larger arteries and their ability to recoil and keep the pressure on the blood as it flows off into the circulation. To illustrate this, think of a garden hose with relatively hard walls. When the water is turned on, the water spurts out under high pressure (because the hose walls don't expand); however, when the water faucet is suddenly turned off, the flow of water stops just as abruptly. This is because the walls of the hose cannot recoil and keep exerting pressure on the water; therefore, the pressure drops, and the flow of water stops. The importance of the elasticity of the arteries is best appreciated when it is lost, as happens in *arteriosclerosis* (ar-te"re-o sklĕ-ro'sis). In arteriosclerosis (also called hardening of the arteries), the blood vessels first become clogged with fats (*atherosclerosis* [ath"er-o" skle-ro'sis]), and then become hardened with calcium deposits. When hard, they lose their ability to stretch and recoil, leading to high blood pressure and all its associated problems (for example, thrombi).

Because the heart alternately contracts and relaxes, the off-and-on flow of blood into the arteries causes the blood pressure to rise and fall during each beat. Thus, two arterial blood pressure measurements are usually made: **systolic pressure** (the pressure in the arteries at the peak of ventricular contraction) and **diastolic pressure** (the pressure when the ventricles are relaxing). Blood pressures are reported in millimeters of mercury (mm Hg), with the systolic pressure written first—120/80 translates to 120 over 80, or a systolic pressure of 120 mm Hg and a diastolic pressure of 80 mm Hg. Normal blood pressure varies considerably from one person to another.

Pulse

The term **pulse** refers to the expansion and then recoil of an artery that occurs with each beat of the left ventricle. Normally the pulse rate (pressure surges/minute) equals the heart rate (beats/minute), and the pulse averages 70–76 beats/minute in a normal resting person.

The pulse may be felt easily on any superficial artery when the artery is compressed over a bone or firm tissue. A few pulse or pressure points are listed next. Attempt to palpate each of them on yourself by placing the fingertips of the first two or three fingers of one hand over the artery. It generally helps to compress the artery firmly as you begin, and then immediately ease up on your pressure slightly. In each case, notice the regularity of the pulse and its relative strength.

- **Carotid artery.** Place your fingertips at the side of your neck.

- **Temporal artery.** Palpate anterior to your ear, in the temple region.

- **Radial artery.** Press at the lateral aspect of your wrist, just above the thumb.

- **Brachial artery.** Place your fingertips in the antecubital fossa at the point where the brachial artery splits into the radial and ulnar arteries.

- **Femoral artery.** Place your fingertips over your groin.

- **Popliteal artery.** Apply pressure at the back of your knee.

When you know the pressure points, you know what region of the body to compress or apply pressure to when blood is spurting from an externally visible wound.

Effects of Various Factors on Blood Pressure and Heart Rate

Arterial blood pressure is directly related to cardiac output (the amount of blood pumped out of the left ventricle per minute) and peripheral resistance to blood flow (Figure 10-10). **Peripheral resistance** is the amount of friction encountered by the blood as it flows through the blood vessels. It is increased by many factors, but probably the most important factor is the constriction, or narrowing, of the blood vessels (mostly the arterioles) as a result of sympathetic nervous system activity or arteriosclerosis. Increased blood volume or blood viscosity (thickness) also raises the peripheral resistance. Any factor that increases either the cardiac output or peripheral resistance causes an almost immediate reflex rise in blood pressure. Many factors can alter blood pressure—age, weight, time of day, exercise, body position, emotional state, and various drugs, to name a few. The influence of a small number of these factors is discussed next.

Autonomic Nervous System. The parasympathetic division has little or no effect on blood pressure, but the sympathetic division is important and is responsive to many different factors. The major action of the sympathetic nerves on the general body circulation is to cause *vasoconstriction* (vas"o-kon-strik'shun), or narrowing, of the blood vessels, which increases the blood pres-

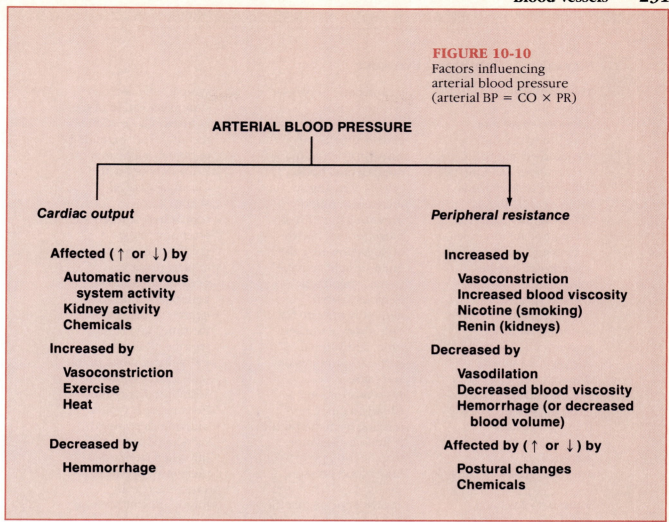

ARTERIAL BLOOD PRESSURE

Cardiac output

Affected (↑ or ↓) by
   **Automatic nervous
     system activity**
   Kidney activity
   Chemicals

Increased by
   Vasoconstriction
   Exercise
   Heat

Decreased by
   Hemmorrhage

Peripheral resistance

Increased by
   Vasoconstriction
   Increased blood viscosity
   Nicotine (smoking)
   Renin (kidneys)

Decreased by
   Vasodilation
   Decreased blood viscosity
   **Hemorrhage (or decreased
     blood volume)**

Affected by (↑ or ↓) by
   Postural changes
   Chemicals

sure. The sympathetic nervous system causes vasoconstriction in many different circumstances. For example, when we stand up suddenly after lying down, the effect of gravity causes blood to pool in the vessels of the legs and feet. Vasoconstriction increases the blood pressure and speeds venous return to the heart and blood delivery to the brain. In the elderly, the sympathetic nervous system is often less efficient than it is in younger people. Therefore, elderly people tend to have a condition called *orthostatic* (or"tho-stat′ik) *hypotension,* which is low blood pressure (*hypotension*) due to position changes. When affected individuals stand up quickly, they tend to become dizzy or lightheaded because not enough blood is reaching their brain temporarily. This condition can usually be dealt with if the elderly person remembers to dangle his or her legs on the side of the bed for a few minutes before standing up. This gives the "age-slowed" sympathetic nervous system time to adjust and prepare to cause the required vasoconstriction.

When blood volume suddenly decreases, as in hemorrhage, blood pressure drops and the heart begins to beat more rapidly (as it tries to compensate). However, since the venous return has been decreased by blood loss, the heart also beats weakly and inefficiently. In such cases, the sympathetic nervous system causes vasoconstriction to increase the blood pressure so that (hopefully) the venous return is increased and the circulation can be continued.

The final example illustrating how the sympathetic nervous system helps maintain homeostasis of the circulatory system concerns its activity when we exercise vigorously or are frightened and have to make a hasty retreat. Under these conditions, there is a generalized vasoconstriction *except* in the skeletal muscles. The vessels of the skeletal muscles vasodilate to increase the blood flow to the working muscles. (It should be noted that the sympathetic nerves *never* cause vasoconstriction of the heart or brain vessels.)

BOX 10-2
Stress and High Blood Pressure

The relationship between arteriosclerosis and high blood pressure is well known and based on the understanding that rigid tubes (that is, blood vessels in this case) cannot expand to accommodate increases in pressure on their walls as can flexible tubes. Dietary factors, when leading to the buildup of fatty materials on blood vessel walls, effectively narrow the internal dimensions of blood vessels and are believed to be important. The presence of fatty accumulations sets the stage for arteriosclerosis.

Stress also has been implicated as a contributor to high blood pressure, but there has been much controversy about *how* it does so. It is known that stress activates the sympathetic nervous system (fight-or-flight response) and releases adrenal cortical hormones; these hormones activate the body's defenses during periods of prolonged stress. Activation of the sympathetic nervous system causes widespread body effects. One is vasoconstriction in nonessential body organs (causing an increase in blood pressure) to redirect blood to vital body organs. However, the flight-or-fight activities of the sympathetic nervous system generally operate for only short periods of time, and adrenal cortical hormones do *not* cause vasoconstriction. Clearly, a piece is missing from the puzzle if stress causes more than fleeting periods of high blood pressure.

It now appears that the missing piece may have been found. Earlier studies on stressed rats indicated that they (a) retained salt (and, consequently, fluids, leading to increased blood pressure), and (b) the salt retention ability was markedly higher in rats that later developed chronic hypertension. These investigations also found that salt retention is caused by sympathetic nervous system activation of the kidneys. Those studies recently were extended to human subjects at the University of North Carolina Medical School where researchers studied salt retention in type A (individuals considered to have driving personalities and be generally under some degree of stress most of the time) and type B (normal to low-stress) individuals. None of the individuals studied had high blood pressure. Because only type A individuals retained substantial amounts of salt when stressed, it presently appears that stress may be an important contributing factor to high blood pressure in individuals *already at risk* for hypertension (due to their type A personality) but is relatively unimportant in those with less driving life-styles. To support this conclusion, follow-up studies are needed to see if those same type A individuals ultimately develop high blood pressure.

KIDNEYS.   The kidneys also play a role in the regulation of arterial blood pressure. As blood pressure (and/or blood volume) increases over what is normal, the kidneys allow more water to leave the body in the urine. Since the source of this water is the blood stream, the blood volume decreases, which in turn decreases the blood pressure. However, when arterial blood pressure decreases, the kidneys retain body water, increasing blood volume, and blood pressure rises.

In addition, when arterial blood pressure is low, the kidney cells also release **renin** to the blood, which activates blood proteins that cause vasoconstriction.

TEMPERATURE. In general, cold has a vasoconstricting effect. This is why your exposed skin feels cold to the touch on a winter day, and why cold compresses are often used to prevent swelling of a bruised area. On the other hand, since heat has a vasodilating effect, warm compresses are used to speed the circulation into an inflammed area.

CHEMICALS. The effects of chemical substances, many of which are drugs, on blood pressure are widespread and well known in many cases. Just a few examples will be given here. Epinephrine increases both heart rate and blood pressure. Nicotine increases blood pressure by causing vasoconstriction. Conversely, both alcohol and histamine cause vasodilation and decrease the blood pressure. The reason why a person who has "one too many" becomes flushed is that the skin vessels have been dilated by alcohol.

DIET. Although medical opinions tend to change and be at odds from time to time, it is generally believed that a diet low in salt, saturated fats, and cholesterol helps to prevent *hypertension,* or high blood pressure. The cause of many cases of hypertension is unknown, but in some cases, it is definitely known to be a result of body-fluid retention, which leads to increased blood volume. In such cases, salt (which holds body fluid) should be restricted. In other cases, hypertension is a result of atherosclerosis or arteriosclerosis. Since saturated fats and cholesterol are deposited in the arteriosclerotic *plaques* (fatty patches on the arterial walls), perhaps it would be wise if these foods were limited in our diet. Hypertension is often called the "silent killer," because it progresses without symptoms in many cases.

LYMPHATIC SYSTEM

The **lymphatic system** is a pumpless system consisting of lymphatic vessels and lymph nodes. Since the closed cardiovascular system is a high-pressure system, it tends to be a little "leaky."

Fluid is forced out of the blood at the arterial ends of the capillary beds and reabsorbed into the blood stream at the venous end. However, *not all* of the lost fluid is returned to the blood at the venous end. The excess fluid that lags behind in the tissue spaces must eventually return to the blood if the vascular system is to be able to operate properly and have a sufficient blood volume. If it does not, fluid accumulates in the tissues, producing *edema.*

The function of the **lymphatic vessels** is to pick up this excess tissue fluid (now called **lymph**) and return it to the blood stream. **Lymph nodes** help protect the body by removing foreign material such as bacteria and tumor cells from the lymphatic stream and by producing agranular WBCs.

The lymphatic vessels form a one-way system, and lymph flows only toward the heart. The microscopic blind-ended lymph capillaries spiderweb through all the tissues of the body and absorb the leaked fluid (primarily water and a small amount of dissolved proteins). The lymph is then transported through successively larger and larger lymphatic vessels—lymph venules and veins—and is finally returned to the venous system through one of the two large ducts in the thoracic region. The **right lymphatic duct** drains the lymph from the right arm and the right side of the head and thorax. The large **thoracic duct** receives lymph from the rest of the body as shown in Figure 10-11. Both ducts empty the lymph into the subclavian vein on their own side of the body.

Like the veins of the cardiovascular system, the lymph veins are thin-walled, and the larger vessels have valves. Since the lymphatic system is a pumpless system, it was assumed that transport of lymph depends entirely on the milking action of the skeletal muscles and the pressure changes in the thorax during breathing. However, recent research has revealed that the smooth muscle in the vessel walls contracts rhythmically and actually helps "pump" the lymph along.

As lymph is transported toward the heart, it is filtered through the thousands of bean-shaped lymph nodes, which cluster along the lymphatic vessels. Particularly large clusters are found in the inguinal, axillary, and cervical regions of the body. Within the lymph nodes are phagocytic

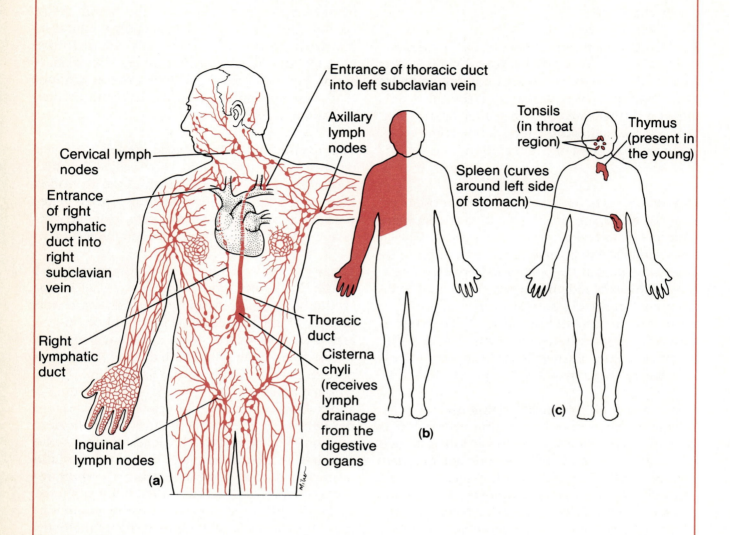

Cervical lymph nodes

Entrance of right lymphatic duct into right subclavian vein

Right lymphatic duct

Inguinal lymph nodes

(a)

Entrance of thoracic duct into left subclavian vein

Axillary lymph nodes

Thoracic duct

Cisterna chyli (receives lymph drainage from the digestive organs

(b)

Tonsils (in throat region)

Thymus (present in the young)

Spleen (curves around left side of stomach)

(c)

FIGURE 10-11
Lymphatic system.
(a) Distribution of lymphatic vessels and lymph nodes.
(b) Darkened area represents body area drained by the right lymphatic duct; the rest of the body is drained by the thoracic duct. **(c)** Body location of the tonsils, spleen, and thymus.

cells, *macrophages* (mak'ro-fāj-ez), which destroy bacteria, viruses, and other foreign substances in the lymph before it is returned to the blood. Agranular leukocytes (see page 203) are also formed in the lymph nodes and released into the lymphatic stream. Although we are not usually aware of the protective nature of the lymph nodes, most of us have had swollen glands during an active infection. This swelling is a result of the trapping function of the nodes.

The tonsils, thymus, and spleen are generally considered to be lymphoid organs. Although all three have roles to play in protecting the body, their role is not that of filtering lymph. The **tonsils** are masses of lymphatic tissue located in the throat where they are found just deep to the mucosa. Their job is to filter and remove any bacteria or other foreign pathogens entering the throat. They carry out this function so efficiently that sometimes they become congested with bacteria and become red, swollen, and sore.

The **thymus,** which functions only during the early years of life, is a lymphatic mass found low in the throat overlying the heart. As described in Chapter 11, the thymus functions in the programming of lymphocytes so that they can carry out their protective functions in the body.

The **spleen** is a blood-rich organ that filters blood. It is located in the left lateral aspect of the abdominal cavity and extends to curl around the anterior aspect of the stomach. Its most important function is to destroy wornout RBCs and return some of their breakdown products to the liver. For example, iron is used again for making hemoglobin, and the rest of the hemoglobin molecule is secreted in bile. Other functions of the spleen include synthesizing WBCs, and acting as a blood reservoir (as does the liver). If hemorrhage occurs, both the spleen and liver contract and empty their contained blood into the circulation to try to bring the blood volume back to normal levels.

DEVELOPMENTAL ASPECTS OF THE CIRCULATORY SYSTEM

The heart begins as a simple tube in the embryo; it is beating and pumping blood by the fourth week of pregnancy. During early development, the heart continues to change and mature, finally becoming a four-chambered structure capable of acting as a double pump by the seventh week. During fetal life, the collapsed lungs and non-functional liver are mostly bypassed by the blood through special shunts. After the seventh week of development, few changes (other than growth) occur in the fetal circulation until birth. Shortly after birth, the bypass structures become blocked, and the special umbilical vessels become nonfunctional.

Congenital heart defects account for about half of all infant deaths resulting from congenital defects. Environmental interferences, such as maternal infection and ingested drugs during the first 3 months of pregnancy when the embryonic heart is forming, seem to be the major causes of such problems. Congenital heart defects may include a ductus arteriosus that does not close, septal openings, and other structural abnormalities of the heart. Such problems are usually corrected surgically.

In the absence of congenital heart problems, the heart usually pumps day-in and day-out and year after year. Homeostatic mechanisms are so effective that we rarely are aware when the heart is working more vigorously unless we can actually hear our pulse pounding in our ears. The heart will hypertrophy and its cardiac output will increase substantially if we exercise regularly and vigorously enough to force it to beat at a higher than normal rate for extended periods of time. This type of exercise is *aerobic exercise.* Not only does the efficiency of the heart increase, but it does so with a decreased pulse rate and blood pressure. Another added benefit of aerobic exercise is that it helps clear fatty deposits from the blood vessel walls, thus helping to slow the progress of atherosclerosis. However, let's raise a caution flag here. The once-a-month or once-a-year tennis player or snow shoveler has not built up this type of heart endurance and strength. When such an individual pushes his or her heart too much, it may not be able to cope with the sudden (and unexpected) demand. This is why many weekend athletes are myocardial infarct victims.

As we begin to get older, more and more signs of circulatory system disturbances start to appear. In some, weakness of the valves of the veins, *varicose veins,* is a major problem. Varicose veins are common in people who stand for long periods of time (for example, dentists and hairdressers) and in obese (or pregnant) individuals. The common factor is the pooling of blood in the feet and legs and a less efficient venous return resulting from inactivity or pressure on the veins. In any case, the overworked valves give way and the veins become twisted and dilated; edema and muscle cramps may also occur. A serious complication of varicose veins is *thrombophlebitis* (throm″bo-fle-bi′-tis), an inflammation of a vein, which results from a clot forming in a vessel with poor circulation or hampered venous return. Since all venous blood must be shunted to the pulmonary circulation before traveling through the body tissues again, a common consequence of thrombophlebitis is clot detachment and *pulmonary embolism,* which is a life threatening condition.

While not everyone has varicose veins, all of us have progressive arteriosclerosis. Some say the process begins at birth, and there's an old saying that goes "You are only as old as your arteries," which refers to this degenerative process. The gradual loss in elasticity in the blood vessels leads to hypertension and hypertensive heart disease. The insidious filling of the blood vessels with the fatty-calcified deposits leads most commonly to *coronary artery disease.* Also, as described earlier, the roughening of the vessel walls encourages thrombus formation. It has been estimated that at least 30% of the population in the United States have hypertension by the time they are age 65, and that cardiovascular disease causes more than one-half of the deaths in individuals over age 65. Although the aging process itself contributes to changes in the walls of the blood vessels that can lead to strokes or myocardial infarcts, most researchers feel that diet, not aging, is the single most important contributing factor to cardiovascular diseases. Thus, there seems to be some agreement that cardiovascular disorders can be decreased if individuals eat less animal fat, cholesterol, and salt. It has also been suggested that regular exercise throughout life, maintaining your diastolic blood pressure under 75 mm Hg,

quitting smoking (if you're a smoker), and reducing the amount of stress in your life are still other ways to reduce your chances of having cardiovascular disease.

Lymphatic system problems are relatively uncommon. Unhappily, some of us become well acquainted with the normal role of the lymphatic system when it malfunctions (as when the lymphatic vessels are blocked) or when parts of it are removed (as in radical breast surgery). In such instances, severe edema is the result.

IMPORTANT TERMS*

Pericardium *(per″ĭ-kar′de-um)*

Myocardium *(mi″o-kar′de-um)*

Atrium *(a′tre-um)*

Ventricles *(ven′trĭ-k′ls)*

Pulmonary circulation

Systemic circulation

Atrioventricular *(a″tre-o-ven-trik′u-lar)* **valves**

Semilunar *(sem″e-lu′nar)* **valves**

Cardiac cycle

Purkinje *(pur-kin′jē)* **system**

Artery

Vein

Capillary *(kap′ĭ-lar″e)*

Cardiac output

Diastolic *(di″ah-stol′ik)* **pressure**

Systolic *(sis-tol′ik)* **pressure**

Pulse

Peripheral *(pĕ-rif′er-al)* **resistance**

Vasoconstriction *(vas″o-kon-strik′shun)*

Lymph

*For definitions, see Glossary.

SUMMARY

A. HEART

1. The heart, located in the thorax, is flanked laterally by the lungs and enclosed in a two-layered pericardium.

2. The walls of the heart (myocardium) are composed of cardiac muscle. The heart has four hollow chambers, the two atria (receiving chambers), and two ventricles (discharging chambers). It is divided longitudinally by a septum.

3. The heart functions as a double pump. The right heart is the pulmonary pump (right heart to lungs to left heart); the left heart is the systemic pump (left heart to body tissues to right heart).

4. Four valves prevent backflow of blood in the heart. The AV valves (mitral and tricuspid) prevent backflow into the atria when the ventricles are contracting. The semilunar valves prevent backflow into the ventricles when the heart is relaxing. The valves open and close in response to pressure changes in the heart.

5. The myocardium is nourished by its own circulation, the coronary circulation, which consists of the right and left coronary arteries, the coronary veins, and the coronary sinus.

6. The time and events occurring from one heartbeat to the next is the cardiac cycle.

7. As the heart beats, sounds resulting from the closing of the valves (''lup-dup'') can be heard. Faulty valves reduce the efficiency of the heart as a pump and result in abnormal heart sounds (murmurs).

8. Cardiac muscle is able to initiate its own contraction in a regular way, but its rate is influenced by both intrinsic and extrinsic factors. The Purkinje system (intrinsic system) increases the rate of heart contraction and ensures that the heart beats as a unit. The SA node is the pacemaker for the heart.

9. Extrinsic factors influencing heart rate are the nerves of the autonomic nervous system

drugs (and other chemicals), and ionic levels in the blood.

B. BLOOD VESSELS

1. Arteries that transport blood from the heart and veins, which conduct blood back to the heart, are conducting vessels. Only capillaries play a role in the actual exchanges with the tissue cells.

2. Except for capillaries, blood vessels are composed of three tunics: the tunica intima forms a friction-reducing lining for the vessel, the tunica media is the bulky middle layer of muscle and elastic tissue, and the tunica externa is the protective outermost connective tissue layer. Capillary walls are formed of endothelium only.

3. Artery walls are thick and strong to withstand pressure fluctuations. They expand and recoil as the heart beats. Vein walls are thinner, their lumen is larger, and they are equipped with valves. These modifications reflect the fact that veins are low pressure vessels.

4. The major arteries of the systemic circulation are all branches of the aorta that leaves the left ventricle. They branch into smaller arteries and then into the arterioles, which feed the capillary beds of the body tissues. For naming and location of the systemic arteries, see p. 221.

5. The major veins of the systemic circulation ultimately converge on one of the vena cavae. All veins above the diaphragm drain into the superior vena cava, and all veins below the diaphragm drain into the inferior vena cava. Both vena cavae enter the right atrium of the heart. See p. 223 for naming and location of the systemic veins.

6. The arterial circulation of the brain is formed by the branches of the paired vertebral and internal carotid arteries. The circle of Willis provides alternate routes for blood flow in cases of a blockage in the brain circulation.

7. The hepatic portal circulation is formed by the veins draining the digestive organs,

which empty into the hepatic portal vein. The hepatic portal vein carries the nutrient-rich blood to the liver where it is processed by the liver cells before the blood is allowed to enter the systemic circulation.

8. The fetal circulation is a temporary circulation seen only in the fetus. It consists primarily of three special vessels: the single umbilical vein that carries nutrient- and oxygen-laden blood to the fetus from the placenta, and the two umbilical arteries that carry carbon dioxide and waste-laden blood from the fetus to the placenta. Special shunts bypassing the lungs and liver are also present.

9. Blood pressure is the pressure that blood exerts on the walls of the blood vessels. It is the force that causes blood to continue to flow in the blood vessels. It is high in the arteries, drops in the capillaries, and lowest in the veins. Blood is forced along a descending pressure gradient. Both systolic and diastolic pressures are recorded.

10. The pulse is the alternate expansion and recoil of a blood vessel wall (the pressure wave) that occurs as the heart beats. It may be felt easily over any superficial artery; such sites are called pressure points.

11. Arterial blood pressure is directly influenced by heart activity (increased heart rate leads to increased blood pressure) and by resistance to blood flow. The most important factors increasing the peripheral resistance are a decrease in the diameter, or stretchiness, of the arteries and an increase in blood viscosity.

12. Many factors influence blood pressure. Some of these factors are the activity of the sympathetic nerves and kidneys, drugs, and diet.

C. LYMPHATIC SYSTEM

1. The lymphatic system consists of the lymphatic vessels, lymph nodes, and certain other lymphoid organs in the body.

2. Blind-ended lymphatic capillaries pick up excess tissue fluid leaked from the blood vessels. The fluid (lymph) flows into the larger lymph veins and finally into the blood vascular system through the right lymphatic duct and the left thoracic duct.

3. Lymph nodes are clustered along lymphatic vessels, and the lymphatic stream flows through them. Lymph nodes form agranular WBCs, and phagocytic cells within them remove bacteria, viruses, and the like from the lymph stream before it is returned to the blood.

4. Other lymphatic organs include the tonsils (located in the throat), which remove bacteria trying to enter the digestive or respiratory tracts; the thymus, a programming region for lymphocytes of the body; and the spleen, RBC graveyard and blood reservoir.

D. DEVELOPMENTAL ASPECTS OF THE CIRCULATORY SYSTEM

1. The heart begins as a tubelike structure; it is beating and pumping blood by the fourth week of embryonic development.

2. Congenital heart defects account for half of all infant deaths resulting from congenital problems.

3. Varicose veins, a structural defect owing to incompetent valves, is a common vascular problem especially in the obese and in people who stand long hours. It is a predisposing factor for thrombophlebitis.

4. Arteriosclerosis is to be expected as a consequence of aging. The gradual loss of elasticity of the arteries leads to hypertension and hypertensive heart disease, and the clogging of the vessels with fatty substances leads to coronary artery disease and strokes. Cardiovascular disease accounts for more than one-half of the deaths in individuals over age 65.

5. Modifications in diet (decreased fats, cholesterol, and salt) and regular exercise may help to reverse the arteriosclerotic process and prolong life.

REVIEW QUESTIONS

1. Describe the location and position of the heart in the thorax.

2. Draw a diagram of the heart showing the three layers composing its wall and its four chambers. *Label* each. Also show where the AV and semilunar valves would be. Now show and label all blood vessels entering and leaving the heart chambers.

3. Trace one drop of blood from the time it enters the right atrium of the heart until it enters the left atrium of the heart. What name is given to this circuit?

4. Clearly explain the difference in *function* of the systemic and pulmonary circulations.

5. Why are the heart valves important? Can the heart function with leaky valves?

6. Why might a thrombus in a coronary artery cause sudden death?

7. What is the function of the fluid that fills the pericardial sac?

8. Define systole, diastole, and cardiac cycle.

9. To which heart chamber(s) do the terms systole and diastole most often apply?

10. How does the heart's ability to contract differ from that of other muscles of the body?

11. What is the function of the Purkinje system of the heart? Name its elements, *in order,* beginning with the pacemaker.

12. What monosyllables are used to describe heart sounds? What causes the heart sounds?

13. Name three different factors that increase heart rate.

14. Name and describe the three tunics making up the wall of arteries and veins from inside out and give the important function of each layer.

15. Describe the structure of capillary walls. How is their structure related to their function in the body?

16. Why are artery walls so much thicker than those of corresponding veins?

17. Since veins are low pressure vessels, blood pressure is not very important in helping to get the blood returned to the heart. Name three factors that are important in promoting venous return.

18. Arteries are often described as vessels that carry oxygen-rich blood while veins are said to carry oxygen-poor (carbon dioxide-rich) blood. Name the two sets of exceptions to this rule that were studied in this chapter.

19. Trace a drop of blood from the left ventricle of the heart to the wrist of the right hand and back to the heart. Now trace it to the dorsum of the right foot and back to the right heart.

20. What two paired vessels provide arterial blood to the brain? What name is given to the communication between them?

21. What is the function of the hepatic portal circulation? Why is a portal circulation a "strange" circulation?

22. The liver and lungs are nearly entirely bypassed in a fetus. Why is this? Name the vessel that bypasses the liver. Name two lung bypasses. Three vessels travel in the umbilical cord—which carries oxygen- and nutrient-rich blood?

23. Define pulse.

24. Which artery is palpated at the following pressure points: At the wrist? At the front of the ear? At the side of the neck?

25. Define blood pressure, systolic pressure, and diastolic pressure.

26. What vital role does blood pressure play?

27. Two elements determine blood pressure— the cardiac output of the heart and the peripheral resistance or friction in the blood vessels. Name two factors that increase cardiac output. Name two factors that increase peripheral resistance.

28. What is the effect of hemorrhage on blood pressure? Why? In which position—sitting, lying down, or standing—is the blood pressure normally the highest? The lowest?

29. Name the most important function of the lymphatic vessels. Of the lymph nodes.

30. Where are the lymph nodes most dense?

31. What is the special role of the tonsils? The spleen?

32. What are varicose veins? What factors seem to promote their formation?

33. Define hypertension and arteriosclerosis. How are they often related? Why is hypertension called the "silent killer"? Name three changes in your life-style that might help to prevent cardiovascular disease in your old age.

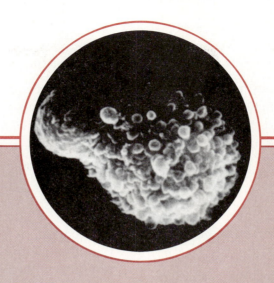

CHAPTER 11

Immunology

Chapter Contents

Function of the immune system • to protect the body against disease by destroying "foreign" cells, and by inactivating toxins and other foreign chemicals with its antibodies

After completing this chapter, you should be able to:

- Name the two arms of the immune response and relate each to a specific lymphocyte (B- or T-cell) population.

- Define antigen and hapten, and name substances that act as complete antigens.

- Compare and contrast the development of B- and T-lymphocytes.

- Explain the clonal selection theory of active immunity.

- Describe the roles of helper, killer, and suppressor T-cells, and plasma cells.

- Indicate the importance of the interactions that occur between macrophages and lymphocytes, T-cells and B-cells, and the various types of T-cells.

- Explain what constitutes immunological memory.

- Explain the importance of lymphocyte circulation through the body.

- Make an appropriately labeled diagram of an antibody, and describe the unique roles of its variable and constant regions.

- Compare and contrast the five classes of antibodies relative to their chief body locations and specific roles in immunity.

- Describe several ways in which antibodies act against antigens.

- Distinguish between active and passive immunity.

- Define monoclonal antibodies, explain how they are produced, and describe their current importance in clinical medicine.

- Compare and contrast immunodeficiencies, allergies, and autoimmune diseases relative to their causes and consequences.

- Describe the effects of aging on immunity.

Most of us would find it wonderfully convenient if we could walk into a single clothing store and buy a complete wardrobe—hat to shoes—that fit us "to a T," regardless of any special figure problems. Yet we take for granted our **immune system**, a built-in defense system that stalks and eliminates, with nearly equal precision, almost any type of pathogen that intrudes in our bodies. When our immune system is operating effectively, it silently and swiftly protects us from most bacteria and their toxins, viral infections, transplanted organs or grafts, and even our own cells that have turned against us. Our resulting resistance to disease is called **immunity**. When our immune system fails, malfunctions, or is disabled, some of the most devastating diseases, such as cancer, rheumatoid arthritis, and AIDS, may result.

OVERVIEW OF THE IMMUNE RESPONSE

Although the study of immunity is a recent science, even the ancient Greeks knew that once a person had suffered through certain diseases, such as mumps and measles, he was unlikely to have that disease again. The basis of this immunity was revealed in the late 1800s when it was shown that animals that survived a serious bacterial infection had "factors" in their blood that protected them from future attacks by the same pathogen. (These factors are now known to be unique proteins, called *antibodies*.) Furthermore, it was demonstrated that if blood serum from these animals was transfused or injected into other animals, the recipient animals would also be protected, even though they had never been exposed to the bacteria themselves.

These landmark experiments revealed three important aspects of the immune response:

1. It is very *specific*—it recognizes, and is directed against, *particular* pathogens or foreign substances.

2. It is *systemic*—immunity is not restricted to the initial infection site.

3. It has *memory*—it recognizes, and mounts an attack on, pathogens previously encountered.

This was exciting news, but in the mid-1900s it was discovered that injection of immune (antibody-containing) serum was not always able to protect a recipient from diseases the donor had survived but that in such cases injection of the donor's lymphocytes would.

Thus, the pieces began to fall into place; and two unique, but overlapping, arms of immunity have been recognized. **Antibody-mediated immunity**, originally called **humoral immunity**, reflects the work of antibodies present in the body's "humors," or fluids (blood, lymphatic fluid, saliva, etc.). Where lymphocytes are required to provide protection, the immunity is called **cellular** or **cell-mediated immunity**, because the protective factor is the living cells.

Structures of the Immune System

Although certain organs of the body (notably lymphatic organs) are intimately involved with the immune response, our immune system is a *functional system,* rather than an organ system in the true sense. Its "structures" are trillions of individual cells that inhabit lymphatic tissues and circulate in body fluids, and a diverse array of molecules.

The most important of the immune cells are *lymphocytes,* which exist in two major "flavors"—**T-lymphocytes** and **B-lymphocytes** (or simply **T-cells** and **B-cells**). T-lymphocytes mediate cel-

lular immunity, whereas B-lymphocytes are involved in the formation of antibody molecules that are released into body fluids, that is, they oversee antibody-mediated immunity.

However, most immune responses involve both arms of immunity and require the services of still another group of cells, the phagocytic *macrophages.* As you will see, cellular interactions between these cells (macrophage to T-cell or B-cell, T-cell to T-cell, and T-cell to B-cell) underlie virtually every phase of the immune response. Besides antibodies, other proteins present in the blood (complement) and various chemicals released by lymphocytes and macrophages also play crucial roles in immunity. Table 11-1 provides a brief summary of the roles of the cells and molecules involved in the immune response.

Bacteria, cancer cells, and other harmful substances use both blood and lymphatic fluid as a means of dispersal within the body. Consequently, the structures of the lymphatic system, such as the lymphatic vessels that transport lymph, and lymph nodes or other lymphoid organs that filter lymph, produce lymphocytes, and house macrophages are intimately intertwined with the operation of the immune system.

Response to Foreign Substances

To summarize then, the immune system is a two-armed defensive system that uses macrophages, sensitized lymphocytes, and specific molecules to identify and destroy all substances—both alive and inert (nonliving)—that are in the body but are not recognized as being part of the body, or as being *self.* Lacking any central controlling organ to respond to such threats, the immune system depends on the ability of its individual cells to recognize foreign substances (antigens) in the body and to communicate with one another so that the system as a whole can mount a response that is specific to those pathogens. In the discussion that follows, we will put the "flesh" on these bones by investigating the nature of antigens, and by describing the roles of the cells and molecules involved in immunity.

TABLE 11-1 Functions of Cells and Molecules Involved in Immunity

Element	*Function in the immune response*
CELLS	
B-cell	Lymphocyte that resides in the lymph nodes or spleen, where it is induced to replicate by antigen-binding and macrophage and helper T-cell interactions; its progeny (clone members) form memory cells or plasma cells
Plasma cell	Antibody-producing "machine"; produces huge numbers of the same antibody (immunoglobulin); represents further specialization of B-cell descendants
Helper T-cell	A *regulatory* T-cell that binds with a specific antigen presented by a macrophage; upon circulating into the spleen and lymph nodes, it stimulates the production of other cells (killer T-cells and B-cells) that help fight the invader; acts both directly and indirectly by releasing lymphokines
Killer T-cell	Also called a cytotoxic T-cell; recruited and activated by helper T-cells; its specialty is killing virus-invaded body cells, as well as body cells that have become cancerous
Suppressor T-cell	Slows or stops the activity of B- and T-cells once the infection (or onslaught by foreign cells) has been conquered
Memory cell	May be a descendant of an activated B- or T-cell; generated during the initial immune response (primary response); may exist in the body for years thereafter, enabling it to respond quickly and efficiently to subsequent infections or meetings with the same antigen
Macrophage	Engulfs and digests antigens that it encounters, and presents parts of them on its plasma membrane for recognition by T-cells bearing receptors for the same antigen; this function, antigen presentation, is essential for normal helper T-cell function; also releases chemicals that activate the T-cells
MOLECULES	
Antibody (immunoglobulin)	Protein produced by a B-cell or its plasma cell offspring, and released into the body fluids (blood, lymph, saliva, mucus, etc.), where it attaches to antigens and acts to neutralize them (by precipitation or agglutination), or "tags" them for destruction by phagocytes (neutrophils) or for lysis by chemicals (complement)
Lymphokines	Chemicals, including the following, released by sensitized T-cells: • Migration inhibitory factor (MIF)—inhibits migration of macrophages, thus keeping them in the immediate area • Macrophage-activating factor (MAF)—"activates" macrophages to become killers • Interleukin II—stimulates T-cells to proliferate • Helper factors—enhance antibody formation by plasma cells • Suppressor factors—suppress antibody formation or T-cell-mediated immune responses • Chemotactic factors—attract leukocytes (neutrophils, eosinophils, and basophils) into the inflamed area

TABLE 11-1 Functions of Cells and Molecules Involved in Immunity

Element	Function in the immune response
	• Lymphotoxin (LT)—a growth inhibitor and cell toxin; causes cell lysis • Gamma interferon—helps make tissue cells resistant to viral infection; also released by macrophages
Complement	Group of blood-borne proteins that are activated when they become bound to antibody-covered antigens; when activated, complement causes lysis of the microorganism and enhances the inflammatory response
Monokines	Chemicals, including the following, released by activated macrophages: • Interleukin I—stimulates T-cells to proliferate and causes fever • Interferon—helps protect tissue cells from virus particles that have invaded them by preventing replication of the virus particles within them

ANTIGENS

Antigens are substances that are capable of mobilizing our immune system and provoking an immune response. Most antigens are large, complex molecules (macromolecules) that are not normally present in our bodies. Consequently, as far as our immune system is concerned, they are intruders, or *nonself*.

An almost limitless variety of substances can act as antigens. Nucleic acids, many large carbohydrates (polysaccharides), and nearly all foreign proteins, including modified proteins such as glycoproteins, are antigenic. Of these, proteins appear to be the strongest, or most potent, antigens. Microorganisms such as bacteria, fungi, and virus particles are antigenic because their surface membranes (or coat, in the case of a virus) bear such foreign macromolecules.

It is also important to remember that our own cells are richly studded with proteins and other large molecules, which are sometimes referred to as "self-antigens." Somehow, as our immune system develops, it takes an inventory of all of these body chemicals so that, thereafter, they are recognized as self and do not trigger an immune response. However, our cells *are* antigenic to other people, just as their cells represent antigens to us. This helps explain why our bodies reject cells of transplanted organs or foreign grafts unless special measures (drugs and others) are taken to cripple the immune response.

As a rule, small molecules are not antigenic; but when they link up with our own proteins, the combination may be recognized by our immune system as foreign. In such cases, the troublesome small molecule is called a *hapten* (haptein = to grasp), or *incomplete antigen*. Perhaps the most dramatic and familiar example of this phenomenon involves the binding of penicillin to blood proteins, which causes a *penicillin reaction* in some people. In such cases, the immune system mounts such a vicious attack that the person's life is often at risk. Besides certain other drugs, chemicals that act as haptens are found in poison ivy and animal dander, as well as in some detergents, soaps, hair dyes, cosmetics, and other commonly used household and industrial products.

ORIGIN, DIFFERENTIATION, AND ACTIVITY OF LYMPHOCYTES

Like all blood cells, lymphocytes originate from the same stem cells (hemocytoblasts) in red bone marrow that produce all other formed elements. However, after birth, lymphocytes are produced from the division of their ancestral cells that have migrated to the lymphatic tissues of the body from the fetal bone marrow. The lymphocytes formed in the fetal bone marrow are immature and essentially identical. Their subsequent transformation, or *differentiation,* into mature T- or B-lymphocytes is a two-stage process, involving (1) the development of immunocompetence and (2) clonal selection, that occurs later in the lymphatic tissues of the body (Figures 11-1 and 11-2).

Development of Immunocompetence

Lymphocytes become **immunocompetent** when they become capable of mounting an immune response against a specific antigen. This readiness is signaled by the appearance of specific proteins on their external surface that enables them to recognize a particular antigen by binding to it. For example, receptors of one lymphocyte can recognize the hepatitis A virus, those of another lymphocyte can bind to pneumococcus bacteria, and so forth.

Although the details of this maturation process are still beyond our grasp, we do know that lymphocytes become immunocompetent *before* meeting the antigens they may later attack. An incredibly large number of possible receptors appear on the surfaces of the different lymphocytes (from the shuffling of their "immunity genes"), which allows us to respond to an equally enormous number of antigens. Only some of the possible antigens our lymphocytes are programmed to resist will ever invade our bodies. Consequently, only some of our total army of immunocompetent cells will be mobilized during our lifetime; the others will be forever idle. As usual, our bodies have done their best to protect us. *Our genes, not antigens, deter-*

mine what specific foreign substances our immune system will be able to recognize and resist.

In humans, the "programming" site where B-cells develop immunocompetence has not yet been identified; and very little is known about how our B-cells accomplish this initial stage of development, other than that they form antibody molecules that they insert into their surface membranes. The bound antibodies serve as the antigen-specific receptors and are ready to bind to that antigen should it invade the body.

B-lymphocytes were first identified in the *bursa of Fabricius,* a digestive organ found only in birds, and thus they were named B- (for bursa) cells. Humans lack a bursa, but the bursa-equivalent organ has been "guestimated" to be the collections of lymphatic tissue called *GALT (gut-associated lymphatic tissues)* in the wall of the small intestine or, perhaps, the bone marrow itself. At any rate, by the time an infant is a few months old, immunocompetent B-cells have been distributed to lymphatic tissues (primarily the lymph nodes and spleen), where they await the antigen-challenge (Figure 11-1). It is in these secondary lymphatic organs that the second stage of the maturation process, clonal selection, occurs.

T-Cells

T-cells, by definition, are lymphocytes that become immunocompetent in the thymus shortly before and after birth (Figure 11-2). The thymus, a lymphoid organ located in the throat and upper thorax, is particularly large during this time of life, and then gradually wastes away and is replaced by connective tissue as we age. The development process appears to be directed by thymic hormones (*thymosin* and others); and while the T-cell stem cells are in the thymus, they divide rapidly, and their numbers increase enormously. The receptors that appear on their surfaces as they become immunocompetent look much like "half-antibodies."

Actually, most of the immature lymphocytes that enter the thymus die there. It seems that the thymus selects only those developing T-cells with the sharpest abilities for identifying foreign substances: It vigorously weeds out and destroys those that are capable of binding with self-antigens and

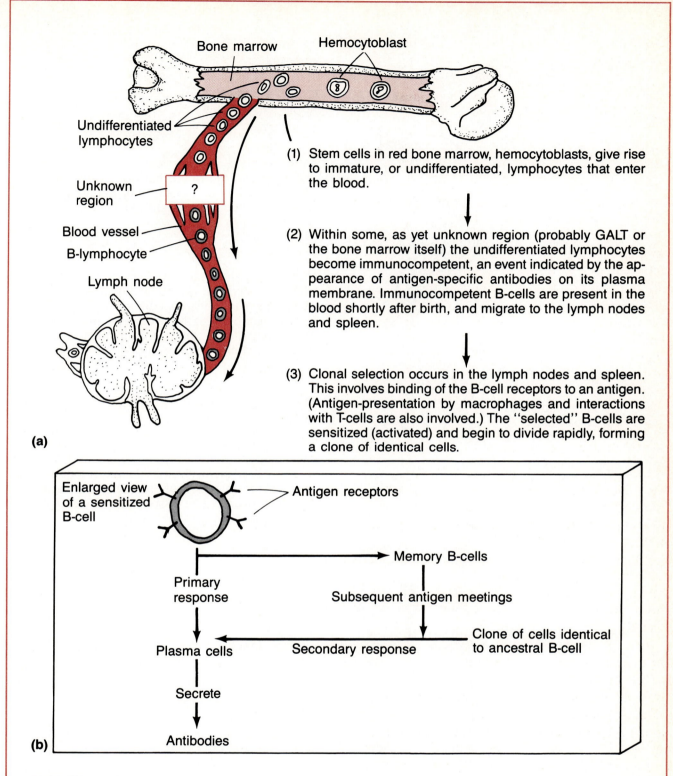

(1) Stem cells in red bone marrow, hemocytoblasts, give rise to immature, or undifferentiated, lymphocytes that enter the blood.

(2) Within some, as yet unknown region (probably GALT or the bone marrow itself) the undifferentiated lymphocytes become immunocompetent, an event indicated by the appearance of antigen-specific antibodies on its plasma membrane. Immunocompetent B-cells are present in the blood shortly after birth, and migrate to the lymph nodes and spleen.

(3) Clonal selection occurs in the lymph nodes and spleen. This involves binding of the B-cell receptors to an antigen. (Antigen-presentation by macrophages and interactions with T-cells are also involved.) The "selected" B-cells are sensitized (activated) and begin to divide rapidly, forming a clone of identical cells.

(a)

(b)

FIGURE 11-1

B-cell development and roles in the immune response. **(a)** Sequence of developmental events. **(b)** Primary and secondary responses of sensitized B-cells.

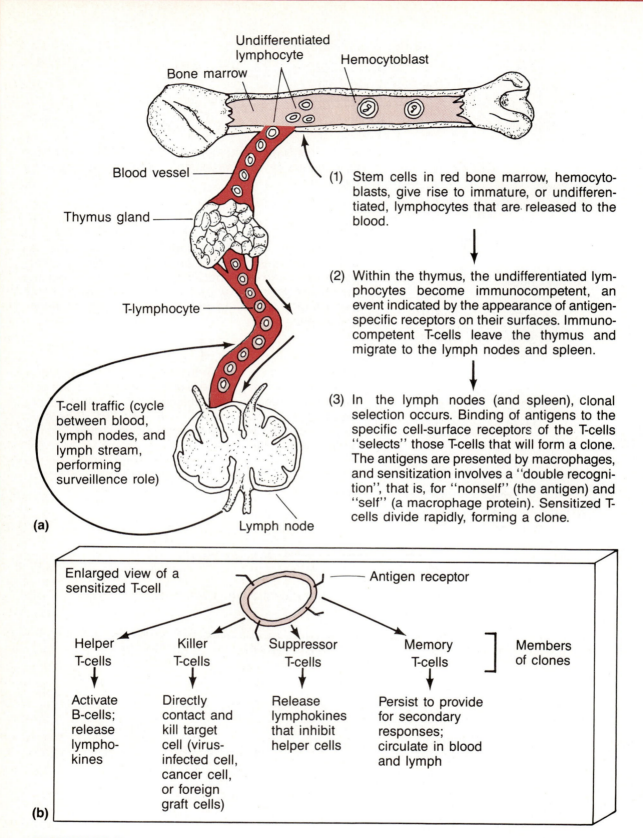

(1) Stem cells in red bone marrow, hemocyto-blasts, give rise to immature, or undifferen-tiated, lymphocytes that are released to the blood.

(2) Within the thymus, the undifferentiated lym-phocytes become immunocompetent, an event indicated by the appearance of antigen-specific receptors on their surfaces. Immuno-competent T-cells leave the thymus and migrate to the lymph nodes and spleen.

(3) In the lymph nodes (and spleen), clonal selection occurs. Binding of antigens to the specific cell-surface receptors of the T-cells "selects" those T-cells that will form a clone. The antigens are presented by macrophages, and sensitization involves a "double recogni-tion", that is, for "nonself" (the antigen) and "self" (a macrophage protein). Sensitized T-cells divide rapidly, forming a clone.

FIGURE 11-2

T-cell development and roles in the immune response. **(a)** Sequence of developmental events. **(b)** Members of a T-cell clone and their roles.

of acting against the body. Thus, the development of *tolerance* for self-antigens is an integral part of the maturation process. (This is true not only for T-cells, but also for B-cells.) Then, two to three days later, the T-cell survivors stream from the thymus into the bloodstream, and circulate to and through other lymphoid organs of the body (lymph nodes or spleen) while waiting for "their" antigen to appear.

Clonal Selection

Clonal selection occurs when antigens bind to the surface receptors of the immunocompetent (but still immature) T-cells and B-cells. This binding event *sensitizes,* or activates, the lymphocyte to "switch on"; that is, it begins to grow, and then multiplies rapidly to form an army of cells all exactly like itself and bearing the same antigen-specific receptors. The resulting family of identical cells descended from the *same* ancestor cell is called a **clone**, and clone formation represents the **primary immune response** to that antigen. Because the trigger for clone formation is antigen binding, the antigens "select" the lymphocytes that will be activated to form clones.

B-cell Clones

As described above, a B-cell proliferates following sensitization (Figure 11-1), and most of its clone members or descendants become **plasma cells**. After an initial lag period, these antibody-producing "factories" swing into action, producing the same highly specific antibodies at an unbelievable rate of about 2000 antibody molecules per second. However, this flurry of activity is short-lived and lasts for only four or five days. Then the plasma cells begin to die. Antibody levels in the blood during this initial response peak in about three weeks and then slowly begin to decline (Figure 11-3).

B-cell clone members that do not become plasma cells remain behind to form a memory bank of sensitized cells that can respond to the same antigen at later meetings with it. These later immune responses, called **secondary responses**, are both much faster and more effective because all of the preparations for this attack have already been made (Figure 11-3). Huge numbers of antibodies literally flood into the bloodstream within hours after recognition of the old-enemy antigen. As noted earlier, because B-cells (and their offspring) act indirectly and at long distance via antibodies released into the blood, they provide for humoral immunity. The manner in which these antibodies function to protect the body is described later in this chapter.

T-cell Clones

The members of a T-cell clone, which provide for cell-mediated immunity, are a much more diverse lot and produce their deadly effects in a variety of ways (Figure 11-2). Some become **killer** (or **cytotoxic**) **T-cells**, effector cells that specialize in killing virus-infected, cancer, or foreign (for example, transplanted or grafted) cells. They accomplish this by binding to them (delivering the so-called "kiss of death") and by releasing toxic chemicals called *lymphotoxins*. Shortly thereafter, the target cells rupture. By that time, the killer cell is long gone and is seeking other foreign prey to attack.

Helper T-cells are regulatory cells that act as the managers of the immune system. When they identify an antigen, they circulate through the body, recruiting other cells to fight the invader. For example, helper T-cells interact directly with B-cells (that have already attached to antigens), prodding them into division (clone production) and then, like the foreman of an assembly line, signaling for antibody formation to begin. They also release chemicals called *lymphokines* that act indirectly to rid the body of antigens by (1) stimulating killer T-cells and B-cells to grow and divide; (2) attracting other types of protective white blood cells, such as neutrophils, into the area; and (3) enhancing the ability of macrophages to ingest and destroy microorganisms. (Actually, the macrophages are pretty effective phagocytes even in the absence of lymphokines, but in their presence they develop an "insatiable appetite.") As the released lymphokines summon more and more cells into the battle, the immune response gains momentum, and the antigens are overwhelmed by the sheer numbers of immune elements acting against them.

Another regulatory T-cell population, the **suppressor T-cells**, releases suppressor chemicals that inhibit the helpers and thus indirectly suppress antibody formation and/or killer T-cell activity. Suppressor T-cells are vital for winding down and finally stopping the entire range of immune responses once an antigen has been successfully

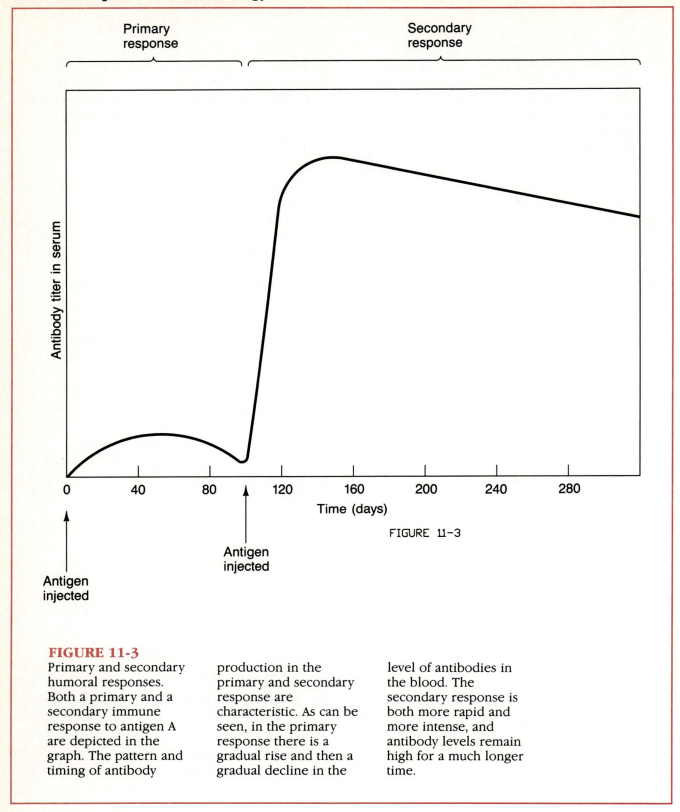

FIGURE 11-3

FIGURE 11-3
Primary and secondary humoral responses. Both a primary and a secondary immune response to antigen A are depicted in the graph. The pattern and timing of antibody production in the primary and secondary response are characteristic. As can be seen, in the primary response there is a gradual rise and then a gradual decline in the level of antibodies in the blood. The secondary response is both more rapid and more intense, and antibody levels remain high for a much longer time.

inactivated and/or destroyed. This helps prevent uncontrollable or unnecessary immune system activity.

Although most of the T-cells enlisted to fight a particular immune response are dead within a few days, a few members of each clone become long-lived **memory cells** that remain to provide the immunologic memory for each antigen encountered and to enable the body to respond more quickly to its subsequent invasions. AIDS, a disease that cripples the immune system by interfering with the normal activity of the T-lymphocytes, will be described later in this chapter.

ORIGIN AND ACTIVITY OF MACROPHAGES

Macrophages arise from monocytes formed in the red bone marrow, and then become broadly distributed throughout the lymphatic organs and connective tissues of the body. Once in the tissues, the macrophages (literally, big eaters) undergo additional changes that allow them to act as front-line body defenders by phagocytizing not only antigens but virtually any type of debris.

The macrophages also perform several functions that are crucial to the immune response, including:

1. Antigen-presentation. Macrophages engulf and process antigens, and then display parts of them, like signal flags, on their own surfaces, where they can be recognized by T-cells (Figure 11-4). Apparently, T-cell activation requires a *double recognition*; the T-cell must not only recognize "nonself," the antigen presented by the macrophage, but also recognize "self" by coupling with a specific protein on the macrophage's surface at the same time. Thus, it seems that antigen binding alone is not enough to sensitize T-cells—they must be "spoon fed" the antigens by macrophages, and a "double handshake" must occur. Although this idea seemed preposterous when it was first suggested, there is no longer any question that *antigen-presentation* is a major role of macrophages, and is essential for activation and clonal selection of the T-cells.* Without macrophage "presenters," the immune response is severely impaired.

2. Secretion of monokines. Macrophages secrete soluble proteins, called *monokines*, that are similar to the lymphokines released by T-cells and, like them, have a variety of functions. For example, macrophages that have engulfed viruses release *interferon*, which helps protect as yet uninvaded body cells from viral effects,

*Recent evidence indicates that macrophage antigen-presentation to B-cells also occurs and is necessary for B-cell clonal selection.

and *interleukin I*, which activates T-cells that have recognized viruses. These two monokines cause much of the fever and malaise that accompany many viral infections. In turn, as noted above, lymphokines released by T-cells activate macrophages. This is a classic example of one hand washing the other.

LYMPHOCYTE TRAFFIC AND IMMUNE SURVEILLANCE

Although many macrophages remain fixed in lymphatic tissues, patiently waiting for antigens to be brought to them, some (such as those in connective tissues) do quite a bit of wandering, patrolling the body as they move. Lymphocytes are extremely mobile and circulate continuously through the body, moving in cycles from the blood into the lymphoid organs and then out again in the blood or lymphatic stream (Figure 11-5).

T-cells account for the bulk (75% to 85%) of the lymphocytes present in blood, where they are recognizable as the *small lymphocytes*. B-cells are much less mobile, circulate more slowly, and tend to remain in the lymphatic organs for extended periods. If you keep in mind that (1) each lymphocyte can respond to only one specific antigen and (2) cellular interactions are required for *all* immune responses, the continual recirculation of lymphocytes between the lymph nodes and the body fluids makes a good deal of sense. For example, this recirculation allows the lymphocytes to perform a surveillance role, and it increases their chances of coming into contact both with (their) antigens and with huge numbers of macrophages and other lymphocytes. Also, since lymph capillaries pick up proteins and pathogens from nearly all body tissues, the lymph nodes are in a strategic position for collecting a wide variety of antigens for the lymphocytes to sample.

ANTIBODIES

Antibodies, also called **immunoglobulins (Igs)**, constitute the *gamma globulin* part of blood proteins. As mentioned, antibodies are soluble proteins secreted by sensitized B-cells, or more spe-

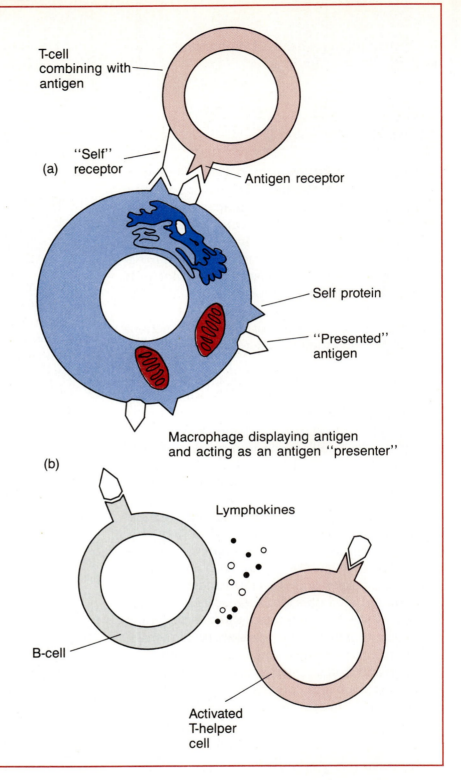

FIGURE 11-4
Cellular interactions in the immune response. **(a)** Macrophages, prevalent in connective tissues and lymphatic tissues of the body, are important both as phagocytes and as antigen-presenters. After they have ingested an antigen, they display parts of it on their surface membranes, where it can be recognized by a T-cell (or B-cell) bearing receptors for the same antigen. During the binding process, the T-cell binds both to the antigen and to macrophage (self) receptors, which leads to T-cell activation. **(b)** Once activated, helper T-cells can interact with B-cells both directly (by binding) and indirectly (by releasing lymphokines that stimulate the B-cells to proliferate). The latter situation is illustrated.

cifically by their plasma-cell offspring, in response to an antigen and are capable of binding specifically with that antigen.

Even though antibodies are formed in response to a staggering number of different antigens, and despite their variety, they all have a similar basic anatomy that allows them to be grouped into five Ig classes, each is slightly different in structure and function. We will consider ways in which the Ig classes differ, but first we will look at how all antibodies are alike.

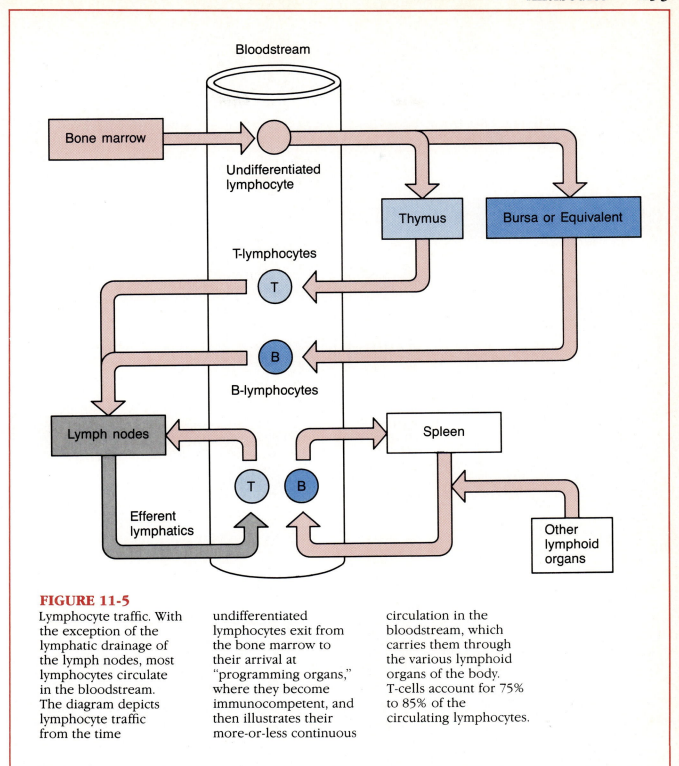

FIGURE 11-5

Lymphocyte traffic. With the exception of the lymphatic drainage of the lymph nodes, most lymphocytes circulate in the bloodstream. The diagram depicts lymphocyte traffic from the time undifferentiated lymphocytes exit from the bone marrow to their arrival at "programming organs," where they become immunocompetent, and then illustrates their more-or-less continuous circulation in the bloodstream, which carries them through the various lymphoid organs of the body. T-cells account for 75% to 85% of the circulating lymphocytes.

Basic Antibody Structure

Regardless of its class, each antibody has a basic structure consisting of four amino-acid (polypeptide) chains linked together by disulfide (sulfur-to-sulfur) bonds (Figure 11-6). Two of the four chains are identical and contain approximately 400 amino acids each; these are called the *heavy chains*. The other two polypeptide chains are also identical to each other, but contain only half as many amino acids as the heavy chains; these are the *light*

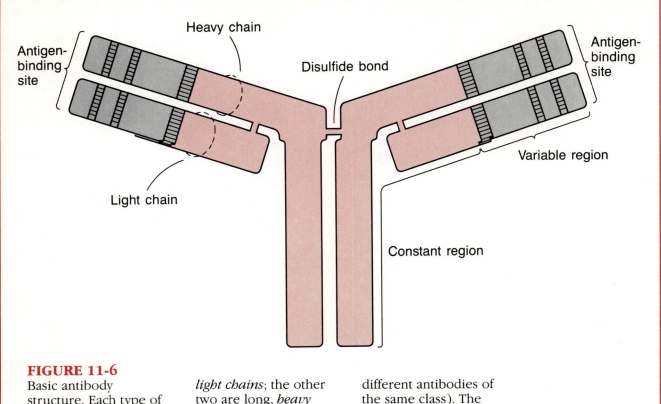

FIGURE 11-6
Basic antibody structure. Each type of antibody is formed of four amino-acid (polypeptide) chains that are joined together by disulfide bonds. Two of the chains are short, *light chains*; the other two are long, *heavy chains*. Each chain has a variable region (which is different in different antibodies) and a constant region (which is essentially identical in different antibodies of the same class). The variable regions are the antigen-binding sites of the antibody; hence, each antibody has two antigen-binding sites.

chains. When the four chains are combined, the antibody molecule formed has two identical halves, each consisting of a heavy and a light chain, and the molecule as a whole is T- or Y-shaped.

When researchers began to investigate the structure of antibodies, they discovered something very peculiar. Each of the four chains forming an antibody had a *variable (V) region* confined entirely to one end and a much larger *constant (C) region.* The naming of these antibody regions reflected the observation that the sequence of the amino acids in the variable regions was quite dissimilar in antibodies responding to different antigens, but their constant regions were the same (or nearly so). This began to make sense when it was discovered that the variable regions of the heavy and light chains in each arm combine their efforts to

form an *antigen-binding site* (Figure 11-6) uniquely shaped to fit its specific antigen. Hence, each antibody has two such antigen-binding regions.

The constant regions of the antibody chains can be compared to the handle of a key. A key handle serves a common function in all keys: It allows you to hold the key and place its tumbler-moving portion into the lock. Likewise, the constant regions of antibody chains serve common functions in all antibodies; that is, they determine the type of the antibody formed (antibody class), how the antibody class will carry out its immune roles in the body, and with what cell types or chemicals the antibody can bind. For example, some, but not all, antibodies can fix complement, some circulate in blood whereas others are found primariiy in mucus and saliva, and some cross placental barriers.

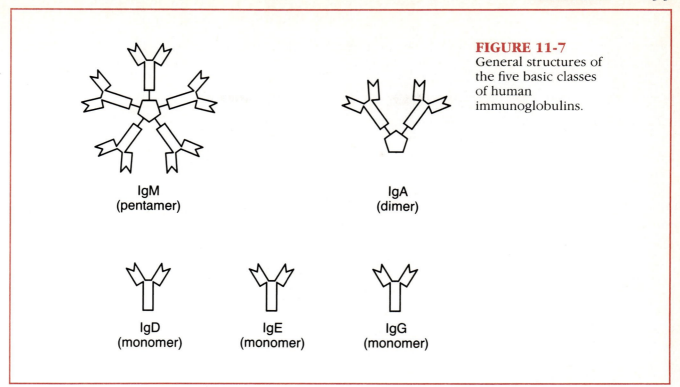

IgM
(pentamer)

IgA
(dimer)

IgD
(monomer)

IgE
(monomer)

IgG
(monomer)

Antibody Classes

There are five major immunoglobulin classes, designated as IgD, IgM, IgG, IgA, and IgE. As illustrated in Figure 11-7, members of the IgD, IgG, and IgE classes have the same basic Y-shaped structure described earlier, and are referred to as *monomers*. IgA antibodies occur in both monomer and *dimer* (two linked monomers) forms. (Only the dimer form is shown in Figure 11-7). Compared to the other antibodies, those of the IgM class are huge. Because they are constructed of five linked monomers, IgM antibodies are called *pentamers* (penta = five).

The antibodies of each class have slightly different biological roles and locations in the body:

Immunoglobulin D (IgD) antibodies are virtually always found attached to B-cells. Although their precise function is not known, they are believed to be identical to the surface receptors that emerge as B-cells become immunocompetent. Thus, IgD antibodies are probably important in the sensitization of B-cells and in the initial events of antibody formation.

Immunoglobulin M (IgM) antibodies account for about 10% of the gamma globulins found in blood plasma. These antibodies, like IgD antibodies, are found inserted in the membranes of B-cells. Additionally, IgM antibodies are the first *released* to the bloodstream by activated B-cells and their progeny plasma cells.

Immunoglobulin G (IgG) antibodies are the most abundant antibodies found in the blood and account for 75% to 85% of all circulating antibodies. These antibodies, like IgM antibodies, fix complement, and they are the only antibody type that can cross the placenta. Hence, the passive immunity that a mother transfers to her fetus or newborn is "with the compliments" of her IgG antibodies. (The importance of complement in antibody activity will be described later in this chapter.)

Immunoglobulin A (IgA) antibodies are released by plasma cells located in epithelial membranes, and are found primarily in mucus and in other secretions, such as saliva, intestinal juice, and tears, that bathe body surfaces. Their major role is to prevent pathogens from gaining entry into the body.

As a general rule, *immunoglobulin E (IgE)* antibodies are found in very low levels in the blood. You might think of these as "troublemaker antibodies," because they are involved in overzealous immune responses called allergies. IgE antibodies attach to mast cells or basophils, and then when the IgE antibodies bind to an antigen, histamine is released from their cell "carriers." This so-called acute allergic response will be described more fully.

Although some may argue the point, it is unlikely that IgE antibodies exist to "bedevil us." Some believe that IgE antibodies play an important role in defending the body against certain types of parasites; and the fact that IgE antibody levels are high in peoples living in the tropics, where parasitic infections are common, seems to support this notion.

Antibody Function

Antibodies defuse, or inactivate, antigens in a number of ways (Figure 11-8)—by complement fixation, neutralization, agglutination, and precipitation—but of these, complement fixation and neutralization are most important.

COMPLEMENT FIXATION

The term **complement** refers to a group of eighteen blood proteins that normally circulate in the inactive state. However, when complement becomes attached, or fixed, to antigen–antibody complexes, it is activated and becomes a major factor in the fight against the foreign cells.

When antibodies bind to surface antigens of foreign cells such as bacteria or mismatched red blood cells, their shape changes in such a way that *complement binding sites* are exposed on their constant regions. This triggers, in rapid-fire sequence, a number of events that result in **complement fixation** to the antigenic cell's surface. The final result is that some of the bound complement proteins assemble themselves so that doughnut-shaped lesions, complete with holes, are formed in the foreign cell's surface (Figure 11-9). This allows water to rush into the cell, causing it to burst.

Besides its ability to rupture, or lyse, invading microorganisms or other foreign cells, activated complement also enhances the inflammatory response. Some of the molecules formed and released during the complement activation process are vasodilators, and some are *chemotaxis chemicals* that attract neutrophils and macrophages into the region. Others alter the cell membranes of the foreign cells so that they become sticky and more

easily phagocytized. This enhancement of phagocytosis is called *opsonization*.

NEUTRALIZATION

Neutralization occurs when antibodies bind to specific sites on bacterial exotoxins (toxic chemicals secreted by bacteria) or on viruses that can act to cause cell injury. In this way, they block the harmful effects of the exotoxin or virus.

AGGLUTINATION AND PRECIPITATION REACTIONS

Because antibodies have more than one antigen-binding site, they can bind to more than one antigen at a time; consequently, antigen–antibody complexes can be crosslinked into large lattices. When the crosslinking involves cell-bound antigens, the process causes clumping of the foreign cells and is called **agglutination**. This type of antigen–antibody reaction occurs when mismatched blood is transfused (the foreign red blood cells are clumped) and is the basis of tests used for blood typing. When the crosslinking process involves large numbers of soluble antigenic molecules and the resulting antigen–antibody complexes are so large that they become insoluble and settle out of solution, the reaction is more precisely called **precipitation**.

There is little question that agglutinated bacteria and immobilized (precipitated) antigen molecules are much more easily captured and engulfed by the body's phagocytes than are freely moving antigens.

TYPES OF IMMUNITY

Active Immunity

When your immune cells encounter antigens and produce antibodies against them, you are exhibiting **active immunity**. Active immunity is *naturally acquired* during bacterial and viral infections, during which we may become symptomatic and suffer a little (or a lot). However, it makes little difference whether the antigen invades the body under its own power or is introduced in the form of a *vaccine*; the response of the immune system is pretty much the same. Indeed, once it was recognized that secondary responses are so much more

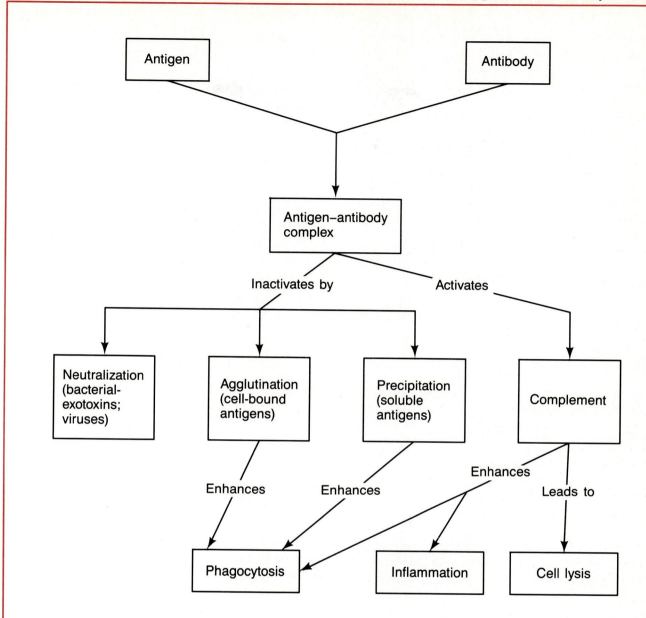

FIGURE 11-8

Mechanism of antibody action. Antibodies provide humoral immunity by forming complexes with specific antigens, and then inactivating or immobilizing them by clumping them together (agglutination) or by causing them to become insoluble (precipitation). In either case, phagocytosis is enhanced. Antibody binding to specific sites on bacterial exotoxins or viruses that are responsible for causing cell damage (neutralization) effectively blocks the harmful effects of those antigens. Another antibody mechanism, complement fixation and activation, is more lethal because it causes lysis of foreign cells. In addition, complement activation enhances the inflammatory response and phagocytosis. (This latter process is called opsonization.)

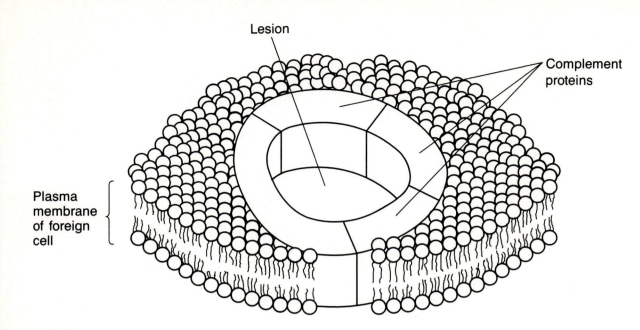

Lesion

Complement
proteins

Plasma
membrane
of foreign
cell

FIGURE 11-9

Lesion induced by
complement fixation.
When complement
proteins are activated
by binding to antibodies
attached to the surface
of foreign cells, a series
of events occurs that
results in some of the
complement proteins
becoming inserted into
the foreign cells' plasma
membranes so that
doughnut-shaped
lesions are formed.
Once the membranes
have been penetrated,
the rapid entry of water
into the cells causes
their rupture (lysis).

vigorous, the race was on to develop vaccines to
"prime" the immune response by providing a first
meeting with the antigen. Most vaccines contain
dead or attenuated (living, but extremely weak-
ened) pathogens. Because the immune attack is
mounted against only *parts* of antigenic molecules,
the pathogen need not be alive to cause the desired
response. Thus, we receive two benefits from vaccines: (1) We are spared most of the signs and
symptoms (and discomfort) of the disease which
would otherwise occur during the primary
response, and (2) the antigens are still able to stimulate antibody production and promote immunologic memory. Note that a person who has been
vaccinated is said to have *artificially acquired*
immunity.

Vaccines are currently available against microorganisms that cause pneumonia, small pox, polio,
tetanus, rabies, diphtheria, and many other diseases. Many potentially serious childhood diseases
have been virtually wiped out in the United States
by active immunization programs. A summary of
the currently recommended schedule for administering vaccines to American children is provided
in Table 11-2.

Passive Immunity

Passive immunity is quite different from active
immunity, both in the antibody source and in the
degree of protection it provides. Instead of being
made by your own plasma cells, the antibodies are
obtained from the serum of an immune human or
animal donor. As a result, immunologic memory
is *not* established, and the protection provided by
the "borrowed antibodies" ends when they naturally degrade in the body.

Passive immunity is conferred *naturally* on a fetus
when the mother's IgG antibodies cross the placenta and enter the fetal circulation. Because human
milk contains IgA and IgG antibodies, a nursing
infant receives similar immunity benefits from the
mother. Essentially, the baby is protected from all
of the antigens to which the mother has been

TABLE 11-2   Recommended Immunization Schedule for U.S. Children

Recommended age	Immunizing agent in vaccine
2 to 3 months	DPT vaccine (diphtheria toxoid, pertussis (whooping cough) vaccine, and tetanus toxoid)
4 to 5 months	DPT vaccine; OPV (oral poliomyelitis vaccine)
6 to 7 months	DPT vaccine; OPV
15 to 19 months	DPT vaccine; OPV; MMR vaccine (combined mumps vaccine, measles vaccine, and Rubella vaccine) or individual mumps, measles, and Rubella vaccines
4 to 6 years	DPT vaccine; OPV
12 to 14 years	TD vaccine (tetanus and diphtheria toxoid)

exposed, and this protection lasts from a few weeks to a few months.

Passive immunity is said to be *artificially* conferred when one receives an infusion of immune serum or gamma globulin. Most often this protection is provided to someone bitten by a poisonous snake (in which case the immune serum is called *antivenom*), or to those exposed to diseases such as botulism or tetanus (in the form of an *antitoxin*). In such cases, the person must be treated quickly if death is to be prevented. The infused antibodies provide immediate protection, but their effect is very short-lived (2 to 3 weeks).

In addition to their use to provide passive immunity, antibodies are prepared commercially for use in research and in clinical laboratory testing, and for treating certain cancers. **Monoclonal antibodies** used for such purposes are pure antibody preparations that exhibit specificity for one and only one antigen. The preparation and current uses of these unique antibodies are described in Box 11-1.

DISORDERS OF IMMUNITY

The most important disorders of the immune system are immunodeficiencies, allergies, and autoimmune diseases.

Immunodeficiencies

The **immunodeficiencies** include a large number of congenital and acquired conditions in which the production or function of the immune cells (or complement) is abnormal. By far the most devastating of the congenital conditions is *congenital thymic aplasia,* in which the thymus fails to develop. Because T-cells normally produced there are required for normal operation of both arms of the immune response, afflicted children have essentially no protection against pathogens. Thus common infections that are easily shrugged off by most people are lethal to those with this condition. Transplants of fetal thymic tissue have been successfully performed for such children, but barring such a transplant, the only hope for survival is living behind protective barriers that keep out all infectious agents.

Almost as tragic is *severe combined immunodeficiency disease (SCID),* in which there is a marked depletion of both B- and T-cells. SCID victims become deathly ill and are tragically wasted by infections that begin when they are about 3 months old. Today many of these children are being kept alive and healthy by new bone marrow transplant techniques.

Currently, the most important and most devastating of the acquired immunodeficiencies is *acquired immune deficiency syndrome (AIDS)*. First identified in 1981 among homosexuals and intravenous

BOX 11-1
Monoclonal Antibodies

One of the most interesting research fields in biology today is genetic engineering, including the production of monoclonal antibodies. Even those not scientifically inclined can hardly escape exposure to this burgeoning and exciting field of inquiry. Many of our ill citizens have benefited from bioengineering methods for the diagnosis of hepatitis and rabies. Currently, research is focusing on the clinical use of monoclonal antibodies for the diagnosis and treatment of cancer.

Although just *what* triggers normal cells to become cancerous still eludes us, we do know that something is amiss in the normal regulation of cell division and that neoplastic (cancerous) cells proliferate in an unruly way. This, in turn, has been traced to alterations in the genes (DNA blueprints) of normal body cells, which in some way causes them to become *oncogenes* (cancer-causing agents).

Just where do monoclonal antibodies fit into this picture? In some cases, normal cells that have become

cancer cells are recognized as foreign (as bacteria and virus particles are) by the body's immune system. B-lymphocytes (B-cells) then mount the attack by producing antibodies against the alien "invaders." If the antibodies are specific enough and produced in large enough numbers, the cancer cells will be destroyed.

However, in many cases, cancer cells somehow manage to escape recognition. Because they are so different from normal cells, it is difficult to understand how this could happen. Part of the problem, undoubtedly, is due to the fact that they have derived from normal cells and have many of the normal, or self-cell, markers. If normal immune response fails, cancer cells proliferate and spread, leading to clinical signs of cancer. Researchers concluded that if large amounts of identical antibodies specifically directed against certain types of cancer cells could be produced, medicine would have a potent weapon. In the 1970s, scientists set about to determine how to meet this challenge. The technique that evolved for the production of monoclonal antibodies

is diagrammed on the facing page.

The isolated, human-engineered *hybridomas* divide potentially forever, making exact copies, or *clones*, of themselves that produce one specific and pure type of antibody known as a monoclonal antibody. It is an ironic situation—making cancer cells that fight cancer cells.

Since 1980, monoclonal antibodies have been used clinically in isolated cases to treat human fluid cancers such as leukemia and lymphomas. As these cancers are present in the circulation, they are readily attacked by injected monoclonal antibodies. These studies have shown promising results, and at least partial remissions have been reported in patients with advanced cancers that have stopped responding to other forms of treatment.

Ongoing animal studies are investigating the use of monoclonal antibodies attached to cell-killing (chemotherapeutic) drugs for cancer therapy. Chemotherapy, as currently employed in humans, uses a "shotgun" approach.

That is, these toxic drugs are injected into the blood and roam freely throughout the bloodstream, killing not only cancerous but also rapidly dividing, normal body cells. The attachment of these drugs to monoclonal antibodies, which specifically recognize cancer cells, provides a "rifle-shot" approach to deliver these drug-tagged antibodies to only diseased tissue. This therapy, when completely developed for human use, will be of unimaginable importance in tracking down and destroying cancer cells that have metastasized.

In other human experiments, monoclonal antibodies are being used as diagnostic probes. Antibodies, tagged with radioactive chemicals (such as radioactive iodine), are injected into the bloodstream; a radioactivity scanner is used; and cancerous body areas are shown as bright patches. For the first time in history, it appears that it may be possible to track the full extent of a cancer within the body.

Another exciting possibility for the use of monoclonal antibodies is as cancer-diagnosing agents to be used on blood samples taken during a routine physical examination. Many cancers shed characteristic proteins into the bloodstream as they grow. If matching antibodies are mixed with a blood sample containing such proteins, it will be possible to detect hidden cancers at the earliest possible stages. Many pharmaceutical companies are funding research to develop such diagnostic tests for stomach, pancreatic, colonic, and prostatic cancers. The ultimate victory would be to completely eliminate cancer with anticancer vaccines, but monoclonal antibodies currently provide the most promise for the diagnosis and treatment of cancer.

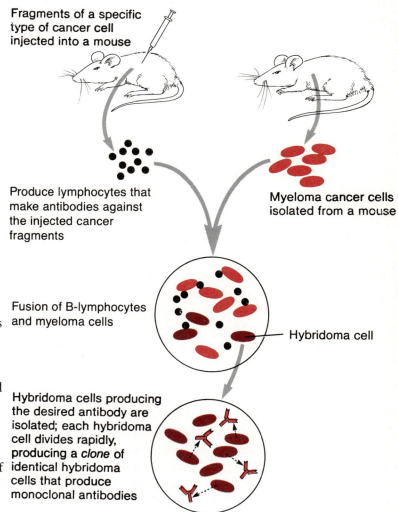

Fragments of a specific type of cancer cell injected into a mouse

Produce lymphocytes that make antibodies against the injected cancer fragments

Myeloma cancer cells isolated from a mouse

Fusion of B-lymphocytes and myeloma cells

Hybridoma cell

Hybridoma cells producing the desired antibody are isolated; each hybridoma cell divides rapidly, producing a *clone* of identical hybridoma cells that produce monoclonal antibodies drug users, AIDS is characterized by severe weight loss, swollen lymph nodes, increasingly frequent opportunistic infections, and bizarre malignancies (that is, Karposi's sarcoma) in people with no previous history of immune system disease. Many of its victims develop slurred speech and severe dementia. Its course is grim and thus far inescapable, finally ending in complete disability, wasting, and death from cancer or overwhelming infection.

It is now known that AIDS is caused by a rapidly mutating virus, that is, one that changes its surface characteristics so often that the immune system has barely responded to one "face" before the virus has changed again. Named the HTLV-III virus by American researchers, it has more recently been named HIV (human immunodeficiency virus). The virus is transmitted principally in blood or semen. Although it has also been found in saliva and tears, the virus is not believed to be transmitted by those secretions or by any form of casual contact.

The virus targets, invades, and then insidiously destroys human helper T-cells. In the absence of these cells, all aspects of cell-mediated immunity are depressed, and there is a profound deficit of normal antibodies. At the same time, suppressor T-cell activity appears to be enhanced, and abnormal antibodies are produced. The whole immune system is short-circuited, or thrown topsy-turvy. The virus also invades brain cells, which accounts for the dementia seen in some AIDS patients.

The years since 1981 have seen the birth of a swift and deadly epidemic. By the end of 1986, 30,000 Americans had been diagnosed as AIDS victims, and half of those had already died. Even more alarming is the assumption that for each case diagnosed, there were probably 100 carriers that did not yet exhibit disease symptoms because of an insufficient incubation period. (Six months to several years may pass before symptoms become apparent.) Consequently, the current prediction is that over 250,000 Americans will have been diagnosed by 1991, with over half of that number dead. Not only have the identified cases jumped exponentially in the at-risk populations, but victims now include hemophiliacs (who must receive the clotting factors of blood) and members of the general heterosexual population. Although diagnostic blood tests are currently available to identify carriers of

the AIDS virus (those exposed to the virus form antibodies against it), no cure has yet been found.

Allergies

At first, the immune response was thought to be purely protective. However, it was not long before its dangerous potentials were discovered. *Allergies* or *hypersensitivities* are overzealous or abnormally vigorous immune responses during which the immune system frequently causes tissue damage as it fights off a perceived "threat" that would otherwise be harmless to the body. The term *allergen* is used to distinguish this type of antigen from those producing essentially normal responses. People rarely die of allergies; they are just miserable with them.

Although there are several different types of allergic response, the most common type is the *acute allergic response*. This type of response, also called *immediate hypersensitivity,* is triggered by the binding of IgE antibodies to mast cells (Figure 11–10). The Ig-coated mast cells release a flood of histamine, an important inflammatory chemical, that causes small blood vessels in the area to become dilated and leaky, and is largely to blame for the best-recognized symptoms of allergy, that is, a runny nose, watery eyes, and itching, reddened skin (hives). When the allergen enters the lungs, the symptoms of asthma also appear, resulting from the stimulation, by histamine, of the smooth muscle in the walls of the bronchioles. As the muscle contracts, the passages narrow and restrict air flow. (Over-the-counter (OTC) antiallergy drugs contain antihistamines that counteract these effects.) Most of these reactions begin within minutes of exposure to the allergen and only last about 30 minutes.

Fortunately, bodywide or systemic acute allergic responses, known as *anaphylaxis*, are fairly rare. Anaphylaxis occurs when the allergen directly enters the blood and circulates rapidly through the body, as might happen with certain insect (bee or wasp) stings or spider bites, and/or following an injection of a foreign substance (such as horse serum, penicillin, or other drugs) into susceptible individuals. The mechanism of anaphylaxis is essentially the same as that of local responses; but when the entire body is involved, the outcome may

FIGURE 11-10
Mechanism of an acute
allergic (immediate
hypersensitivity)
response.

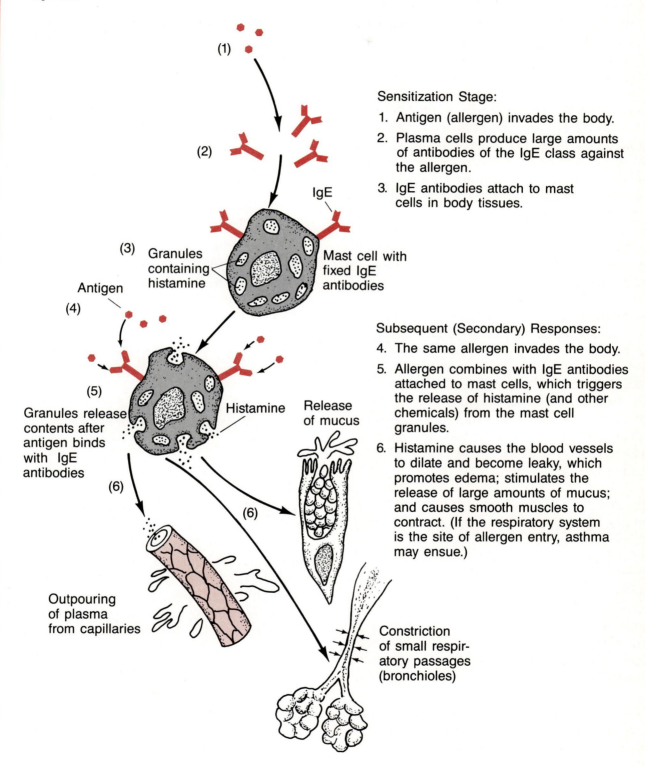

Sensitization Stage:

1. Antigen (allergen) invades the body.
2. Plasma cells produce large amounts of antibodies of the IgE class against the allergen.
3. IgE antibodies attach to mast cells in body tissues.

Subsequent (Secondary) Responses:

4. The same allergen invades the body.
5. Allergen combines with IgE antibodies attached to mast cells, which triggers the release of histamine (and other chemicals) from the mast cell granules.
6. Histamine causes the blood vessels to dilate and become leaky, which promotes edema; stimulates the release of large amounts of mucus; and causes smooth muscles to contract. (If the respiratory system is the site of allergen entry, asthma may ensue.)

IgE

(3) Granules containing histamine

Mast cell with fixed IgE antibodies

Antigen

(4)

(5)

Granules release contents after antigen binds with IgE antibodies

Histamine

Release of mucus

(6)

(6)

Outpouring of plasma from capillaries

Constriction of small respiratory passages (bronchioles)

be life-threatening. For example, smooth muscles of the lung passages contract, making it difficult to breathe; and, unless treated immediately, the sudden vasodilation (and fluid loss) may precipitate circulatory collapse and death. Epinephrine is routinely injected to reverse these histamine-mediated effects.

Delayed reaction allergies (delayed hypersensitivity reactions) mediated by T-cells take much longer to appear (hours to days) than do the acute reactions produced by antibodies. Lymphokines released by activated T-cells are the major mediators of this type of allergic response; thus antihistamine drugs are *not* helpful against the delayed types of allergies. (Only the corticosteroid drugs seem to provide some relief.)

The best known examples of delayed hypersensitivity reactions are those classed as *allergic contact dermatitis* reactions, which follow skin contact with poison ivy, some heavy metals, and certain cosmetic and deodorant chemicals. These types of agents act as haptens—that is, after diffusing through the skin and attaching to body proteins, they are perceived as foreign by the immune system. The *Mantoux* and *tine tests*, skin tests for detection of tuberculosis, also depend on delayed hypersensitivity reactions. When the tubercle antigens are injected just under (or scratched into) the skin, a small hard lesion forms if the person has been sensitized to the antigen.

Autoimmune Diseases

Sometimes the body's ability to tolerate self-antigens, while still recognizing and attacking foreign antigens, fails. When this happens, antibodies (autoantibodies) and sensitized T-cells are produced that attack and damage our own tissues. This puzzling phenomenon is called *autoimmune disease,* because it is one's own immune system that produces the disorder.

How does the normal state of self-tolerance break down? Although there are probably hundreds of triggers for autoimmune disease, most types seem to reflect one of the following:

1. Some change in the structure of self-antigens—that is, body proteins may suddenly become

"intruders" if they are distorted or altered in any way, such as by hapten attachment or partial breakdown due to infectious (bacterial or more importantly viral) disease. For example, damage to self-proteins by viruses is thought to underlie symptoms of multiple sclerosis.

2. Appearance of self-proteins in the circulation that were not previously exposed to the immune system. Not all body proteins are exposed to immune surveillance during fetal development, and such "hidden" antigens are found in sperm cells, the eye lens, and certain proteins in the thyroid gland. *Hashimoto's disease* is a thyroid-destroying autoimmune disease that follows thyroid trauma (for example, as might occur in an automobile accident) in which thyroid antigens escape into the bloodstream.

3. Cross-reaction with self-antigens of antibodies produced against foreign antigens. For instance, antibodies produced during an initial infection caused by streptococcus bacteria are known to cross-react with heart antigens, thus causing damage to both the heart muscle and its valves. This age-old disease is called *rheumatic fever.*

DEVELOPMENTAL ASPECTS OF THE IMMUNE SYSTEM

Stem cells that form lymphocytes originate in the liver during the first month of embryonic development. Later, the bone marrow becomes the predominant source of the stem cells (hemocytoblasts) from which the lymphoid cells arise. As described earlier, in late fetal life and shortly after birth, immature lymphocytes migrate from the bone marrow to populate the thymus or bursa-equivalent organ, where they become immunocompetent.

The thymus is the first lymphoid organ to appear in the developing embryo. Other lymphoid organs, particularly the lymph nodes and the spleen, are poorly developed before birth; but shortly thereafter they become heavily populated by B- and T-cells, and their development parallels that of the immune system. It has been suggested that the release of the hormone thymosin by the embryonic thymus controls the development of the other lymphatic tissues during infancy.

Although the ability of our immune system to recognize foreign substances is genetically determined, the nervous system seems to play a role in both the control and activity of the immune response. The immune response is definitely impaired in individuals who are under severe stress—for example, in those mourning the death of a beloved family member or friend. Our immune system normally serves us well throughout our lifetime, until old age. During the later years, its efficiency begins to wane, and the body's ability to fight infections declines. Additionally, we become more "open" to both autoimmune and immune-deficiency diseases.

The greater incidence of cancer in elderly people is assumed to reflect the increasing failure of the immune system. Just why this happens is not really known, but it is assumed that "genetic aging" and its consequences are probably at fault. However, as some recent studies suggest, there may be additional problems. The primary defense of the respiratory and gastrointestinal tracts lies in their mucosal linings, which provide both mechanical and antibody (IgA) barriers. Although IgA levels in an elder's body are normal, those immunoglobins cannot get to the mucosal surface, where they normally carry out their protective sentinel role. This may help to explain why respiratory and digestive tracts of the elderly are more prone to infection and there is an increased incidence of bowel cancer in the aged.

IMPORTANT TERMS*

Immune response

Humoral *(hu'mor-al)* **immunity**

Cellular immunity

B-lymphocyte (B-cell)

T-lymphocyte (T-cell)

Immunocompetence

Clonal selection

Plasma cell

Primary response

Secondary response

Killer T-cell

Helper T-cell

Suppressor T-cell

Memory cell

Macrophage *(mak'ro-fāj)*

Immunoglobulin *(im"mu-no-glob'u-lin)*

Complement fixation

Neutralization

Active immunity

Passive immunity

Immunodeficiency

Allergy (hypersensitivity)

Autoimmune disease

*For definitions, see Glossary.

SUMMARY

A. OVERVIEW OF THE IMMUNE RESPONSE

1. Resistance to disease is called immunity.

2. Immune responses are specific, are systemic, and have memory.

3. The two arms of the immune response are humoral immunity, mediated by antibodies, and cellular immunity, mediated by living cells (lymphocytes).

4. Structures of the Immune System
   a. The immune system consists of individual cells—T- and B-lymphocytes, and macrophages—and requires the services of a number of molecules (antibodies, complement, and others).

b. Because both lymphocytes and macrophages reside in lymphatic tissues, and antigens and lymphocytes travel through the body in lymph, the structures of the lymphatic system are important to the immune response.

5. Response to foreign substances—cellular interactions between lymphocytes and between lymphocytes and macrophages and a variety of molecules are necessary for all immune responses.

B. ANTIGENS

1. Antigens are large, complex molecules (or parts of them) that are recognized as foreign by the body. Foreign proteins are the strongest antigens.

2. Complete antigens provoke an immune response and bind with the products of that response (antibodies or sensitized lymphocytes).

3. Incomplete antigens, or haptens, are small molecules that are unable to cause an immune response by themselves, but do so when they bind to body proteins and the complex is recognized as foreign.

C. ORIGIN, DIFFERENTIATION, AND ACTIVITY OF LYMPHOCYTES

1. All lymphocytes arise from hemocytoblasts of bone marrow.

2. Development of Immunocompetence
   a. After leaving the marrow, immature lymphocytes circulate to the specific lymphoid organs, where they become immunocompetent, an event indicated by the appearance of antigen-specific receptors on their surfaces.
   b. T-lymphocytes become immunocompetent in the thymus. B-lymphocytes differentiate in the bursa-equivalent organ(s), which as yet is unidentified but is believed to be GALT or bone marrow.

3. Clonal Selection
   a. Clonal selection occurs when an antigen binds to the surface receptors of a B- or T-cell. This initiates the primary response, in which the lymphocyte divides rapidly, forming a clone of like cells. The various clone cells then specialize, some becoming effector or regulatory cells and others becoming memory cells that mediate secondary responses.

b. B-cells mount the humoral response via antibodies (immunoglobulins) released to the circulation. Most B-cell clone offspring become antibody-producing plasma cells. Antibodies protect against toxins, and against some bacteria and viruses.
c. T-cells mount the cellular, or cell-mediated, response that acts against most bacteria and viruses, tumor cells, and foreign organs or grafts. Members of a T-cell clone differentiate into killer T-cells and two types of regulatory cells—the helpers and the suppressors.
   (1) Helper T-cells release helper factors and interact directly with B-cells bound to antigens. They also liberate lymphokines, chemicals that enhance the killing activity of macrophages, attract other leukocytes into the area, or act as helper factors that stimulate the activity of B-cells and killer T-cells.
   (2) Killer T-cells are effector T-cells that directly attack virus-infected cells (and other cellular antigens) and promote their lysis.
   (3) Suppressor T-cells terminate the normal immune response by releasing suppressor factors.

D. ORIGIN AND ACTIVITY OF MACROPHAGES

1. Macrophages arise from monocytes formed in bone marrow.

2. A major role of macrophages in immunity is phagocytosis of antigens and antigen-presentation. Antigen-presentation is required for activation and clonal selection of T-cells and B-cells. Recognition of self and nonself must occur at the same time.

3. Macrophages also secrete monokines, chemicals that activate T-cells (interleukin) or protect body cells from the destructive effects of viruses (interferon).

E. LYMPHOCYTE TRAFFIC AND IMMUNE SURVEILLANCE

1. Many macrophages remain fixed in lymphoid organs. Lymphocytes (particularly T-cells) circulate continually between the blood, lymph, and lymphoid organs, performing their surveillance role.

2. Lymphocyte circulation facilitates both the localization and identification of pathogens

in the body, and the cellular interactions required for the immune response.

F. ANTIBODIES

1. Basic Antibody Structure
   a. Antibodies are proteins produced by sensitized B-cells or plasma cells in response to an antigen, and are capable of binding with that antigen.
   b. An antibody is composed of four polypeptide chains (two heavy and two light) connected by disulfide bonds to form a Y-shaped molecule.
   c. Each polypeptide chain has both a variable and a constant region. The variable regions on the heavy and light chains cooperate to form two antigen-binding sites, one on each arm of the Y. The constant regions determine antibody function and class.

2. Antibody Classes
   a. There are five antibody classes:
      (1) IgD—believed equivalent to the B-cell antigen receptor.
      (2) IgM—the first antibody released to the blood during the primary response; fixes complement; the largest of the antibodies.
      (3) IgG—the most abundant Ig in plasma and the major antibody of secondary responses; fixes complement; crosses the placenta.
      (4) IgA—found in mucus and other body secretions; protects mucosal surfaces.
      (5) IgE—found primarily in the mucosa of the gastrointestinal and respiratory tracts; attaches to mast cells or basophils, and when it binds with its antigens, histamine is released by the mast cell; involved in immediate types of allergies.
   b. Monoclonal antibodies are pure preparations of a single antibody type that are used in the diagnosis and treatment of certain cancers.

3. Antibody Function
   a. Antibodies act by binding to antigens.
   b. Some antibodies (IgG and IgM) fix (activate) complement, which lyses foreign cells, and enhances inflammation and phagocytosis.
   c. Antibody binding to bacterial exotoxins and viruses blocks their harmful effects on body cells; this is neutralization.
   d. By crosslinking antigens into large lattices (agglutination or precipation), antibodies cause antigens to become immobile targets for phagocytes.

G. TYPES OF IMMUNITY

1. Active Immunity
   a. Active immunity occurs when the immune system is exposed to an antigen and clonal selection occurs (as during viral/bacterial infection or when a vaccine is injected).
   b. Active immunity provides immunologic memory.

2. Passive Immunity
   a. Passive immunity is produced by the injection, into a recipient, of antibodies produced by another person or by an animal. The borrowed antibodies provide short-lived protection.
   b. Because the recipient's system is not primed by antigen exposure, immunologic memory does not occur.

H. DISORDERS OF IMMUNITY

1. Immunodeficiencies
   a. Immunodeficiencies result from abnormalities in any of the immune elements.
   b. Congenital thymic aplasia and severe combined immunodeficiency disease are congenital diseases in which the victim is virtually unprotected from pathogens.
   c. AIDS is an acquired immunodeficiency caused by a virus that attacks and cripples the helper T-cells. Victims succumb to overwhelming infection or cancer.

2. Allergies
   a. Allergy or hypersensitivity is a condition in which the immune system overresponds to an otherwise harmless antigen and tissue destruction occurs.
   b. Acute allergy (immediate hypersensitivity), as seen in hayfever, hives, and anaphylaxis, is due to IgE antibodies.
   c. Delayed hypersensitivity (for example, contact dermatitis) reflects the activity of T-cells and lymphokines, and nonspecific killing by activated macrophages.

3. Autoimmune Diseases
   a. Autoimmune disease occurs when the body's self-tolerance breaks down, and antibodies and/or T-cells attack the body's own tissues.
   b. Most forms of autoimmune disease result from changes in the structure of self-antigens, appearance of formerly hidden self-antigens in the blood, and cross-reactions

with self-antigens and antibodies formed against foreign antigens.

I. DEVELOPMENTAL ASPECTS OF THE IMMUNE SYSTEM

1. Development of the immune response occurs around the time of birth.

2. The thymus gland is the first lymphoid organ to appear in the embryo. Other lymphoid organs remain relatively undeveloped until after birth.

3. The ability of immunocompetent cells to recognize foreign antigens is genetically determined and controlled, in part, by the nervous system. Stress appears to interfere with the normal immune response.

4. The efficiency of the immune response wanes in old age, and infections, cancer, immunodeficiencies, and autoimmune diseases become more prevalent.

REVIEW QUESTIONS

1. Define immune response.

2. Define antigen. What is the difference between a complete antigen and an incomplete antigen (hapten)?

3. Differentiate clearly between humoral and cell-mediated immunity, and between the roles of B-lymphocytes and T-lymphocytes.

4. Although the immune system has two arms, it has been said, "No T-cells, no immunity." How is this so?

5. Define immunocompetence. What event or observation signals that a B-cell or T-cell has developed immunocompetence? Where does the "programming phase" occur in the case of T-cells? In the case of B-cells?

6. Binding of antigens to the receptors of immunocompetent lymphocytes leads to clonal selection. Describe the process of clonal selection. What nonlymphocyte cell is a central actor in this process, and what is its function?

7. Name the cell types that would be present in a B-cell clone. List the function of each type.

8. Describe the specific roles of helper, killer, and suppressor T-cells in cell-mediated immunity. Which population is thought to be disabled in AIDS?

9. Compare and contrast a primary and a secondary immune response. Which is more rapid, and why?

10. Describe the structure of an antibody, and note the importance of its variable and constant regions.

11. Name the five classes of immunoglobulins. Which is most likely to be found attached to a B-cell membrane? Which is most abundant in plasma? Which is important in allergic responses? Which is the first Ig to be released during the primary response? Which can cross placental barriers?

12. Antibodies help to defend the body. How?

13. What is complement, and how does it cause bacterial lysis?

14. Define allergy, and distinguish between acute (immediate) types of allergy and delayed allergic reactions relative to cause and consequences.

15. What events can result in the loss of self-tolerance or autoimmune disease?

16. Declining efficiency of the immune system with age is thought to reflect what event?

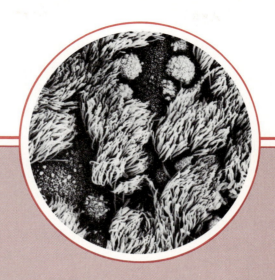

CHAPTER 12

Respiratory System

Chapter Contents

Function of the respiratory system • to exchange respiratory gases in the blood, oxygen supplied and carbon dioxide removed

After completing this chapter, you should be able to:

- Define the following terms: cellular respiration, external respiration, ventilation, expiration, and inspiration.

- Name the major respiratory system organs (or to identify them on a diagram or model) and describe the function of each.

- Describe protective mechanisms in the respiratory system.

- Explain how the respiratory muscles cause volume changes that lead to air flow into and out of the lungs (breathing).

- Define and compare the following respiratory volumes: tidal volume, vital capacity, expiratory reserve volume, inspiratory reserve volume, and residual air.

- Describe several nonrespiratory air movements and explain how they modify or differ from normal respiratory air movements.

- Describe the process of gas exchange in the lungs and the tissues.

- Name the brain areas involved in the control of respiration.

- Name several physical factors that may modify respiratory rate.

- Explain the relative importance of the respiratory gases (oxygen and carbon dioxide) in modifying the rate and depth of breathing.

- Explain why it is not possible to voluntarily stop breathing.

- Define apnea, dyspnea, hyperventilation, and hypoventilation.

The millions of cells in the body require an abundant and continuous supply of oxygen to carry out their vital processes. There is no such thing as being able to "do without oxygen" for a little while the way you can do without food, water, or sleep. As the cells use oxygen, they give off carbon dioxide, a waste product the body must get rid of. The cellular processes that actually use oxygen are referred to as *internal* (or *cellular*) *respiration.*

The circulatory and respiratory system are both intimately involved in taking oxygen into the body, seeing that it is delivered to the cells, and eliminating carbon dioxide. The respiratory system organs are responsible for the gas exchanges that occur between the blood and the external environment; this is *external respiration.* The transportation of respiratory gases between the lungs and the tissue cells is accomplished by the organs of the circulatory system with blood as the transporting fluid. If either system fails, the cells begin to die from oxygen starvation and the accumulation of carbon dioxide. If this situation is not corrected, the result is death.

ANATOMY

The organs of the respiratory system include the nasal cavities, pharynx, larynx, trachea, bronchi (and their smaller branches) and the lungs that contain the **alveoli** (al-ve'o-li), or **air sacs.** Since gas exchanges with the blood happen only in the alveoli, the rest of the respiratory system structures are really just passageways to bring air into the lungs. However, these passageways *do* have another very important job. They purify (or air-condition) the incoming air so that air reaching the lungs has many fewer irritants (such as dust or bacteria) than the air that entered the system. This may not seem too important in a quiet seashore or rural community, but *all* air contains dust and bacteria. In addition, if you live in the city, soot and auto exhaust particles add to the debris in the air. Despite all these efforts, the lungs become dull gray and mottled as we grow older—a very undesirable state as compared to the bright healthy pink of an infant's lungs. Sneezes and coughs are explosive reactions of the respiratory organs to irritants that manage to get into the passageways. As the respiratory system organs are described in detail next, locate each on Figure 12-1.

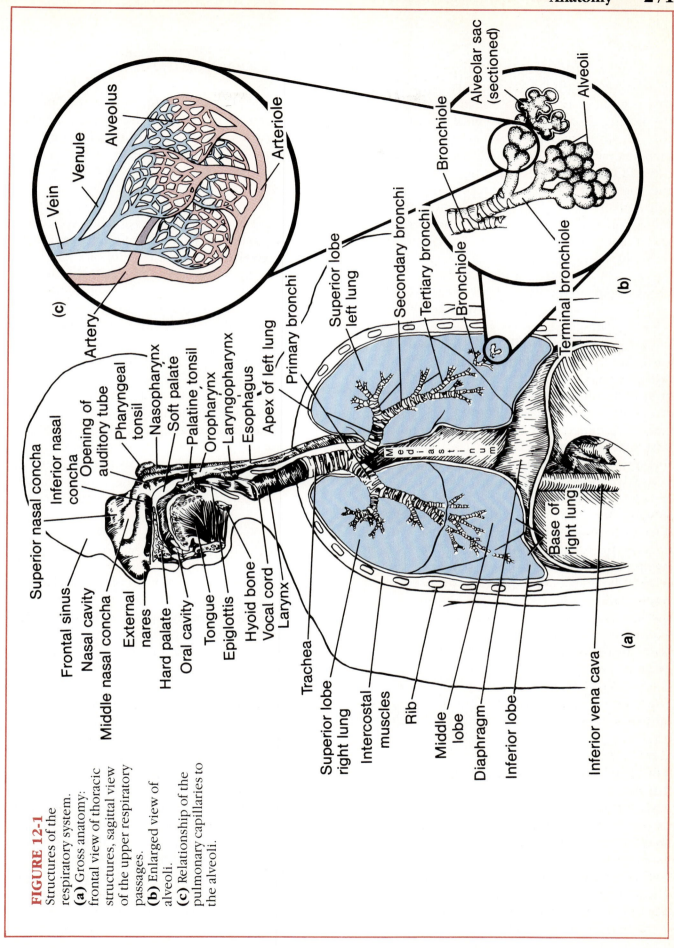

FIGURE 12-1
Structures of the respiratory system.
(a) Gross anatomy: frontal view of thoracic structures, sagittal view of the upper respiratory passages.
(b) Enlarged view of alveoli.
(c) Relationship of the pulmonary capillaries to the alveoli.

Superior nasal concha
Frontal sinus
Nasal cavity
Middle nasal concha
Inferior nasal concha
Opening of auditory tube
Pharyngeal tonsil
Nasopharynx
Soft palate
Palatine tonsil
External nares
Oropharynx
Hard palate
Laryngopharynx
Oral cavity
Esophagus
Tongue
Apex of left lung
Epiglottis
Primary bronchi
Hyoid bone
Vocal cord
Larynx
Superior lobe left lung
Trachea
Secondary bronchi
Tertiary bronchi
Superior lobe right lung
Bronchiole
Intercostal muscles
Rib
Middle lobe
Diaphragm
Inferior lobe
Base of right lung
Inferior vena cava
Mediastinum

(a)

Bronchiole
Alveolar sac (sectioned)
Alveoli
Bronchiole
Terminal bronchiole

(b)

Vein
Venule
Alveolus
Arteriole
Artery

(c)

Nasal Cavities

The nose, whether "pug" or "ski-jump" in nature, is the only externally visible part of the respiratory system. During breathing, air enters the nose by passing through the **external nares,** or nostrils. The interior of the nose consists of two **nasal cavities,** which are separated from each other by the midline **nasal septum.** Since the mucosa lining the nasal cavities is blood-rich, the air is warmed as it flows over it. In addition, the sticky mucus produced by the mucosa moistens the air and traps incoming bacteria and other foreign particles.

Projecting from the lateral wall of each nasal cavity are three lobes called **conchae** (kong′ke), which greatly increase the surface area of the mucosa that is exposed to the air. The nasal cavity is surrounded by the **paranasal sinuses** in the frontal, sphenoid, ethmoid, and maxillary bones. The sinuses lighten the skull, and it is believed that they act as resonance chambers in speech. Because the mucosa of the sinuses is continuous with that of the nasal cavities, nasal infections often spread to the sinuses, causing *sinusitis* (si″nŭ-si′tis), which is difficult to treat. The nasolacrimal ducts, which drain tears from the eyes, also empty into the nasal cavities (which explains why you have to blow your nose when you cry). The nasal cavities are separated from the oral cavity below by a partition, the **palate** (pal′et). Anteriorly, where the palate is supported by bone, is the **hard palate;** the unsupported posterior part is the **soft palate.** The genetic defect *cleft palate* (failure of the bones forming the palate to fuse medially) results in difficulty in breathing as well as oral cavity functions such as chewing and speech. The olfactory receptors for the sense of smell are located in the mucosa in the superior part of the nasal cavities.

Pharynx

The **pharynx** (far′inks) is a muscular passageway about 5-inches long and is commonly called the throat. It is a common passageway for air and food. Air enters its superior portion (**nasopharynx** [na″zo-far′inks]) from the nasal cavities anteriorly, and then descends through the **oropharynx** (o″ro-far′inks) and **laryngopharynx** (lah-ring″go-far′inks) to enter the **larynx** below. (Food travels from the mouth through the oro-

pharynx and laryngopharynx, but instead enters the esophagus posteriorly.)

The eustachean tubes, which drain the middle ear, open into the nasopharynx. Since the mucosae of these two regions are continuous, ear infections (for example, *otitis media* (o-ti′tis me′de-ah)) often follow a sore throat or other types of pharyngeal infections.

Clusters of lymphatic tissue called tonsils are also found in the pharynx. The **pharyngeal** (fah-rin′je-al) **tonsils,** often called adenoids, are located high in the nasopharynx; the **palatine tonsils** are found in the oropharynx at the end of the soft palate. If the pharyngeal tonsils become seriously inflamed and swollen (as during a bacterial infection), they obstruct the nasopharynx and force the person to breathe through his or her mouth. In mouth breathing, air is not properly moistened, warmed, or filtered before reaching the lungs. Many children seem to have almost continuous *tonsillitis.* Years ago the belief was that, in such cases, the tonsils were more trouble than they were worth, and they were routinely removed. Presently, because of the widespread use of antibiotics, this is no longer necessary (or true) today.

Larynx

The **larynx** (lar′inks), or voicebox, is located inferior to the pharynx. It is formed by nine cartilages. The largest of these cartilages is the **thyroid cartilage,** which protrudes anteriorly and is commonly called the Adam's apple. The superior opening of the larynx is protected by a flap of elastic cartilage, the **epiglottis** (ep″ĭ-glot′is). The epiglottis is sometimes referred to as the "guardian of the airways." When we swallow food or fluids, the larynx rises, and the epiglottis forms a tight lid over its opening. This routes food into the **esophagus** (ĕ-sof′ah-gus), or food tube, posteriorly. If anything other than air enters the larynx, a cough reflex attempts to expel the substance and to prevent any type of foreign material from continuing down into the lungs. Because this protective reflex does *not* work when we are unconscious, it is never a good idea to try to give fluids to an unconscious person when attempting to revive them. When we are not swallowing, the larynx is more inferior, and the epiglottis does

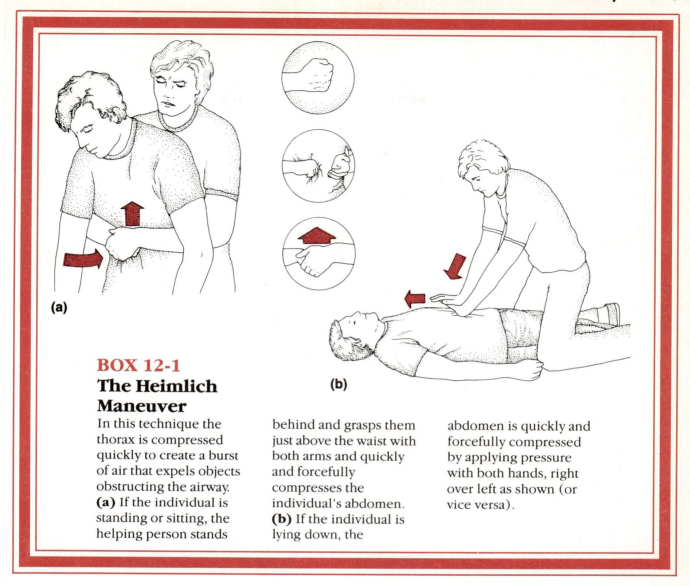

BOX 12-1
The Heimlich Maneuver

In this technique the thorax is compressed quickly to create a burst of air that expels objects obstructing the airway. **(a)** If the individual is standing or sitting, the helping person stands behind and grasps them just above the waist with both arms and quickly and forcefully compresses the individual's abdomen. **(b)** If the individual is lying down, the abdomen is quickly and forcefully compressed by applying pressure with both hands, right over left as shown (or vice versa).

---

not restrict the passage of air into the lower respiratory passages.

Palpate your larynx by placing your hand midway on the anterior surface of your neck. Swallow. Can you feel the larynx rising as you swallow?

Part of the mucous membrane of the larynx forms a pair of folds, or cordlike, structures—**vocal folds,** or **vocal cords,** which vibrate with expelled air. This ability of the vocal cords to vibrate allows us to speak. The slit-like passageway between the vocal folds is the **glottis.**

Trachea

Air entering the **trachea** (tra′ke-ah), or windpipe, from the larynx travels down its length (a little over 4 inches) to the level of the fifth thoracic vertebra, which is approximately midchest. The trachea is lined with ciliated mucosa. The cilia beat in unison and in a direction opposite to that of the incoming air. They propel mucus, loaded with dust particles and other debris, away from the lungs toward the throat where it can be swallowed or spat out. The trachea is fairly rigid, because its walls are reinforced with C-shaped cartilage rings. The open parts of the rings face posteriorly toward the esophagus. These cartilages serve a double purpose. The incomplete parts allow the esophagus to expand anteriorly when swallowing a large food bolus; the solid portions support the trachea walls and keep it *patent,* or open, in spite of the pressure changes occurring during breathing. Since the trachea is the only way air can enter the lungs, it is vital that the trachea remain open. Although it is difficult to collapse the trachea, it can become obstructed or

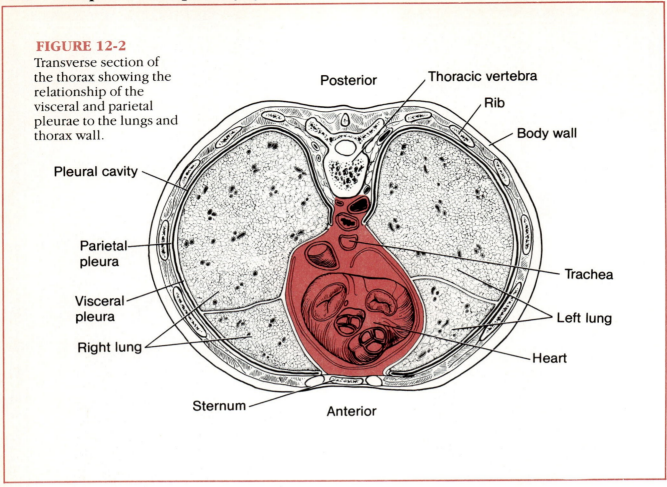

FIGURE 12-2

Transverse section of the thorax showing the relationship of the visceral and parietal pleurae to the lungs and thorax wall.

Posterior

Thoracic vertebra

Rib

Body wall

Pleural cavity

Parietal pleura

Visceral pleura

Right lung

Trachea

Left lung

Heart

Sternum

Anterior

blocked, for example, by the growth of a tumor. Also, many people have suffocated after misswallowing a piece of food, frequently meat, which suddenly closed off the trachea. This is often referred to as a "cafe coronary." Today many restaurants are training their employees to use the Heimlich maneuver in which the air in a person's own lungs is used to "pop out" or expel an obstructing piece of food (Box 12-1). In some cases, where breathing has been obstructed, an emergency *tracheotomy* ([tra"ke-ot'o-me] surgical opening into the trachea) is done to provide an alternate route for air to reach the lungs. Individuals with tracheostomy tubes in place tend to form very large amounts of mucus the first few days because of irritation to the trachea; thus, they must be suctioned frequently during this time to prevent the mucus from pooling in their lungs.

Primary Bronchi

The right and left **primary bronchi** (brong'ki) are formed by the division of the trachea into two branches. Each primary bronchus plunges into

the medial depression (hilus) of the lung on its own side. The right primary bronchus is larger in diameter and straighter than the left one. For this reason, the right bronchus is the more common site for an inhaled foreign object to become lodged. By the time air reaches the bronchi, the incoming air has been warmed and filtered. The primary bronchi and their smaller subdivisions within the lungs are direct routes to the air sacs.

Lungs

The paired **lungs** are fairly large organs. They occupy all of the thoracic cavity except for the most central area, the **mediastinum** (me"de-as-ti'num) where the heart is found. The narrow superior portion of each lung (located just deep to the clavicle) is the **apex;** the broad lung area which rests on the diaphragm is the **base.** Each lung is, in turn, divided into lobes. The left lung has two lobes while the right lung has three.

The surface of each lung is covered with a visceral serosa called the **pulmonary,** or **visceral, pleu-**

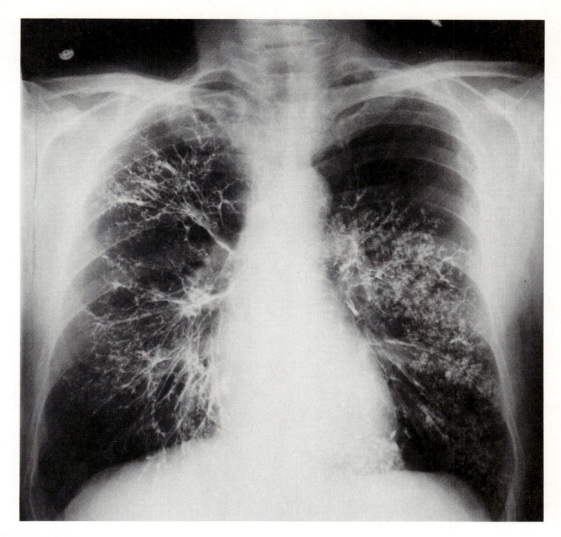

FIGURE 12-3
Bronchogram showing the extensive branching of the respiratory tree.

ra, and the walls of the thoracic cavity are lined by the **parietal pleura.** The pleural membranes produce a slippery serous secretion, which allows the lungs to glide easily over the thorax wall during breathing movements. Serous fluid also causes the two pleural layers to cling tightly together so that the "pleural space" is more of a potential space than an actual one. Figure 12-2 shows the position of the pleura on the lungs and the thorax wall. In *pleurisy* (ploor'ĭ-se), an inflammation of the pleura, the production of pleural fluid often decreases. The result is extremely uncomfortable breathing, and the affected individual feels a stabbing pain with each breath taken. Conversely, the pleurae may produce too much rather than too little fluid, and fluid accumulates between the pleural membranes. The accumulated fluid exerts pressure on the lungs and makes breathing difficult, but this condition is much less painful. Fluid is usually removed by aspiration with a needle through an intercostal space.

Other than a background of elastic connective tissue, the lungs are mostly air spaces. After the primary bronchi enter the lungs, they subdivide into smaller and smaller branches (secondary and tertiary bronchi, and so on), finally ending in the smallest passageways, **bronchioles** (brong'ke-ōls) (see Figure 12-1). Because of this extreme branching and rebranching of the respiratory passageways within the lungs, the network formed is often referred to as the *respiratory tree.* A bronchogram, which shows this extensive branching, is reproduced in Figure 12-3. All but the smallest

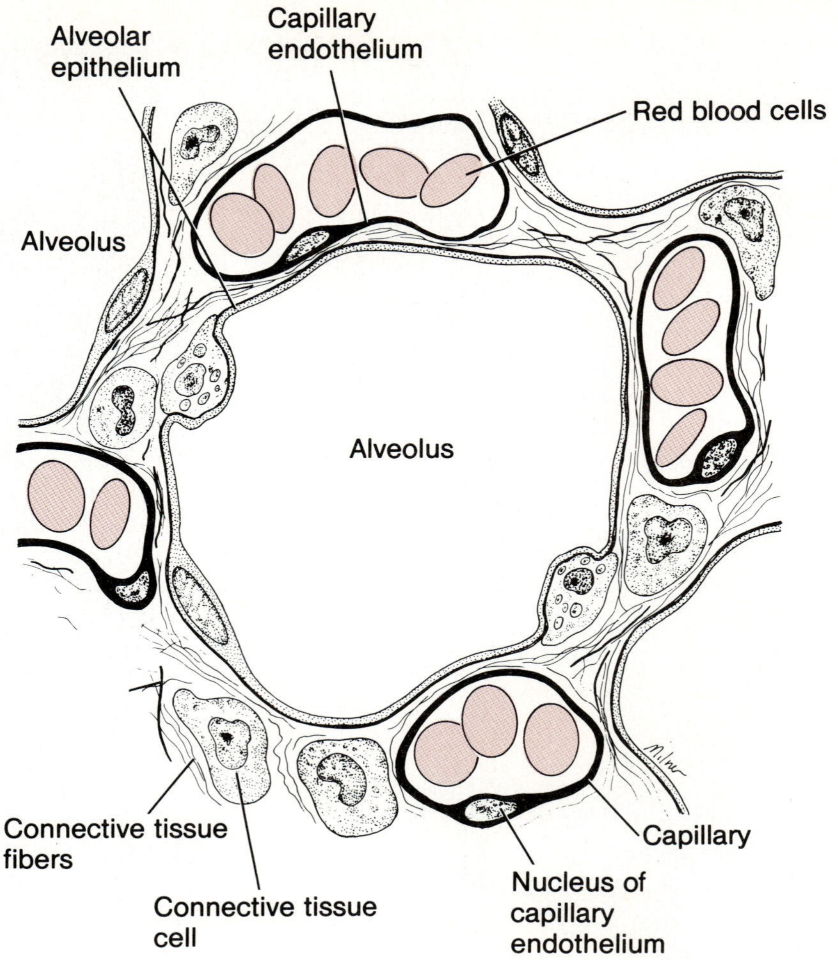

FIGURE 12-4
Diagrammatic view of the thin respiratory membrane that separates the alveolar air from the pulmonary capillary blood.

branches have cartilage reinforcements in their walls. The bronchioles eventually terminate in **alveoli,** or air sacs, that resemble clusters of grapes (see Figure 12-1). There are millions of alveoli. They make up the bulk of the lungs and make them feel soft and spongy. Thus, in spite of their relatively large size, the lungs weigh only about 2½ pounds. The final line of defense for the respiratory system is in the alveoli. Phagocytic cells, called "dust cells," wander in and out of the alveoli picking up bacteria, carbon particles, and other debris.

The walls of the alveoli are composed of a single, thin layer of squamous epithelium. These walls are so thin it is hard to imagine their thinness, but a sheet of notepaper is much thicker. The external surfaces of the alveoli are spider-webbed with pulmonary capillaries, which surround each cluster (Figure 12-4). It is through these thin walls, the alveolar and capillary walls, that the gas exchanges occur by simple diffusion—the oxygen

passing from the alveolar air into the capillary blood, and the carbon dioxide leaving the blood to enter the alveolar air. It has been estimated that the total gas exchange surface provided by the alveolar walls of a healthy man is about 600 square feet, or approximately equal to the area of a tennis court.

PHYSIOLOGY

The main purpose of respiration is to supply the body with oxygen and dispose of carbon dioxide. To do this, three processes must occur.

1. Air must move into and out of the lungs so that the air in the alveoli is continuously changed; this is called **ventilation,** or breathing.

2. The exchange of gases between the blood and the alveoli must occur; this is called **external respiration.**

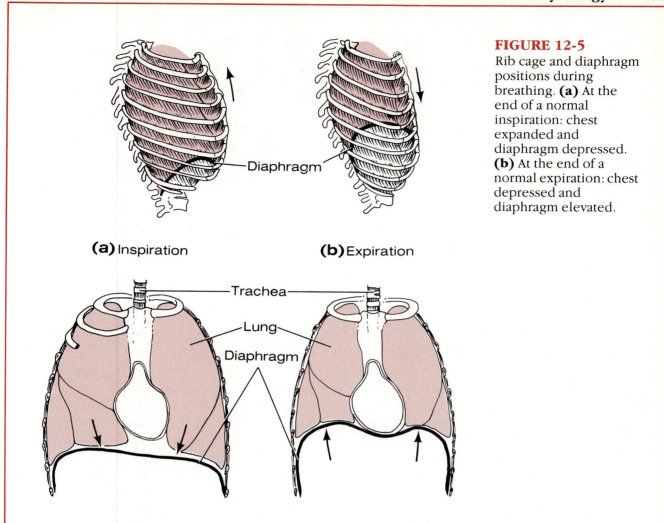

FIGURE 12-5
Rib cage and diaphragm positions during breathing. **(a)** At the end of a normal inspiration: chest expanded and diaphragm depressed. **(b)** At the end of a normal expiration: chest depressed and diaphragm elevated.

Diaphragm

(a) Inspiration   **(b)** Expiration

Trachea

Lung

Diaphragm

3. Gas exchanges must be made between the blood and body tissue cells, which then use the oxygen in their metabolic activities. This process is **internal respiration.**

Ventilation and external respiration are functions of the lungs and are discussed here. Internal respiration will be considered in the discussion of metabolism in Chapter 13.

Mechanics of Respiration

Ventilation, or breathing, has two phases: **inspiration,** when air is taken into the lungs, and **expiration,** when air passes out of the lungs. As the inspiratory muscles (external intercostals and diaphragm) contract during inspiration, the size of the thoracic cavity increases. The diaphragm moves from its relaxed dome shape to become flattened; this increases the superoinferior vol-

ume. The contraction of the external intercostals lifts the rib cage and increases the anteroposterior and lateral dimensions of the thorax (Figure 12-5). Since the lungs cling to the thorax walls like flypaper (due to the attraction of the pleural membranes for each other), they are pulled to the larger size of the thorax. As the volume within the lungs (intrapulmonary volume) increases, the gases in the lungs spread out to fill the larger space. The result is a decrease in the gas pressure within the lungs, and a partial vacuum (pressure less than atmospheric pressure) is formed. The partial vacuum sucks air into the lungs, and air continues to move into the lungs until the intrapulmonary pressure becomes equal to atmospheric pressure. This series of events is inspiration.

Expiration is a much more passive process than inspiration. During expiration, the inspiratory muscles relax, and the elastic lung tissues recoil. As a result, both the intrathoracic and intrapulmonary volumes decrease. As the intrapulmonary volume decreases, the gases inside the lungs are forced more closely together, and the intrapulmonary pressure rises to a point higher than atmospheric pressure. This causes the gases to flow out of the lungs to equalize the pressure inside and outside the lungs. Ordinarily expiration is an effortless process, but if the respiratory passageways are narrowed by spasms of the bronchioles (as in *asthma*) or clogged with mucus or fluid (as in *chronic bronchitis* or *pneumonia*), expiration becomes a much more active process. In such cases, the internal intercostal muscles are activated to help depress the rib cage, and the abdominal muscles contract and help to force air from the lungs by squeezing the abdominal organs upward.

The pressure within the pleural space (intrapleural pressure) is *always* negative. This is important, because it is the major factor preventing collapse of the lungs. If for any reason the intrapleural pressure becomes equal to the atmospheric pressure, the lungs will immediately recoil completely and collapse. This phenomenon is seen when air enters the pleural space (*pneumothorax* [nu"mo-tho'raks]) through a chest wound, or when a lung bleb ruptures allowing air to enter the pleural space through the respiratory tract. A **bleb** is an air-filled space beneath the pleura. They are common in conditions of bronchiole obstruction. Generally speaking, a pneumothorax is treated by drawing the air out of the intrapleural space with chest tubes, which allows the lungs to reinflate. Pneumothorax due to the rupture of a lung bleb is relatively common in individuals who have *emphysema* (em"fi-se'mah), one of the chronic pulmonary diseases. In emphysema, the elasticity of the lungs decreases as fibrosis of the lungs occurs. In addition, the walls of the alveoli break through, making the gas exchange chambers larger. In this case, larger is not better, because the larger chambers provide less surface area for gas exchange. Affected individuals have *hypoxia* (hi-pok'se-ah), or chronic oxygen deficiency. They retain carbon dioxide and have *dyspnea* (disp'ne-ah), or difficult or labored breathing. In addition, since their lungs are no longer elastic, they must use a great deal of energy to exhale, and they are chronically tired. Studies have indicated that smoking seems to be a predisposing factor to emphysema.

Nonrespiratory Air Movements

There are many situations other than breathing that move air into or out of the lungs and that may modify the normal respiratory rhythm. For the most part, these nonrespiratory air movements are a result of reflex activity, but some may be produced voluntarily. Examples of the most common nonrespiratory air movements are given next.

- **Cough**   A cough involves taking a deep breath, closing the glottis, and forcing the air superiorly from the lungs against the glottis. Then suddenly the glottis opens, and a blast of air passes upward. Coughs act to clear the lower respiratory passageways.

- **Sneeze**   A sneeze is similar to a cough except expelled air is directed into the nasal cavities instead of the oral cavity. The **uvula** ([u'vu-lah], a tag of tissue hanging from the soft palate) becomes depressed, closing the oral cavity off from the pharynx and routing the air upward through the nasal cavities. Sneezes clear the upper respiratory passageways.

- **Crying**   Crying involves inspiration followed by the release of air in a number of short breaths.

- **Laughing**   Laughing is essentially the same as crying. The respiratory patterns of crying and laughing are so similar that sometimes we have to look at a person to tell whether they are laughing or crying.

- **Hiccups**   Hiccups are believed to be initiated by irritation of the diaphragm or the phrenic nerve, which serves the diaphragm. Essentially hiccups are sudden inspirations due to spasms of the diaphragm. The sound occurs when inspired air hits the vocal folds of the closed glottis.

- **Yawn**   A yawn is a very deep breath, which is believed to be triggered by the need to increase the amount of oxygen in the blood. A yawn ventilates all of the alveoli. (This is not the case in normal quiet breathing.)

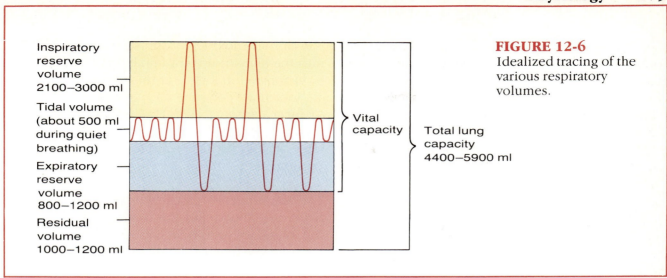

FIGURE 12-6
Idealized tracing of the various respiratory volumes.

Respiratory Volumes and Capacities

Many factors affect respiratory capacity—for example, a person's size, sex, age, and physical condition. Normal quiet breathing moves about 500 ml of air into and out of the lungs with each breath. This respiratory volume is referred to as the **tidal volume (TV).**

As a rule, a person can inhale much more air that is inhaled during a normal, or tidal, breath. The amount of air that can be taken in forcibly over the tidal volume is the **inspiratory reserve volume (IRV).** Normally the inspiratory reserve volume is between 2100–3000 ml.

Similarly, after a normal expiration, more air can be forcibly exhaled. The amount of air that can be forcibly exhaled after a tidal expiration is approximately 1000 ml and is the **expiratory reserve volume (ERV).**

Even after the most strenuous expiration, there are still about 1100 ml of air, which cannot be voluntarily expelled, in the lungs. This is the **residual volume.** The residual volume air is important, because it allows gas exchange to go on continuously even between breaths.

The total amount of exchangeable air is around 4500 ml, and this respiratory volume is the **vital capacity (VC).** The vital capacity is the sum of the TV + IRV + ERV.

Obviously, much of the air that enters the respiratory tract remains in the respiratory passageways and never reaches the alveoli. During a normal tidal breath, this amounts to about 150 ml and is called the **dead space volume.** The functional volume (air that actually reaches the alveoli and contributes to gas exchange) is about 350 ml. The respiratory volumes are summarized in Figure 12-6.

Respiratory capacities are measured with a *spirometer* (spi-rom′ĕ-ter). As a person breathes, the volumes of air exhaled can be read on an indicator, which moves to indicate the changes in air volume inside the apparatus. Spirometer testing is useful for evaluating losses in respiratory functioning and in following the course of some respiratory diseases. For example, in diseases such as chronic bronchitis and pneumonia in which inspiration is obstructed, the IRV decreases. In diseases such as emphysema where expiration is hampered, the ERV is much lower than normal.

Respiratory Sounds

As air flows in and out of the respiratory tree, it produces two recognizable sounds that can be picked up with a stethoscope. The **bronchial sounds** are produced by air rushing through the large respiratory passageways (trachea and bronchi). The second sound type, **vesicular** (vĕ-sik′u-lar) **breathing sounds,** occurs as air fills the alveoli. The vesicular sounds are soft and resemble a muffled breeze sound. Diseased respiratory tissue, mucus, or pus can produce abnormal chest sounds such as *rales* (a rasping sound) and *wheezing* (a whistling sound).

External Respiration, Gas Transport, and Gas Exchange in the Tissues

As explained earlier, external respiration is the actual exchange of gases between the alveoli and the blood. It is important to remember that all gas exchanges are made according to the laws of diffusion; that is, movement occurs *toward* the area of lower concentration of the diffusing substance. There is always more oxygen in the alveoli than there is in the blood, because the body cells continually remove oxygen from the blood. Thus, oxygen always tends to move from the air of the alveoli through the alveolar-capillary walls into the oxygen-poor blood of the pulmonary capillaries. On the other hand, as tissue cells remove oxygen from the blood, they release carbon dioxide into the blood. Since the concentration of carbon dioxide is much higher in the pulmonary capillaries than it is in the air of the alveoli, it will leave the blood to pass into the alveoli and be flushed out of the lungs during expiration. Blood draining from the lungs into the pulmonary veins is oxygen-rich and carbon dioxide-poor, and is ready to be pumped to the systemic circulation.

The pickup of oxygen and the unloading of carbon dioxide from the blood go hand in hand. Most of the oxygen attaches to hemoglobin molecules inside the RBCs to form *oxyhemoglobin* **(ok″se-he″mo-glo′bin)(Figure 12-7a). Also, a small amount of oxygen is carried dissolved in the plasma.**

Most carbon dioxide is carried in plasma as the bicarbonate ion (HCO_3^-). A smaller amount is carried inside the RBCs bound to hemoglobin. Carbon dioxide carried inside the RBCs attaches to hemoglobin at a different site than oxygen does, and so, it does not interfere in any way with oxygen transport. (However, *carbon monoxide* binds at the same site on hemoglobin as oxygen and competes vigorously with oxygen for the binding spots—so much so, it crowds oxygen out. This is why carbon monoxide poisoning is lethal.)

For carbon dioxide to diffuse out of the blood into the alveoli, it must first be released from its bicarbonate ion form. For this to occur, bicarbonate ions must combine with hydrogen ions (H^+) to form carbonic acid (H_2CO_3). Carbonic acid then quickly splits to form water and carbon dioxide. Carbon dioxide then diffuses from the blood and enters the alveoli.

The exchange of gases, which takes place between the blood and the tissue cells and makes the blood oxygen-poor/carbon dioxide-rich, is the opposite of what occurs in the lungs. This process in which oxygen is unloaded and carbon dioxide loaded into the blood is shown in Figure 12-7b. In this process, carbon dioxide diffuses out of tissue cells to enter plasma. In the plasma, it combines with water to form carbonic acid, which quickly releases the bicarbonate ions. (It should be mentioned that most conversion of carbon dioxide to bicarbonate ions occurs inside the RBCs, and *then* the bicarbonate ions diffuse out into plasma where they are transported.) At the same time, oxygen is released from hemoglobin, and the oxygen diffuses quickly out of the blood to enter the tissue cells.

Control of Respiration

Nervous Control

The activity of the respiratory muscles, diaphragm, and external intercostals, is regulated by nerve impulses transmitted to them from the brain by the phrenic and intercostal nerves. The neural centers that control respiratory rhythm and depth are located in the medulla and pons. The medulla contains the **inspiratory** and **expiratory centers.** The pons centers are the **pneumotaxic** (nu-mo-tak′sik) **center** (an expiratory center) and the **apneustic** (ap-nu′stik) **center,** which provides the inspiratory drive. Impulses going back and forth between the pons and medulla centers normally maintain a rate of 14–18 respirations/minute. This normal respiratory rate is *eupnea* (ūp-ne′ah).

In addition, the bronchioles and alveoli have stretch receptors that respond to extreme overinflation (which might damage the lungs) as well as extreme deflation (which indicates that an individual is not ventilating properly). In the case of overinflation, impulses are sent from the stretch

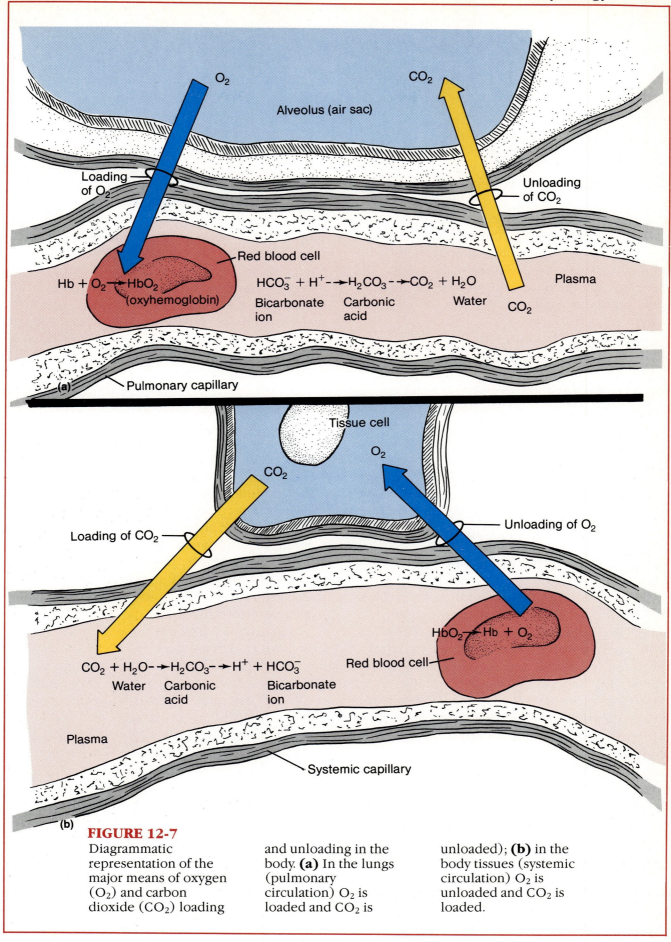

FIGURE 12-7

Diagrammatic representation of the major means of oxygen (O_2) and carbon dioxide (CO_2) loading and unloading in the body. **(a)** In the lungs (pulmonary circulation) O_2 is loaded and CO_2 is unloaded); **(b)** in the body tissues (systemic circulation) O_2 is unloaded and CO_2 is loaded.

receptors to the medulla by the vagus nerves, and the expiratory center is activated. In extreme deflation, impulses traveling from the lungs to the medulla excite the inspiratory center, and inspiration occurs.

During exercise, we breathe more deeply and at a faster rate because the brain centers send more rapid impulses to the respiratory muscles. After strenuous exercise, expiration becomes active and the abdominal muscles and any other muscles capable of lifting the ribs are used to aid expiration. If the medulla centers are completely suppressed (as with an overdose of sleeping pills, morphine, or alcohol), respiration stops completely, and death occurs.

FACTORS INFLUENCING THE RATE AND DEPTH OF RESPIRATION

PHYSICAL FACTORS.   Although the brain centers set the basic rhythm of breathing, there is no question that physical factors such as talking, coughing, and exercise can modify both the rate and the depth of breathing. Some of these have already been examined in the earlier discussion of nonrespiratory air movements. Increased body temperature results in an increase in the rate of breathing.

VOLITION (CONSCIOUS CONTROL).   We all have consciously controlled our breathing pattern at one time or another. During singing and swallowing, breath control is extremely important, and many have held their breath for short periods to swim under water. However, voluntary control of breathing is limited, and the respiratory centers will simply ignore messages from the cortex (your wishes) when the oxygen supply in the blood is getting low. All you need to do to prove this is to try to talk normally or to hold your breath after running at breakneck speed for a few minutes. It simply cannot be done. Many toddlers try to manipulate their parents by holding their breath "to death." Even though this threat causes many parents to become anxious, they need not worry because the involuntary controls take over—normal respiration will begin again.

EMOTIONAL FACTORS.   Emotional factors can also modify the rate and depth of breathing. Have you ever watched a horror movie with baited (held)

breath or been so scared by what you saw that you were nearly panting? Have you ever touched something cold and clammy and gasped? All of these are a result of reflexes, which can be initiated by emotional stimuli.

CHEMICAL FACTORS.   Although many factors can modify respiratory rate and depth, the *most* important factors are chemical—that is, the levels of carbon dioxide and oxygen in the blood. Increased levels of carbon dioxide and decreased blood pH are the most important stimuli leading to an increase in the rate and depth of breathing. (Actually, an increase in carbon dioxide levels and decreased blood pH are the same thing in this case, because increased carbon dioxide retention leads to increased levels of carbonic acid that decreases blood pH.) Changes in carbon dioxide concentrations in the blood seem to act directly on the medulla centers.

Conversely, changes in oxygen concentrations in the blood are detected by chemoreceptor regions in the aorta (aortic arch) and carotid artery (carotid body), which in turn send impulses to the medulla when blood oxygen levels are dropping. This is an interesting fact—although every cell in the body must have oxygen to live, the body's need to rid itself of carbon dioxide (not to take in oxygen) is the most important stimulus for breathing in a healthy person. Decreases in oxygen levels only become important stimuli when the oxygen levels are dangerously low. However, in people that retain carbon dioxide (as in chronic lung diseases like emphysema and chronic bronchitis), increased levels of carbon dioxide are no longer recognized as important by the brain, and decreasing oxygen levels become the respiratory stimulus. (This is why such patients are always given low levels of oxygen. If they are given high levels, they would stop breathing because their respiratory stimulus [low oxygen levels] would be gone.)

In healthy individuals, respiratory system homeostatic mechanisms are obvious. As carbon dioxide starts to accumulate in blood and blood pH begins to drop, you begin to breathe deeper and more rapidly. This blows off more carbon dioxide and decreases the amount of carbonic acid, which returns blood pH to the normal range. On the other hand, when blood starts to become slightly alkaline, or basic (for whatever reason), your breathing will become slower and more shallow. Slower breathing allows carbon dioxide to accumulate in the blood and brings pH back into the normal range.

In cases of extremely slow or shallow breathing *(hypoventilation)* or fast deep breathing *(hyperventilation),* the amount of carbonic acid in the blood can be dramatically changed or modified. Carbonic acid increases greatly during hypoventilation and decreases substantially during hyperventilation. In both situations, the buffering ability of the blood is likely to be overwhelmed, so that *acidosis* or *alkalosis* occurs. Hyperventilation, often brought on by anxiety, frequently leads to brief periods of *apnea* (ap-ne'ah), cessation of breathing, until the carbon dioxide builds up in the blood again. If breathing stops for an extended time, *cyanosis* (si"ah-no'sis) may occur. Cyanosis is a bluish cast of the skin, especially visible in the nail beds and lips, due to insufficient oxygen in the blood. In addition, dizziness and even fainting may occur during hyperventilation as a result of alkalosis. These symptoms can be decreased by having the hyperventilating individual breathe into a paper bag.

DEVELOPMENTAL ASPECTS OF THE RESPIRATORY SYSTEM

During fetal life, the lungs are filled with fluid, and all respiratory exchanges are made by the placenta. At birth the fluid-filled pathway is drained, and the respiratory passageways become filled with air. The alveoli inflate and begin to function in gas exchange. The success of this change—that is, from nonfunctional-to-functional respiration—depends on the presence of a fatty molecule, **surfactant** (sur-fak'tant), made by the alveolar cells. Surfactant lowers the surface tension of the water-film lining each alveolar sac so that the alveoli do not collapse between each breath. Surfactant is not usually present in large enough amounts to accomplish this function until late in pregnancy—that is, between 28 and 32 weeks. Infants born prematurely (before week 28) or those in which surfactant production is inadequate for other reasons (as in many infants born to diabetic mothers) have *infant respiratory distress syndrome*

(RDS). These infants have dyspnea (labored breathing) within a few hours after birth and have to use tremendous amounts of energy just to keep reinflating their alveoli, which collapse after each breath. RDS accounts for over 20,000 newborn deaths a year. However, the odds for these babies surviving are now increasing, because of the current use of equipment that supplies a positive pressure continuously to the lungs and keeps the alveoli open and working in gas exchange.

In normal newborn infants, the respiratory rate is about 40 respirations/minute and it takes nearly 2 weeks for the lungs to become fully inflated. The respiratory rate continues to drop through life—in the infant it is around 30/minute, at 5 years it is around 25/minute, and in adults it is between 12–18/minute. However, it often begins to increase again in old age. The lungs continue to mature throughout childhood, and more alveoli are formed until young adulthood. A recent medical discovery found that when smoking is begun during the early teens, complete maturation of the lungs never occurs, and those additional alveoli are lost forever.

Under normal conditions the respiratory system works so efficiently and smoothly that we are not even aware of it. Most problems that do occur are a result of external factors—for example, the obstruction of the trachea by a piece of food, aspiration of food particles or vomitus (which leads to *aspiration pneumonia*), or viral and bacterial infections. The common cold blocks the upper respiratory passageways with mucus. *Bronchitis* causes the lining of the bronchial tubes to swell and produce excessive mucus; both obstruct the respiratory passageways. *Asthma,* which may be caused by various factors such as allergy or anxiety, also obstructs the respiratory passageways by causing the bronchioles to constrict, causing the affected person to wheeze and gasp for air. By far the most damaging diseases are those that destroy lung tissue (such as *emphysema* and *tuberculosis*) or obstruct the alveoli (such as *pneumonia*). For many years, tuberculosis and pneumonia were the worst killers in the United States. Antibiotics have decreased their lethality to a large extent, but they are still dangerous diseases.

As we age, the chest wall becomes more rigid and the lungs begin to lose their elasticity. Both of these factors result in a slowly decreasing ability to ventilate the lungs. Vital capacity decreases by about one-third by the age of 70. In addition, just as the overall effectiveness of the immune system decreases as we age, many of the respiratory system's protective mechanisms become much less efficient. Ciliary activity of the mucosa decreases, and the phagocytes in the lungs become sluggish. The net result is that the elderly population is more susceptible to respiratory tract infections, particularly pneumonia.

IMPORTANT TERMS*

Alveolar (al-ve'o-lar)	**Internal respiration**
Nasal cavities	**Inspiration**
Pharynx (far'inks)	**Expiration**
Larynx (lar'inks)	**Oxyhemoglobin** (ok"se-he"mo-glo'bin)
Vocal cords	**Bicarbonate ion**
Primary bronchi (brong'ki)	**Inspiratory center**
Pleura	**Expiratory center**
Bronchioles (brong'ke-ōls)	**Surfactant** (sur-fak'tant)
External respiration	

*For definitions, see Glossary.

SUMMARY

A. ANATOMY

1. The nasal cavities, which contain receptors for sense of smell, are two chambers within the nose. They are divided by the nasal septum and separated from the oral cavity by the palate. The nasal cavities are lined with a mucosa, which warms, filters, and moistens the incoming air. Paranasal sinuses and nasolacrimal ducts drain into the nasal cavities.

2. The pharynx (throat) is a mucosa-lined, muscular tube. The pharynx has three regions—nasopharynx, oropharynx, and laryngopharynx. The nasopharynx functions in respiration only; the other two regions serve both respiration and digestive functions. The pharynx contains tonsils (pharyngeal and palatine), which act as part of the body's defense system.

3. The larynx (voicebox) is a cartilage structure; the most prominent cartilage is the thyroid cartilage (Adam's apple). The larynx connects the pharynx with the trachea below. The opening (glottis) is hooded by the epiglottis, which prevents aspiration of food or drink into the respiratory passages when swallowing. The larynx contains the vocal cords, which produce sounds used in speech.

4. The trachea (windpipe) extends from the larynx to the primary bronchi. The trachea is a smooth muscle tube lined with a ciliated mucosa and reinforced with C-shaped cartilage rings, which keep the trachea patent.

5. Primary bronchi have right and left members, resulting from the subdivision of the trachea. Each plunges into the hilus of the lung on its side.

6. The lungs are paired organs in the thoracic cavity, which flank the mediastinum. The lungs are covered with visceral pleura; the thorax wall is lined with parietal pleura. Pleural secretions decrease friction during

breathing. The lungs are primarily elastic tissue plus the passageways of the respiratory tree. The smallest passageways (bronchioles) end in clusters of alveoli. Alveoli have thin walls through which gas exchanges are made with the pulmonary capillary blood.

B. PHYSIOLOGY

1. Mechanics of respiration. Gas travels from an area of high pressure to an area of low pressure. Pressure outside the body is atmospheric pressure; pressure inside the lungs is intrapulmonary pressure; pressure in the intrapleural space is intrapleural pressure. Intrapleural pressure is always negative. Movement of air into and out of the lungs is called ventilation, or breathing. When the inspiratory muscles contract, the intrapulmonary volume increases, its pressure decreases, and air rushes in (inspiration). When the inspiratory muscles relax, the lungs recoil and air rushes out (expiration). Expansion of the lungs is helped by the cohesion between the pleurae and the presence of surfactant in the alveoli.

2. Nonrespiratory air movements. Nonrespiratory air movements are voluntary or reflex activities that move air into or out of the lungs. These include coughing, sneezing, laughing, crying, hiccuping, and yawning.

3. Respiratory volumes and capacities. Air volumes exchanged during breathing are TV, IRV, ERV, and VC (see p. 279 for values). A nonexchangeable respiratory volume is the residual volume, which allows gas exchange to go on continually.

4. Respiratory sounds. Bronchial sounds are the sounds of air passing through the large respiratory passageways. Vesicular breathing sounds are made as air fills the alveoli.

5. External respiration, gas transport, and exchange in the tissues. Gases move according to the laws of diffusion. Oxygen moves from the alveolar air into the pulmonary blood. Most of the oxygen is transported bound to hemoglobin inside the RBCs. Carbon dioxide moves from the pulmonary blood into the alveolar air. Most carbon dioxide transport is in the form of the bicarbonate ion in the plasma. At the body tissues, oxygen moves from the blood to the tissues, whereas carbon dioxide moves from the tissues to the blood.

6. Control of respiration
   a. Nervous control. The neural centers for control of respiratory rhythm are in the medulla and pons. Reflex arcs initiated by stretch receptors in the lungs also play a role in modifying respiration by notifying the neural centers of excessive overinflation or overdeflation.
   b. Physical factors. Increased body temperature, exercise, speech, singing, yawning, and other nonrespiratory air movements modify the rate and depth of breathing.
   c. Volition. Breathing may be consciously controlled, if it does not interfere with homeostasis.
   d. Emotional factors. Some emotional stimuli interfere with or modify breathing. Examples are fear, anger, and excitement.
   e. Chemical factors. Increases or decreases in the blood levels of carbon dioxide are the most important stimuli affecting respiratory rhythm and depth. Carbon dioxide acts directly on the medulla. High levels of carbon dioxide in the blood result in faster deeper breathing while decreased carbon dioxide levels lead to shallow slow breathing. Hyperventilation may result in apnea and dizziness, due to the alkalosis produced. Oxygen is less important as a respiratory stimulus in normal healthy people. It *is* the stimulus for those whose systems have become accustomed to high levels of carbon dioxide.

C. DEVELOPMENTAL ASPECTS OF THE RESPIRATORY SYSTEM

1. Premature infants have problems keeping their lungs inflated due to the lack of surfactant in their alveoli. (Surfactant is formed late in pregnancy.)

2. The lungs continue to mature and form more alveoli until young adulthood.

3. During youth and middle age most respiratory system problems are a result of external factors, such as substances that

physically block the respiratory passageways and infections.

4. In old age, the thorax becomes more rigid and the lungs become less elastic; these changes lead to a decreased vital capacity. Protective mechanisms of the respiratory system decrease in effectiveness in elderly persons predisposing them to more respiratory tract infections.

REVIEW QUESTIONS

1. What is the most basic function of respiration?

2. Clearly explain the difference between external and internal respiration.

3. Trace the route of air from the external nares to an alveolus.

4. *Why* is it important that the trachea is reinforced with cartilage rings? What is the advantage of the fact that the rings are incomplete posteriorly?

5. Where in the respiratory tract is the air filtered, warmed, and moistened?

6. The trachea has cilia and goblet cells which produce mucus. What is the specific protective function of each of these?

7. Which primary bronchus is the more likely site for an inspired object to become lodged? Why?

8. In terms of general health, what is the importance of the fact that the eustachean tubes and the sinuses drain into the nasal cavities and nasopharynx?

9. The lungs are mostly passageways and elastic tissue. What is the role of the elastic tissue? Of the passageways?

10. What is it about the structure of the alveoli that makes them an ideal site for gas exchange?

11. What do TV, IRV, ERV, and VC mean? Which of these values is the largest? Why?

12. Name several nonrespiratory air movements and explain how each differs from normal breathing.

13. The contraction of the diaphragm and the external intercostal muscles begins inspiration. Explain exactly what happens, in terms of volume and pressure changes in the lungs, when these muscles contract.

14. What causes air to flow out of the lungs during expiration?

15. What is the major way oxygen is transported in the blood? Carbon dioxide?

16. What determines in which direction carbon dioxide and oxygen will diffuse in the lungs? In the tissues?

17. Name the two major brain areas involved in the nervous control of breathing.

18. Name three physical factors that can modify respiratory rate or depth.

19. Name two chemical factors that modify respiratory rate and depth. Which is usually more important?

20. Define hyperventilation. If you hyperventilate, do you retain or expel more carbon dioxide? What effect does hyperventilation have on blood pH?

21. Why doesn't Mom have to worry when 3-year-old Johnny threatens "to hold his breath til he dies"?

CHAPTER 13

Digestive System

Chapter Contents

Function of the digestive system
- to break down ingested food into particles small enough to be absorbed into the blood

Function of metabolism
- to produce energy (ATP) in the cells; involves constructive and degradative activities

The above electron micrograph from *Tissues and Organs: A Text-Atlas of Scanning Electron Microscopy* by Richard G. Kessel and Randy H. Kardon. W. H. Freeman and Co. Copyright © 1979.

After completing this chapter, you should be able to:

- Identify the overall function of the digestive system as digestion and absorption of food stuffs.

- Name the alimentary canal and accessory digestive organs and identify each on an appropriate diagram or torso model.

- Describe the general functions of each of the digestive system organs.

- Describe the composition and function(s) of saliva.

- Name the deciduous and permanent teeth and describe the anatomy of the generalized tooth.

- Explain how villi aid absorption in the small intestine.

- Describe the mechanisms of swallowing, vomiting, and defecation.

- Describe how foodstuffs in the digestive tract are mixed and moved along the tract.

- Describe the function of local hormones.

- List the major enzymes or enzyme groups produced by the various digestive organs or accessory organs and name the foodstuffs on which they act.

- Name the end products of protein, fat, and carbohydrate digestion.

- State the function of bile in the digestive process.

- Define enzyme, metabolism, anabolism, and catabolism.

- Describe the metabolic roles of the liver.

- Recognize the uses of carbohydrates, fats, and proteins in cell metabolism.

- Describe the effect of aging on the digestive system.

The treatment that food undergoes, once it has been eaten, is a rough one. Once food is swallowed, it passes totally out of our voluntary control—thus, our only digestive decision is to eat or not to eat.

The digestive system provides the body with nutrients, water, and electrolytes needed for good health. The organs of this system are responsible for food *ingestion, digestion, absorption,* and the *elimination* of the undigestible remains as feces. Essentially the digestive system consists of a hollow tube extending from the mouth to the anus. Various glands outside this tube produce substances that aid the digestive process, and these are secreted into the tube at various points along the way. Food material within this tube is technically outside the body because it has contact only with the cells lining the tract. For the food to become available to the body cells, it must first be digested. Digestion involves food breakdown by physical means (churning and chewing) and chemical means (breakdown by protein molecules called enzymes) into smaller molecules. The end products can then pass through the epithelial cells lining the tract into the blood, a process called absorption.

ANATOMY

The organs of the digestive system are separated into two major groups, the **alimentary** (al″e-men′tar-e) **canal (gastrointestinal tract)** and the **accessory digestive organs.** The alimentary canal is coiled and twisted and is about 30 feet long in a living person. Accessory structures consist primarily of salivary glands, gall bladder, liver, and pancreas, which secrete their products into the alimentary canal. The organs of the alimentary canal are described in sequence next. As each organ is identified and described, find it on Figure 13-1.

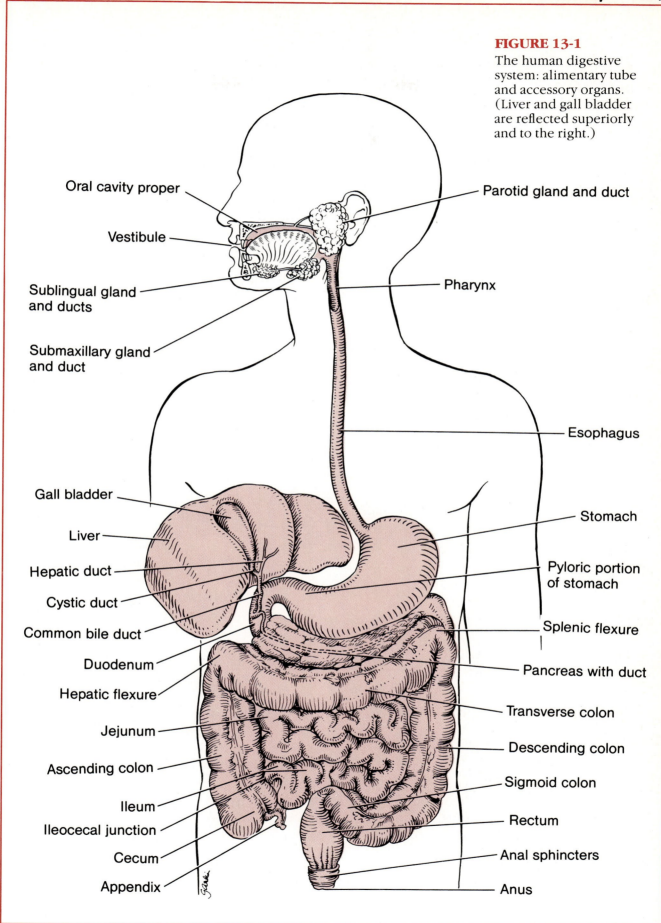

FIGURE 13-1
The human digestive system: alimentary tube and accessory organs. (Liver and gall bladder are reflected superiorly and to the right.)

Oral cavity proper

Vestibule

Sublingual gland and ducts

Submaxillary gland and duct

Gall bladder

Liver

Hepatic duct

Cystic duct

Common bile duct

Duodenum

Hepatic flexure

Jejunum

Ascending colon

Ileum

Ileocecal junction

Cecum

Appendix

Parotid gland and duct

Pharynx

Esophagus

Stomach

Pyloric portion of stomach

Splenic flexure

Pancreas with duct

Transverse colon

Descending colon

Sigmoid colon

Rectum

Anal sphincters

Anus

FIGURE 13-2
Oral and pharyngeal cavities. **(a)** Sagittal view of the head. **(b)** Anterior view of the oral cavity.

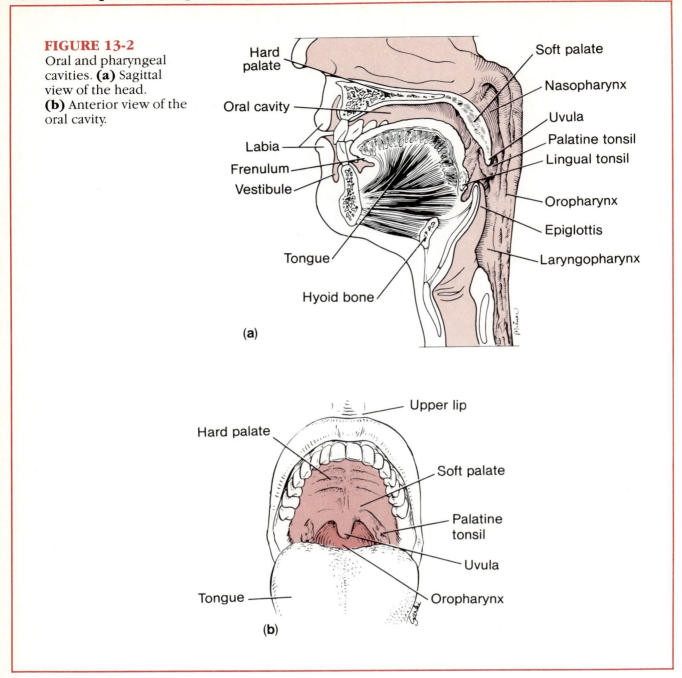

(a)

(b)

Organs of the Alimentary Canal

MOUTH

Food enters the digestive tract through the **mouth,** or **oral cavity,** a mucous membrane–lined cavity. The **labia** (la′be-ah) **(lips)** protect its anterior opening, the *cheeks* form its lateral walls, the *hard palate* forms its anterior roof, and the *soft palate* forms its posterior roof. The **uvula** (u′vu-lah) is a fingerlike projection of the soft palate, which extends downward from its posterior edge.

The muscular **tongue** occupies the floor of the mouth. The tongue has several attachments—two of these are to the hyoid bone and the styloid processes of the skull. The **frenulum** (fren′u-lum), a membrane, secures the tongue to the floor of the mouth. The space between the lips, cheeks, and the teeth is the **vestibule;** the area contained by the teeth is the **oral cavity proper** (Figure 13-2).

On each side of the mouth at its posterior end are masses of lymphatic tissue, the *palatine tonsils.* Other masses of lymphatic tissue, the *lingual* (ling′gwal) *tonsils,* cover the base of the tongue

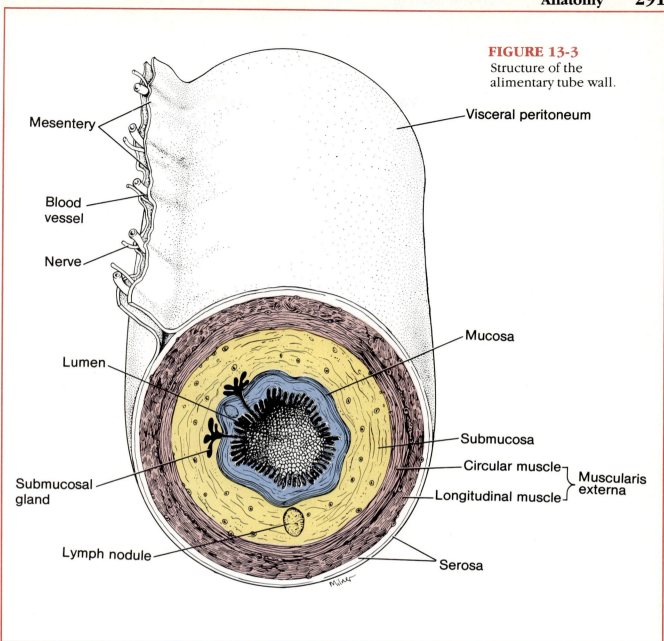

FIGURE 13-3
Structure of the alimentary tube wall.

Mesentery

Blood vessel

Nerve

Visceral peritoneum

Lumen

Submucosal gland

Lymph nodule

Mucosa

Submucosa

Circular muscle ⎫
Longitudinal muscle ⎦ Muscularis externa

Serosa

just posterior to the oral cavity proper. The tonsils, along with other lymphatic tissues, are part of the body's defense system. When the tonsils become inflamed and enlarge, the entrance into the throat (pharynx) is partially blocked, which makes swallowing difficult and painful.

As food enters the mouth, it is mixed with saliva and *masticated* (chewed). The cheeks and closed lips hold the food between the teeth during chewing, and the nimble tongue continually mixes food with saliva during chewing and initiates swallowing. Thus, the physical and chemical breakdown of food begins before the food has even left the mouth. As noted in Chapter 7, *papil-*

lae containing taste buds, or taste receptors, are found on the tongue surface. Thus, besides its manipulating function, the tongue allows us to enjoy and appreciate food as it is eaten.

PHARYNX

From the mouth, food passes posteriorly into the **pharynx,** which is a common passageway for food, fluids, and air. As explained in Chapter 12, the pharynx is subdivided into the *nasopharynx* (part of the respiratory passageway), the *oropharynx,* and *laryngopharynx,* which is continuous with the esophagus below.

The walls of pharynx contain two skeletal muscle layers. The cells of the inner layer run longitudinally; those of the outer layer (the constrictor muscles) run around the wall in a circular fashion. Alternating contractions of these two muscle layers propel food through the pharynx into the esophagus below. This propelling mechanism is *peristalsis* (per″ĭ-stal′sis).

ESOPHAGUS

The **esophagus** (ĕ-sof′ah-gus), or *gullet,* runs from the pharynx through the diaphragm to the stomach. It is approximately 25-cm (10-in.) long and is essentially a passageway that conducts food to the stomach in a wavelike peristaltic motion.

Because the tissue arrangement in the alimentary-tube walls changes in various areas along its length to serve special functions, it might be helpful to describe the general wall structure for future reference.

The walls of the alimentary canal organs from the esophagus to the large intestine have four characteristic layers (Figure 13-3):

- **Mucosa** is the innermost layer, which lines the cavity or lumen of the organ. It consists primarily of a surface epithelium, plus a small amount of connective tissue (lamina propria) and a scanty smooth muscle layer.

- **Submucosa** is found just beneath the mucosa. It consists primarily of connective tissue containing blood vessels, nerve endings, and lymphatic vessels.

- **Muscularis externa** is a muscle layer commonly made up of a circular inner layer and a longitudinal outer layer of smooth muscle cells.

- **Serosa** is the outermost layer of the wall. It consists of a single layer of flat serous fluid–producing cells, the **visceral peritoneum** (per″i-to-ne′um). The visceral peritoneum is continuous with the slick slippery **parietal peritoneum,** which lines the abdominopelvic cavity by way of a membrane extension, the **mesentery** (mes′en-ter″e). These relationships are illustrated in Figure 13-4.

STOMACH

The **C**-shaped **stomach** (Figure 13-5, p. 294 is on the left side of the abdominal cavity nearly hidden by the liver and diaphragm. Different regions of the stomach have been named. The *cardiac re-*

gion surrounds the **cardioesophageal** (kar″de-o-ĕ-sof″ah-je′al) **sphincter** through which food enters the stomach from the esophagus. The *fundus* is the expanded part of the stomach lateral to the cardiac region. The *body* is the midportion, and the *pyloris* (pi-lo′rus) is the terminal part of the stomach. The pylorus is continuous with the small intestine through the **pyloric sphincter.** The stomach is approximately 25-cm (10-in.) long, but its diameter depends on how much food it contains. When it is full, it can hold over ½ gallon of food. When empty, it seems to collapse inward on itself, and its mucosa is thrown into large folds called **rugae** (roo′je). The convex lateral surface of the stomach is the *greater curvature;* its concave medial surface is the *lesser curvature.* The **lesser omentum** (o-men′tum), a double layer of peritoneum, extends from the liver to the lesser curvature. The **greater omentum,** another extension of the peritoneum, drapes downward and covers the abdominal organs like an apron before attaching to the posterior body wall (see Figure 13-4). The greater omentum is riddled with fat deposits, which help to insulate, cushion, and protect the abdominal organs.

The stomach acts as a temporary "storage tank" for food as well as a site for food breakdown. In addition to longitudinal and circular layers, its wall contains a third obliquely arranged layer of muscle in the muscularis externa. This muscle allows it not only to move food along the tract but also to churn, mix, and pummel the food, physically breaking it down to smaller fragments. The chemical breakdown of proteins begins in the stomach. The mucosa of the stomach has deep pits *(gastric pits),* which lead into *gastric glands.* The gastric glands secrete many substances that contribute to the stomach secretion called **gastric juice.** For example, some cells produce *intrinsic factor,* a substance needed for the absorption of vitamin B_{12} from the stomach. The *chief cells* produce protein-digesting enzymes, and the *parietal cells* produce hydrochloric acid, which makes the stomach contents acid and activates the enzymes. The *mucous neck cells* produce a sticky mucus, which clings to the stomach mucosa and protects the stomach wall itself from being digested.

Most digestive activity occurs in the pyloric region of the stomach. After food has been pro-

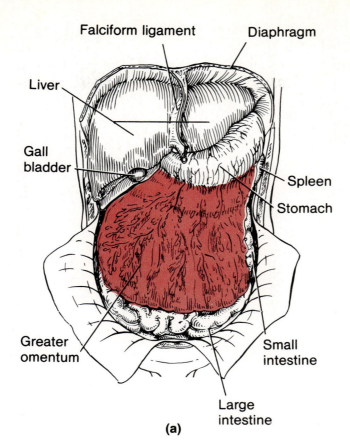

Falciform ligament   Diaphragm

Liver

Gall
bladder

Spleen

Stomach

Greater
omentum

Small
intestine

Large
intestine

(a)

FIGURE 13-4
Peritoneal attachments
of the abdominal
organs. **(a)** Anterior
view, omentum in
place. **(b)** Sagittal view
of a female torso.

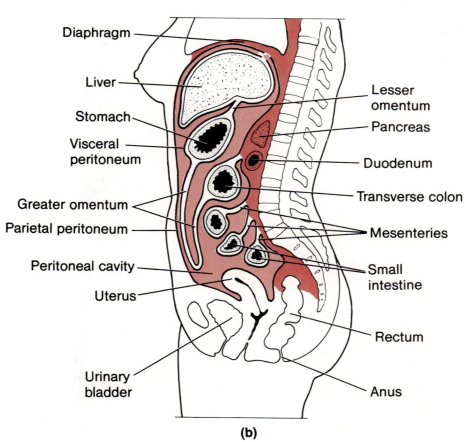

Diaphragm

Liver

Stomach

Visceral
peritoneum

Greater omentum

Parietal peritoneum

Peritoneal cavity

Uterus

Urinary
bladder

Lesser
omentum

Pancreas

Duodenum

Transverse colon

Mesenteries

Small
intestine

Rectum

Anus

(b)

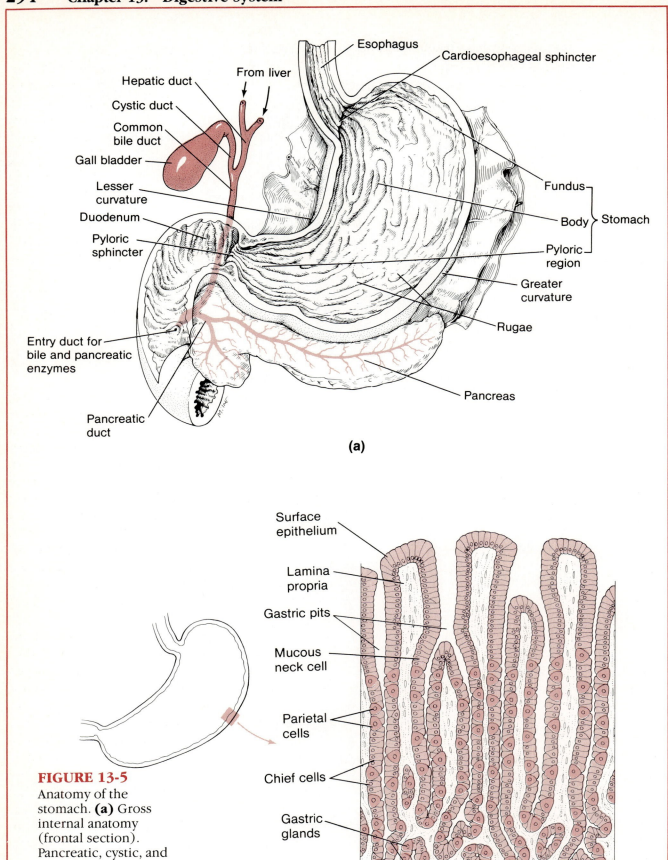

FIGURE 13-5
Anatomy of the stomach. **(a)** Gross internal anatomy (frontal section). Pancreatic, cystic, and hepatic ducts also shown. **(b)** Enlarged view of gastric pits (longitudinal section).

cessed in the stomach, it resembles heavy cream and is called *chyme* (kīm). The chyme enters the small intestine through the pyloric sphincter.

SMALL INTESTINE

The **small intestine** is a tube extending from the pyloric sphincter to the **ileocecal** (il"e-o-se'kal) **valve** (see p. 297). It is the longest section of the alimentary tube. Its length varies, but its average measurement is approximately 3 m (10 ft.) in a living person. The small intestine hangs into the abdominal cavity suspended by the fan-shaped mesentery from the posterior abdominal wall (see Figure 13-4). The large intestine encircles and frames it in the abdominal cavity. The small intestine has three subdivisions: the **duodenum** (do"o-de'num), which curves around the head of the pancreas, is about 30-cm (1-ft) long; the **jejunum** (je-oo'num) is 0.8–1.1-m (3–4-ft) long and extends from the duodenum to the ileum; the **ileum** (il'e-um), about 2-m (7-ft) long, is the terminal part of the small intestine (see Figure 13-1). The ileum joins the large intestine at the ileocecal valve.

Chemical digestion of foods begins in earnest in the small intestine, which is able to process only small amounts of foods at one time. The pyloric sphincter (literally, gatekeeper) controls food movement into the small intestine from the stomach and prevents the small intestine from being overwhelmed. Enzymes, produced by the intestinal cells and (more importantly) by the pancreas and ducted into the duodenum through the **pancreatic duct,** complete the chemical breakdown of foods in the small intestine. Bile (formed by the liver) also enters the duodenum through the **common bile duct** in the same area (see Figure 13-5). The pancreatic and common bile ducts join and then empty their products (in common) into the duodenum.

Nearly all food absorption occurs in the small intestine, where three structures that increase the absorptive area appear—the **microvilli, villi,** and **circular folds (plicae circulares** [pli'se ser-ku-la'res]) (Figure 13-6). Microvilli are tiny projections of the plasma membrane of the mucosa cells. Villi are fingerlike projections of the mucosa, which give it a velvety appearance and feel (much like the soft nap of a turkish towel). Within each villus is a rich capillary bed and a modified

lymph capillary, a **lacteal.** The digested foodstuffs are absorbed through the mucosa cells into both of these structures; the absorption process is discussed later in the chapter. Circular folds are deep folds of both mucosa and submucosa layers. Unlike the rugae folds of the stomach, the circular folds do not disappear when food fills the small intestine. All of these structural modifications, which increase the surface area, decrease in number toward the end of the small intestine. On the other hand, local collections of lymphatic tissue called **Peyer's patches** found in the submucosa increase toward the end of the small intestine. This reflects the fact that the remaining (undigested) food residue in the intestine contains huge numbers of bacteria, which must be prevented from entering the blood stream if at all possible.

LARGE INTESTINE

The **large intestine** (Figure 13-7) is about 15-m (5-ft) long and extends from the ileocecal valve to the anus. It frames the small intestine on three sides and has the following subdivisions: **cecum** (se'kum), **appendix, colon, rectum,** and **anal canal.** The saclike cecum is the first part of the large intestine. Hanging from the cecum is the blind wormlike appendix. The appendix is a potential trouble spot. Since it is usually twisted, it is an ideal location for bacteria to accumulate and multiply. Inflammation of the appendix, *appendicitis,* is the result. The colon is divided into several distinct regions. The **ascending colon** travels up the right side of the abdominal cavity and makes a turn (the *hepatic flexure*) to travel across the abdominal cavity as the **transverse colon.** It then turns again at the *splenic flexure* and continues down the left side as the **descending colon** and enters the pelvis where it becomes the S-shaped **sigmoid** (sig'moid) **colon.** The sigmoid colon, rectum, and anal canal lie in the pelvis. The anal canal ends at the **anus** (a'nus) which has an external *voluntary sphincter* and an internal *involuntary sphincter.* The sphincters, which act rather like purse strings to open and close the anus, are ordinarily closed except during defecation when the undigested residue is eliminated from the body as feces.

Because most of the absorption has been done before the large intestine is reached, no villi are seen in the large intestine, but there are tremen-

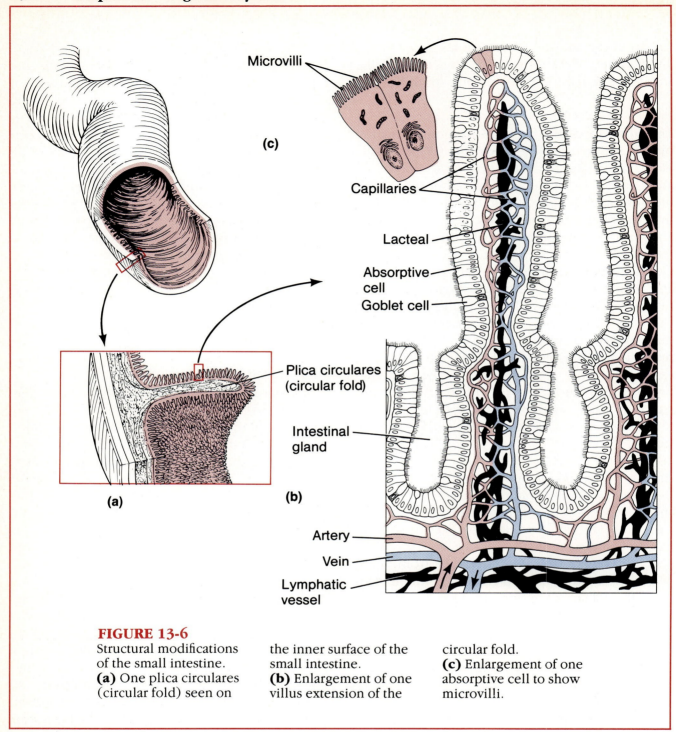

Microvilli

(c)

Capillaries

Lacteal

Absorptive
cell

Goblet cell

Plica circulares
(circular fold)

Intestinal
gland

(b)

Artery

Vein

Lymphatic
vessel

(a)

FIGURE 13-6
Structural modifications
of the small intestine.
(a) One plica circulares
(circular fold) seen on

the inner surface of the
small intestine.
(b) Enlargement of one
villus extension of the

circular fold.
(c) Enlargement of one
absorptive cell to show
microvilli.

dous numbers of *goblet cells* (see Figure 13-6)
there that produce mucus. The mucus acts as a
lubricant to ease the passage of feces to the end of
the digestive tract.

In the large intestine, the longitudinal muscle layer of the muscularis externa is reduced. The short
bands of muscle cause the wall to pucker into
small pocketlike sacs, **haustra** (haws'trah).

Accessory Digestive Organs

PANCREAS

The **pancreas** is a soft, pink, triangular gland that
extends across the abdomen from the spleen to
the duodenum (see Figures 13-1 and 13-5). The
pancreas is important to the digestive process, because it produces a wide spectrum of enzymes

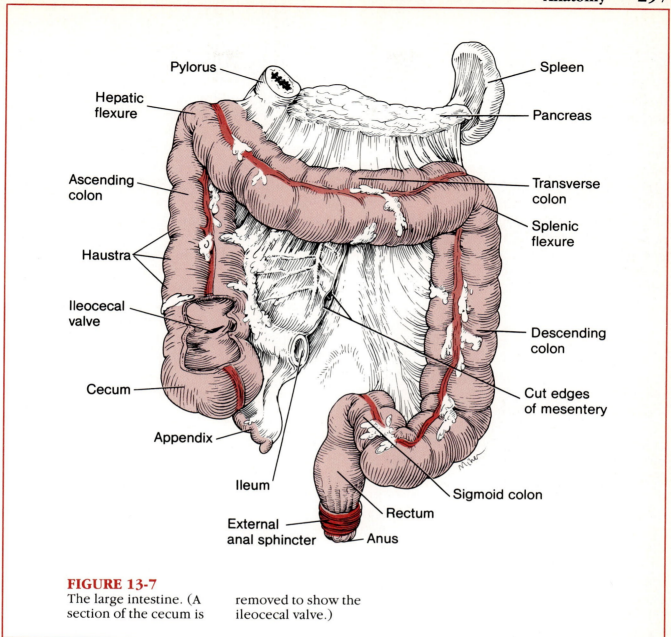

Pylorus

Spleen

Hepatic
flexure

Pancreas

Ascending
colon

Transverse
colon

Splenic
flexure

Haustra

Ileocecal
valve

Descending
colon

Cecum

Cut edges
of mesentery

Appendix

Ileum

Sigmoid colon

Rectum

External
anal sphincter

Anus

FIGURE 13-7
The large intestine. (A
section of the cecum is
removed to show the
ileocecal valve.)

that break down all categories of digestible foods. The pancreatic enzymes are secreted into the duodenum in an alkaline fluid, which neutralizes the acid chyme coming in from the stomach. The pancreas also has an endocrine function; it produces the hormones insulin and glucagon as explained in Chapter 8.

LIVER AND GALL BLADDER

The **liver** is the largest gland in the body. It is located under the diaphragm, more to the right side of the body (see Figure 13-1). As described earlier, the liver overlies and almost completely covers the stomach. The liver has four lobes and is suspended from the diaphragm and abdominal wall by a delicate mesentery cord, the **falciform** (fal'sĭ-form) **ligament** (see Figure 13-4a).

There is no question that the liver is one of the body's most important organs; it has many metabolic and regulatory roles. However, its digestive function is to produce **bile,** which leaves the liver through the **hepatic duct** and enters the duodenum through the common bile duct (see Figure 13-1). Bile is *not* an enzyme. It emulsifies fats by physically breaking apart large fat globules into smaller ones, which provides a larger surface area for the fat-digesting enzymes to work on.

The **gall bladder** is a small, green sac embedded in the inferior surface of the liver (see Figures 13-1 and 13-5). When digestion is not occurring in the digestive tract, bile backs up the **cystic duct** and enters the gall bladder to be stored. In the gall bladder, bile is concentrated by the removal of water. Later, when fatty food enters the duodenum, a hormonal stimulus causes the gall bladder to contract and spurt out the stored bile, making it available to the duodenum. If bile is stored in the gall bladder for too long or too much water is removed, solid particles of bile substances called *gall stones* begin to form. Since gall stones tend to be quite sharp, the individual may have severe pain (the typical gall bladder attack) when the gall bladder contracts.

If the hepatic or common bile ducts are blocked (for example, by wedged gall stones), bile is prevented from entering the small intestine, and it begins to accumulate and eventually backs up into the liver. This exerts pressure on the liver cells, and bile begins to enter the blood stream. As it circulates through the body, the tissues begin to become yellow, or *jaundiced* (jawn′dist). Blockage of the ducts is just one cause of jaundice; more often it results from actual liver problems such as *hepatitis* (an inflammation of the liver) or *cirrhosis* (sir-ro′sis), a condition in which the liver is severely damaged and becomes hard and fibrous. Cirrhosis is almost guaranteed when one drinks alcoholic beverages in excess and for many years.

SALIVARY GLANDS

Three pairs of **salivary glands** empty their secretions into the mouth. The large **parotid** (pah-rot′id) **glands** are anterior to the ear. *Mumps,* a common childhood disease, is an inflammation of the parotid glands. If you look at the location of the parotid glands in Figure 13-1, you can readily understand why people with mumps complain that it hurts to open their mouth or chew.

The **submaxillary glands** and the small **sublingual** (sub-ling′gwal) **glands** empty their secretions into the floor of the mouth through tiny ducts. The product of the salivary glands, *saliva,* is a mixture of mucus and serous fluids. The mucus moistens and helps bind food together into a mass called a *bolus* (bo′lus), which makes chewing and swallowing easier. The clear serous fluid

portion contains an enzyme, *salivary amylase* (am′ĭ-lās), which begins the process of starch digestion in the mouth. Saliva also contains substances that inhibit bacteria; therefore, it has a protective function as well.

TEETH

Ordinarily, by the age of 21, two sets of teeth have been formed (Figure 13-8). The first set is the **deciduous** (de-sid′u-us) **teeth,** or **milk teeth.** The deciduous teeth begin to erupt around 6 months, and a baby has a full set (20 teeth) by the age of 2 years. The first teeth to appear are the lower central incisors, an event which is usually anxiously awaited by the child's parents. The child begins to lose the milk teeth around the age of 6, and a second set of teeth, the **permanent teeth,** gradually replaces them. As the deeper permanent teeth enlarge and develop, the roots of the milk

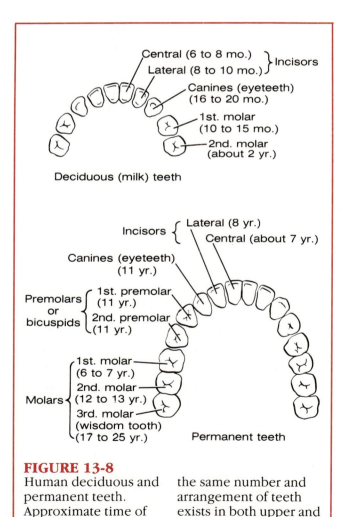

FIGURE 13-8

Human deciduous and permanent teeth. Approximate time of tooth eruption is shown in parentheses. (Since the same number and arrangement of teeth exists in both upper and lower jaws, only one jaw is shown in each case.)

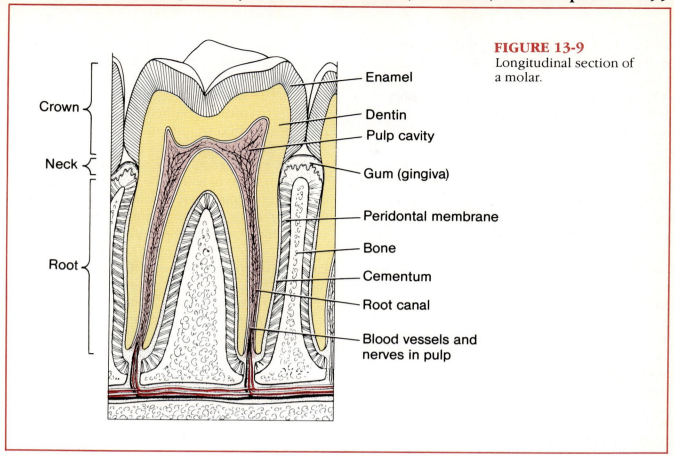

FIGURE 13-9
Longitudinal section of a molar.

Labels: Crown, Neck, Root; Enamel, Dentin, Pulp cavity, Gum (gingiva), Peridontal membrane, Bone, Cementum, Root canal, Blood vessels and nerves in pulp

teeth are resorbed; they then begin to loosen and are finally lost by the age of 12. The teeth are classed according to shape and function as incisors, canines (eyeteeth) premolars (bicuspids), and molars.

Although 32 is the normal number of permanent teeth, not all of us develop a full set. In many, the number three molars, commonly called wisdom teeth, never erupt.

A tooth consists of two major regions—the **crown** and the **root** as shown in Figure 13-9. The crown is the superior part of the tooth visible above the **gingiva** (jin-jī′vah), or **gum.** It is covered by enamel. **Enamel** is the hardest substance in the body and is fairly brittle, because it is heavily mineralized with calcium salts. The portion of the tooth embedded in the jawbone is the root; the root and crown are connected by the tooth region called the **neck.** The outer surface of the root is covered by a substance called **cementum.** Cementum attaches the teeth to the **periodontal** (per″e-o-don′tal) **membrane,** a fibrous membrane that holds the tooth in place in the bony jaw. **Dentin,** a bonelike material, forms the bulk

of the tooth. The dentin surrounds the **pulp cavity,** which contains a number of structures (connective tissue, blood vessels, and nerve fibers) collectively called the pulp. **Pulp** supplies nutrients to the tooth tissues and provides for tooth sensations. Where the pulp cavity extends into the root, it becomes the **root canal,** which provides a route for blood vessels, nerves, and other pulp structures to enter the tooth.

FUNCTIONS OF THE DIGESTIVE SYSTEM: FOOD BREAKDOWN, MOVEMENT, AND ABSORPTION

As described earlier, before foods can be absorbed by the cells of the digestive tract, they must first be broken down into their building blocks. The breakdown process is accomplished both by mechanical or physical means such as chewing and churning and by chemical means in which enzymes (proteins that increase the rates of chemical reactions) break the bonds holding the food molecules together. The digestive process occurs bit by bit; no one digestive organ does it all. Thus, in one sense, the digestive tract can be viewed as a "disassembly line" in which food is carried from one stage of its digestive processing to the next, and its nutrients are made available to the cells in the body en route.

The building blocks, or units, of *carbohydrate* foods are *monosaccharides* (mon"o-sak'ah-rīds), or simple sugars. We need to remember only three of these that are common in our diet—*glucose, fructose,* and *galactose.* Glucose is by far the most important, and when we talk about blood sugar levels, glucose is the "sugar" being referred to. Fructose is the sugar most abundant in fruits, and galactose is found in milk. Essentially, the only carbohydrates that our digestive system is able to digest, or to break down to simple sugar units, are *sucrose* (table sugar), *lactose* (milk sugar), *maltose* (malt sugar), and *starch.* Sucrose, maltose, and lactose are referred to as *disaccharides,* or double sugars, because each consists of two simple sugars that are linked together. Starch is a *polysaccharide* (literally, many sugars) formed of hundreds of glucose units linked together. Although we *do* eat foods containing other polysaccharides, such as cellulose, we do not have enzymes capable of breaking them down. While the indigestible polysaccharides do not provide us with any nutrients, they do help move the foodstuffs along the gastrointestinal tract by providing bulk, or fiber, to our diet.

Proteins are digested to their building blocks, which are *amino* (am'in-oh) *acids.* Intermediate products of protein digestion are polypeptides, proteoses (pro'te-os-es), and peptides. When fats (lipids) are digested, they yield two different types of building blocks—*fatty acid chains* and an alcohol called *glycerol* (glis'er-ol). The breakdown of carbohydrates, proteins, and fats is summarized in Figure 13-10.

Activities Occurring in the Mouth, Pharynx, and Esophagus

Food Breakdown

Both mechanical and chemical digestion begin in the mouth. First the food is physically broken down into smaller particles by chewing. Then, as the food is mixed with saliva, salivary amylase begins the chemical digestion of starch, breaking it down into maltose.

Saliva is normally secreted in response to the presence of food in the mouth, but simple pressure in the mouth such as occurs when rubber bands or sugarless gum is chewed will also stimulate the release of saliva. We also know that psychologic or emotional stimuli can cause salivation. For example, the mere thought of a hot fudge sundae will make many a mouth water.

Essentially no absorption of food occurs in the mouth. (However, some drugs such as nitroglycerine are absorbed easily through the oral mucosa.) The pharynx and esophagus have no digestive function; they simply provide passageways to carry food to the next processing site, the stomach.

Food Movement—Swallowing and Peristalsis

In order for food to be sent on its way from the mouth, it must first be swallowed. *Deglutition* (deg"loo-tish'un), or swallowing, is a complicated process that involves the coordinated activity of several structures (tongue, soft palate, pharynx, and esophagus). It has two major phases. The first, or *buccal phase,* occurs in the mouth and is voluntary. Once the food has been chewed and well mixed with saliva, the bolus is forced into the pharynx by the tongue. As food enters the pharynx, it passes out of our control and into the realm of reflex activity.

The second, or *involuntary phase,* transports food through the pharynx and esophagus. The parasympathetic division of the autonomic nervous system (primarily the vagus nerves) controls this phase and acts to promote the digestive process from this point on. Muscular contractions block off all routes that the food might take except for the desired route, that is, into the digestive pathways. The tongue blocks off the mouth, the soft palate rises to block off the nasal passages, and the larynx rises so that its opening (into the respiratory passageways) is covered by the flap-like epiglottis. Food is moved through the pharynx and then into the esophagus below by wavelike peristaltic contractions (first the longitudinal muscles contract, and then the circular muscles

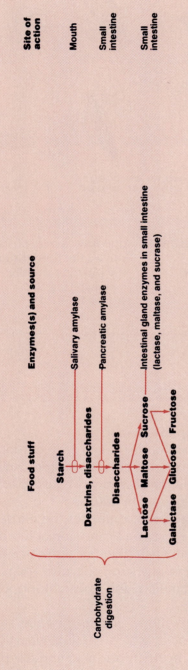
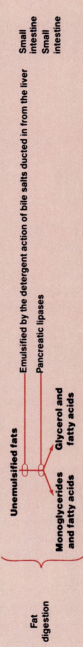

FIGURE 13-10
Flow sheet of digestion and absorption of foodstuffs.

Food stuff   **Enzyme(s) and source**   **Site of action**

Carbohydrate digestion

Starch — Salivary amylase — Mouth

Dextrins, disaccharides — Pancreatic amylase — Small intestine

Disaccharides — Intestinal gland enzymes in small intestine (lactase, maltase, and sucrase) — Small intestine

Lactose   Maltose   Sucrose

Galactase   Glucose   Fructose

Absorption: The monosaccharides (glucose, galactose, and fructose) are absorbed into the capillary blood in the villi and transported to the liver via the hepatic portal v.

Protein digestion

Proteins — Pepsin (stomach glands) in the presence of HCl — Stomach

Proteoses, peptones — Pancreatic enzymes (trypsin, chymotrypsin, carboxypeptidase) — Small intestine

Small polypeptides, dipeptides — Intestinal gland enzymes (aminopeptidases and dipeptidases) — Small intestine

Amino acids

Absorption: The amino acids are absorbed into the capillary blood in the villi and transported to the liver via the hepatic portal v.

Fat digestion

Unemulsified fats — Emulsified by the detergent action of bile salts ducted in from the liver — Small intestine

Monoglycerides and fatty acids — Pancreatic lipases — Small intestine

Glycerol and fatty acids

Absorption: Absorbed primarily into the lacteals of the villi (glycerol and short-chain fatty acids are absorbed into the capillary blood in the villi). Transported to the liver via the systematic circulation (hepatic artery), which receives the lymphatic flow from the thoracic duct or via the hepatic portal v.

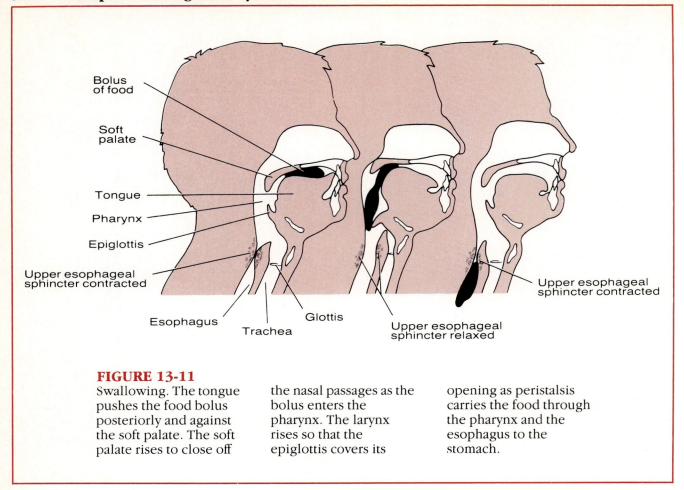

FIGURE 13-11

Swallowing. The tongue pushes the food bolus posteriorly and against the soft palate. The soft palate rises to close off the nasal passages as the bolus enters the pharynx. The larynx rises so that the epiglottis covers its opening as peristalsis carries the food through the pharynx and the esophagus to the stomach.

contract) of their muscular walls. The events of the swallowing process are illustrated in Figure 13-11.

If we try to talk or inhale while swallowing, these protective mechanisms may be "short-circuited," and food may manage to enter the respiratory passages. This triggers still another protective reflex (*coughing*), which involves the upward rush of air from the lungs in an attempt to expel the food.

Once food has reached the end of the esophagus, it presses against the cardioesophageal sphincter, causing it to open, and the food enters the stomach. The movement of food through the pharynx and esophagus is so automatic that a person can swallow, and food will reach the stomach even when he is standing on his head. Gravity plays no part in food transport once it has left the mouth, which explains why our astronauts (in zero gravity of outer space) can still swallow and get nourishment.

Activities of the Stomach

FOOD BREAKDOWN

The sight, smell, and taste of food stimulate parasympathetic nervous system reflexes, which increase the secretion of gastric juice by the stomach glands. In addition, the presence of food in the stomach stimulates the release of a local hormone, *gastrin,* by the stomach cells. Gastrin prods the stomach glands to produce still more of the protein-digesting enzymes (pepsinogen and rennin), mucus, and hydrochloric acid.

Hydrochloric acid makes the stomach contents very acid, which is (somewhat) dangerous since both hydrochloric acid and the protein-digesting enzymes have the ability to digest the stomach itself. However, as long as the mucus production is adequate, the stomach is "safe" and will remain unharmed by the local conditions. However, the cardioesophageal sphincter is often not as efficient as it should be, and if so, gastric juice will back up into the esophagus, which has little mucus protection. This results in a characteristic pain known as *heartburn,* which uncorrected leads to inflammation of the esophagus (*esophagitis* [ĕ-sof″ah-ji′tis]) and perhaps even to the ulceration of the esophagus.

However, the extremely acid environment that hydrochloric acid provides *is* necessary, because it is needed to activate pepsinogen to *pepsin,* the active protein-digesting enzyme. Rennin, the second protein-digesting enzyme produced by the stomach, works primarily on milk and converts it to a curdy-appearing substance (much like sour milk). Many mothers mistakenly think that when their infants spit up a curdy substance after having their bottle that the milk has soured in their stomach. Rennin is produced in large amounts in infants, but it is not believed to be produced in adults.

Other than beginning protein digestion, little digestion occurs in the stomach. With the exception of alcohol (which seems somehow to have a "special pass"), virtually no absorption occurs through the stomach walls.

FOOD MIXING AND MOVEMENT

As food enters and fills the stomach, its wall begins to stretch (at the same time the gastric juices are secreted in increasing amounts, as just described). Churning, or mixing, movements then begin to occur in the stomach. These mixing movements involve all three muscle layers of the stomach wall; they compress, pummel, and continually mix the food with the gastric juice so that the semifluid chyme is formed. Perhaps it would help if you visualize this process like the mixing of a cake mix in which the floury mixture is continually folded on itself and mixed with the liquid until it reaches uniform texture.

Once the food has been well mixed, peristalsis begins in the lower half of the stomach and forces the chyme through the pyloric sphincter (see Figure 13-5) into the duodenum. Because the pyloric sphincter is only partially opened, only small amounts of food are allowed to enter it at one time from the stomach. When the duodenum is filled with chyme and its wall is stretched, a nervous reflex, the *enterogastric* (en″ter-o-gas′trik)

reflex, occurs. This reflex inhibits the vagus nerves from stimulating the stomach muscles and slows the emptying of the stomach as long as food remains in the duodenum. Generally, it takes about 4 hours for the stomach to completely empty after eating a meal.

Activities of the Small Intestine

FOOD BREAKDOWN AND ABSORPTION

Food reaching the small intestine is only partially digested. Carbohydrate and protein digestion have been started, but virtually no fats have been digested to this point. Here the process of food digestion is accelerated. Having been churned and partially digested by enzymes, the food now takes a rather wild 4–8 hour journey through the looping coils and twists of the small intestine. By the time the food reaches the end of the small intestine, digestion is completed. In addition, nearly all food absorption occurs in the small intestine.

Although the small intestine cells produce a few important enzymes (for example, enzymes that break down double sugars into simple sugars), **intestinal juice** is relatively enzyme-poor. Protective mucus is probably the most important intestinal gland secretion. However, foods entering the small intestine are literally deluged with enzymes ducted in from the pancreas. **Pancreatic juice** contains enzymes that digest proteins (trypsin, peptidases), carbohydrates (amylase), fats (lipases), and nucleic acids (nucleases). In addition to enzymes, pancreatic juice contains a rich supply of bicarbonate, which makes it very basic. Thus, when pancreatic juice reaches the small intestine, it neutralizes the acid chyme coming in from the stomach and provides the proper environment for activity of the intestinal and pancreatic digestive enzymes.

The release of pancreatic juice into the duodenum is stimulated by both the vagus nerves and local hormones produced by the small intestine cells. When chyme enters the small intestine, it stimulates the mucosa cells to produce two hormones, *secretin* (se-kre′tin) and *cholecystokinin* (ko"le-sis"to-kin'in). The hormones enter the blood and circulate to their target organs, the pancreas, liver, and gall bladder. Both hormones work together to stimulate the pancreas to release its enzyme- and bicarbonate-rich product. In addition, secretin causes the liver to increase its output of bile. Cholecystokinin causes the gall bladder to contract and release stored bile into the common bile duct so that both bile and pancreatic juice enter the small intestine at the same time. As mentioned before, bile is not an enzyme. Instead, it acts like a detergent to emulsify, or mechanically break down, large fat globules into thousands of tiny ones, which provides a much greater surface area for the lipases to work on. Bile is also necessary for the absorption of fats and fat-soluble vitamins (vitamins K, D, A) from the intestinal tract. If *either* bile or pancreatic juice are absent, essentially no fat digestion or absorption goes on, and fatty bulky stools are the result. In such cases, blood-clotting problems also occur because the liver needs vitamin K to make prothrombin, one of the clotting factors.

Absorption of water and of the end products of digestion occurs all along the length of the small intestine. Most substances are absorbed through the intestinal cell walls by the process of *active transport*. They then enter the capillary bed in the villus to be transported in the blood to the liver. The exception seems to be the fats, or lipids, which are absorbed passively by the process of *diffusion*. Lipids enter both the capillary bed and the lacteal in the villus and are carried to the liver by both blood and lymphatic fluids.

At the end of the ileum, all that remains is some water, undigestible food materials (plant fibers such as cellulose), and large amounts of bacteria. It is this debris that enters the large intestine through the ileocecal valve. The complete process of food digestion and absorption is summarized in Figure 13-10.

FOOD MIXING AND MOVEMENT

While foods are in the small intestine, two types of movement occur (Figure 13-12). *Segmental movements* are local constrictions of the intestine, and they occur rhythmically. They are not important in moving food along the tract. Instead, they *mix* chyme with digestive juices and increase the rate of absorption by continually moving different portions of the chyme over adjacent regions of the intestinal wall.

As mentioned previously, peristaltic movements are the major means of propelling food through the digestive tract. They involve waves of contraction that move along the length of the intestine, followed by waves of relaxation. The net effect is that the food is moved through the small intestine much in the same way that toothpaste is squeezed from a tube.

Local irritation of the small intestine or the stomach, such as might occur with bacterial food poisoning, may stimulate the *emetic* (ĕ-met'ik) *center* in the brain (medulla). The emetic center, in turn, causes vomiting. Vomiting is essentially a reverse peristalsis occurring in the stomach (and perhaps the small intestine) that is accompanied by contraction of the abdominal muscles and the diaphragm, which increases the pressure on the abdominal organs. The emetic center may also be activated through other pathways; disturbance of the equilibrium apparatus of the inner ear during a boat ride on a rough ocean is one such example.

Activities of the Large Intestine

FOOD BREAKDOWN AND ABSORPTION

What is finally delivered to the large intestine contains few nutrients, but the residue still has about 12 hours more to spend in the digestive tract. With the exception of a small amount of digestion of the residue by the "resident" bacteria, virtually no further food breakdown goes on in the large intestine.

Bacteria residing in the large intestine also make some vitamins (vitamin K and some B vitamins). Absorption by the large intestine is limited to the absorption of these vitamins, some ions, and most of the remaining water. The more or less solid product delivered to the rectum is called *feces* (fe' sēz), or the stool.

MOVEMENT OF THE RESIDUE AND DEFECATION

Peristalsis and *mass movements* are the two major types of movements occurring in the large intes-

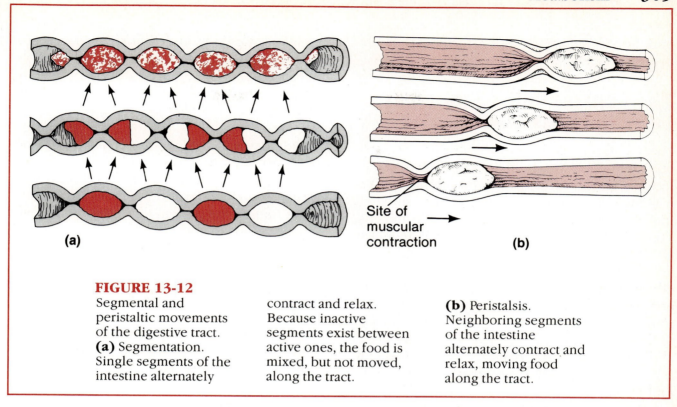

FIGURE 13-12
Segmental and peristaltic movements of the digestive tract. **(a)** Segmentation. Single segments of the intestine alternately contract and relax. Because inactive segments exist between active ones, the food is mixed, but not moved, along the tract. **(b)** Peristalsis. Neighboring segments of the intestine alternately contract and relax, moving food along the tract.

Site of muscular contraction

(a)

(b)

tine. Mass movements are long, slow-moving waves (mass peristalsis) that move over large areas three to four times daily and force the contents of the large intestine toward the rectum. The rectum is generally empty. When feces is rushed into it by mass movements and its wall is stretched, the *defecation reflex* is initiated. The defecation reflex is a cord-mediated (sacral region) reflex that causes the walls of the sigmoid colon and the rectum to contract and the anal sphincters to relax. As the feces is forced through the anal canal, messages reach the brain giving us time to make a decision as to whether the external voluntary sphincter should remain open or be constricted to stop the passage of feces. If it is not convenient, defecation (or moving the bowels) can be delayed temporarily; within a few seconds the reflex contractions end, and the rectal walls relax. With the next mass movement, the defecation reflex is initiated again.

Watery stools, or *diarrhea* (di"ah-re′ah), result from any condition that rushes food residue through the large intestine before it has had sufficient time to absorb the water (as in irritation of the colon by bacteria). Because fluids and ions are lost from the body, prolonged diarrhea may result in dehydration and electrolyte imbalance. Conversely, when food residue remains in the large intestine for extended periods, too much water is absorbed, and the stool becomes hard and difficult to pass *(constipation)*.

METABOLISM

Metabolism (mĕ-tab′o-lizm) is a broad term referring to all chemical reactions that are necessary to maintain life. It involves processes in which substances are broken down to simpler substances (*catabolism* [kah-tab′o-lizm]) and processes in which larger molecules or structures are built from smaller ones (*anabolism* [ah-nab′o-lizm]). During catabolic reactions, energy is released and captured to make ATP, which is the energy-rich molecule used by body cells to energize all their activities.

We find that not all foodstuffs are treated in the same way by body cells. For example, carbohydrates, particularly glucose, are usually broken down to make ATP. Fats are used to build cell membranes, make myelin sheaths, and insulate the body with a fatty cushion. They are also used for making ATP when there are inadequate carbohydrates in the diet. Proteins tend to be carefully conserved (even hoarded) by the body cells. This is easy to understand when you recognize that proteins are the major building materials for making cell structures.

The Central Role of the Liver in Metabolism

The liver is one of the most versatile organs in the body. Without it we would die within 24 hours. Its role in digestion (that is, manufacture of bile) is important to the digestive process to be sure, but it is only one of the many functions of liver cells. The liver cells detoxify drugs and other chemicals (for example, alcohol), make many substances vital to the body as a whole (blood proteins such as albumin and clotting proteins, and cholesterol), and play a central role in metabolism. Because of its key roles, nature has provided us with a surplus of liver tissue. We have much more than we need, and even if part of it is damaged or removed, it is one of the few body organs that can regenerate rapidly and easily.

A unique circulation, the hepatic portal circulation, brings nutrient-rich blood draining from the digestive viscera directly to the liver. Since the liver is the major metabolic organ of the body, this detour that nutrients take through the liver reflects the wisdom of the body; it ensures that the liver's needs will be met first. As blood loaded with nutrients circulates slowly through the liver, liver cells remove amino acids, fatty acids, and glucose from the blood, and phagocytic cells remove and destroy bacteria that have managed to get through the digestive tract walls and into the blood.

The liver is extremely important in helping to maintain blood glucose levels within normal range (around 100 mg glucose/100 ml of blood). After a carbohydrate-rich meal, thousands of glucose molecules are removed from the blood and combined together to form large polysaccharide molecules called *glycogen* (gli′ko-jen) *molecules,* which are then stored in the liver. This process is *glycogenesis* (gli″ko-jen′ĕ-sis). Later, as the body cells remove glucose from the blood to meet their needs, blood glucose levels begin to drop. At this time, liver cells will break down the stored glycogen (*glycogenolysis* [gli″ko-jĕ-nol′ĭ-sis]) and release glucose bit by bit to the blood to maintain homeostasis of blood glucose levels. If necessary, the liver can also form glucose from noncarbohydrate substances such as fats and proteins. This

process is *gluconeogenesis* (glu″ko-ne″o-jen′ĕ-sis), which means formation of new sugar. As described in Chapter 8, hormones such as thyroxine, insulin, and glucagon are important in controlling the handling of glucose in all body cells.

All blood proteins made by the liver are built from the amino acids its cells pick up from the blood. The completed proteins are then released back into the blood to travel throughout the circulation. Albumin, the protein that is most abundant in blood, holds fluids in the bloodstream. When insufficient albumin is present in blood, fluid leaves the bloodstream and accumulates in the tissue spaces causing edema. Possible causes are liver damage or kidney disease (in which the kidney "filters" are damaged so that albumin is allowed to pass out of the bloodstream into the urine product). The role of the clotting proteins made by the liver is discussed in Chapter 9.

Finally, some of the fats and fatty acids picked up by the liver cells are broken down to make ATP for use by the liver cells. The rest of the fats are broken down to simpler substances such as acetic acid and acetoacetic acid (two acetic acids linked together) and released into the blood. The liver also forms cholesterol.

Nutrients not needed by the liver cells, as well as the products of liver metabolism, are released into the blood and drain from the liver in the hepatic vein to enter the systemic circulation where they become available to other body cells.

Carbohydrate, Fat, and Protein Metabolism in Body Cells

CARBOHYDRATE METABOLISM

Glucose, the major breakdown product of carbohydrate digestion, is the major fuel used for making ATP in most cells of the body. Just as an oil furnace uses oil (its fuel) to produce heat, the cells of the body use their preferred fuel (glucose) to produce cellular energy (ATP). The liver is an exception; it routinely uses fats as well, thus saving glucose for other body cells. Essentially, glucose is broken apart piece by piece. Hydrogen removed is combined with oxygen to form water, and the carbon leaves the cells as carbon dioxide (Figure 13-13). The actual chemical pathways

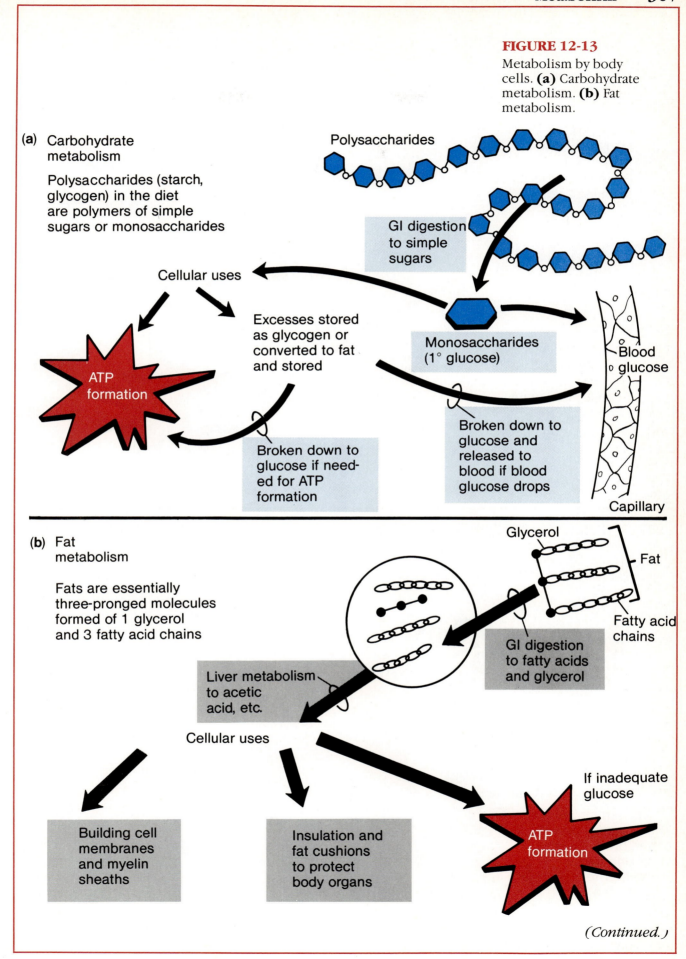

FIGURE 12-13
Metabolism by body cells. **(a)** Carbohydrate metabolism. **(b)** Fat metabolism.

(a) Carbohydrate metabolism

Polysaccharides (starch, glycogen) in the diet are polymers of simple sugars or monosaccharides

Polysaccharides

GI digestion to simple sugars

Cellular uses

Excesses stored as glycogen or converted to fat and stored

Monosaccharides (1° glucose)

Blood glucose

ATP formation

Broken down to glucose if needed for ATP formation

Broken down to glucose and released to blood if blood glucose drops

Capillary

(b) Fat metabolism

Fats are essentially three-pronged molecules formed of 1 glycerol and 3 fatty acid chains

Glycerol

Fat

Fatty acid chains

GI digestion to fatty acids and glycerol

Liver metabolism to acetic acid, etc.

Cellular uses

Building cell membranes and myelin sheaths

Insulation and fat cushions to protect body organs

If inadequate glucose

ATP formation

(Continued.)

FIGURE 13-13
(continued)
(c) Protein metabolism.
(d) ATP production.

PROTEIN METABOLISM

Proteins are polymers
of amino acids

CELLULAR USES

Build and repair
body tissues
(membrane proteins,
muscle proteins,
etc.)

Protein

GI digestion
to amino acids

RARE

ATP
formation

ATP formation if
inadequate glucose
and fats

(c)

(d)

ATP FORMATION — "FUELING THE METABOLIC PUMP"
As shown below and described in a–c, all categories
of food *can* be oxidized to provide energy (ATP); however,
glucose is the preferred molecule for this purpose

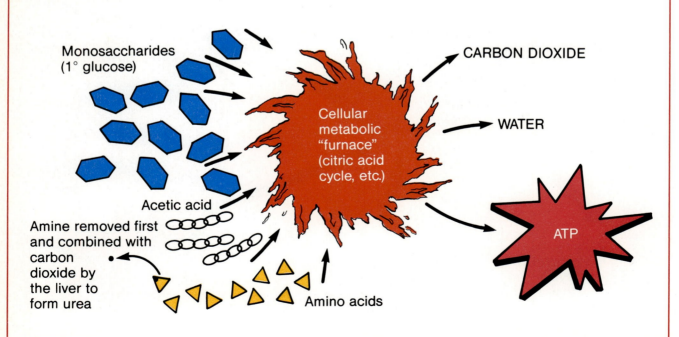

Monosaccharides
(1° glucose)

CARBON DIOXIDE

Cellular
metabolic
"furnace"
(citric acid
cycle, etc.)

WATER

ATP

Acetic acid

Amine removed first
and combined with
carbon
dioxide by
the liver to
form urea

Amino acids

(the citric acid cycle and others) go on inside the mitochondria, but we need not concern ourselves with the details of these pathways here. The overall reaction can be summed up in a simple statement (see Figure 13-13*d*):

Glucose + Oxygen =
> Carbon dioxide + Water + ATP

This helps explain why homeostasis of blood glucose levels is so critically important. If there are excessively high levels of glucose in the blood (*hyperglycemia* [hi″per-gli-se′me-ah]), some of this excess will be stored in body cells (particularly liver and muscle cells) as glycogen. If blood glucose levels are still too high, excesses will be converted to fat and stored in adipose tissue. There is no question that eating large amounts of empty calorie foods such as candy and other sugary sweets causes a rapid deposit of fat in the body tissues. When blood glucose levels are too low (*hypoglycemia*), stored glycogen is broken down by the liver, and glucose is released to the blood as just explained. These various fates of carbohydrates are also shown in Figure 13-13*a*.

FAT METABOLISM

Fats are somewhat important for the building of cell structures. All cell membranes have fat, or lipid, molecules as part of their structure, and the myelin sheaths of neurons (see Chapter 6) are formed from fatty substances.

Actually, the liver handles most of the fat metabolism that goes on in the body. After it has taken what it needs to make ATP for its own use, it releases relatively small, fat-breakdown products to the blood as previously explained. Body cells remove the fat products from the blood and build them into their membrane structures as needed. Fats are also used to form fatty cushions around body organs (for example, around the kidneys and behind the eyeballs). In addition, a small amount is burned, or broken down, for energy. For fat products to enter the citric acid cycle reactions, they must first be broken down to acetic acid. Within the mitochondria, the acetic acid is then completely oxidized and carbon dioxide, water, and ATP are formed. When there is not enough glucose to fuel the needs of the cells for energy, much larger amounts of fats are used to produce ATP. Because fat breakdown and metab-

olism is not nearly as efficient as carbohydrate metabolism, many of the intermediate products begin to accumulate in the blood (such as acetoacetic acid and acetone). These cause the blood to become acidic (acidosis, or ketosis), and the breath takes on a fruity odor as acetone diffuses from the lungs. Ketosis is a common consequence of "no-carbohydrate diets," uncontrolled diabetes mellitus, and starvation in which the body is forced to rely almost totally on dietary or stored body fats to fuel its energy needs.

Excess fats are stored in fat depots such as the hips, abdomen, breasts, and subcutaneous tissue. Fat in subcutaneous tissue is important as insulation for the deeper body organs. However, excessive subcutaneous fat restricts movement and places much greater demands on the circulatory system as a whole. The metabolism and uses of fats are shown in Figure 13-13*b*.

PROTEIN METABOLISM

Proteins make up the bulk of cellular structures, and they are carefully conserved by the body cells. Proteins taken in the diet are broken down to amino acids. Once the liver has completed its processing of the blood draining the digestive tract, the remaining amino acids circulate to the body cells in the systemic circulation. The cells remove amino acids from the blood and use them to build proteins both for their own use (enzymes, membranes, mitotic spindle proteins, muscle proteins) and for export (mucus, hormones, and others). The cells do not take too many chances with their amino acid supply. They use ATP to actively transport amino acids into their interior even though (in many cases) there may be more of those amino acids inside the cell than there are in the blood flowing past them. Even though this may appear to be cellular greed, there is an important reason for this active uptake of amino acids. Cells cannot build their proteins unless *all* of the needed amino acids, which number around 20 types, are present. Since 9 of these amino acids can not be made by the cells, they are available to the cells only through dietary pathways; such amino acids are called *essential amino acids*. This helps explain the avid accumulation of amino acids, which ensures that all amino acids needed will be available for present and at least some of the future protein-building needs of the cells (see Figure 13-13*c*).

TABLE 13-1 Factors Determining the Basal Metabolic Rate (BMR)

Factor	Variation	Effect on BMR
Surface area	Large surface area in relation to body volume, as in thin/small individuals	Increased
	Small surface area in relation to body volume, as in large/heavy individuals	Decreased
Sex	Male	Increased
	Female	Decreased
Thyroxine production	Increased	Increased
	Decreased	Decreased
Age	Young/rapid growth	Increased
	Aging, elderly	Decreased
Strong emotions (anger or fear) and infections		Increased

Amino acids are used to make ATP only when proteins are overabundant and/or when carbohydrates and fats are not available for this purpose. When it is necessary to oxidize amino acids for energy, their amine groups are removed as *ammonia,* and the rest of the molecule enters the citric acid cycle pathways in the mitochondria. The ammonia that is released during this process is toxic to body cells, especially nerve cells. Again, the liver comes to the rescue by combining the ammonia with carbon dioxide to form *urea* (u-re'ah). Urea, which is not harmful to the body cells, is then flushed from the body in the urine.

Metabolic Rate and Body Temperature Regulation

BASAL METABOLIC RATE

When nutrients are broken down to produce cellular energy (ATP), they yield different amounts of energy. The energy value of foods is measured in a unit called the kilocalorie (kcal). In general, carbohydrates yield 4 kcal/g, fats yield 9 kcal/g, and proteins yield 4 kcal/g when they are broken

down completely for energy production. Most meals, and even many foods, are mixtures of carbohydrates, fats, and proteins. Thus, to determine the caloric value of a meal, we must know how many grams of each type of foodstuff it contains. For most of us, this is a difficult chore indeed, but approximations can easily be made with the help of a simple, calorie-values guide available in any drugstore.

The amount of energy used by the body is also measured in kilocalories. The *basal metabolic rate* (BMR) is the amount of energy used by the body when it is under basal conditions, that is, at rest. It reflects the energy supply a person's body needs just to perform essential life activities such as breathing, maintaining the heartbeat, and kidney function.

As shown in Table 13-1, small/thin/male people tend to have a higher BMR than large/obese/female people. Age is also important; children and adolescents require large amounts of energy for growth and have relatively high BMRs. In old age the BMR decreases dramatically as the muscles begin to atrophy.

The amount of thyroxine produced by the thyroid gland is probably the most important factor in determining a person's BMR. The more thyroxine produced, the higher the metabolic rate will be. Today, thyroid activity is easily determined with blood tests such as the PBI (protein-bound iodine), T_3, and T_4 blood determinations.

TOTAL METABOLIC RATE

When we are active, much more glucose must be oxidized to provide energy for these additional activities. Digesting food and even modest physical activity increases the body's caloric requirements dramatically. These additional fuel requirements are above and beyond the energy required to maintain the body in the basal state. *Total metabolic rate* (TMR) refers to the total amount of kilocalories the body must consume to fuel *all* ongoing activities. Muscular work is the major body activity that increases the TMR.

When the total amount of calories consumed is equal to the TMR, homeostasis is maintained, and our weight remains constant. However, if we eat more than we need to sustain our activities, excess calories appear in the form of fat deposits. Conversely, if we are extremely active and do not properly feed the "metabolic furnace," we begin to catabolize fat reserves and even tissue proteins to satisfy our TMR. This principle is used in every good weight-loss diet. The total amount of calories needed are calculated on the basis of body size and age. Then, 20% or more of the caloric requirements are cut from the daily diet. If the dieting person exercises regularly, weight drops off even more quickly because the TMR is being increased over and above the person's former rate.

MAINTAINING BODY TEMPERATURE

Although we have been emphasizing that foods are burned to produce ATP, ATP is not the only product of cell catabolism. Most of the energy released as foods are oxidized escapes as heat. Less than 40% of available food energy is actually captured to form ATP. The heat released warms the tissues and, more importantly, the blood, which circulates to all body tissues keeping them at homeostatic temperatures.

It is important that a constant body temperature (36.1°–37.8°C [97–100°F]) be maintained regardless of the heat-producing processes occurring within the body or cold temperatures outside the body. The body's thermostat is in the hypothalamus of the brain. Through autonomic nervous system pathways, the hypothalamus continually regulates body temperature by initiating various heat-liberating or heat-promoting mechanisms.

When the environmental temperature is cold, body heat must be conserved. The skin capillaries constrict to restrict the flow of blood to deeper (and more vital) body organs, and the skin is temporarily bypassed by the blood. When this happens, the temperature of the exposed skin drops to that of the external environment. This situation is not a problem for a brief period of time, but if it is extended, the skin cells, which are deprived of oxygen and nutrients, begin to die. This condition, *frostbite*, is extremely serious. When the core body temperature (the temperature of the deep organs) drops to the point beyond which simple constriction of skin capillaries can handle the situation, *shivering* begins. Shivering, involuntary shudderlike contractions of the voluntary muscles, is very effective in increasing the body temperature, because muscle activity produces large amounts of heat.

An extremely low body temperature resulting from prolonged exposure to cold is *hypothermia*. In hypothermia, the individual's vital signs (respiratory rate, blood pressure, heart rate) decrease. Drowsiness sets in and the person becomes oddly comfortable even though previously he or she felt extremely cold. Uncorrected, the situation progresses to coma and finally death as metabolic processes stop in such low temperatures.

Just as the body must be protected from becoming too cold, it must also be protected from excessively high temperatures. Most heat loss occurs through the skin. When the temperature of the body increases over what is desirable, the capillary beds in the skin become flushed with warm blood, and heat is allowed to radiate from the skin surface. However, if the external environment is as hot or hotter than the body, heat cannot be lost by radiation and the only means of getting rid of excess heat is by the evaporation of perspiration off the skin surface. This is an efficient means of body-heat loss as long as the air is dry; if it is very humid, evaporation occurs at a much slower rate.

Under such circumstances, our heat-liberating mechanisms cannot work well, and we feel miserable and become irritable. This is why the hot humid days of August are often called "dog days." When normal heat-liberating processes become ineffective, body temperature increases to levels that threaten homeostasis. This condition, *heat stroke,* can be fatal.

Fever is a special type of *hyperthermia,* or elevated body temperature. Most often fever results from infection somewhere in the body, but certainly other conditions may cause it (for example, cancer, allergic reactions, CNS injuries). Injured tissue cells and bacteria release chemical substances called *pyrogens* (pi′ro-gens) that act directly on the hypothalamus, causing its thermostat to be set to a higher temperature. It cannot be emphasized strongly enough that fever is controlled hyperthermia. After the thermostat resetting, heat promoting mechanisms are initiated (constriction of the skin capillaries and shivering), and the body temperature is allowed to rise until it reaches the new setting. Body temperature is then maintained at the "fever setting" until natural body-defense processes or antibiotics are effective in breaking the disease process. At that point, the thermostat is reset again to a lower or normal level, which causes heat-loss mechanisms to swing into action—the individual begins to sweat, and the skin becomes flushed and warm. Physicians have long recognized that these events signaled a turn for the better in their patients and said that the patient had passed the crisis.

DEVELOPMENTAL ASPECTS OF THE DIGESTIVE SYSTEM

The very young embryo is flat and pancakelike in shape. However, soon this flattened cell mass folds to form a cylindrical body. Its internal cavity becomes the cavity of the digestive tract. By the fifth week of development, the alimentary canal is a tubelike structure extending from the mouth to the anus; shortly after, the glandular organs (salivary glands, liver, and pancreas) bud out, or outpocket, from the mucosa of the alimentary tube. Since these glands retain their connections (ducts), they can easily empty their secretions into the digestive tract to promote its digestive functions.

The digestive system is susceptible to many congenital defects, which interfere with feeding, and a number of genetically caused problems interfere with normal metabolism. The most common congenital defect is the *cleft palate/cleft lip* defect. Of the two, cleft palate is more serious because it interferes with feeding, and the child is unable to suck properly. Additionally, if the feeding is aspirated (drawn into the lungs) pneumonia results. Another relatively common congenital defect is a *tracheoesophageal fistula* (tra″ke-o-e-sof′ah-je-al fis′tu-lah). In this condition, there is an opening or connection between the esophagus and the trachea; in addition, the esophagus often (but not always) ends in a blind sac and does not connect to the stomach. The baby chokes, drools, and becomes cyanotic during feedings, because food is entering the respiratory passageways. All three defects are corrected surgically.

There are many types of inborn errors of metabolism or genetically based problems, but perhaps the two most common are *cystic fibrosis* and *phenylketonuria* (fen″il-ke″to-nu′re-ah) (PKU). Cystic fibrosis primarily affects the lungs, but it also significantly impairs the activity of the pancreas. In cystic fibrosis, huge amounts of mucus are produced, which block the passages of involved organs. Blockage of the pancreatic duct prevents pancreatic fluid from reaching the small intestine. As a result, fats and fat-soluble vitamins fail to be digested or absorbed in any significant amounts, and bulky fat-laden stools result. This condition is usually handled by administering pancreatic enzymes with meals.

PKU involves an inability of tissue cells to use one particular amino acid (phenylalanine [fen″il-al′ah-nīn]), which is present in all protein foods. In such cases, brain damage and retardation occur unless the infant is put on a special diet low in phenylalanine until about the age of 6.

During fetal life, the developing infant receives all of its nutrients through the placenta. At least at this period of life, obtaining and processing nutrients is no problem (assuming the mother is adequately nourished). In the newborn baby, obtaining nutrition is the most important activity. Several reflexes present at this time help in this

BOX 13-1
Ulcers

Because of its activity in protein digestion, it hardly seems likely that the stomach can prevent its own digestion. Although a layer of thick mucus produced by the stomach glands performs very efficiently to protect the stomach wall, this line of defense may fail occasionally. For example, if too much hydrochloric acid (or too little mucus) is produced or if drugs (such as aspirin) are taken that interfere with the mucous lining, the stomach does in fact begin to digest itself and ulcers appear. (The photograph shows a peptic ulcer in the mucosa of the gastrointestinal tract.) If the ulcer is very severe and fails to respond to medication, a

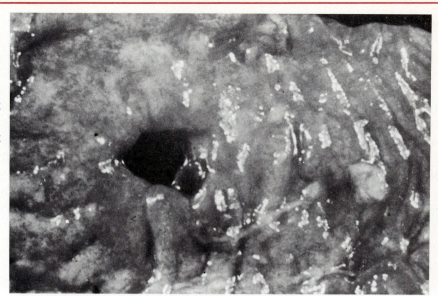

gastrectomy (surgical removal of all or part of the stomach) may be required to prevent the possibility of perforation of the stomach wall and hemorrhage. After a gastrectomy, particularly one which removes all or most of the stomach, inadequate amounts of intrinsic factor may be made.

Under such conditions, the individual may need to receive vitamin–B$_{12}$ injections for the rest of his or her life to prevent metabolic problems (particularly pernicious anemia). (Photo from Purtilo, D.T. 1978. *A Survey of Human Diseases.* Menlo Park, Calif.: Addison-Wesley Publishing Co.)

---

activity; for example, the rooting reflex helps the infant find the nipple (mother's or bottle), and the sucking reflex helps him or her to hold on to the nipple and swallow. The stomach of a newborn infant is very small, which means that feeding must be frequent (every 3–4 hours). Peristalsis is rather inefficient at this time, and vomiting of the feeding it not at all unusual.

Teething begins around age 6 months and continues until about the age of 2. During this interval, the infant progresses to more and more solid foods and usually is eating the adult diet by toddlerhood. Appetite decreases in the school-age child and then increases again during adolescence when the growth spurt is ongoing. (Parents of adolescents usually bewail their high food costs.)

All through childhood and into adulthood the di-

gestive system operates with relatively few problems unless there are abnormal interferences such as contaminated food or extremely spicy or irritating foods (which may cause inflammation of the gastrointestinal tract, or *gastroenteritis* [gas″-tro-en-ter-i′tis]). Inflammation of the appendix, appendicitis, is particularly common in teenagers (for some unknown reason). Between middle age and early old age, the metabolic rate decreases by 5%–8% in every 10-year period. This is the time of life when the weight seems to creep up, and obesity often becomes a fact of life. If a desired weight is to be maintained, we must be aware of this gradual change and be prepared to reduce caloric intake. Two distinctly middle age digestive system problems are *ulcers* and *gall bladder problems* (inflammation of the gall bladder and gallstones). Ulcers are seen more frequently in people who are constantly rushing and under pressure, but they can also result from overindulgence in spicy foods, drugs, or alcohol. Ulcers and gall bladder problems are handled conventionally (diet and drugs), if possible.

During old age, gastrointestinal tract activity declines. Fewer digestive juices are produced, and peristalsis slows. Taste and smell become less acute and periodontal disease (inflamed gums and loosening teeth) become endemic. Many elderly individuals live alone or are living on a reduced income. These factors, along with increas-

ing physical disability, tend to make eating less appealing, and nutrition is often inadequate in many of our elderly citizens. Cancer of the gastrointestinal tract is a fairly common problem in the elderly. Cancer of the stomach and colon rarely have early signs, and the disease often progresses to an inoperable stage (that is, it has spread to distant parts of the body as well) before the individual seeks medical attention. It has been suggested that diets high in plant fiber might help to decrease the incidence of colon cancer, but this has still not been proven.

IMPORTANT TERMS*

Alimentary *(al″ĕ-men′tar-e)* **canal**	**Villi**
Oral cavity	**Large intestine**
Peristalsis *(per″ĭ-stal′sis)*	**Rectum**
Esophagus *(ĕ-sof′ah-gus)*	**Pancreas**
Mucosa	**Liver**
Peritoneum *(per″ĭ-to-ne′um)*	**Gallbladder**
Mesentery *(mes′en-ter″e)*	**Bile**
Stomach	**Salivary glands**
Gastric juice	**Intestinal juice**
Chyme *(kīm)*	**Pancreatic juice**
Small intestine	**Metabolism**

*For definitions, see Glossary.

SUMMARY

A. ANATOMY

1. The digestive system consists of a hollow tube extending from the mouth to the anus (alimentary canal) and several accessory digestive organs. The wall of the alimentary canal has four main tissue layers—mucosa, submucosa, muscularis externa, and serosa. The serosa (visceral peritoneum) is continuous with the parietal peritoneum, which lines the abdominal cavity wall.

2. Organs of the alimentary tube include:
   a. The mouth or oral cavity contains the teeth and tongue and is bounded by the lips,

cheeks, and palate. Tonsils guard the posterior margin of the oral cavity.
   b. The pharynx is a muscular tube that provides a passageway for food and air.
   c. The esophagus is a muscular tube that completes the passageway from the pharynx to the stomach.
   d. The stomach is a C-shaped organ located on the left side of the abdomen beneath the diaphragm. Food enters the stomach through the cardioesophageal sphincter and leaves it to enter the small intestine through the pyloric sphincter. The stomach has a third oblique layer of muscle in its wall that allows it to perform mixing, or

churning, movements. Gastric glands produce hydrochloric acid, pepsin, rennin, mucus, and intrinsic factor. Mucus protects the stomach lining itself from being digested.

  e. The small intestine is a tubelike organ suspended from the posterior body wall by the mesentery. Its subdivisions are the duodenum, jejunum, and ileum. Food digestion and absorption are completed here. Pancreatic juice and bile enter the duodenum through a sphincter at the distal end of the common bile duct. Microvilli, villi, and circular folds increase the surface area of the small intestine for enhanced absorption. Food leaves the small intestine to enter the large intestine through the ileocecal valve.

  d. The large intestine forms a U-shaped frame around the small intestine. Subdivisions are the cecum; appendix; ascending, transverse, and descending colon; sigmoid colon; rectum; and anal canal. The large intestine delivers undigested food residue (feces) to the body exterior.

3. Accessory organs duct substances into the alimentary tube.
  a. The pancreas is a soft gland lying in the mesentery between the stomach and small intestine. Pancreatic juice contains enzymes that digest all categories of foods in an alkaline fluid.
  b. The liver is a four-lobed organ overlying the stomach. Its digestive function is to produce bile, which it ducts into the small intestine.
  c. The gall bladder is a muscular sac that serves only as a storage organ. When fat digestion is not ongoing, the continuously made bile backs up the cystic duct and enters the gall bladder. The gall bladder stores and concentrates bile.
  d. The salivary glands (three pairs—parotid, submaxillary, and sublingual) duct saliva into the oral cavity. Saliva contains mucus and serous fluids. The serous component contains the enzyme, salivary amylase.

4. Two sets of teeth are formed. The first set, deciduous teeth, consists of 20 teeth that begin to appear at 6 months and are lost by the age of 12. The permanent teeth (32) begin to replace the deciduous teeth around the age of 7. A typical tooth consists of a crown covered with enamel and a root covered with cementum. The bulk of the

tooth is bonelike dentin. The pulp cavity contains blood vessels and nerves.

B. FUNCTIONS OF THE DIGESTIVE SYSTEM: FOOD BREAKDOWN, MOVEMENT, AND ABSORPTION

1. Foods must be broken down to their building blocks to be absorbed. The building blocks of carbohydrates are simple sugars, or monosaccharides; the building blocks of proteins are amino acids; the building blocks of fats, or lipids, are fatty acids and glycerol.

2. Both mechanical (chewing) and chemical food breakdown begin in the mouth. The mucus component of saliva helps bind food together into a bolus; salivary amylase in the serous component begins the chemical breakdown of starch to maltose. Saliva is secreted in response to the presence of food in the mouth, mechanical pressure, and psychic stimuli. Essentially no food absorption occurs in the mouth.

3. Swallowing involves the activity of the tongue, palate, and larynx. The buccal phase is voluntary; the tongue pushes the bolus into the pharynx. The involuntary pharyngeal/esophageal phase involves the closing of nasal and respiratory passages and the conduction of the food to the stomach by peristalsis.

4. When food enters the stomach, gastric secretion is stimulated by the vagus nerves and by gastrin (a local hormone). Hydrochloric acid activates the protein-digesting enzymes, pepsin and rennin, and the chemical digestion of proteins is begun. Food is also mechanically broken down by the churning activity of the stomach muscles. Movement of chyme into the small intestine is controlled by the enterogastric reflex.

5. Chemical digestion of fats, proteins, and carbohydrates is completed in the small intestine by intestinal enzymes, and more importantly, pancreatic enzymes. Alkaline pancreatic juice neutralizes the acid chyme and provides the proper environment for the operation of enzymes. Both pancreatic juice (the only source of lipases) and bile

(formed by the liver) are necessary for normal fat breakdown and absorption. Bile acts as a fat emulsifier. Local hormones (secretin and cholecystokinin) produced by the small intestine stimulate the release of bile and pancreatic juice. Segmental movements mix foods while peristaltic movements move foodstuffs along the length of the small intestine.

Most nutrient absorption occurs by active transport into the capillary blood of the villi. Fats are absorbed by diffusion into both the capillary blood and the lacteals in the villi.

6. The large intestine receives bacteria-laden indigestible food residue. Activities of the large intestine are absorption of water, salts, and vitamins made by resident bacteria. The feces is delivered to the anus by peristalsis and mass peristalsis. When feces enters the rectum, the defecation reflex is initiated.

C. METABOLISM

1. Metabolism includes all chemical breakdown (catabolic) and building (anabolic) reactions needed to maintain life.

2. The liver is the key metabolic organ of the body. Its cells remove simple sugars, amino acids, and fats from the hepatic portal blood. The liver performs glycogenesis, glycogenolysis, and gluconeogenesis to maintain homeostasis of blood glucose levels. Its cells make blood proteins and other substances and release them to the blood. Fats are burned by the liver cells to provide some of their energy (ATP); excesses are released to the blood in simpler forms that can be used by other tissue cells of the body. Phagocytic cells cleanse the hepatic portal blood by removing bacteria from it.

3. Carbohydrates, most importantly glucose, are the body's major energy fuel. As glucose is oxidized, carbon dioxide, water, and ATP are formed. During hyperglycemia, glucose is stored as glycogen or converted to fat for storage. In hypoglycemia, glycogenolysis and fat breakdown occur to restore normal blood glucose levels.

4. Fats insulate the body, protect organs, build some cell structures (membranes and myelin sheaths), and provide a reserve energy source. When carbohydrates are not available, fats are oxidized to produce ATP. Excessive fat breakdown causes the blood to become acid. Excess dietary fat is stored in subcutaneous tissue and other fat depots.

5. Proteins form the bulk of cell structure and most functional molecules. They are carefully conserved by body cells. Amino acids are actively taken up from the blood by tissue cells. Amino acids that cannot be made by body cells are called essential amino acids. Amino acids are burned or oxidized to form ATP only when other fuel sources are not available. Ammonia, released as amino acids are catabolized, is detoxified by the liver cells that combine it with carbon dioxide to form urea.

6. When the three major types of foods are oxidized for energy, they yield different amounts of energy. Carbohydrates yield 4 kcal/g; fats yield 9 kcal/g; and proteins yield 4 kcal/g. The basal metabolic rate (BMR) is the total amount of energy used by the body when one is in a basal state. Age, sex, body surface area, and most importantly, the amount of thyroxine produced influence an individual's BMR.

7. The total metabolic rate (TMR) is the number of calories used by the body to accomplish all ongoing daily activities. It increases dramatically as muscle activity increases. When TMR equals total caloric intake, weight remains constant.

8. As foods are catabolized to form ATP, more than 50% of energy released escapes as heat, which warms the body. The hypothalamus initiates heat-liberating processes (radiation of heat from the skin surface and sweating) or heat-promoting processes (vasoconstriction of skin blood vessels and shivering) as necessary to maintain body temperature within normal limits. Fever (hyperthermia) represents a body temperature that is regulated at higher than normal levels.

D. DEVELOPMENTAL ASPECTS OF THE DIGESTIVE SYSTEM

1. The alimentary tract forms as a hollow tube. The accessory glands form as outpocketings from this tube.

2. Common congenital defects include cleft palate, cleft lip, and tracheoesophageal fistula, which interfere with normal nutrition. Inborn errors of metabolism that are often seen are phenylketonuria (PKU) and cystic fibrosis.

3. Various inflammatory conditions plague the digestive system through life. Appendicitis is common in adolescents, gastroenteritis and food poisoning may occur at any time (given the proper irritating factors), ulcers and gall bladder problems increase in the middle-age years. Obesity and diabetes mellitus are other (noninflammatory) conditions that are bothersome during the later middle-age years.

4. The efficiency of all digestive system processes decreases in the elderly, and periodontal disease is common. Gastrointestinal cancers, such as stomach and colon cancer, appear with increasing frequency in an aging population.

REVIEW QUESTIONS

1. Make a simple line drawing of the organs of the alimentary tube and label each organ.

2. Add three labels to your drawing—salivary glands, liver, and pancreas—and use arrows to show where each of these organs empties its secretion into the alimentary tube.

3. Name the layers of the alimentary tube wall from the lumen out.

4. What is the mesentery? The peritoneum?

5. Name the subdivisions of the small intestine in an anterior-to-posterior direction. Do the same for the subdivisions of the large intestine.

6. The digestive system has many structural modifications. Describe the structure and function of villi.

7. What is the normal number of permanent teeth? Of deciduous teeth? What substance covers the tooth crown? What substance makes up the bulk of a tooth? What is pulp, and where is it?

8. Name the three pairs of salivary glands. Name two functions of saliva.

9. Assume you have been chewing a piece of bread for 5 or 6 minutes. How would you expect its taste to change during this time? Why?

10. Name two regions of the digestive tract where mechanical food breakdown occurs and explain how it is accomplished in those regions.

11. Name the organ where protein digestion first begins.

12. Why is it necessary for the stomach contents to be so acidic? How does the stomach protect itself from digestion?

13. Only one organ produces enzymes capable of digesting all groups of foodstuffs. What organ is this?

14. Explain why fatty stools result from the absence of bile and/or pancreatic juice.

15. Define emulsify.

16. What is the function of gastrin? Of secretin?

17. Describe the two phases of swallowing.

18. How do segmental and peristaltic movements differ?

19. What are the end products of protein digestion? Of fat digestion? Of carbohydrate digestion?

20. Where does most nutrient absorption occur?

21. What substances are absorbed in the large intestine?

22. What is the composition of feces?

23. Define defecation reflex, constipation, and diarrhea.

24. Define metabolism, anabolism, and catabolism.

25. Define glyconeogenesis, glycogenolysis, and glycogenesis.

26. Which food group is *most* important as a fuel source (that is, for catabolism and ATP production)? Which is most important for building cell structures?

27. What is the harmful result when excessive amounts of fats are burned to produce ATP? Name two conditions that might lead to this result.

28. Define BMR and name two factors that are important in determining an individual's BMR.

29. If your total caloric intake exceeds your TMR, what can you expect to happen?

30. How many calories are produced when 1 g of carbohydate is oxidized? With 1 g of protein? With 1 g of fat? If you just ate 100 g of food that was 20% protein, 30% carbohydrate, and 10% fat, how many calories did you consume?

31. Some of the energy released as foods are oxidized is captured to make ATP. What happens to the rest of it?

32. Where is the body's thermostat?

33. Name two ways in which heat is lost from the body. Name two ways in which heat is retained or generated.

34. What is a fever? What does it indicate?

35. Name three digestive system problems common to middle-aged adults. Name one common to teenagers. Name three common in elderly persons.

CHAPTER 14

Urinary System

Chapter Contents

Function of the urinary system • to rid the body of nitrogenous wastes while regulating electrolyte, water, and acid-base balance of the blood

The above electron micrograph from *Tissues and Organs: A Text-Atlas of Scanning Electron Microscopy* by Richard G. Kessel and Randy H. Kardon. W. H. Freeman and Co. Copyright © 1979.

After completing this chapter, you should be able to:

- Name and identify the organs of the urinary system on a diagram or torso model.

- Identify the following regions of a kidney (longitudinal section): hilus, cortex, medulla, medullary pyramids, calyces, pelvis, and renal columns.

- Recognize that the nephron is the structural and functional unit of the kidney and describe its anatomy.

- Describe the process of urine formation, identifying the areas of the nephron that are responsible for filtration, reabsorption, and secretion.

- Describe the function of the kidneys as excretion of nitrogen-containing wastes and maintenance of water/electrolyte/acid–base balance of the blood.

- Explain the role of antidiuretic hormone (ADH) in the regulation of water balance by the kidney.

- Explain the role of aldosterone in sodium and potassium balance of the blood.

- Define: polyuria, anuria, oliguria, and diuresis.

- Describe the normal composition of urine.

- List substances that are abnormal urinary components.

- Describe the general function of the ureters, bladder, and urethra.

- Define micturition.

- Describe the difference in control of the external and internal bladder sphincters.

- Compare the course and length of the male urethra to that of a female.

- Name three common urinary tract problems.

- Describe the effect of aging on urinary system functioning.

Cellular metabolism produces various wastes such as carbon dioxide and nitrogen-containing wastes (urea, ammonia, and others), as well as imbalances of water and essential ions. Although several other organs (lungs and skin) have roles to play in excretion of body wastes, the kidneys rid the body of most nitrogenous wastes. These small, dark red organs with their kidney-bean shape are probably the second most complex organs in the body. (Most people would agree that the brain deserves first place for complexity.) Ridding the body of wastes is only one job of the kidneys; they also regulate the chemical makeup of the blood so that the proper balance between water and salts is continuously assured.

To perform this task, each kidney serves as a blood filter, and each day gallons of fluid filter out of the blood stream and flow through tiny tubules in the kidneys. Although it might seem that the kidneys could be overwhelmed by this "flood," most often they deal with it efficiently and selectively. Toxins, metabolic wastes, and excess ions are allowed to leave the body in the urine; but needed substances such as water, glucose, and amino acids are retained and returned to the blood. Because of this role, the kidneys are considered major, if not *the* major, homeostatic organs of the body. Malfunction of the kidneys leads to dangerous changes in blood composition, which are always serious and sometimes fatal.

Organs of the urinary system include the paired kidneys and ureters and the single urinary bladder and urethra (Figure 14-1). The kidneys alone perform the functions just described and manufacture urine in the process. The other organs of this system provide temporary storage reservoirs for urine or carry it from one body region to another.

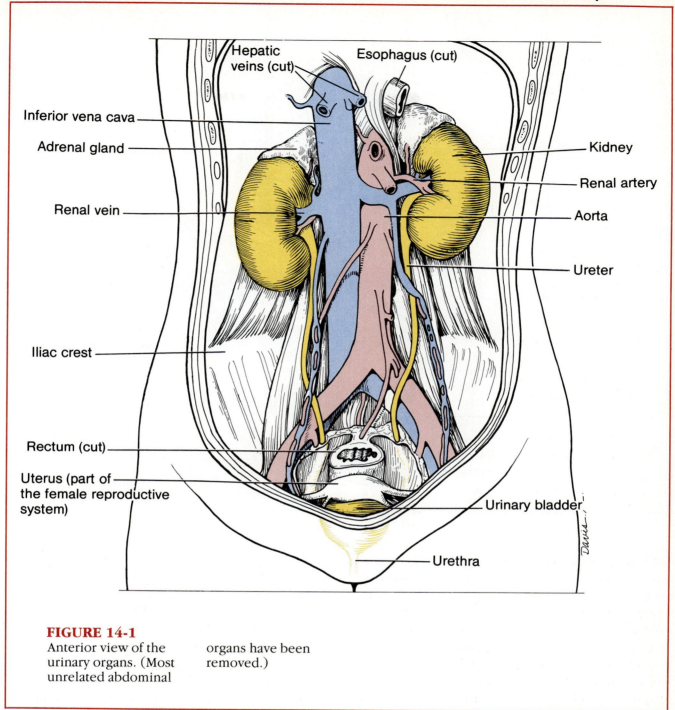

FIGURE 14-1
Anterior view of the urinary organs. (Most unrelated abdominal organs have been removed.)

KIDNEYS

Location and Structure

Although many believe that the kidneys are located in the lower back, this is not their location.

Instead, they lie against the dorsal body wall beneath the parietal peritoneum in the *superior* lumbar region where they receive some protection from the lower part of the rib cage. Because it is crowded by the liver, the right kidney is positioned slightly lower than the left kidney. Each kidney (about 12.5-cm [5-in.] long, 7.5-cm [3-in.] wide, and 2.5-cm [1-in.] thick) has a medial indentation, the **hilus** (hi′lus). A fibrous *renal capsule* encloses each kidney, and in a living person an additional capsule of fat surrounds the kidneys

FIGURE 14-2
Longitudinal section of
a kidney showing the
larger arteries supplying
kidney tissue.

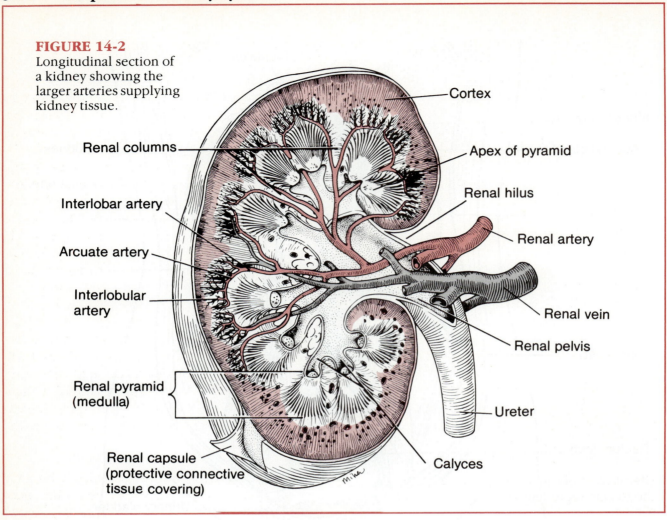

Renal columns

Interlobar artery

Arcuate artery

Interlobular
artery

Renal pyramid
(medulla)

Renal capsule
(protective connective
tissue covering)

Cortex

Apex of pyramid

Renal hilus

Renal artery

Renal vein

Renal pelvis

Ureter

Calyces

and helps hold them in place against the muscles of the trunk wall. If the amount of fatty tissue decreases or is scanty (as with rapid weight loss or in very thin individuals), the kidneys are not as securely held in place and may drop to a lower position in the abdominal cavity; this is *ptosis* (to′sis). Ptosis can result in problems if the ureters, which drain urine from the kidneys, become kinked. When this happens, urine, which can no longer pass through the ureters, begins to back up into the kidney and exert pressure on kidney tissue. This condition is *hydronephrosis* (hi″dro-nĕ-fro′sis), and it can be severely damaging to the kidney.

Several structures are seen at the hilus region of each kidney. Two **renal arteries** branch off the descending aorta, and each plunges into the hilus of a kidney. A **renal vein,** which receives the venous blood draining from the kidney, and a **ureter,** which drains urine from the kidney, exit at the hilus region (Figure 14-2).

When a kidney is cut lengthwise, many more structural details become apparent as can be seen by looking at Figure 14-2. The outer kidney region, which is lighter in color, is the **renal cortex.** (The word cortex comes from the Latin word meaning bark.) Deep to the cortex is a darker reddish-brown area, the **renal medulla.** The medulla has many basically triangular regions, which have a striped appearance, **medullary** (med′u-lār″e) **pyramids.** The broader **base** of each pyramid faces toward the cortex; its tip, the **apex,** points toward the inner region of the kidney. The pyramids are separated by areas of lighter-staining tissue, the **renal columns.** The renal columns appear to be extensions of cortex tissue, which dip downward into the medulla.

Medial to the hilus is a flat, basinlike cavity, the **renal pelvis.** As Figure 14-2 shows, the pelvis is continuous with the ureter. Extensions of the pelvis, **calyces** (kal′ĭ-sēz), form cup-shaped areas that enclose the tips of the pyramids. Calyces collect urine, which continuously drains from the tips of the pyramids into the pelvic area. Urine then flows out of the pelvis into the ureter, which transports it to the bladder to be stored.

Since the kidney cells must continuously cleanse the blood and adjust its composition, it is not surprising that the kidneys have a very rich blood supply. Approximately one-quarter of the total blood supply of the body passes through the kidneys each minute. The arterial supply of each kidney is the renal artery. Once inside the pelvis, the renal artery breaks up into several branches called **interlobar arteries,** which travel through the renal columns to reach the cortex. At the top of the medulla, interlobar arteries give off branching arteries called **arcuate** (ar'ku-āt) **arteries,** which curve over the medullary pyramids. Small **interlobular arteries** then branch off the arcuate arteries and run upward to supply the cortex tissue. Venous blood draining from the kidney flows through veins that trace the pathway of the arterial supply but in a reverse direction (that is, interlobular veins to arcuate veins to interlobar veins to the renal vein in the pelvis).

Nephrons and Urine Formation

NEPHRONS

Each kidney contains approximately 1 million tiny structures, **nephrons** (nef'rons). Nephrons are responsible for the processes of filtration, reabsorption, and secretion that go on in the kidney to form the urine product. Figure 14-3 shows the anatomy and relative positioning of nephrons in each kidney.

Each nephron consists of two main structures: a **glomerulus** (glo-mer'u-lus), which is a knot of capillaries, and a **renal tubule.** The closed end of the renal tubule is enlarged and cup-shaped and completely surrounds the glomerulus. This portion of the renal tubule is called **Bowman's capsule.** The inner layer of Bowman's capsule is made up of very specialized cells, *podocytes* (pod'o-sīts). Podocytes have long branching processes that intertwine with those of other podocytes and cling to the capillary walls of the glomerulus. The podocytes form a porous, or holey, membrane surrounding the glomerulus.

The rest of the tubule is about 5-cm (approximately 2-in.) long. As it extends from Bowman's capsule, it coils and twists before dropping down into a long hairpin loop and then again becomes coiled and twisted before entering a collecting duct. These different regions of the tubule have specific names; in order from Bowman's capsule they are the **proximal convoluted tubule, loop of Henle** (hen'le), and the **distal convoluted tubule.** The lumen surfaces (surface exposed to the filtrate) of the tubule cells in the proximal convoluted tubules are covered with dense microvilli, which increases their surface area tremendously. Microvilli also occur on tubule cells in other parts of the tubule but in much reduced numbers.

Most of the nephron is located in the cortex; only portions of the loops of Henle dip into the medulla. The **collecting ducts** (or **tubules**), each of which receives urine from many nephrons, run downward through the medullary pyramids, giving them their striped appearance. They deliver the final urine product into the calyces and pelvis of the kidney.

Each and every nephron is associated with two capillary beds—the glomerulus (mentioned earlier) and the **peritubular** (per″ĭ-tu'bu-lar) **capillary bed.** The glomerulus is both fed and drained by *arterioles*. The *afferent* arteriole is the "feeder vessel," and the *efferent* arteriole receives blood that has passed through the glomerulus. The efferent arteriole then breaks up to form the peritubular capillary bed, which closely clings to the whole length of the tubule. The peritubular capillaries then drain into an interlobular vein that leaves the cortex. The glomerulus is unlike any other capillary bed in the entire body. Because it is fed *and* drained by arterioles, which are high-resistance vessels, blood pressure in the glomerulus is ordinarily very high. This extremely high pressure forces fluid and solutes (smaller in size than proteins) out of the blood into Bowman's capsule.

As noted earlier, urine formation is a result of three processes—filtration, reabsorption, and secretion. Each of these processes is illustrated in Figure 14-4 and described in more detail next.

FILTRATION. As just described, the glomerulus acts as a filter. Filtration is a nonselective, passive

FIGURE 14-3
Structure of the nephron. **(a)** Wedge-shaped section of kidney tissue indicating the positioning of nephrons in the kidney. **(b)** Detailed anatomy of a nephron and its associated blood supply.

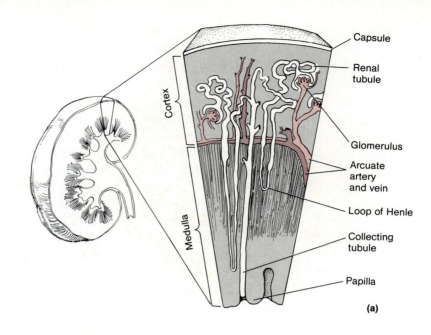

Capsule

Renal tubule

Glomerulus

Arcuate artery and vein

Loop of Henle

Collecting tubule

Papilla

Cortex

Medulla

(a)

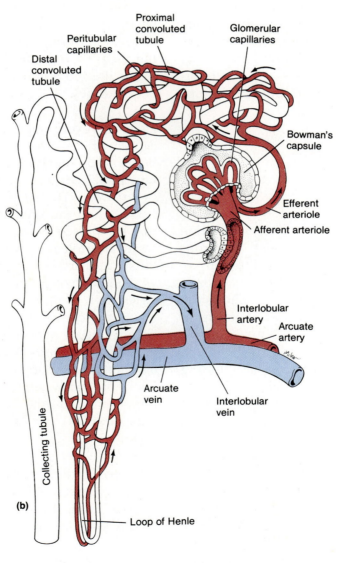

Distal convoluted tubule

Peritubular capillaries

Proximal convoluted tubule

Glomerular capillaries

Bowman's capsule

Efferent arteriole

Afferent arteriole

Interlobular artery

Arcuate artery

Arcuate vein

Interlobular vein

Collecting tubule

Loop of Henle

(b)

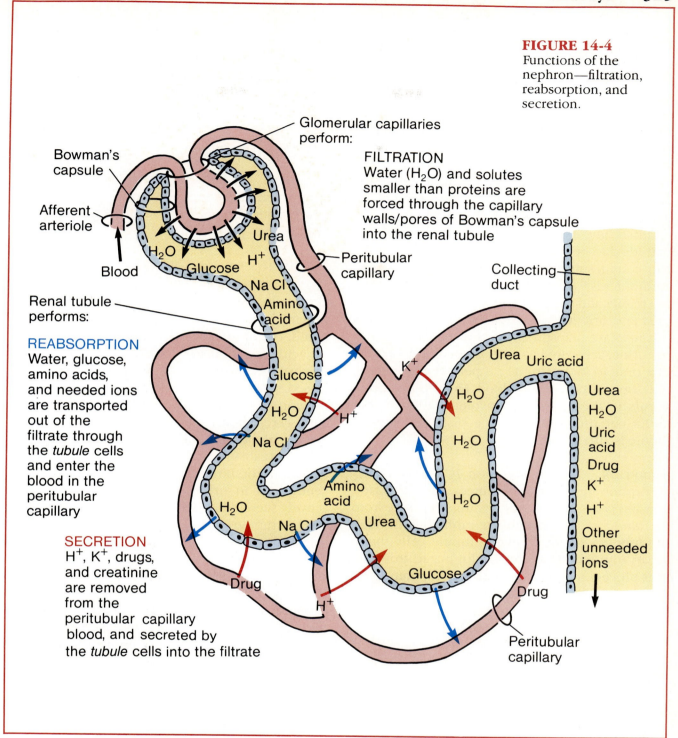

FIGURE 14-4
Functions of the nephron—filtration, reabsorption, and secretion.

Glomerular capillaries perform:

FILTRATION
Water (H_2O) and solutes smaller than proteins are forced through the capillary walls/pores of Bowman's capsule into the renal tubule

Bowman's capsule

Afferent arteriole

Blood

Peritubular capillary

Collecting duct

Renal tubule performs:

Urea

H^+

H_2O
Glucose
Na Cl
Amino acid

REABSORPTION
Water, glucose, amino acids, and needed ions are transported out of the filtrate through the *tubule* cells and enter the blood in the peritubular capillary

Glucose

H_2O

Na Cl

H^+

K^+

H_2O

H_2O

H_2O

Urea   Uric acid

Urea
H_2O
Uric acid
Drug
K^+
H^+
Other unneeded ions

SECRETION
H^+, K^+, drugs, and creatinine are removed from the peritubular capillary blood, and secreted by the *tubule* cells into the filtrate

Amino acid

Na Cl   Urea

H_2O

Glucose

Drug

H^+

Drug

Peritubular capillary

process. The filtrate that is formed is essentially blood plasma without blood proteins. Both proteins and blood cells are normally too large to pass through the filtration membrane, and when either of these appear in the urine, it is a pretty fair bet that there is some problem with the glomerular filters. As long as the systemic blood pressure is normal, filtrate will be formed. However, if arterial blood pressure drops too low, the glomerular pressure becomes inadequate to force substances out of the blood into the tubules, and filtrate formation stops.

REABSORPTION. Besides wastes and excess ions that must be removed from the blood, the filtrate contains many useful substances (including water, glucose, amino acids, and ions), which must be reclaimed from the filtrate and returned to the blood. Reabsorption begins as soon as the filtrate enters the proximal convoluted tubule. The tubule cells are transporters, taking up needed substances from the filtrate and then passing them out their posterior aspect into the peritubular capillary blood. Some reabsorption is done passively (for example, water passes by osmosis), but the reabsorption of most substances depends on active transport processes, which use membrane carriers and are very selective. There is an abundance of carriers for substances that need to be retained, and few or no carriers for substances that need to be eliminated from the body. This is why certain substances (for example, glucose and amino acids) are nearly entirely removed from the filtrate and metabolic waste products (such as urea and uric acid) are poorly reabsorbed. Various ions are reabsorbed or allowed to go out in the urine, according to what is needed at a particular time to maintain the proper pH and electrolyte composition of the blood. Most reabsorption occurs in the proximal convoluted tubules, but under certain conditions, the distal convoluted tubule is also active.

Secretion.   Tubular secretion is essentially reabsorption in reverse. Substances, such as hydrogen and potassium ions and ammonia, move from the blood of the peritubular capillaries through the tubule cells or from the tubule cells themselves into the filtrate to be eliminated in urine. This process seems to be important for getting rid of substances not already in the filtrate or as an additional means for controlling blood pH.

Control of Blood Composition by the Kidneys

Blood composition depends on three major factors: diet, cellular metabolism, and urine output. In 24 hours, the kidney's 2 million nephrons filter approximately 150–180 L of blood plasma through their glomeruli into the tubules, which process the filtrate by taking substances out of it (reabsorption) and adding substances to it (secretion). In the same 24 hours, only about 1.0–1.8 L of urine are produced. Obviously, urine and filtrate are different. Filtrate contains everything that blood plasma does (except proteins), but by the time the collecting ducts have been reached,

the filtrate has lost most of its water and just about all of its nutrients and necessary ions. What remains (urine) contains most of the waste or unneeded substances. Assuming we are healthy, our kidneys can keep our blood composition fairly constant despite wide variations in diet and cell activity.

In general, the kidneys have four major roles to play, which help keep the blood composition relatively constant. Each of these roles is discussed briefly next.

Excretion of Nitrogen-Containing Wastes

Urea, uric acid, and creatinine are the most important nitrogen-containing wastes found in blood. *Urea* (u-re′ah), which is formed when amino acids are used to produce energy, is an end product of protein breakdown. *Uric acid* is released when nucleic acids are metabolized, and *creatinine* (kre-at′ĭ-nin) is associated with creatine (kre-′ah-tin) metabolism in muscle tissue. Because the tubule cells have few membrane carriers to reabsorb these substances, they tend to remain in the filtrate and are found in high concentrations in urine. In addition, creatinine is actively secreted into the filtrate by proximal convoluted tubule cells.

Water and Electrolyte Balance

If the body is to remain properly hydrated, we cannot lose more water than we take in. Most water intake is a result of fluids and foods we take in in our diet; however, a small amount is produced during cellular metabolism as explained in Chapter 14. There are several routes for water to leave the body. Some water vaporizes out of the lungs, some is lost in perspiration, and some leaves the body in the stool. The job of the kidneys is like that of a juggler. That is, if large amounts of water are lost in other ways, they compensate by putting out less urine to conserve body water. On the other hand, when water intake is excessive, the kidneys excrete generous amounts of urine and the anguish of a too-full bladder becomes very real. Just how the kidneys accomplish this balancing act is explained in more detail next.

Reabsorption of water and electrolytes by the kidneys is regulated by hormones. When blood volume drops for any reason (for example owing to hemorrhage or excessive water loss through sweating or diarrhea), the arterial blood pressure drops, which in turn decreases the amount of filtrate formed by the kidneys. In addition, highly sensitive cells in the hypothalamus called *osmoreceptors* (oz″mo-re-cep′tors) react to the change in blood composition (that is, less water and more solutes) by becoming very irritable and sending nerve impulses to the posterior pituitary. The posterior pituitary then releases antidiuretic hormone (ADH.) (The term antidiuretic is derived from *diuresis* (di″u-re′sis), which means flow of urine from the kidney, and *anti,* which means against.) As one might guess, this hormone prevents excess water loss in the urine. ADH travels in the blood to its target, the kidney tubule cells. Its major effect is to cause the cells of the distal convoluted tubules and collecting ducts to increase their rate of water reabsorption. As water is returned to the blood stream, blood volume and blood pressure increase to their normal levels, and a small amount of very concentrated urine is formed. ADH is released more or less continuously unless the solute concentration of the blood drops too low; when this happens, the osmoreceptors become "quiet," and excess water is allowed to leave the body in the urine.

When ADH is *not* released (as perhaps owing to injury or destruction of the hypothalamus or posterior pituitary gland), huge amounts of very dilute urine (up to 25 L/day) flush from the body day after day. This condition, *diabetes insipidus* (in-sip′e-dus), can lead to severe dehydration and electrolyte imbalances. Affected individuals are continually thirsty and have to drink fluids almost continuously to maintain normal fluid balance.

A second hormone that helps to regulate blood composition and blood volume by acting on the kidney is *aldosterone* (al″do-ster′ōn). Aldosterone is released by the adrenal cortex in response to several conditions—decreased blood volume, decreased blood sodium ions (Na^+) or increased potassium ions (K^+) in the blood. Sodium is the most important ion in the plasma that must be regulated. When too little sodium is in the blood, the blood becomes too dilute, and water tends to leave the blood stream and flow out into the tissue spaces, causing edema and possible circulatory collapse. Normally, sodium is reabsorbed ac-

tively along the entire length of the nephron tubules. In the presence of aldosterone, tubular reabsorption of sodium increases even more, especially in the ascending loop of Henle and the distal convoluted tubule. For each sodium ion that is reabsorbed, a potassium ion is secreted into the filtrate. Thus, as sodium concentration in the blood increases, potassium decreases, bringing these two ions back to their normal balance in the blood. A second effect of aldosterone is to increase water reabsorption by the tubule cells, because as sodium is reabsorbed, water follows it back into the blood.

ACID–BASE BALANCE OF THE BLOOD

For the cells of the body to function properly, the blood pH must be maintained between 7.35 and 7.45, a very narrow range. Cellular metabolism continuously adds substances to the blood, which tend to disturb its acid–base balance. Many acids are produced (for example, phosphoric acid, sulfuric acid, and many types of fatty acids). In addition, carbon dioxide, which is released during energy production, forms carbonic acid. (However, in this case, the lungs have the chief responsibility for eliminating carbon dioxide from the body.) Ammonia and other basic substances are also released to the blood.

Unneeded acids and bases (including bicarbonate ions) tend not to be reabsorbed, as discussed earlier. If selective reabsorption of bases or acids by the kidney tubule cells is not enough to correct the blood pH, the tubule cells actively secrete hydrogen ions and ammonia (or whatever is necessary) into the filtrate to help clear the undesirable substances from the blood by a different route. Urine pH varies from 4.5–8.0, which reflects the ability of the renal tubules to excrete basic or acid ions to maintain blood pH homeostasis.

Characteristics of Urine

Freshly voided urine is generally clear and pale to deep yellow. The normal yellow color is due to *urochrome* (u′ro-krōm), a pigment that results from the body's destruction of hemoglobin. The more solutes there are in the urine, the deeper yellow its color; on the other hand, dilute urine is a pale, straw color. At times, urine may have an abnormal color; this might be a result of eating

TABLE 14-1   Abnormal Urinary Constituents

Substance	Name of condition	Possible causes
Glucose	Glycosuria (gli"ko-su'-re-ah)	Nonpathologic: Excessive intake of sugary foods Pathologic: Diabetes mellitus
Proteins	Proteinuria (pro"te-in-u're-ah) (also called albuminuria)	Nonpathologic: Physical exertion, pregnancy Pathologic: Glomerulonephritis, hypertension
WBCs and bacteria	Pyuria (pi-u'-re-ah)	Urinary tract infection
RBCs	Hematuria	Bleeding in the urinary tract (owing to trauma, kidney stones, infection)
Hemoglobin	Hemoglobinuria (he"mo-glo-bĭ-nu'-re-ah)	Various: Transfusion reaction, hemolytic anemia
Bile pigment	Bilirubinuria (bil"ĭ-roo-bi-nu'-re-ah)	Liver disease (hepatitis)

certain foods (beets) or the presence of bile or blood in the urine.

The odor of fresh urine is slightly aromatic, but if it is allowed to stand, it takes on an ammonia odor caused by bacterial action on the urine solutes. Some drugs, vegetables (such as asparagus), and various diseases (such as diabetes mellitus) alter the usual odor of urine.

Urine pH is usually slightly acid (around 6), but changes in body metabolism and certain types of diet may cause it to be much more acidic or basic. For example, a diet that contains large amounts of protein (eggs and cheese) and whole wheat products causes urine to become quite acid—thus, such foods are called acid-ash foods. Conversely, a vegetarian diet is called an alkaline-ash diet, because it causes urine to become quite basic as the kidneys excrete the excess bases. Bacterial infection of the urinary tract also may cause the urine to be quite alkaline.

Since urine is water plus solutes, urine weighs more, or is more dense, than distilled water. The term used to compare how *much* heavier urine is than distilled water is *specific gravity*. Whereas the specific gravity of water is 1.0, the specific gravity of urine usually ranges from 1.001–1.030 (dilute to concentrated urine, respectively). Urine is generally dilute (that is, it has a low specific gravity) when a person drinks excessive fluids, uses diuretics (drugs that increase urine output), or has chronic renal failure. (In chronic renal failure, the kidney loses its ability to concentrate urine.) Conditions that produce urine with a high specific gravity include inadequate fluid intake, fever, and a kidney inflammation called *pyelonephritis* (pi"e-lo-nĕ-fri'-tis). When the urine becomes extremely concentrated, solutes such as uric acid begin to form crystals (kidney stones). Kidney stones tend to be sharp and may cause severe pain, particularly if they become wedged in a ureter.

Solutes normally found in urine include sodium and potassium ions, urea, uric acid, creatinine, ammonia, bicarbonate ions, and various other ions depending on blood composition. With certain diseases, urine composition can change dramatically, and the presence of abnormal substances in urine is often helpful in diagnosing the problem. This is why a routine urinalysis should always be part of any good physical examination.

Substances *not* normally found in urine are glucose, blood proteins (primarily albumin), red blood cells, hemoglobin, white blood cells (pus), and bile. Naming and possible causes of conditions in which abnormal urinary constituents and volumes might be seen are summarized in Table 14-1.

URETERS

The **ureters** (u-re′ters) are slender tubes each about 25–30-cm (10–12-in.) long and 3.3 mm (¼ in.) in diameter. Each ureter runs behind the peritoneum from the hilus of a kidney to enter the posterior aspect of the bladder at a slight angle (see Figure 14-1). The superior end of each ureter is continuous with the pelvis of the kidney, and its mucosa lining is continuous with that lining the renal pelvis and the bladder below.

Essentially, the ureters are passageways to carry urine from the kidneys to the bladder. Although it might appear that urine could simply drain to the bladder below by gravity, the ureters do play an active role in urine transport. Smooth muscle layers in their walls contract at the rate of one to five times per minute to force urine into the bladder by peristalsis. Once urine has entered the bladder, it is prevented from flowing back into the ureters by small valvelike folds of bladder mucosa that flap over the ureter openings.

URINARY BLADDER AND MICTURITION

The **urinary bladder** is a smooth collapsible muscular sac located anteriorly to the small intestine and just posterior to the pubic symphysis. If the interior of the bladder is scanned, three openings are seen—the two ureter openings and the single opening of the **urethra** (u-re′thrah), which drains the bladder (Figure 14-5). The triangular region of the bladder, which is outlined by these three openings, is the **trigone** (tri′gōn). In males, the *prostate gland* (part of the male reproductive system) surrounds the neck of the bladder where it empties into the urethra.

The bladder wall contains three layers of smooth muscle, and its mucosa is a special type of epithelium, *transitional epithelium*. Both of these structural features make the bladder uniquely suited for its function of urine storage. When the bladder is empty, it is collapsed and is perhaps 5–7.5-cm (2–3-in.) long at most, and its walls are thick and thrown into folds. As urine begins to accumulate, the bladder expands. The smooth muscle wall stretches and thins, and the transitional epithelial cells slide over one another to increase the internal volume of the bladder by decreasing the thickness of its wall. A moderately full bladder is about 12.5-cm (5-in.) long and holds about 500 ml (1 pint) of urine, but it is capable of holding more than double that amount. When the bladder is distended (or really stretched by urine), it becomes firm and may be felt just above the pubic symphysis. Although urine is formed continuously by the kidneys, it is usually removed from the body when the time is convenient. In the meantime, the bladder provides a temporary storage tank for the urine.

Voiding, or *micturition* (mik″tu-rish′un), is the act of emptying the bladder. Two sphincters or valves, the **internal urethral sphincter** (more superiorly located) and the **external urethral sphincter** (more inferiorly located) control the release of urine from the bladder. Ordinarily, the bladder continues to collect urine until about 300 ml have accumulated. At about this point, stretching of the bladder wall activates stretch receptors. Impulses transmitted to the sacral region of the spinal cord, in turn, cause the bladder to go into reflex contractions. As the contractions become stronger, stored urine is forced past the internal sphincter (a smooth muscle, involuntary sphincter) into the upper part of the urethra. It is then that a person feels the urge to void. The lower external sphincter is skeletal muscle and is voluntarily controlled. If it is not convenient to void, this sphincter can be kept closed. On the other hand, if it is convenient, the external sphincter can be relaxed so that urine is flushed from the body. When one chooses not to void, the reflex contractions of the bladder will stop temporarily and urine will continue to accumulate in the bladder. After 200–300 ml more have been collected, the micturition reflex will again occur.

A condition called *incontinence* (in-kon′tĭ-nes) occurs when we are unable to voluntarily control the external sphincter. Incontinence is normal in children 2 years old or younger, because they have not yet gained control over their voluntary sphincter. Past that age, incontinence is usually a result of emotional problems, pressure (as in

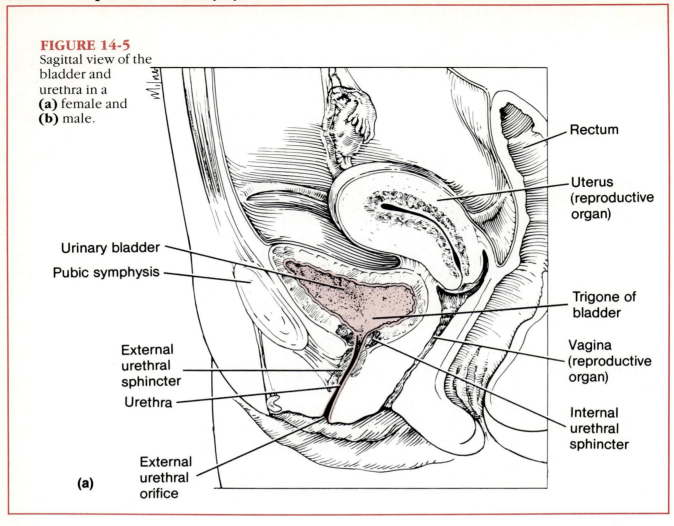

FIGURE 14-5
Sagittal view of the bladder and urethra in a **(a)** female and **(b)** male.

Rectum

Uterus (reproductive organ)

Urinary bladder

Pubic symphysis

Trigone of bladder

Vagina (reproductive organ)

External urethral sphincter

Urethra

Internal urethral sphincter

(a)

External urethral orifice

pregnancy), or nervous system problems (stroke or spinal cord injury).

Urinary retention is essentially the opposite of incontinence. It is a condition in which the bladder is unable to expel its contained urine. There are various causes for urinary retention, but it often occurs after surgery in which general anesthesia has been given. It appears that it takes a little time for the smooth muscles to regain their activity. Another cause of urinary retention, which occurs primarily in elderly men, is enlargement, or *hypertrophy,* of the prostate gland, which surrounds the neck of the bladder. As it enlarges, it narrows the urethra, making it very difficult to void.

URETHRA

The **urethra** is a thin-walled tube that carries urine by peristalsis from the bladder to the outside of the body. The length and relative function of the urethra differs in the two sexes. In females, it is about 3–4-cm (1½-in.) long and its external orifice, or opening, lies anteriorly to the vaginal opening. Its function is to conduct urine to the body exterior. Since the female urinary orifice is so close to the posterior anal opening (see Figure 14-5*a*) and feces contains a good deal of bacteria, improper toileting habits (that is, wiping from back to front rather than from front to back) can easily carry bacteria into the urethra. Moreover, since the mucosa of the urethra is continuous with that of the rest of the urinary tract organs, an inflammation of the urethra (*urethritis* [u″rĕ-thri′tis]) can easily ascend the tract to cause bladder inflammation (*cystitis* [sis-ti′tis]) and even kidney inflammation.

In males, the urethra is approximately 20-cm (8-in.) long. It opens at the tip of the penis after traveling down its length. The urethra of the male has a double function: it carries urine out of the body, and it provides the passageway through

FIGURE 14-5
(continued)

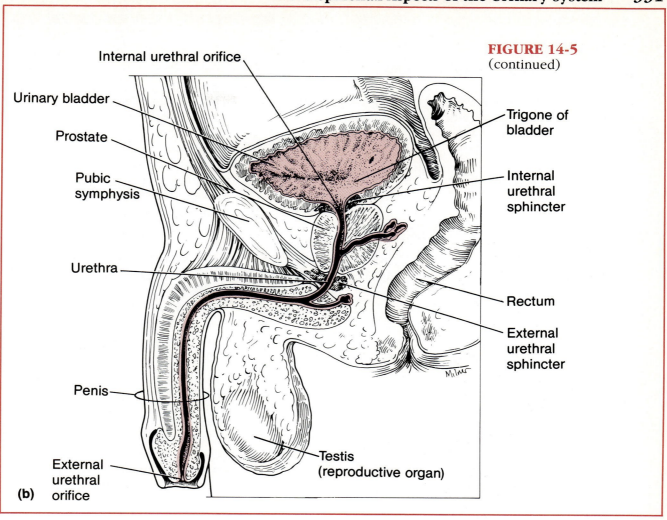

Internal urethral orifice

Urinary bladder

Prostate

Pubic symphysis

Urethra

Penis

External urethral orifice

(b)

Trigone of bladder

Internal urethral sphincter

Rectum

External urethral sphincter

Testis (reproductive organ)

which sperm is ejected from the body. Thus, in males, the urethra is part of both the urinary and reproductive systems.

DEVELOPMENTAL ASPECTS OF THE URINARY SYSTEM

When you trace the development of the kidneys in a young embryo, it almost seems as if they can't make up their mind about whether to come or go. The first tubule system forms and then begins to degenerate as a second, lower set appears. The second set, in turn, degenerates as a third set makes its appearance. This third set develops into the functional kidneys, which are excreting urine by the third month of fetal life. However, it is important to remember that the fetal kidneys do not work nearly as hard as they will after birth, be-

cause exchanges with the mother's blood through the placenta allow her system to clear many of the undesirable substances from the fetal blood.

There are many congenital abnormalities of this system, but three of the most common are polycystic kidneys, exstrophy of the bladder, and hypospadias.

Polycystic (pol″e-sis′tik) *disease* is a degenerative disease that appears to run in families. In this disease, one or both kidneys are enlarged and have many blister-like sacs containing blood, mucus, or urine. These cysts interfere with renal function. There is not too much that can be done for this condition except to prevent further kidney damage by avoiding infection.

In *exstrophy* (ek′stro-fe) *of the bladder,* the bladder and distal ends of the ureters protrude through an abnormal opening in the abdominal wall and are exposed to the external environment. This com-

BOX 14-1
Renal Failure and Dialysis

Like that vast unseen and unappreciated army of sanitation workers who keep our water drinkable and tend to the city's waste disposal, the kidneys continually maintain the purity of the fluids of our internal environment. Without their continual efforts, body fluids become contaminated with nitrogen-containing wastes, and *uremia* sets in, which totally disrupts life processes. Signs and symptoms of uncontrolled uremia include diarrhea, vomiting, labored breathing, convulsions, coma, and finally death.

Although not common, renal failure may occur when the number of functioning nephrons becomes too low to carry out the normal urinary system functions. Possible causes of renal failure include:

1. Repeated damaging infections of the kidneys.

2. Physical trauma to the kidneys.

3. Chemical poisoning of the tubule cells by heavy metals (mercury or lead) or organic solvents (dry cleaning fluids).

4. Inadequate blood delivery to the tubule cells (as sometimes happens with arteriosclerosis).

In renal failure, filtrate formation decreases or stops completely. Because toxic wastes accumulate quickly in the blood when the tubule cells are not working, *dialysis* must be performed to cleanse the blood while the kidneys are shut down. In dialysis, the patient's blood is passed through a membrane tubing that is permeable only to certain substances, and the tubing is immersed in a bathing solution

that is similar to normal "cleansed" plasma. As blood circulates through the tubing, substances present in the blood but not present in the bath diffuse out of the blood through the tubing to enter the bathing solution. In this way, needed substances are retained in the blood while substances such as nitrogenous wastes and ion excesses are removed.

If the damage is nonreversible (as in chronic, slowly progressing renal failure), the disease progresses to total inability of the kidneys to concentrate or process urine, and kidney transplant is the only answer. Unhappily, the signs and symptoms of this serious kidney problem are obvious to the affected person only after about 75% of renal function has been lost.

plication is usually corrected surgically between the ages of 3 and 4 years.

Hypospadias (hi"po-spa'de-as) is a condition found in male babies only. It occurs when the urethral orifice is located on the ventral surface of the penis. Surgery is generally done around 12 months of age to correct this problem.

Because the bladder is very small and the kidneys are unable to concentrate urine for the first 2 months, the newborn baby voids from 5 to 40

times per day depending on the amount of fluids taken in. By 2 months the infant is voiding approximately 400 ml/day, and the amount steadily increases until about the age of 8 when urine output reaches 1000 ml. By adolescence, adult urine output (1400 ml/day) is achieved.

Control of the voluntary sphincter goes hand in hand with nervous system development. By 15 months, most toddlers are aware when they have voided, and by 18 months, they can hold urine in their bladder for about 2 hours, which is the first

BOX 14-1
(continued)

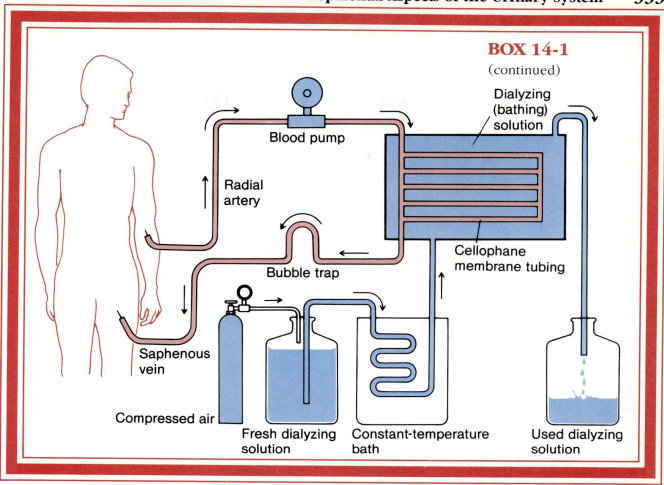

Dialyzing (bathing) solution

Blood pump

Radial artery

Cellophane membrane tubing

Bubble trap

Saphenous vein

Compressed air

Fresh dialyzing solution

Constant-temperature bath

Used dialyzing solution

sign that toilet training (for urination) can begin. Daytime control usually occurs well before nighttime control is accomplished. It is unrealistic to expect that complete nighttime control will occur before 4 years of age.

During childhood and through late middle age, most urinary system problems are infectious, or inflammatory, conditions. Many types of bacteria may invade the urinary tract to cause urethritis, cystitis, or pylonephritis. *Escherichia coli* (esh"er-i'ke-a ko'li), a resident of the digestive tract (where it generally causes no problems), acts as a pathogen (disease causer) in the sterile environment of the urinary tract. *E. coli* infections of the urinary tract are common in females, because of the closeness of the urinary and digestive tract external openings.

Venereal diseases (primarily reproductive tract infections) may also invade and cause inflammation in the urinary tract, which leads to the clogging of some of its ducts.

Streptococcal (strep"to-kok'al) infections such as strep throat and scarlet fever, which occur during childhood, may cause inflammatory damage to the kidneys if the original infections are not treated promptly or properly. A common sequel to untreated childhood strep infections is *glomerulonephritis* (glo-mer"u-lo-ně-fri'tis). In glomerulonephritis, the glomerular filters become clogged with antigen–antibody complexes, which result from the strep infection. Common symptoms of this disease are albuminuria (al"bu-mĭ-nu're-ah), hematuria (hem"ah-tu're-ah), and decreased urine output.

As we age, there is a progressive decline in kidney function. By age 70, the rate of filtrate formation is only about half that of the middle-aged adult. This is believed to be a result of faulty renal circulation owing to arteriosclerosis, which affects the entire circulatory system of the aging person. In addition to a decrease in the number of functional nephrons, the tubule cells become less efficient in their ability to concentrate urine.

Another consequence of aging is a loss of bladder tone, causing many elderly individuals to experience *urgency* (a feeling that it is necessary to void) and *frequency* (frequent voiding of small amounts of urine). In many, incontinence is the final outcome of the aging process. This loss of control is a tremendous blow to the pride of many of our aging people. Urinary retention is another common problem; most often it is a result of hypertrophy of the prostate gland in males. Some of the problems of incontinence and retention can be avoided by a regular regimen of activity that keeps the body as a whole in optimum condition and promotes alertness to elimination signals.

IMPORTANT TERMS*

Renal cortex

Renal medulla

Renal pelvis

Nephrons *(nef'rons)*

Glomerulus *(glo-mer'u-lus)*

Renal tubule

Bowman's capsule

Collecting ducts

Peritubular *per"ĭ-tu'bu-lar)* **capillaries**

Arterioles *(ar-te're-ōls)*

Filtration

Reabsorption

Secretion

Nitrogen-containing wastes

Water balance

Electrolyte balance

Diuresis *(di"u-re'sis)*

Acid–base balance

Ureters *(u-re'ters)*

Urinary bladder

Urethra *(u-re'thrah)*

Micturition *(mik"tu-rish'un)*

---

*For definitions, see Glossary.

SUMMARY

A. KIDNEYS

1. The paired kidneys are retroperitoneal in the superior lumbar region. Each kidney has a medial indentation (hilus) where the renal artery, renal vein, and ureter are seen. A fatty cushion holds the kidneys in position against the trunk wall.

2. A longitudinal section of the kidney reveals an outer cortex, a deeper medulla, and a medial pelvis. Extensions of the pelvis (calyces) surround the tips of the medullary pyramids and collect the urine draining from their tips.

3. The renal artery, which enters the kidney, breaks up into interlobar arteries that travel outward through the medulla. The interlobar arteries split into arcuate arteries that in turn branch to produce the interlobular arteries, which serve the cortex.

4. Nephrons are the structural and functional units of the kidneys. Each consists of a glomerulus and a renal tubule. Subdivisions of the renal tubule (from the glomerulus) are Bowman's capsule, proximal convoluted tubule, loop of Henle, and distal convoluted tubule. Besides a glomerulus, a second capillary bed, the peritubular capillary bed, is associated with each nephron.

5. Functions of the nephron include filtration, reabsorption, and secretion. Filtrate formation is the role of the high-pressure glomerulus Filtrate is essentially plasma

without the blood proteins. Reabsorption is done by the tubule cells. In reabsorption, needed substances are removed from the filtrate (amino acids, glucose, water, and some ions) and returned to the blood. Secretion, a second role of the tubule cells, involves the secretion of additional substances into the filtrate by the tubule cells. Secretion is important in ridding the body of drugs and excess ions and in maintaining the acid–base balance of the blood.

6. Blood composition depends on diet, cellular metabolism, and urinary output. To maintain blood composition, the kidneys:
    a. Allow nitrogen-containing wastes (urea, ammonia, creatinine, and uric acid) to go out in the urine.
    b. Maintain water and electrolyte balance by absorbing more or less water and ions in response to hormones. ADH increases water reabsorption and conserves body water. Aldosterone increases tubular reabsorption of sodium and water and decreases tubular reabsorption of potassium.
    c. Maintain acid–base balance by failing to reabsorb or by actively secreting excess acids and bases.

7. Urine is clear, yellow, and usually slightly acid, but its pH value can vary widely. Substances normally found in urine are nitrogen-containing wastes, water, and various ions (always sodium and potassium). Substances not normally found in urine include glucose, albumin (or other blood proteins), blood, pus (WBCs), and bile.

B. URETERS

1. The ureters are slender tubes running from each kidney to the bladder. They conduct urine by peristalsis from the kidney to the bladder.

C. URINARY BLADDER AND MICTURITION

1. The bladder is a collapsed sac anterior to the small intestine. It has two inlets (ureters) and one outlet (urethra). In males, the prostate gland surrounds its outlet.

2. The function of the bladder is to store urine. As the bladder fills, its muscular wall stretches.

3. Micturition is the emptying of the bladder. The micturition reflex causes the involuntary internal sphincter to open when stretch receptors in the bladder wall are stimulated. Since the external sphincter is voluntarily controlled, micturition can ordinarily be temporarily delayed. Incontinence is the inability to control micturition.

D. URETHRA

1. The urethra is a tubule that conducts urine from the bladder to the body exterior. In females, it is 3–4-cm long and conducts only urine. In males, it is 20-cm long and conducts both urine and sperm.

E. DEVELOPMENTAL ASPECTS OF THE URINARY SYSTEM

1. The kidneys begin to develop in the first few weeks of embryonic life and are excreting urine by the third month.

2. The more common congenital abnormalities include polycystic kidneys, exstrophy of the bladder, and hypospadias.

3. The most common urinary system problems in children and young to middle-aged adults are infections caused by digestive system microorganisms, venereal disease–causing microorganisms, and streptococcus.

4. Renal failure is an uncommon, but extremely serious, problem of this system. In renal failure, the kidneys are unable to concentrate urine.

5. Renal function decreases with age. The filtration rate decreases and the tubule cells become less efficient at concentrating urine, leading to urgency, frequency, and incontinence. Urinary retention is another common problem of the elderly individual.

REVIEW QUESTIONS

1. Name the organs of the urinary system and describe the general function of each organ.

2. Describe the location of the kidneys in the body.

3. Make a diagram of a longitudinal section of a kidney. Identify and label the cortex, medulla, medullary pyramids, renal columns, and pelvis.

4. Name the structural and functional unit of the kidney.

5. Trace the pathway a uric acid molecule takes from a glomerulus to the urethra. Name every microscopic or gross structure it passes through on its journey.

6. What is the function of the glomerulus? What two functions do the renal tubules perform?

7. Besides ridding the body of wastes formed during cell metabolism, the kidney continually adjusts blood chemistry in other ways. What are these three other ways?

8. Explain the important difference between filtrate and urine.

9. How does aldosterone modify the chemical composition of urine?

10. What hormone has a name that means "against urine flow"? What condition happens if it is not secreted?

11. Name three substances normally found in blood that are not normally found in urine. Give the name of the condition when each of the named substances *is* found in urine.

12. Why is a urinalysis a routine part of any good physical examination?

13. Define micturition and then describe the micturition reflex.

14. What sometimes happens when urine becomes too concentrated or remains too long in the bladder?

15. Define incontinence.

16. How is the female urethra different from that of the male in structure and function?

17. Why is cystitis more common in females?

18. What type of problem most commonly affects the urinary system organs?

19. Describe the changes that occur in kidney and bladder function in old age.

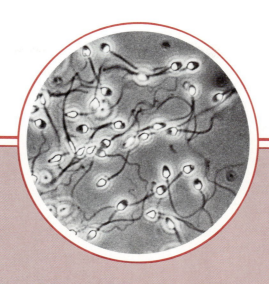

CHAPTER 15

Reproductive System

Chapter Contents

Functions of the reproductive system • to ensure the continuity of the species and to produce offspring

After completing this chapter, you should be able to:

- Discuss the general function of the reproductive system.

- Identify organs of the male reproductive system when provided with a model or diagram and discuss the general function of each.

- Name the endocrine and exocrine products of the testes.

- Discuss the composition of semen and name the glands that produce it.

- Trace the pathway followed by a sperm from the testis to the body exterior.

- Define erection, ejaculation, and circumcision.

- Identify the organs of the female reproductive system when provided with an appropriate model or diagram and discuss the general function of each.

- Describe the functions of the graafian follicle and corpus luteum of the ovary.

- Define endometrium, myometrium, and ovulation.

- Explain the location of the following regions of the female uterus: cervix, fundus, and body.

- Define meiosis, spermatogenesis, and oogenesis.

- Describe the structure of a sperm and relate its structure to function.

- Describe the effect of FSH and LH on testis and ovary functioning.

- Describe the phases and control of the menstrual cycle.

- Define menopause, fertilization, and zygote.

- Describe the process and timing of implantation.

- Distinguish between an embryo and a fetus.

- Name the functions of the placenta.

- Name several ways that pregnancy alters or modifies the functioning of the mother's body.

- List several agents that can interfere with normal fetal development.

- List common reproductive system problems seen in adult and aging males and females.

Although the reproductive system structures of males and females are different, they have a common purpose. Most simply, the unique function of the reproductive system is to produce offspring.

The reproductive role of the male is to manufacture sperm and deliver them to the female reproductive tract. The female, in turn, produces eggs. If the time is suitable, the combination of a sperm and an egg produces a fertilized egg, which is the first cell of a new individual. Once fertilization has occurred, the female uterus provides a protective environment in which the embryo, later called the fetus, develops until birth.

Although the biologic drive to reproduce is strong in all animals, in humans, emotions as well as cultural and social factors, often enhance or inhibit its expression. Every day, we are greeted by some type of emphasis on the sexual difference between males and females. Television, newspapers, and magazines sell products by using advertising that plays on our natural interest in sexual differences. For these reasons, it is often difficult for many of us to view our reproductive abilities and drives as natural traits that enrich our lives.

MALE REPRODUCTIVE SYSTEM

The primary reproductive organs of the male are the testes, which have both an exocrine (sperm-producing) function and an endocrine (testoster-

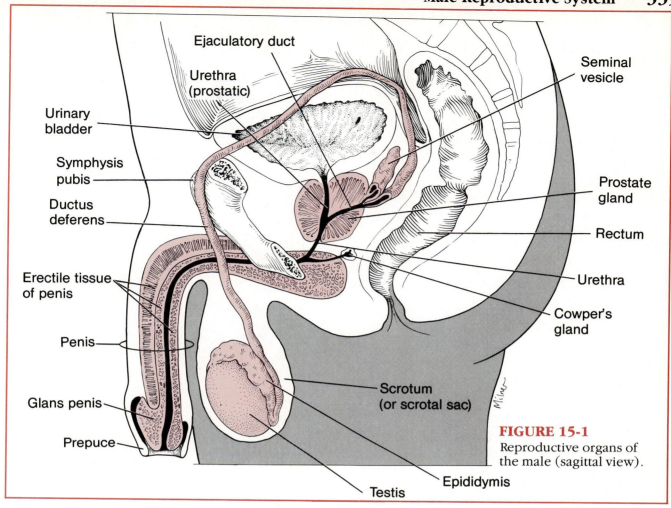

Ejaculatory duct

Urethra
(prostatic)

Urinary
bladder

Symphysis
pubis

Ductus
deferens

Erectile tissue
of penis

Penis

Glans penis

Prepuce

Seminal
vesicle

Prostate
gland

Rectum

Urethra

Cowper's
gland

Scrotum
(or scrotal sac)

FIGURE 15-1
Reproductive organs of
the male (sagittal view).

Epididymis

Testis

one-producing) function. All other reproductive structures are ducts or glands that aid in the safe delivery of sperm to the body exterior or the female reproductive tract (Figure 15-1).

Testes

LOCATION AND GROSS ANATOMY

The paired oval **testes** (tes′tēz), also called the male **gonads,** lie suspended in the **scrotal sac** outside the abdominopelvic cavity. This is a rather exposed location, but apparently the temperature (slightly lower than body temperature) that exists there is necessary for the production of healthy sperm.

Each testis is approximately 4-cm (1½-in.) long and 2.5-cm (1-in.) wide. A connective tissue capsule, the *tunica albuginea* (tu′nĭ-kah al″bu-jin′e-ah) surrounds each testis. Extensions of this capsule *(septa)* plunge into the testis and divide it into a large number of lobes. Each lobe contains one to four tightly coiled **seminiferous** (se″mĭ-

nif′er-us) **tubules,** the actual "sperm-forming factories" (Figure 15-2). Seminiferous tubules of each lobe empty sperm into another set of tubules, the *rete* (re′te) *testis* located at one side of the testis. Sperm travel through the rete testis to enter the first part of the duct system, the epididymis (ep″ĭ-did′ĭ-mis), which is located outside the testis.

FUNCTIONS: SPERMATOGENESIS AND TESTOSTERONE PRODUCTION

SPERMATOGENESIS. Sperm production, or *spermatogenesis* (sper″mah-to-jen′ĕ-sis), begins during puberty and continues throughout life. Every day a man makes millions of sperm. Since only one sperm fertilizes an egg, it seems that nature has made sure that the human species will not be endangered for lack of sperm.

Sperm formation occurs in the seminiferous tubules of the testis as noted earlier. As seen by looking at Figure 15-3, the process is begun by primitive stem cells called **spermatogonia** (sper″mah-to-go′ne-ah) found in the outer region of

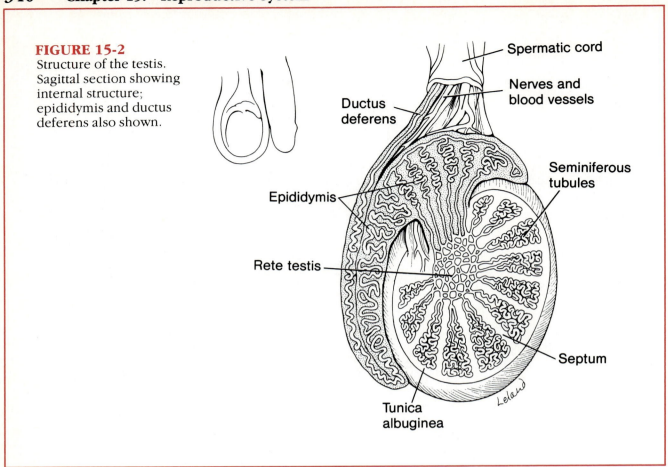

FIGURE 15-2
Structure of the testis. Sagittal section showing internal structure; epididymis and ductus deferens also shown.

Spermatic cord

Nerves and blood vessels

Ductus deferens

Seminiferous tubules

Epididymis

Rete testis

Septum

Tunica albuginea

each tubule. Spermatogonia go through rapid mitotic divisions to build up the stem cell line. Before puberty (generally described as the period between the ages of 13 and 15 years), all such divisions simply produce more stem cells. However, during puberty, follicle-stimulating hormone (FSH) is secreted in increasing amounts by the anterior pituitary gland, and from this time on each division of a spermatogonium produces one stem cell and another cell called a **primary spermatocyte** (sper′mah-to-sīt″), which is destined to undergo *meiosis* (mi-o′sis). Meiosis is a special type of nuclear division that differs from mitosis (studied in Chapter 2) in two major ways. Meiosis consists of two rapid divisions of the nucleus and results in four (instead of two) daughter cells. In spermatogenesis, the daughter cells are called spermatids (sper′mah-tids). Spermatids have only half as much genetic material as other body cells do. (In humans, this is 23 chromosomes rather than the usual 46 chromosomes.) Then, when the sperm and the egg (which also contains only 23 chromosomes) unite, the normal number of 46 chromosomes is reestablished.

As meiosis occurs, the dividing cells are pushed toward the lumen of the tubule. Thus, progress of meiosis can be followed from the tubule periphery to the lumen. The spermatids, which are the products of meiosis, are *not* functional sperm. They are nonmotile cells and have too much excess baggage to function well in reproduction. They must still undergo further changes in which their excess cytoplasm is stripped away and a tail is formed. In this last stage of sperm development, all the excess cytoplasm is sloughed off, and what remains is compacted into the three regions of the mature **sperm**—the *head, midpiece, and tail* (Figure 15-4). The mature sperm is a greatly streamlined cell equipped with a means of propelling itself and a high rate of metabolism, which enable it to move long distances in a short time to get to the egg. It is a prime example of the fit between form and function.

The sperm head contains DNA, or genetic material. Essentially, it *is* the nucleus of the spermatid. Anterior to the nucleus is the **acrosome** (ak′ro-sōm), which is similar to a large lysosome. When

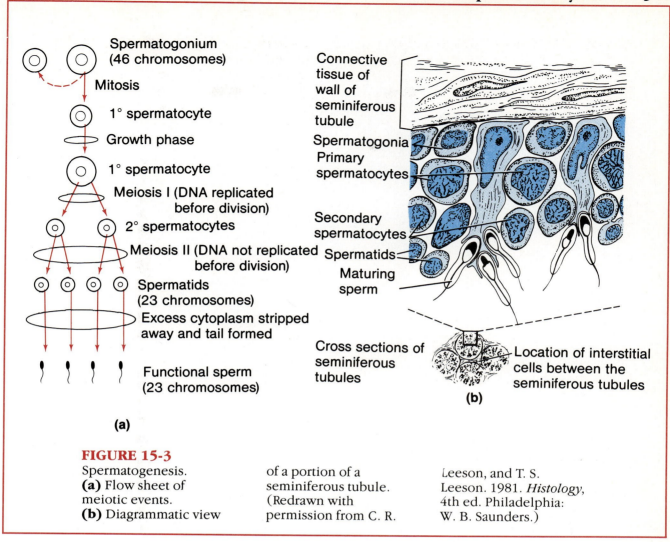

**Spermatogonium
(46 chromosomes)**

Mitosis

1° spermatocyte

Growth phase

1° spermatocyte

Meiosis I (DNA replicated
before division)

2° spermatocytes

Meiosis II (DNA not replicated
before division)

**Spermatids
(23 chromosomes)**

Excess cytoplasm stripped
away and tail formed

**Functional sperm
(23 chromosomes)**

(a)

Connective
tissue of
wall of
seminiferous
tubule

Spermatogonia
Primary
spermatocytes

Secondary
spermatocytes
Spermatids
Maturing
sperm

Cross sections of
seminiferous
tubules

Location of interstitial
cells between the
seminiferous tubules

(b)

FIGURE 15-3
Spermatogenesis.
(a) Flow sheet of
meiotic events.
(b) Diagrammatic view
of a portion of a
seminiferous tubule.
(Redrawn with
permission from C. R.
Leeson, and T. S.
Leeson. 1981. *Histology*,
4th ed. Philadelphia:
W. B. Saunders.)

a sperm comes into close contact with an egg (or more precisely, an oocyte), the acrosomal membrane breaks down and releases enzymes, which help the sperm to penetrate the egg. Filaments, which form the tail, arise from centrioles in the midpiece. Wrapped tightly around these filaments are mitochondria that apparently provide the ATP needed for tail movement. The tail filaments propel the sperm when they are powered by ATP.

TESTOSTERONE PRODUCTION.   Lying between the seminiferous tubules, and softly padded with connective tissues, are **interstitial** (in″ter-stish′al) **cells.** The interstitial cells (see Figure 15-3) produce **testosterone** (tes-tos′tĕ-rōn), the hormonal product of the testes. During puberty, interstitial cells are activated by FSH and luteinizing hormone (LH) (also called interstitial cell-stimulating hormone [ICSH]) released by the anterior pi-

tuitary gland. From this time on, testosterone is produced continuously (more or less) for the rest of a man's life. The rise in testosterone production in the young male stimulates his reproductive organs to develop to their adult size and causes the secondary sex characteristics of the male to appear. *Secondary sex characteristics* typical of males include:

1. Deepening of the voice owing to the enlargement of the larynx

2. Increased hair growth all over the body but particularly in the axillary and pubic regions and the face (the beard)

3. Enlargement of skeletal muscles to produce the heavier muscle mass typical of the male physique

4. Increased heaviness of the male skeleton due to thickening of the bones

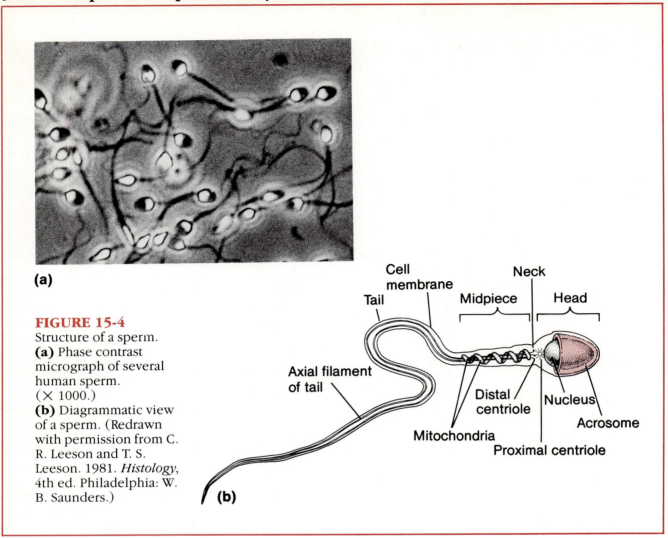

(a)

FIGURE 15-4
Structure of a sperm.
(a) Phase contrast
micrograph of several
human sperm.
(× 1000.)
(b) Diagrammatic view
of a sperm. (Redrawn
with permission from C.
R. Leeson and T. S.
Leeson. 1981. *Histology*,
4th ed. Philadelphia: W.
B. Saunders.)

Because testosterone is responsible for the appearance of these typical masculine characteristics, it is often referred to as the "masculinizing" hormone.

If testosterone is not produced, the secondary sex characteristics never appear in the young man, and his other reproductive organs remain childlike. This is *sexual infantilism.* Castration of the adult male (or the inability of his interstitial cells to produce testosterone) results in a decrease in the size and function of his reproductive organs as well as a decrease in his sex drive. Sterility also occurs because testosterone is necessary for the final stages of sperm production.

Duct System

The accessory structures that form the duct system, which transports sperm from the body, are the epididymis, ductus deferens, and urethra (Figure 15-1).

EPIDIDYMIS

The **epididymis** is a coiled tube about 5–6 m (20-ft.) long, seen capping the superior part of the testis and then running down its posterior side. The epididymis forms the first part of the duct system and provides a temporary storage site for immature sperm that enter it from the testis. While they remain in the epididymis, the sperm complete their maturation process to become fully functional sperm. When a male is sexually stimulated, the walls of the epididymis contract to expel the sperm into the next part of the duct system, the ductus deferens.

DUCTUS DEFERENS

The **ductus deferens** (duk'tus def'er-enz), or **vas deferens,** runs upward from the epididymis through the inguinal canal into the pelvic cavity and arches over the superior aspect of the bladder. This tube is enclosed, along with blood vessels and nerves, in a connective tissue sheath called the **spermatic cord.** The end of the ductus deferens empties into the **ejaculatory** (e-jak′u-lah-to″re) **duct** that carries sperm through the prostate gland to empty them into the urethra, which travels down the length of the penis to the exterior of the body.

As Figure 15-1 illustrates, part of the ductus deferens lies in the scrotal sac, which hangs outside the body cavity. Some men voluntarily opt to take full responsibility for birth control by requesting a *vasectomy* (vah-sek′to-me). In this relatively minor operation, the surgeon makes a small incision into the scrotum and then cuts through (or cauterizes) the ductus deferens. Sperm are still produced but they can no longer reach the body exterior; thus, the man is sterile after this procedure.

Accessory Glands

The accessory glandular structures include the paired seminal vesicles, the single prostate gland, and Cowper's glands (see Figure 15-1). These glands produce **seminal fluid,** or **semen** (se′-men), the fluid that carries sperm out of the body. The relative alkalinity of all the accessory secretions is thought to have a role in protecting the delicate sperm against the acid conditions of the female reproductive tract. The amount of seminal fluid ejected during ejaculation is relatively small (2–6 ml), but there are over 100 million sperm in each milliliter.

SEMINAL VESICLES

The **seminal vesicles** produce about 60% of the fluid volume of semen. Their secretion is rich in sugar (fructose) and other substances, which nourish and activate the sperm passing through the tract. The duct of each seminal vesicle joins that of the ductus deferens on the same side to form the ejaculatory duct; thus, sperm and seminal fluid enter the urethra together during *ejaculation* (propulsion of semen out of the body).

PROSTATE GLAND

The **prostate gland** is a single gland about the size and shape of a chestnut and encircles the upper area of the urethra just below the bladder. It secretes a milky alkaline fluid, which has a role in activating sperm, into the urethra.

Hypertrophy of the prostate gland, which is common in old age, constricts the urethra. This is a troublesome condition that makes urination difficult.

COWPER'S GLANDS

Cowper's glands are tiny pea-sized glands inferior to the prostate gland. They form a thick clear mucus, which drains into the urethra. The secretion of Cowper's glands is believed to serve primarily as a lubricant during sexual intercourse.

External Genitalia

The **external genitalia** (jen″i-tal′e-ah) of the male include the scrotum and the penis (see Figure 15-1). The **scrotum** (skro′tum) is a divided sac, or pouch of skin, that hangs from the midline of the body between the legs and posterior to the penis. Under normal conditions, the scrotum hangs loosely from its attachments allowing the testes to have a temperature that is below body temperature. However, when the external temperature is very cold, the scrotum draws up and pulls the testes closer to the body so that the temperature there does not become too cold for viable sperm production.

The **penis** (pe′nis) is designed to deliver sperm into the female reproductive tract. The skin covered penis consists of a **shaft,** which ends in an enlarged tip, the **glans.** The skin covering the penis is loose and it folds downward to form a cuff of skin, the **prepuce,** or **foreskin,** around the proximal end of the glans. Frequently, the foreskin is surgically removed *(circumcision).*

Internally, the penis contains the **urethra,** which carries both semen and urine to the body exterior. The urethra (see Figure 15-1) is surrounded by three elongated areas of erectile tissue, a spongy tissue that becomes filled with blood during sexual excitement. This causes the penis to become rigid and enlarge so that it may more adequately serve as a penetrating device. This event is called *erection.*

FEMALE REPRODUCTIVE SYSTEM

The role of a female in reproduction is much more complex than that of a male. Not only must she produce the female germ cells (ova), but her body must also nurture and protect a developing

FIGURE 15-5
Sagittal section of the human female reproductive system.

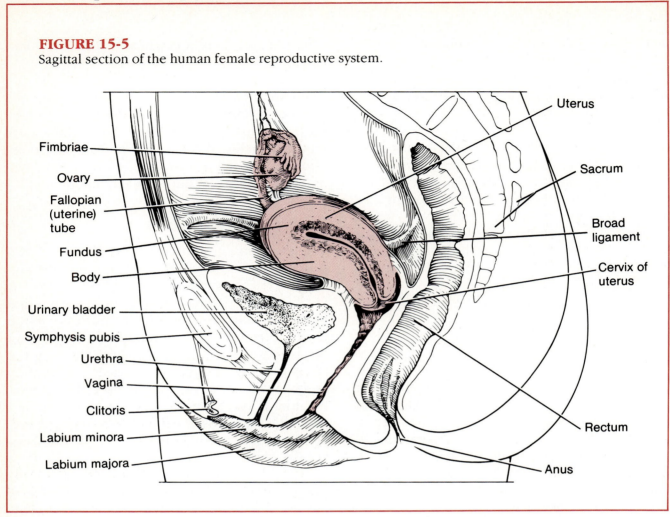

Fimbriae

Ovary

Fallopian
(uterine)
tube

Fundus

Body

Urinary bladder

Symphysis pubis

Urethra

Vagina

Clitoris

Labium minora

Labium majora

Uterus

Sacrum

Broad
ligament

Cervix of
uterus

Rectum

Anus

fetus during the 9 months of pregnancy. Ovaries are the primary reproductive organs of a female. Like the testes of a male, ovaries produce both an exocrine product (the eggs, or ova) and endocrine products (estrogens and progesterone). The other accessory structures of the female reproductive system transport, nurture, or otherwise serve the needs of the reproductive cells and/or the developing fetus.

Ovaries

LOCATION AND GROSS ANATOMY

The paired **ovaries** (o'vah-rez) are pretty much the size and shape of almonds. They lie against the lateral walls of the pelvis where they are enclosed and held in place by a fold of peritoneum, the *broad ligament* (Figure 15-5).

An internal view of an ovary reveals many tiny saclike structures, **follicles** (Figure 15-6). Each folli-

cle consists of an immature egg surrounded by one or more layers of very different cells. *follicle cells*. As a developing egg within a follicle begins to ripen or mature, the follicle enlarges and develops a fluid-filled central region (the antrum). At this stage, the follicle is called a *mature,* or *graafian (graf'e-an) follicle,* and the developing egg is ready to be ejected from the ovary, an event called *ovulation*. Ovulation generally occurs every 28 days, but it can occur more or less frequently in some women. In older women, the surface of the ovaries are scarred and pitted, which attests to the fact that many eggs have been released.

FUNCTIONS: OOGENESIS AND HORMONE PRODUCTION

OOGENESIS AND THE OVARIAN CYCLE. As described earlier, sperm production in males begins at puberty and generally continues throughout

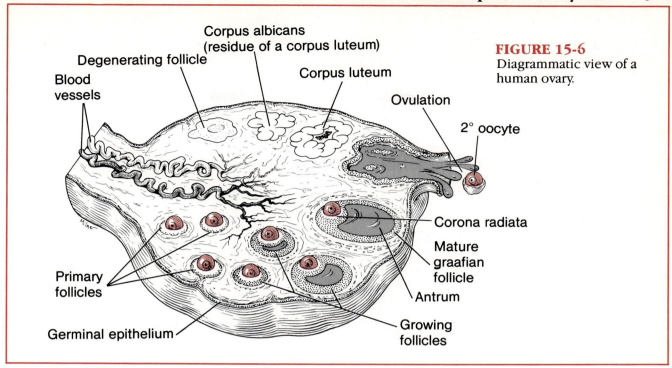

Blood vessels

Degenerating follicle

Corpus albicans (residue of a corpus luteum)

Corpus luteum

Ovulation

2° oocyte

Primary follicles

Germinal epithelium

Growing follicles

Corona radiata

Mature graafian follicle

Antrum

FIGURE 15-6
Diagrammatic view of a human ovary.

life. The situation is quite different in females. The total supply of eggs that a female can release has already been determined by the time she is born. In addition, a female's reproductive ability (that is, her ability to release eggs) usually begins during puberty and ends when or before she is in her 50s. The period in which a woman's reproductive capability gradually declines and then finally ends is *menopause.*

The specialized type of cell division (meiosis), which occurs in the male testes to produce sperm, also occurs in the ovaries of the female. But in this case, female sex cells are produced, and the process is called *oogenesis* (o″o-jen′ĕ-sis). This process is shown in Figure 15-7 and described in more detail next.

In the developing female fetus, **oogonia** (o″o-go′ne-ah), the female stem cells, multiply rapidly to increase their number, and then their daughter cells, **primary oocytes** (o′o-sīts), push into the ovary connective tissue where they become surrounded by a single layer of follicle cells to form the primary follicles. By birth, the oogonia no longer exist, and a female's lifetime supply of primary oocytes (approximately 400,000 of them) is already in place in the ovarian follicles awaiting the chance to begin meiosis to produce functional eggs. The primary oocytes remain in this state of suspended

animation all through childhood; thus, the wait is a long one—a minimum of 10–14 years.

At puberty, the anterior pituitary gland begins to release FSH, which stimulates a small number of primary follicles to grow and mature each month, and one ovulation begins to occur each month. Since the reproductive life of a female is at best about 38 years (from the age of 12 to approximately 50) and there is only one ovulation per month, only 400 to 500 ova out of a potential of 400,000 are released during a woman's lifetime. Again, nature has provided us with a generous oversupply of sex cells.

As a follicle grows larger, prodded on by FSH, the primary oocyte it contains begins meiosis and undergoes the first meiotic division to produce two different cells. The larger cell is a **secondary oocyte** and the second very tiny cell is a **polar body.** By the time follicles have ripened to the mature, or graafian follicle, stage, they contain secondary oocytes and can be seen protruding or bulging blister-like from the surface of the ovaries (see Figure 15-6). Follicle development to this stage takes about 14 days, and ovulation (of a secondary oocyte) occurs just about that time in response to a second hormone, LH, released by the anterior pituitary. Generally speaking, whichever follicle is at the proper stage of maturity when the

FIGURE 15-7
Oogenesis. *Left,* flow sheet of meiotic events. *Right,* correlation with follicle development and ovulation in the ovary.

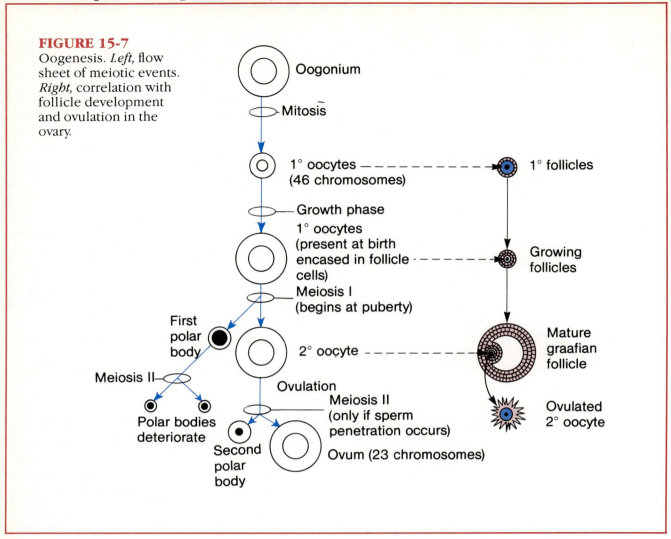

LH stimulus occurs will rupture and release its oocyte into the peritoneal cavity. The mature follicles that are not ovulated become overripe and gradually deteriorate. In addition to triggering ovulation, LH also causes the ruptured follicle to change into a glandular structure, the corpus luteum. (Both the follicles and the corpus luteum produce hormones as will be described later.)

If the ovulated secondary oocyte is penetrated by a sperm, its nucleus will undergo the second meiotic division that produces another polar body and the **ovum nucleus.** Once the ovum nucleus (with 23 chromosomes) has been formed, its chromosomes are combined with those of the sperm to form the fertilized egg, which is the first cell of the yet-to-be offspring. However, if the secondary oocyte is not penetrated by a sperm, it simply deteriorates without ever completing meiosis to form the functional egg. Although meiosis

in males results in four functional sperm, meiosis in females yields only one functional ovum and three tiny polar bodies. Since the polar bodies have essentially no cytoplasm, they deteriorate and die shortly.

Another major difference between males and females concerns the size and structure of their sex cells. Sperm are tiny and equipped with tails for locomotion. They have little nutrient-containing cytoplasm; thus, the nutrients in seminal fluid are vital to their survival. In contrast, the egg is a large nonmotile cell, which is well stocked with nutrient reserves that nourish the developing embryo until it can take up residence in the uterus.

HORMONE PRODUCTION BY THE OVARIES. As the ovaries become active at puberty to produce the ova, the production of ovarian hormones also begins. The follicle cells of the growing and mature follicles produce **estrogen,** which causes the appearance of the *secondary sex characteristics* in the young woman. Such changes include:

1. Enlargement of the other organs of the female reproductive system (fallopian tubes, uterus, vagina, and external genitals)

2. Development of the breasts

3. Increased deposits of fat beneath the skin in general, but particularly in the hips and breasts

4. Widening and lightening of the pelvis

5. Onset of menses, or the menstrual cycle

The second ovarian hormone, **progesterone,** is produced by a special glandular structure of the ovaries, the **corpus luteum** (kor'pus lu'te-um). (The name means "yellow body," which describes the color of this glandular tissue.) After ovulation occurs, the ruptured follicle is converted to the corpus luteum, which looks and acts completely different. Once formed, the corpus luteum produces progesterone (and some estrogen) as long as LH is still present in the blood. Generally speaking, the corpus luteum has stopped producing hormones by 10 to 14 days after ovulation. Except for working with estrogen to establish the menstrual cycle, progesterone does not contribute to the appearance of the secondary sex characteristics. Progesterone is important during pregnancy when it helps maintain the pregnancy and prepare the breasts for milk production. (However, the source of progesterone during pregnancy is the placenta, not the ovaries.)

Duct System

The fallopian tubes, uterus, and vagina form the duct system of the female reproductive tract (see Figure 15-5).

FALLOPIAN (UTERINE) TUBES

The **fallopian** (fal-lo'pe-an) **tubes** form the first part of the duct system. Each of the fallopian tubes is about 10-cm (4-in.) long and extends medially from an ovary to empty into the superior region of the uterus. Unlike the male duct system, which is continuous with the tubule system of the testis, there is little or no actual contact between the fallopian tubes and the ovaries. The distal end of each fallopian tube is expanded and has finger-like projections called *fimbriae* (fim'bre-ah), which partially surround the ovary. As an oocyte is expelled from an ovary during ovulation, the waving fimbriae create fluid currents that act to carry the oocyte into the fallopian tube where it begins its passage to the uterus. (Obviously, many eggs are lost in the peritoneal cavity.) The oocyte is carried toward the uterus by a combination of fallopian tube wall peristalsis and the action of *cilia* (sil'e-ah), which beat toward the uterus.

The usual and most desirable site of fertilization is the fallopian tube, because the journey to the uterus takes about 5 days and the oocyte is usually nonviable after 24–48 hours. To reach the oocyte, the sperm must swim upward through the vagina and uterus to reach the fallopian tubes. This must be a difficult journey, since they must swim against the downward current created by the cilia—rather like swimming against the tide.

UTERUS

The **uterus** (u'ter-us), located between the urinary bladder and rectum, is a hollow organ that is normally about the size and shape of a pear. (However, during pregnancy the uterus increases tremendously in size to accomodate the growing fetus and can be felt well above the umbilicus during the later part of pregnancy.) The uterus is suspended in the pelvis by the same peritoneal fold (the broad ligament) that anchors the ovaries in place.

The major portion of the uterus is referred to as the **body.** Its superior rounded region above the entrance of the fallopian tubes is the **fundus,** and its narrow inferior outlet, which protrudes into the vagina below, is the **cervix.**

The wall of the uterus is thick and composed of three layers. The inner layer or mucosa is the **endometrium** (en-do-me'tre-um). If fertilization occurs, the fertilized egg (actually the young embryo by the time it reaches the uterus) burrows into the endometrium of the uterus (this process is called *implantation*) and resides there for the rest of its development. When a woman is not pregnant, the endometrial lining sloughs off periodically, usually about every 28 days, in response to changes in the levels of ovarian hormones in the blood. This process, called *menses,* is discussed on p. 350.

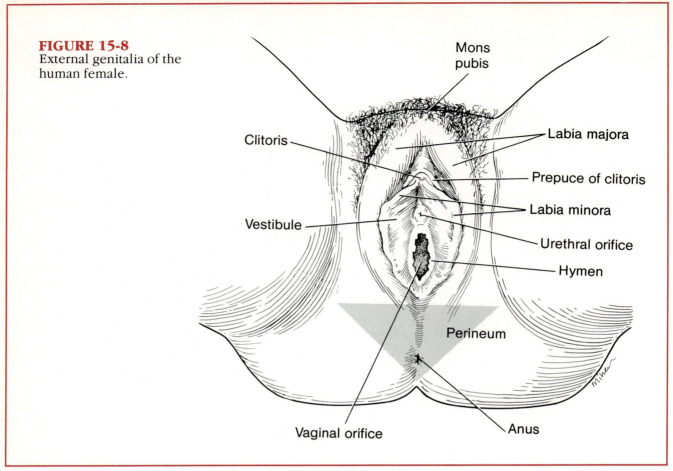

FIGURE 15-8
External genitalia of the human female.

Mons pubis

Clitoris

Labia majora

Prepuce of clitoris

Labia minora

Vestibule

Urethral orifice

Hymen

Perineum

Vaginal orifice

Anus

The **myometrium** (mi-o-me′tre-um) is the thick middle layer of the uterus; it is composed of interlacing bundles of smooth muscle. The myometrium plays an active role during the delivery of a baby when it contracts rhythmically to force the baby out of the body. The outermost layer of the uterus is the **epimetrium** (ep-ĭ-me′tre-um), a serous membrane.

VAGINA

The thin-walled muscular **vagina** (vah-ji′nah) is about 10-cm (4-in.) long; it extends from the body exterior to the uterus superiorly. Often called the birth canal, the vagina provides a passageway for the delivery of an infant and for the menstrual flow to leave the body. It also is the female reproductive organ that receives the penis (and semen) during sexual intercourse.

The distal end of the vagina is partially closed by a thin fold of the mucosa call the *hymen* (hi′men). The hymen is generally ruptured during the first sexual intercourse. However, in isolated cases, it is extremely tough and must be surgically removed.

External Genitalia

The female reproductive structures that are located external to the vagina are the **external genitalia** (Figure 15-8). The external genitalia, also called the **vulva,** include the labia, the clitoris, urethral and vaginal orifices, and the greater vestibular glands.

The *mons pubis* is a fatty rounded area overlying the pubic symphysis. After puberty, this area is covered with pubic hair. Running posteriorly from the mons pubis are two elongated, hair-covered skin folds, the **labia majora** (la′be-ah ma-jo′rah), which enclose two smaller hair-free folds, the **labia minora.** The labia majora correspond to the scrotum of the male. The labia minora enclose a region called the **vestibule,** which contains many structures: the **clitoris** (kli′to-ris), most anteriorly, followed by the external openings of the urethra and the vagina.* A pair of mucus-produc-

*While the male urethra carries both urine and semen, the female urethra has no reproductive function; it is strictly a urine passageway.

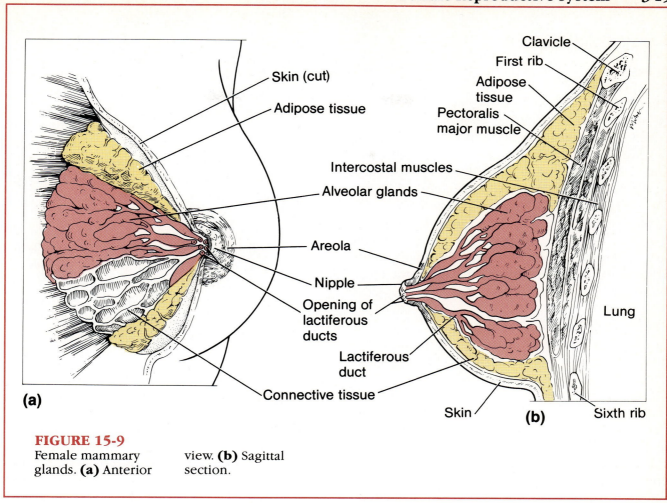

(a)

FIGURE 15-9
Female mammary   view. **(b)** Sagittal
glands. **(a)** Anterior   section.

ing glands, the **greater vestibular glands,** flank the vagina, one on each side. Their secretion lubricates the distal end of the vagina during intercourse. (These glands are not shown in Figure 15-8.) The diamond-shaped region between the anterior end of the labia minora folds and the anus (located most posteriorly) is the **perineum** (per"i-ne'um).

The clitoris is a small protruding structure that corresponds to the male penis. Like the penis, it is composed of very sensitive erectile tissue that becomes swollen with blood during sexual excitement. The clitoris differs from the penis in that it lacks a reproductive duct.

Mammary Glands

The **mammary glands** exist, of course, in both sexes but they have a reproductive function only in females. Since the role of the mammary glands is to produce milk to nourish a newborn baby, they are more important when reproduction has

already been accomplished. Stimulation by female sex hormones, especially estrogens, causes the female mammary glands to increase in size at puberty.

The mammary glands are contained within the rounded skin-covered breasts anterior to the pectoral muscles of the thorax. Slightly below the center of each breast is a pigmented area, the **areola** (ah-re'o-lah), which surrounds a central protruding **nipple** (Figure 15-9).

Internally each mammary gland consists of 15 to 20 **lobes,** which radiate around the nipple. The lobes are padded and separated from each other by connective tissue and fat (adipose tissue). Within each lobe are smaller chambers called **lobules,** which contain clusters of **alveolar glands** that produce the milk when a woman is lactating. The alveolar glands of each lobule pass the milk into the **lactiferous** (lac-tif'er-us) **ducts,** which open to the outside at the nipple.

Menstrual Cycle

As described earlier, the uterus is the female reproductive organ within which the young embryo implants and develops. However, it is important to understand that the uterus is not always prepared to receive an embryo. In fact, it is only receptive for a very short period each month. Not surprisingly, this brief interval coincides exactly to the time at which a fertilized egg would reach the uterus, which is approximately 7 days after ovulation. The events of the *menstrual,* or *uterine, cycle* are the cyclic changes that the endometrium, or mucosa, of the uterus goes through month after month as it responds to changes in the levels of ovarian hormones in the blood.

Since the production of estrogen and progesterone by the ovaries is, in turn, regulated by the anterior pituitary hormones, FSH and LH, it is important to understand how these "hormonal pieces" fit together. Generally speaking, both female cycles are about 28 days long with ovulation occuring midway in the cycles, or on day 14. Figure 15-10 illustrates the events occuring both in the ovary (the ovarian cycle) and in the uterus (menstrual cycle) at the same time. The stages of the menstrual cycle are also described briefly next.

1. **Days 1–5: Menses.** During this interval, the thick endometrial lining of the uterus is sloughing off, or becoming detached, from the uterine wall. This is accompanied by bleeding for 3–5 days.

2. **Days 6–14: Proliferative stage.** Under the influence of estrogens produced by the growing follicles of the ovaries, the endometrium is repaired, glands are formed in the endometrium, and the endometrial blood supply is increased. (Ovulation occurs in the ovary at the end of this stage in response to a sudden burstlike release of LH from the anterior pituitary.)

3. **Days 15–28: Secretory stage.** Rising levels of progesterone production by the corpus luteum of the ovary increases the blood supply to the endometrium even more. It also causes the endometrial glands to increase in size and begin secreting nutrients into the uterine cavity. These nutrients will sustain a developing embryo (if one is present) until it has implanted. If fertilization has occurred, the embryo produces a hormone very similar to LH, which

causes the corpus luteum to continue producing its hormones. If fertilization has not occurred, the corpus luteum begins to degenerate toward the end of this period as LH blood levels decline. Lack of ovarian hormones in the blood causes the blood vessels supplying the endometrium to kink and go into spasms. When deprived of oxygen and nutrients, the endometrial cells begin to die, which sets the stage for menses to begin on day 28.

Although this explantation assumes a classic 28-day cycle, the length of the menstrual cycle is actually highly variable. Sometimes it is as short as 21 days or as long as 38 days. Only one interval is fairly constant in all females; the time from ovulation to the beginning of menses is almost always 14–15 days.

Survey of Pregnancy and Embryonic Development

Because the birth of a baby is such a familiar event, we tend to lose sight of the wonder of the reproductive process. In every instance it begins with a single cell, the fertilized egg, and ends with an extremely complex human being consisting of billions of cells. The development of an embryo is very complex, and details of this process can fill a good-sized book. Thus, our intention here is simply to outline the important events of pregnancy and embryonic development.

As previously described, ovulation occurs on day 14 of a woman's ovarian cycle, which coincides with the onset of the secretory cycle in the uterus. Sperm are motile cells that can propel themselves by lashing movements of their tails. If sperm are deposited in a female's vagina at the approximate time of ovulation, they are attracted to the oocyte by chemicals which act as "homing devices," allowing the sperm to locate the oocyte. It takes 1–2 hours for sperm to complete the journey up the female duct system to the end of the fallopian tubes, and if an oocyte is en route in the tube, fertilization is a distinct possibility.

Once fertilization has been accomplished, changes occur in the fertilized egg, or **zygote** (zi'-gōt), which prevent other sperm from gaining entry. Of the millions of sperm ejaculated by a male, only *one* can penetrate an oocyte. As the zygote

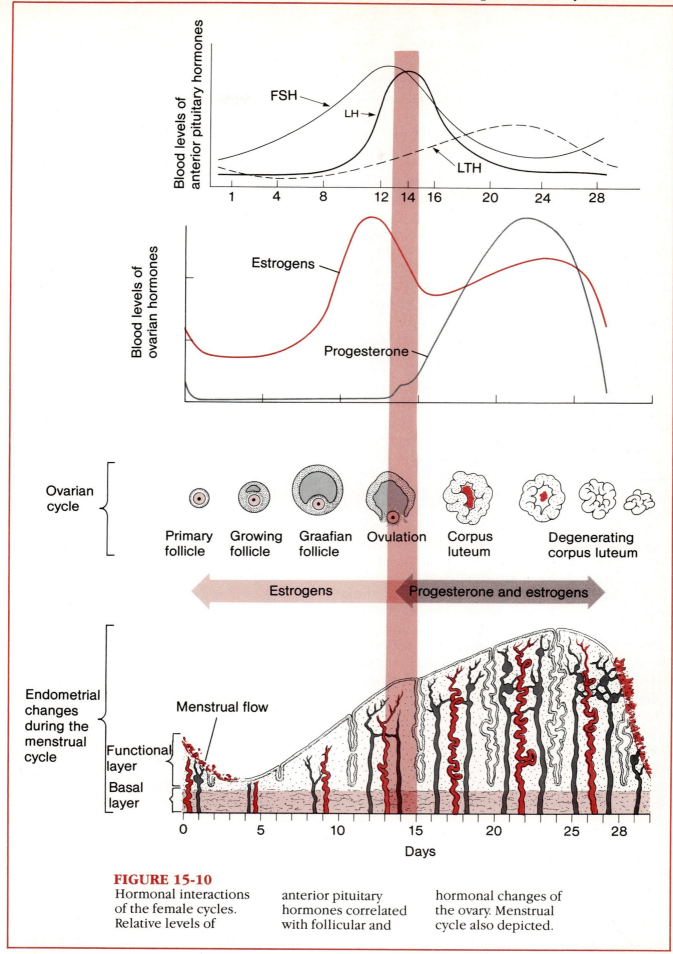

FIGURE 15-10
Hormonal interactions of the female cycles. Relative levels of anterior pituitary hormones correlated with follicular and hormonal changes of the ovary. Menstrual cycle also depicted.

continues its journey down the fallopian tube (propelled along by peristalsis and cilia), it begins to undergo rapid mitotic cell divisions—forming first two cells, then four, and so on. (This early stage of embryonic development, called *cleavage,* is shown in Figure 15-11.) By the time the developing embryo reaches the uterus (about 3 days after ovulation, or on day 17 of the woman's cycle), it is a tiny ball of 16 cells, which looks like a raspberry. The uterine endometrium is still not fully prepared to receive the embryo at this point, so the embryo floats free in the uterine cavity temporarily using the uterine secretions for nutrition. While still unattached, the embryo continues to develop until it has about 100 cells, and then it hollows out to form a ball-like structure, a **blasto-cyst** (blas′to-sist). At the same time, it produces a hormone much like LH, which prods the corpus luteum of the ovary to continue producing its hormones. (If this was not the case, the endometrium would be sloughing off shortly in menses.)

By day 7 after ovulation, the blastocyst has attached to the endometrium and has eroded the lining away in a small area, embedding itself in the thick velvety mucosa. Implantation has usually been completed and the uterine mocosa has grown over the burrowed-in embryo by day 14 after ovulation—the day the woman would ordinarily be expecting to start menses. After it is securely implanted, the blastocyst develops elaborate projections, **chorionic villi** (ko″re-on′ik vil′i), which cooperate with the tissues of the mother's uterus to form a temporary structure, the **placenta** (plah-sen′tah). Once the placenta has formed, the platelike embryonic body, now surrounded by a fluid-filled sac, the **amnion** (am′ne-on), is attached to the placenta by a blood vessel–containing stalk of tissue, the **umbilical cord.** (The special features of the umbilical blood vessels and fetal circulation are discussed on p. 226.) Generally by the third week, the placenta is functioning to deliver nutrients and oxygen to and remove wastes from the embryonic blood. All exchanges are made through the placental barrier. By the end of the second month of pregnancy, the placenta has also become an endocrine organ and is producing estrogen and progesterone to maintain the pregnancy. (At this time, the corpus luteum of the ovary becomes inactive.)

By the eighth week of embryonic development, the groundwork has been completed. All the organ systems have been laid down, at least in rudimentary form, and the embryo looks distinctly human. By the ninth week of development, the embryo is referred to as a **fetus.** From this point on, the major activities are growth and organ specialization. By the end of the tenth lunar month (approximately 270 days after fertilization), the fetus is 45–50-cm (18–20-in.) long and weighs between 2.7–4.1 kg (6-9 lb). At this time, the fetus is said to be full term and is ready to be born.

The later months of pregnancy place great demands on the mother's system. Her blood volume increases, and her kidneys have additional burden of disposing of the fetal wastes. As the fetus grows, the uterus enlarges and pushes superiorly. Since this causes the abdominal organs to be pushed superiorly as well, they inhibit the downward movements of the diaphragm and breathing becomes difficult. Early in pregnancy, frequency of urination may result from pressure of the expanding uterus on the bladder. Some women have additional problems such as backaches (resulting from muscle stress owing to changes in their center of gravity) or nausea and vomiting (the result of chemical and hormonal changes in a woman's system during pregnancy).

Obviously, good maternal nutrition is very necessary all through pregnancy to ensure that the developing fetus has all the building materials (proteins, calcium, iron, and the like) it needs to form its tissues and organs. The old expression "A pregnant woman is eating for two" has encouraged many women to eat *twice* the amount of food actually needed during pregnancy, which, of course, leads to excessive weight gain. Actually, a pregnant woman only needs about 300 additional calories to sustain proper fetal growth. Thus, the emphasis should be on eating high quality food, not just more food.

Since many potentially harmful substances can cross through the placental barrier into the fetal blood, the pregnant woman should be very much aware of what she is taking into her body. Substances that may cause life-threatening conditions (and even fetal death) include alcohol, nicotine, and many types of drugs (anticoagulants, antihypertensives, sedatives, and some antibiotics). Maternal infections, particularly German measles, are another possible source of fetal defects or death. Termination of a pregnancy by loss of a

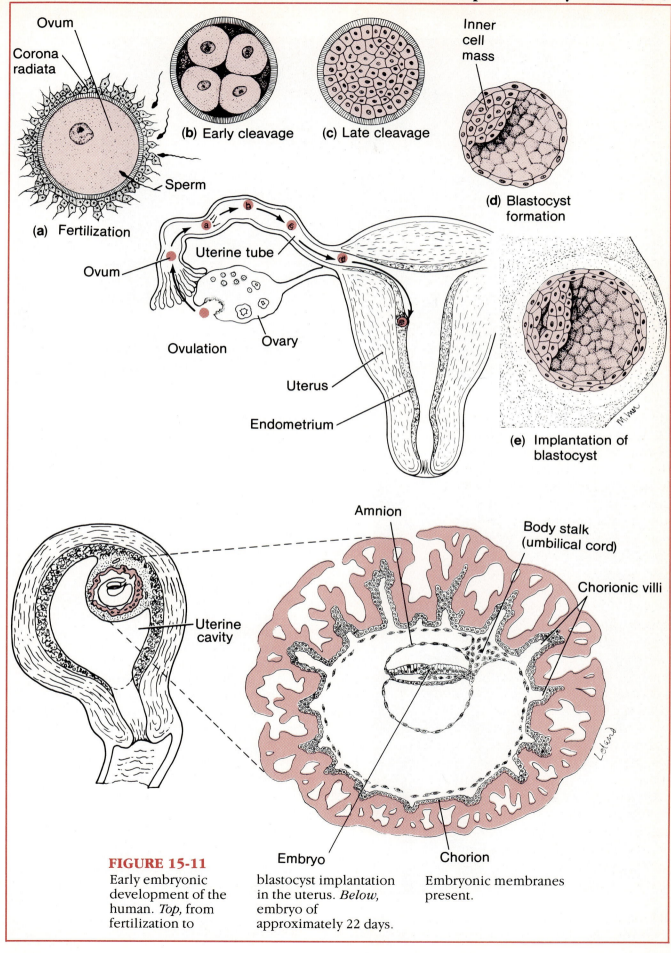

Ovum
Corona radiata
Sperm

(a) Fertilization

(b) Early cleavage

(c) Late cleavage

Inner cell mass

(d) Blastocyst formation

Ovum
Uterine tube
Ovulation
Ovary
Uterus
Endometrium

(e) Implantation of blastocyst

Uterine cavity

Amnion
Body stalk (umbilical cord)
Chorionic villi
Embryo
Chorion

FIGURE 15-11
Early embryonic development of the human. *Top,* from fertilization to blastocyst implantation in the uterus. *Below,* embryo of approximately 22 days. Embryonic membranes present.

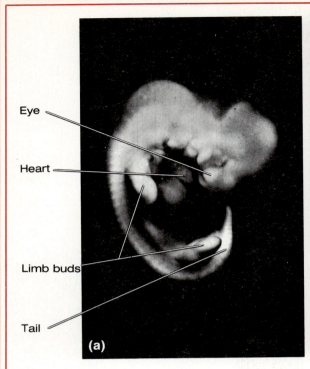

Eye

Heart

Limb buds

Tail

(a)

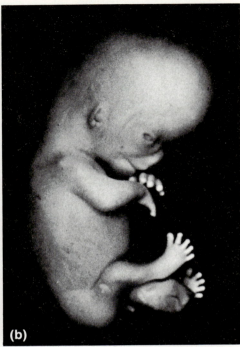

(b)

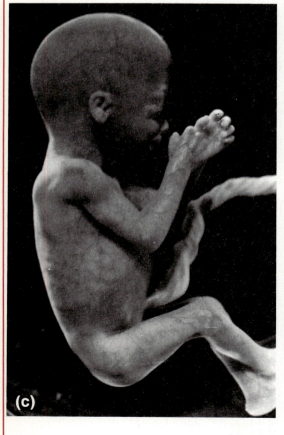

(c)

BOX 15-1
Development of the Human Fetus

(a) The 5-week old embryo has a disproportionately huge brain, which is visible through the thin skin, and its eyes are beginning to form. The heart is also large relative to the embryo's size, and it is functioning to pump blood through the embryonic circulation. A tail, which will eventually disappear, and limb buds are present. The limb buds will form the structures of the arms and legs.

(b) At 8 weeks of gestation, the embryo is a little over 2.5-cm (1-in.) long. The tail can no longer be seen, the face is formed, and the embryo (soon to be called a fetus) looks distinctly human. All the major organs have been formed, and many are functional. The limbs are well developed, and fingers/toes are present. Increasing coordination of the nervous and muscular systems is indicated by reflex movements of the embryo. Beyond this time in development, growth and tissue specialization are the major fetal activities.

(c) At 28 weeks of gestation, the fetus is still approximately 12 weeks away from being full term. It is now 35–40-cm (14–16-in.) long and weighs approximately 1.4 kg (3 lb). The body is covered with fine hair, and fingernails are present. Most body systems are capable of being fully functional except the hypothalamus (which regulates body temperature) and the lungs. A baby born at this time is considered to be premature, and respiration and body temperature are carefully monitored.

fetus that occurs during the first 20 weeks of pregnancy is called *abortion*.

DEVELOPMENTAL ASPECTS OF THE REPRODUCTIVE SYSTEM

Although the sex of an individual is determined at the time of fertilization (that is, males have X and Y sex chromosomes and females are XX), the gonads do not form until the eighth week of embryonic development. Prior to this time, the embryonic reproductive structures of males and females are identical and are said to be in the indifferent stage. After the gonads have formed, the development of the duct system begins and is determined by whether testosterone is present. The usual case is that, once formed, the embryonic testes produce testosterone, and the development of the male duct system and external genitalia then follows. When testosterone is not produced, as is the case in female embryos that form ovaries, the female ducts and external genitalia result. There have been some unusual cases in which fetal testes have failed to produce testosterone, and genetic males have developed the female accessory reproductive structures. However, if a pregnant woman has a masculinizing tumor of her adrenal cortex, a female embryo develops the male ducts and external genitalia. Individuals having accessory reproductive structures that do not match their gonads are called *pseudohermaphrodites* (su″do-her-maf′ro-dīts) to distinguish them from true *hermaphrodites,* rare individuals that possess both ovarian and testicular tissues. In recent years, many pseudohermaphrodites have sought sex change operations.

The male testes, formed in the abdominal cavity at approximately the same location as the female ovaries, descend to enter the scrotum about 1 month before birth. Failure of the testes to make their normal descent leads to a condition called *cryptorchidism* (krip-tor′kī-dizm). Because this condition results in sterility of a male (and also puts him at risk for cancer of the testes), surgery is usually performed during childhood to rectify this problem.

Abnormal separation of chromosomes during meiosis can lead to congenital defects of this system. For example, some males have an extra female sex chromosome and are thus called XXY males. These individuals have the normal male accessory structures, but their testes atrophy, causing them to be sterile. Other abnormalities occur when a child has only one sex chromosome. An XO female appears normal, but lacks ovaries; YO males die during development. Other much less serious conditions affect males primarily; these include *phimosis* (fi-mo′sis), which essentially is a narrowing of the foreskin of the penis, and misplaced urethral openings.

Since the reproductive system organs do not function until puberty, there are few problems with this system during childhood. In adults, the most common problems are infections. Vaginal infections are more common in young and elderly women and in those whose resistance is low. Common infections include *Escherichia coli* (spread from the digestive tract), venereal disease–causing microorganisms, and yeasts (a type of fungus). Untreated vaginal infections may spread throughout the female reproductive tract and cause sterility. This more widespread infection is called pelvic inflammatory disease (PID). Problems involving painful or abnormal menses may result from infection or hormone imbalance.

The most common inflammatory conditions in males are *urethritis, prostatitis,* and *epididymitis* (ep″ĭ-did″ĭ-mi′tis), all of which may follow sexual contacts in which venereal disease microorganisms are transmitted. *Orchiditis* (or″kĭ-di′tis), inflammation of the testes, is rather uncommon but is serious because it can cause sterility. Orchiditis most commonly follows veneral disease or mumps (in an adult male).

Neoplasms represent a major threat to reproductive system organs, particularly in females. Tumors of the breast are the most common cancer type and the leading cause of death in adult women. Since most breast lumps are discovered by women themselves in routine monthly self-breast exams, this simple examination should be a health-maintenance priority in every woman's life. Breast cancers are usually removed surgically; further therapy may include radiation or the use of chemotherapeutic drugs.

Cancer of the cervix of the uterus is the second most common female cancer. It primarily strikes women between the ages of 30 and 50 years. Risk factors include frequent cervical inflammations, veneral disease, multiple pregnancies, and an active sex life with multiple partners. A yearly Pap smear is the single most important diagnostic technique for detecting slowly growing cervical cancer.

Less is known about the effect of aging on the reproductive system than is known about other body systems. In females, a natural decrease in ovarian function occurs in the fourth or fifth decade of life—leading to the cessation of ovulation and ending childbearing ability. When ovarian hormones are no longer produced, the female accessory reproductive structures tend to atrophy. *Senile vaginitis,* which makes a women susceptible to vaginal infections, is common in postmenopausal women. A male's reproductive capability seems unending, and healthy elderly men are able to produce sperm (and father offspring) well into their 80s and beyond. The single most common problem of the reproductive system of elderly men is hypertrophy of the prostate gland (discussed previously), which interferes with both urinary and reproductive system functions. In addition, prostatic hypertrophy often seems to precede cancer of the prostate gland, which is the most common site of reproductive system cancer in males.

IMPORTANT TERMS*

Epididymis *(ep"i-did'i-mis)*

Spermatogenesis *(sper"mah-to-jen'ě-sis)*

Sperm

Interstitial *(in"ter-stish'al)* **cells**

Testosterone *(tes-tos'těr-ōn)*

Ductus deferens *(duk'tus def'er-enz)*

Seminal fluid

Seminal vesicles

Ejaculation

Prostate gland

Cowper's glands

Penis *(pe'nis)*

Follicle

Ovulation

Oogensis *(o"o-jen'ě-sis)*

Estrogen

Progesterone

Corpus luteum *(kor'pus lu'te-um)*

Fallopian *(fal-lo'pe-an)* **tubes**

Fimbriae *(fim'bre-ah)*

Uterus *(u'ter-us)*

Endometrium *(en-do-me'tre-um)*

Vagina *(vah-ji'nah)*

Labia *(la'be-ah)*

Mammary glands

Menstrual cycle

Pregnancy

Implantation

Fertilization

Zygote *(zi'gōt)*

Placenta *(plah-sen'tah)*

Amnion *(am'ne-on)*

Embryo

Fetus

Menopause

Puberty

---

*For definitions, see Glossary.

SUMMARY

A. MALE REPRODUCTIVE SYSTEM

1. The paired testes, male gonads, reside in the scrotum outside the abdominopelvic cavity. The testes have both an exocrine (sperm-producing) and an endocrine (testosterone-producing) function.
   a. Spermatogenesis (sperm production) begins at puberty in the seminiferous tubules in response to FSH. An additional process, which strips excess cytoplasm from the spermatid, is necessary for the production of functional motile sex cells (sperm).
   b. Testosterone production begins in puberty in response to FSH and LH and is produced by the interstitial cells of the testes. Testosterone causes the appearance of the secondary sex characteristics in males and is necessary for sperm maturation.

2. The duct system includes the epididymis, ductus deferens, and the urethra. Sperm maturation occurs in the epididymis. When ejaculation occurs, sperm are propelled through the duct passageways to the body exterior.

3. The accessory glands of the male include the seminal vesicles, prostate gland, and Cowper's glands. These glands produce an alkaline fluid (seminal fluid), which activates and nourishes the sperm.

4. The external genitalia include:
   a. The scrotum is a skin-covered sac that hangs outside the abdominopelvic cavity and provides the proper temperature for producing viable sperm.
   b. The penis consists of three columns of erectile tissue surrounding the urethra. The erectile tissue provides a way for the penis to become rigid so that it may more adequately serve as a penetrating device during sexual intercourse.

B. FEMALE REPRODUCTIVE SYSTEM

1. The ovaries, female gonads, are located against the lateral walls of the pelvis. They produce female sex cells (exocrine function) and hormones (endocrine function).
   a. Oogenesis (production of female sex cells) occurs in the ovarian follicles, which are activated at puberty by FSH and LH to mature and eject oocytes (ovulation) on a cyclic basis. The female egg, or ovum, is formed only if sperm penetration of the secondary oocyte occurs. In females, meiosis produces only one functional ovum (plus three nonfunctional polar bodies) as opposed to the four functional sperm/ meiosis produced by males.
   b. Hormone production: Estrogen is produced by the ovarian follicles in response to FSH. Progesterone is produced by the corpus luteum in response to LH. Estrogen stimulates the development of the secondary sex characteristics of the female.

2. The duct system consists of:
   a. The fallopian tubes extend from the vicinity of an ovary to the uterus. Ends are fringed and "wave" to direct ovulated oocytes into the fallopian tubes, which conduct the oocyte (embryo) to the uterus by peristalsis and ciliary action.
   b. The uterus is a pear-shaped organ in which the embryo implants and develops. Its mucosa (endometrium) sloughs off each month in menses unless an embryo has become embedded in it. The myometrium contracts rhythmically during the birth of a baby.
   c. The vagina is a passageway between the uterus and the body exterior, which allows a baby or the menstrual flow to pass out of the body. It also receives the penis and semen during sexual intercourse.

3. The external genitalia of the female include the labia majora and minora (skin folds), the clitoris, and the urethral and vaginal openings.

4. The mammary glands are milk-producing glands found in the breasts. They produce milk after the birth of a baby in response to hormonal stimulation.

5. The menstrual cycle concerns changes that occur in the endometrium in response to changes in blood levels of ovarian hormones. The phases are:

a. Menses. The endometrium sloughs off and bleeding occurs. Ovarian hormones are at their lowest levels.

b. Proliferative phase. The endometrium is repaired, thickens, and becomes well vascularized in response to increasing levels of estrogen.

c. Secretory phase. Endometrial glands begin to secrete nutrient substances, and the lining becomes more vascular in response to increasing levels of progesterone.

If fertilization does not occur, the phases are repeated about every 28 days.

6. If fertilization occurs, embryonic development begins immediately. By day 14 after ovulation, the young embryo has implanted in the endometrium, and the placenta is being formed. The placenta serves the respiratory, nutritive, and excretory needs of the embryo and produces the hormones of pregnancy. All major organ systems have been laid down by 8 weeks, and at 9 weeks the embryo is called a fetus.

A pregnant woman has increased respiratory, circulatory, and urinary demands placed on her system by the developing fetus. Good nutrition is necessary to produce a healthy baby.

C. DEVELOPMENTAL ASPECTS OF THE REPRODUCTIVE SYSTEM

1. The reproductive system structures of males and females are identical during early development. The gonads develop by the eighth week. The presence or absence of testosterone determines whether male or female accessory reproductive organs will be formed.

2. Important congenital defects result from abnormal separation of sex chromosomes during sex cell formation.

3. The reproductive system is inactive during childhood.

4. The most common reproductive problems during young adulthood are infections of the reproductive tract.

5. Neoplasms of the reproductive tract organs are a major threat to females. Breast cancer is responsible for the greatest number of deaths in adult women; cervical cancer is the second most common female cancer. Prostatic cancer is the most common reproductive system cancer seen in males.

6. After menopause, a female's reproductive capabilities end, and her reproductive organs begin to atrophy. Reproductive capacity does not appear to decline significantly in aging males. The most common problem of elderly males is prostatic hypertrophy.

REVIEW QUESTIONS

1. What are the primary sex organs, or gonads, of males? What are their two major functions?

2. Name the organs forming the male duct system from the male gonad to the body exterior.

3. What is the function of seminal fluid? Name the three types of glands that produce it.

4. The penis contains erectile tissue that becomes engorged with blood during sexual excitement. What term is used to describe this event?

5. Define ejaculation.

6. Why are the male gonads not found in the abdominal cavity? Where are they found?

7. How does enlargement of the prostate gland interfere with a male's reproductive function?

8. What structures in the testes form the sex cells? When does spermatogenesis begin? What causes it to begin?

9. The process of spermatogensis actually forms cells called spermatids. How are the spermatids converted to functional sperm?

10. Testosterone causes the male secondary sex characteristics to appear at puberty. Name three examples of male secondary sex characteristics.

11. Name the female gonad and list its two major functions.

12. Name the structures of the female duct system and describe the important functions of each.

13. Since the fallopian tubes are not continuous with the ovaries, how can you explain the fact that all ovulated "eggs" do not end up in the female's peritoneal cavity?

14. What anterior pituitary hormones cause follicle development and ovulation to occur in the ovary? What is a follicle? What is ovulation?

15. The female cell that is ovulated is not a mature sex cell (ovum). When or under what conditions does it become mature?

16. What hormone can be called the "feminizing" hormone? What ovarian structures produce this hormone? Name the second hormone produced by the ovary.

17. List and describe the events of the menstrual cycle. Why is the menstrual cycle important?

18. Define menopause. What does this mean to a female?

19. What is the role of the mammary glands?

20. Define fertilization. Where does fertilization usually occur? Describe the process of implantation.

21. What are the functions of a placenta?

22. How is a pregnant woman's body functioning altered, or changed, by her pregnancy?

23. Compare the effect of aging on the male and female reproductive systems.

Basic Chemistry

Appendix Contents

After completing this appendix, you should be able to:

- Differentiate clearly between matter and energy.

- List the major energy forms and provide one example (from the body) of the use of each energy form.

- Define chemical element and list the four elements that form the bulk of body matter.

- Explain the relationship between elements and atoms.

- List the subatomic particles, and describe their relative masses, charges, and positions in the atom.

- Define radioisotope, and explain briefly how radioisotopes are used in diagnosis and treatment of disease.

- Recognize that chemical reactions involve the interaction of electrons in the making and breaking of chemical bonds.

- Define molecule and explain its relationship to compounds.

- Differentiate between ionic, polar covalent, and nonpolar covalent bonds, and describe the importance of hydrogen bonds.

- Contrast synthesis, decomposition, and exchange reactions.

- Distinguish between organic and inorganic compounds.

- Differentiate clearly between a salt, an acid, and a base.

- List several salts (or their ions) that are vitally important to body function.

- Explain the importance of water to body homeostasis and provide several examples of the various roles of water.

- Explain the concept of pH and state the pH of blood.

- Compare and contrast carbohydrates, lipids, proteins, and nucleic acids in terms of their building blocks, structures, and general functions in the body.

- Differentiate between structural and functional proteins.

- Compare and contrast the structure and general functions of DNA and RNA.

- Define enzyme and explain the role of enzymes.

- Explain the importance of ATP in the body.

Many short courses in anatomy and physiology lack the time to consider chemistry as a topic, which explains the placement of chemistry concepts in an appendix. So why include it at all? Very simply, the food you eat, the medicines you take when you are ill, and virtually every object in your environment are chemical substances. Indeed, your whole body is made up of chemicals—thousands of them—continuously interacting with one another at a pace that makes a chemical-manufacturing plant appear idle.

Although it is possible to study anatomy without much reference to chemistry, chemical reactions underlie all body processes—movement, digestion, the pumping of your heart, and even your thoughts. Hence, a basic understanding of chem-

istry will enhance your ability to understand body processes. This appendix reviews the basic chemistry and biochemistry (chemistry of living material) involved in body functions, and thus can be used both as a brief introduction to these concepts by students who lack a chemistry background and as a mini-refresher for others.

DEFINITION OF CONCEPTS: MATTER AND ENERGY

Matter

Matter is anything that occupies space and has mass (weight). Chemistry studies the nature of matter—how its building blocks are put together and interact.

Matter exists in *solid, liquid,* and *gaseous states,* and examples of each state are found in the body. Solids, like bones and teeth, have a definite shape and volume. Liquids have a definite volume but they conform to the shape of their container. Examples of body liquids are blood plasma and the interstitial fluid that bathes all body cells. Gases have neither a definite shape nor a definite volume. The air we inhale is a gas.

Matter may be changed both physically and chemically. **Physical changes** do not alter the basic nature of a substance; examples include changes in state such as ice melting to become water and cutting our food into smaller pieces. **Chemical changes** do alter the composition of the substance—often substantially. The fermentation of grapes to make wine and the digestion of food in the body are examples of chemical changes.

Energy

In contrast to matter, **energy** is massless, does not take up space, and can only be measured by its effects on matter. Energy is commonly defined as the ability to do work or to put matter into motion. When energy is actually doing work (putting matter into motion), it is referred to as *kinetic* energy. When it is inactive or stored, it is called *potential* energy. All forms of energy exhibit both kinetic and potential energy work capacities.

Actually, energy is a physics topic, but discussions of matter and energy cannot be completely separated. All living things are built of matter, and in order to grow and function, they require a continuous supply of energy. Because this is so, it is worth taking a brief detour to introduce the energy forms used by the body to do its work.

Forms of Energy

Chemical energy is the energy stored in the bonds of chemical substances. When these bonds are broken, the stored, or *potential, energy* is released for use; that is, it becomes active, or *kinetic, energy* that causes an effect on matter. For example, when gasoline molecules are broken apart in your automobile engine, the energy released powers your car. In like manner, all body activities are "run" by the chemical energy of foods we eat.

Electrical energy is energy that results from the movement of charged particles. In your home, electrical energy reflects the flow of electrons along the household wiring. In your body, an electrical current is generated when charged particles called ions move across cell membranes.

Mechanical energy is *directly* involved in moving matter. When you ride a bicycle, your legs provide the mechanical energy that moves the pedals. We can take this example one step further back: As the muscles in your legs shorten, they pull on your bones, causing your limbs to move (so that you can pedal the bike).

Radiant energy is energy that travels in waves; that is, energy of the electromagnetic spectrum, which includes heat (infrared), light, radio, and cosmic waves. Of these, *heat* and *light* have the greatest influence on body functioning. For example, when matter is heated, its particles begin to move more quickly, that is, their kinetic energy (energy of motion) is increased. This is important to the chemical reactions that occur in the body because the higher the temperature, the faster those reactions occur. Light energy, which stimulates the retinas of your eyes, is important in vision.

Energy Form Conversions

With a few exceptions, energy is easily converted from one form to another. For example, an electrical current carried to a lamp socket is converted into light energy by the bulb, and the chemical energy of foods may ultimately be transformed into the electrical energy of a nerve impulse or the mechanical energy of shortening muscles.

Energy conversions are quite inefficient, and some of the initial energy supply is always "lost" to the environment as heat. (It is not really lost, because the total energy of the system is constant—energy cannot be created or destroyed, but that part given off as heat is *unusable.*) You can demonstrate this very easily by touching a light bulb that has been lit for an hour or so. The fact that it is hot reveals that some of the electrical energy reaching the bulb is producing heat instead of light. Likewise, all energy conversions that occur in the body liberate heat; it is this heat that makes us warm-blooded animals and contributes to our relatively high body temperature.

COMPOSITION OF MATTER

Elements and Atoms

A long time ago, it was established that all matter is composed of a limited number of basic substances called **elements**. The term "element" refers to a unique substance that cannot be decomposed or broken down into simpler substances by ordi-

nary chemical methods. Examples of elements include many commonly known substances, such as oxygen, carbon, gold, copper, and iron.

So far, 108 different elements have been discovered. Approximately 90 of these are naturally occurring; the rest are produced artificially (man-made) in accelerator devices. Only four elements—carbon, oxygen, hydrogen, and nitrogen—make up about 96% of the body, but several others are present in small or trace amounts. The most abundant elements found in the body and their roles are listed in Table A-1.

The building block of an element, the smallest particle that still retains its special properties, is called

TABLE A-1 Common Elements Comprising the Human Body

Element	Atomic symbol	Percentage of body mass	Role
Oxygen	O	65.0	A major component of both organic and inorganic molecules; as a gas, is essential to the oxidation of glucose and other food fuels, during which cellular energy (ATP) is captured
Carbon	C	18.5	The primary elemental component of all organic molecules, including carbohydrates, lipids, and proteins
Hydrogen	H	9.5	A component of most organic molecules; in ionic form, influences the acidity of body fluids
Nitrogen	N	3.2	A component of proteins and nucleic acids (genetic material)
Calcium	Ca	1.5	Found as a salt in bones and teeth; in ionic form, is required for muscle contraction, neural transmission, and blood clotting
Phosphorus	P	1.0	Present as a salt, in combination with calcium, in bones and teeth; also present in nucleic acids and many proteins; forms part of the high-energy compound ATP
Potassium	K	0.4	In its ionic form, is the major intracellular cation; necessary for the conduction of nerve impulses and for muscle contraction
Sulfur	S	0.3	A component of proteins (particularly contractile proteins of muscle)

TABLE A-1   Common Elements Comprising the Human Body (continued)

Element	Atomic symbol	Percentage of body mass	Role
Sodium	Na	0.2	As an ion, is the major extracellular cation; important for water balance, conduction of nerve impulses, and muscle contraction
Chlorine	Cl	0.2	In ionic form, is a major extracellular anion
Magnesium	Mg	0.1	Present in bone; also an important coenzyme in a number of metabolic reactions
Iodine	I	<0.1	Needed to make functional thyroid hormones
Iron	Fe	<0.1	A component of the functional hemoglobin molecule (which transports oxygen within red blood cells) and many respiratory enzymes
Chromium	Cr		
Cobalt	Co		
Copper	Cu		
Fluorine	F		
Manganese	Mn		Referred to as the trace elements because required in very minute amounts; many found as part of enzymes or required for enzyme activation
Molybdenum	Mo		
Selenium	Se		
Silicon	Si		
Tin	Sn		
Vanadium	V		
Zinc	Zr		

an **atom**. Because all elements are unique, the atoms of each element differ from those of all other elements. Each element is designated by a one- or two-letter chemical shorthand called an **atomic symbol**. In most cases, the atomic symbol is simply the first (or first two) letters of the element's name. For example, C stands for carbon, O for oxygen, and Ca for calcium. In a few cases, the atomic symbol is taken from the Latin name for the element. For instance, sodium is indicated by Na (from the Latin word *natrium*).

Atomic Structure

The word "atom" comes from the Greek word meaning "incapable of being divided," and until this century, this characteristic of an atom was accepted as a scientific truth. According to this notion, you could theoretically divide a pure element, such as a block of gold, into smaller and smaller particles until you got down to the individual atoms, and then could subdivide no further. We now know that atoms, although indescribably small, are clusters of even smaller (subatomic) particles and that under very special circumstances atoms can be split into these smaller particles. However, this never happens in normal chemical reactions (or in the body); and when it does happen, the subparticles no longer exhibit the properties of the element. Thus, the "old" idea of the atom as the indivisible building block of matter is still very usable.

The atoms representing the 108 elements are composed of different numbers and proportions of three basic subatomic particles, which differ in their mass, electrical charge, and location within the atom (Table A-2). **Protons** (p^+) have a positive charge, whereas **neutrons** (n^0) are uncharged, or neutral. Protons and neutrons are heavy particles and have approximately the same mass (1 atomic mass unit, 1 Amu). The tiny pointlike **electrons** (e^-) bear a negative charge equal in strength to the positive charge of the proton, but their mass is so small that they are *considered to be* essentially massless.

The electrical charge of a particle is a measure of its ability to attract or repel other charged particles. Particles with the same type of charge ($+$ to $+$ or $-$ to $-$) repel each other, but particles with unlike charges ($+$ to $-$) attract one another. Neutral particles are neither attracted nor repelled by charged particles.

Because all atoms are electrically neutral, the number of protons an atom has must be precisely balanced by its number of electrons (the $+$ and $-$ charges will then cancel the effect of each other). Thus, hydrogen has one proton and one electron, and iron has 26 protons and 26 electrons. For any atom, the number of protons and electrons is always equal.

PLANETARY MODEL OF AN ATOM

The *planetary model* of an atom portrays the atom as resembling a miniature solar system (Figure A-1) in which the protons and neutrons are clus-tered at the center of the atom in the *atomic nucleus*. Because the nucleus contains all of the heavy particles, it is fantastically dense and is positively charged. The tiny electrons orbit around the outside of the nucleus, where they form a haze of negative charge often called the *electron cloud*. The electrons occupy nearly the entire volume of the atom and determine its chemical behavior (that is, its ability to bond with other atoms).

As shown in Figure A-2, the simplest and smallest atom, hydrogen, has 1 proton, 1 electron, and no neutrons. Next in size is the helium atom, consisting of 2 protons and 2 neutrons in the nucleus, and 2 orbiting electrons. The next is lithium, with 3 protons, 4 neutrons, and 3 electrons. If this list were continued, all 108 atoms could be described by adding one proton and one electron at each step. The number of neutrons is not as easy to pin down; but light atoms tend to have equal numbers of protons and neutrons, whereas in larger atoms, neutrons outnumber protons.

ATOMIC NUMBER

Each element is given a number, called its **atomic number**, that is equal to the number of protons its atoms contain. Because each element contains a different number of protons, you could say that elements are "identified" by their atomic numbers. The number of protons is always equal to the number of electrons; hence the atomic number also indicates the number of electrons an atom contains.

TABLE A-2	Subatomic Particles		
Particle	*Position in atom*	*Mass (Amu units)*	*Charge*
Proton (p^+)	Nucleus	1	+
Neutron (n^0)	Nucleus	1	0
Electron (e^-)	Orbitals outside the nucleus	1/1800	−

FIGURE A-1

Planetary model of an atom. The central dense nucleus contains the protons and neutrons. The electrons are found outside the nucleus in orbits where they produce a haze of negative charge called the "electron cloud."

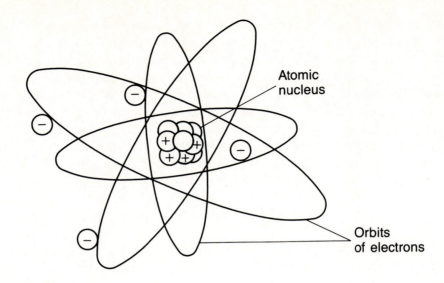

Atomic nucleus

Orbits of electrons

Key: (+) = Proton

⃝ = Neutron

(−) = Electron

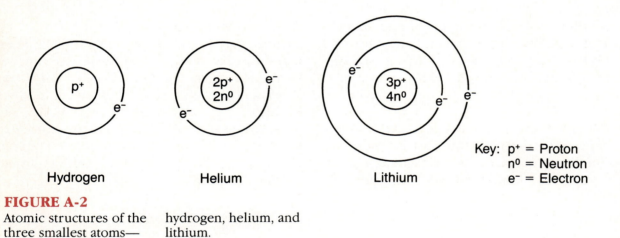

Hydrogen   Helium   Lithium

Key: p⁺ = Proton
n⁰ = Neutron
e⁻ = Electron

FIGURE A-2

Atomic structures of the three smallest atoms—hydrogen, helium, and lithium.

ATOMIC MASS NUMBER

The **atomic mass number** of any atom is the sum of the protons and neutrons contained in its nucleus. (The mass of the electrons is so small that it is ignored.) Hydrogen has one bare proton in its nucleus; thus its atomic number and atomic mass number are the same (1). Helium, with 2 protons and 2 neutrons, has a mass number of 4.

ATOMIC WEIGHT AND ISOTOPES

At first glance, it would seem that the **atomic weight** of atoms should be equal to their atomic mass. This would be so if there were only one type of atom representing each element. However, the atoms of almost all elements exhibit two or more structural variations, and these varieties are called isotopes. **Isotopes** have the same number of protons and electrons, but vary in the number of *neutrons* they contain. Thus, the isotopes of an element have the same atomic number but have different atomic masses. Because all of an element's isotopes have the same number of electrons (and electrons determine bonding properties), their chemical properties are *exactly* the same.

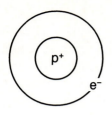 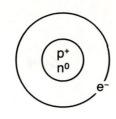 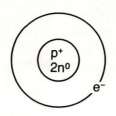

Key: e⁻ = Electron
p⁺ = Proton
n⁰ = Neutron

FIGURE A-3
The isotopes of hydrogen. Notice that all three isotopes have the same number of protons (1) and electrons (1). They vary only in the number of neutrons that they contain.

TABLE A-3   Atomic Structures of the Most Abundant Body Elements

Element	Symbol	Atomic number	Mass number	Atomic weight	Electrons in valence shell
Calcium	Ca	20	40	40.08	2
Carbon	C	6	12	12.011	4
Chlorine	Cl	17	35	35.453	7
Hydrogen	H	1	1	1.008	1
Iodine	I	53	127	126.905	7
Iron	Fe	26	56	55.847	2
Magnesium	Mg	12	24	24.305	2
Nitrogen	N	7	14	14.007	5
Oxygen	O	8	16	15.999	6
Phosphorus	P	15	31	30.974	5
Sodium	Na	11	23	22.99	1
Sulfur	S	16	32	32.064	6

As a general rule, the atomic weight of any element is approximately equal to the mass number of its most abundant isotope. For example, hydrogen has an atomic number of 1, and has isotopes with atomic masses of 1, 2, and 3 (Figure A-3); but its atomic weight is 1.0079, which reveals that its lightest isotope is present in much greater amounts than its heavier forms. The atomic numbers, mass numbers, and atomic weights for elements commonly found in the body are provided in Table A-3.

The heavier isotopes of certain atoms are unstable and tend to decompose to become more stable; such isotopes are called **radioisotopes**. This process of spontaneous atomic decay is called **radioactivity** and can be compared to a tiny explosion. All types of radioactive decay (alpha, beta, and gamma) involve the emission (ejection) of nuclear particles or energy, and are damaging to living cells. Alpha emission tends to be least damaging, and gamma radiation is most harmful.

Radioisotopes are used in minute amounts to tag biological molecules so that they can be followed, or traced, through the body, and are valuable tools for medical diagnosis and treatment. For example, a radioisotope of iodine can be used to check blood circulation through the lungs or to scan the thyroid gland of a patient suspected of having a thyroid tumor. Radium and certain other radioisotopes are used to destroy localized cancers.

MOLECULES AND COMPOUNDS

When atoms of the same, or different, elements chemically combine, **molecules** are formed. When two or more atoms of the *same* element bond together, a molecule of that element is produced. For example, when two hydrogen atoms bond, the product is a molecule of hydrogen gas:

$$H \text{ (atom)} + H \text{ (atom)} \rightarrow H_2 \text{ (molecule)}^*$$

(In the example given, the atoms taking part in the reaction are indicated by their atomic symbols, and the composition of the product is indicated by a *molecular formula* that shows its atomic makeup. The chemical reaction is shown by writing a *chemical equation*. When two or more *different* atoms bind together to form a molecule, the molecule is more specifically referred to as a molecule of a **compound**. For example, four hydrogen atoms and one carbon atom can interact chemically to form methane:

$$4H + C = CH_4 \text{ (methane)}$$

Thus, a molecule of methane is a compound, but a molecule of hydrogen gas is not.

It is important to understand that compounds always

have properties quite different from those of the atoms making them up, and it would be next to impossible to determine the atomic makeup of a compound without analyzing it chemically. Notice that just as an atom is the smallest particle of an element that still retains that element's properties, a molecule is the smallest particle of a compound that still retains the properties of that compound. If you break the bonds between the atoms of the compound, properties of the atoms, rather than those of the compound, will be exhibited.

CHEMICAL BONDS AND CHEMICAL REACTIONS

Chemical reactions occur whenever atoms combine with, or dissociate from, other atoms. When atoms unite chemically, **chemical bonds** are formed, and the properties of the product that results are very different from those of the atoms composing it.

Bond Formation

ROLE OF ELECTRONS

It is important to understand that a chemical bond is not an actual physical structure, like a pair of handcuffs linking two people together, but is instead an energy relationship that involves interactions between the electrons of the reacting atoms. Because this is so, we will devote a few words to the role of electrons in bond formation.

As illustrated in Figure A-4, electrons occupy generally fixed regions of space around the nucleus, called *energy levels* or *electron shells*. The maximum number of energy levels of any atom is 7, and these are numbered from the nucleus outward. The electrons closest to the nucleus are those most strongly attracted to its positive charge, and those farther away are less securely held. As a result, the more distant electrons are likely to interact with other atoms.

Perhaps this situation can be compared to the development of a child. During infancy and the toddler years, the child spends most of its time at home, and is shaped and molded by the ideas and demands of its parents. However, when the child goes to school, it is increasingly influenced by its

---

*Notice that when the number of atoms is written as a subscript, the subscript indicates that the atoms are joined by a chemical bond. Thus, 2H represents two unjoined atoms, but H_2 indicates that the two hydrogen atoms are bonded together to form a molecule.

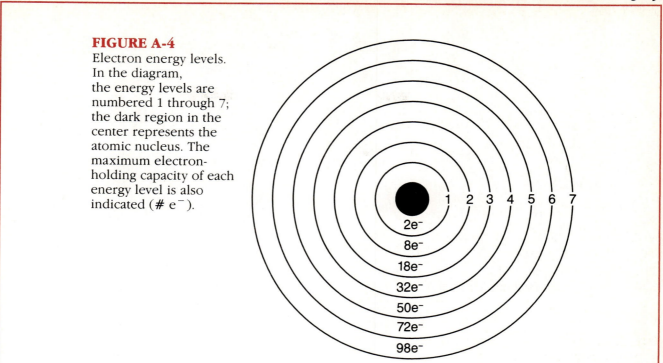

FIGURE A-4
Electron energy levels. In the diagram, the energy levels are numbered 1 through 7; the dark region in the center represents the atomic nucleus. The maximum electron-holding capacity of each energy level is also indicated (# e$^-$).

friends and other adults, such as teachers and coaches. Thus, just as the child is more likely to become involved with "outsiders" as it roams farther from home, electrons are more influenced by outsiders (other atoms) as they get farther and farther away from the positive influence of "home" (the nucleus).

Each electron shell has a maximum number of electrons that it can hold. Shell 1, closest to the nucleus, is small and can accommodate only 2 electrons. Shell 2 holds a maximum of 8. Shell 3 holds 8 electrons in elements up to the atomic number 20 and can accommodate up to 18 electrons in larger atoms. Thus, subsequent shells hold larger and larger numbers of electrons, and in most cases the shells tend to be filled consecutively.

The only electrons important to a consideration of bonding behavior are those in the atom's outermost shell. This shell is called the **valence shell**, and its electrons determine the chemical behavior of the atom. As a general rule, the electrons of inner shells do not take part in bonding.

When the valence shell of an atom contains the maximum number of electrons that it can hold, the atom is completely stable and is chemically inactive (inert). When the valence shell is incom-

pletely filled, an atom will tend to gain, lose, or share electrons with other atoms to reach a stable state. When any of these events occur, chemical bonds are formed.

The key to chemical reactivity is referred to as the "rule of 8s"; that is, atoms interact in such a way that they will have 8 (or a multiple of 8) electrons in their valence shell. The first energy level represents an exception to this rule, as it is "full" when it has 2 electrons. (As you might guess, atoms must approach each other very closely for their electrons to interact—in fact, their outermost energy levels must overlap.)

TYPES OF CHEMICAL BONDS

IONIC BONDS.   **Ionic bonds** form when electrons are completely transferred from one atom to another. Atoms are electrically neutral; but when they gain or lose electrons during bonding, their positive and negative charges are no longer balanced, and charged particles, called **ions**, result. When an atom gains an electron, it acquires a net negative charge because it now has more electrons than protons. Negatively charged ions are more specifically called *anions*. When an atom loses an electron, it becomes a positively charged ion (a *cation*) because it now possesses more protons

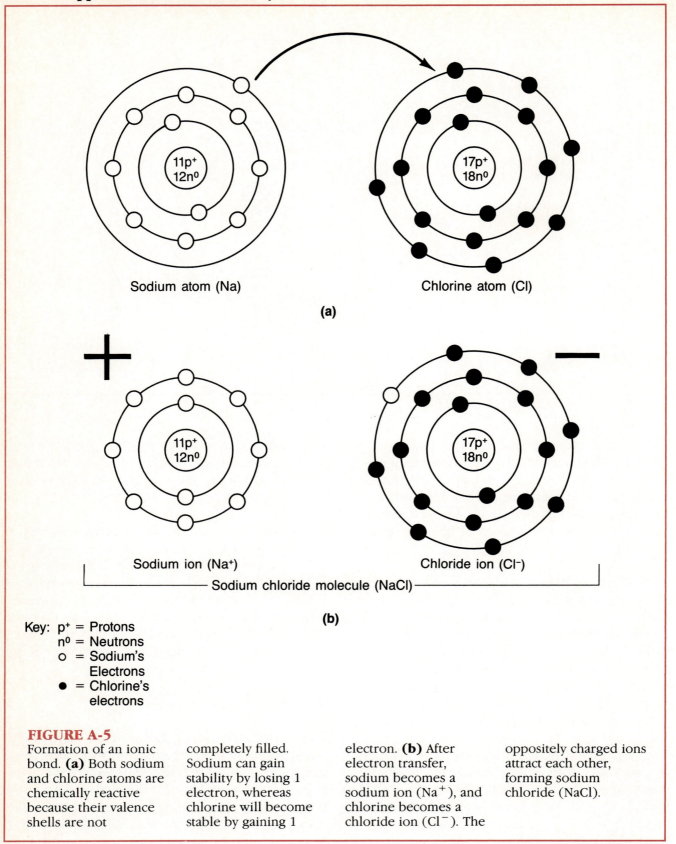

Sodium atom (Na)

Chlorine atom (Cl)

(a)

Sodium ion (Na⁺)

Chloride ion (Cl⁻)

Sodium chloride molecule (NaCl)

(b)

Key: p^+ = Protons
n^0 = Neutrons
o = Sodium's Electrons
● = Chlorine's electrons

FIGURE A-5
Formation of an ionic bond. **(a)** Both sodium and chlorine atoms are chemically reactive because their valence shells are not completely filled. Sodium can gain stability by losing 1 electron, whereas chlorine will become stable by gaining 1 electron. **(b)** After electron transfer, sodium becomes a sodium ion (Na^+), and chlorine becomes a chloride ion (Cl^-). The oppositely charged ions attract each other, forming sodium chloride (NaCl).

than electrons. Both anions and cations result when an ionic bond is formed. The two oppositely charged ions attract each other and tend to stay close together.

The formation of sodium chloride (NaCl), common table salt, provides a good example of ionic bonding. As illustrated in Figure A-5, sodium's valence shell contains only 1 electron and so is

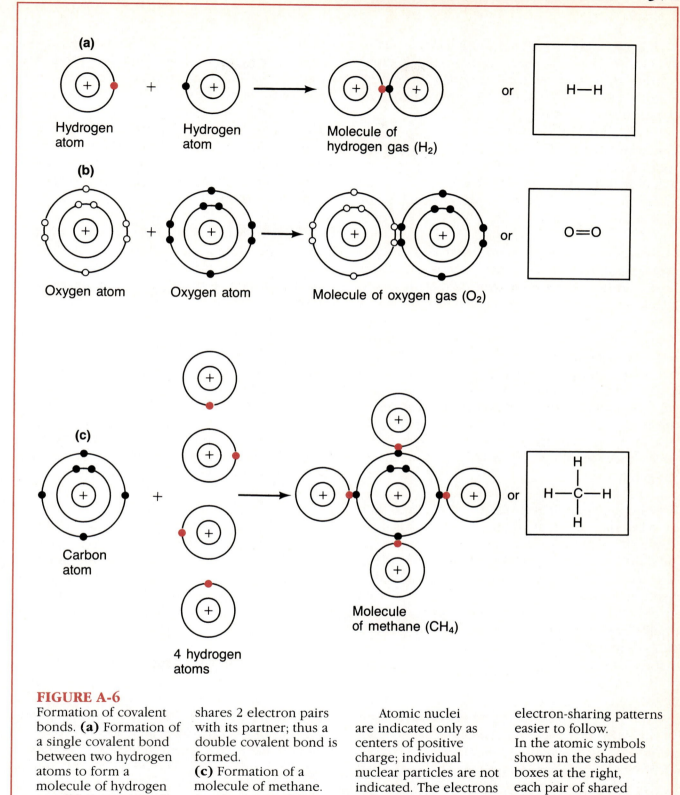

FIGURE A-6

Formation of covalent bonds. **(a)** Formation of a single covalent bond between two hydrogen atoms to form a molecule of hydrogen gas. **(b)** Formation of a molecule of oxygen gas. Each oxygen atom shares 2 electron pairs with its partner; thus a double covalent bond is formed. **(c)** Formation of a molecule of methane. A carbon atom shares 4 electron pairs with 4 hydrogen atoms.

Atomic nuclei are indicated only as centers of positive charge; individual nuclear particles are not indicated. The electrons of interacting atoms are shown in different colors to make the electron-sharing patterns easier to follow. In the atomic symbols shown in the shaded boxes at the right, each pair of shared electrons is indicated by a single line between sharing atoms.

incomplete. However, if this single electron is "lost" to another atom, the valence shell becomes shell 2, which contains 8 electrons; sodium becomes a cation (Na^+) and achieves stability. Chlorine only needs 1 electron to fill its valence shell, and it is much easier to gain 1 electron (forming Cl^-) than it is to try to give away 7. Thus, the ideal situation is for sodium to donate its valence-shell electron to chlorine, and this is exactly what happens in the interaction between these two atoms. Sodium chloride and most other compounds formed by ionic bonding fall into the general category of chemicals called **salts**.

COVALENT BONDS.   Electrons do not have to be completely lost or gained for atoms to become stable. Instead, they can be shared in such a way that each atom is able to fill its valence shell for at least part of the time.

Molecules in which atoms share electrons are called *covalent molecules*, and their bonds are **covalent bonds**. For example, hydrogen, with its single electron, can become stable if it fills its valence shell (level 1) by sharing a pair of electrons—its own and one from another atom. As shown in Figure A-6*a*, a hydrogen atom can share an electron pair with another hydrogen atom to form a molecule of hydrogen gas. The shared electron pair orbits the whole molecule and satisfies the stability needs of both hydrogen atoms. Likewise, two oxygen atoms can share 2 pairs of electrons with each other (Figure A-6*b*) to form a molecule of oxygen gas (O_2).

A hydrogen atom may also share its electron with an atom of a different element. Carbon has 4 valence shell electrons but needs 8 to achieve stability. As shown in Figure A-6*c*, when methane (CH_4) is formed, carbon shares 4 electron pairs with 4 hydrogen atoms (one pair with each hydrogen atom). Because the shared electrons orbit and "belong to" the whole molecule, each atom has a full valence shell enough of the time to satisfy its stability needs.

In the covalent molecules described thus far, electrons have been shared *equally* between the atoms of the molecule. Such molecules are called *nonpolar covalently bonded molecules*. However, electrons are not shared equally in all cases. When covalent bonds are made, the molecule formed always has a definite three-dimensional shape. A molecule's shape plays a major role in determining just what other molecules (or atoms) it can interact with; the shape may also result in unequal electron-pair sharing. The following two examples illustrate this principle (Figure A-7).

Carbon dioxide is formed when a carbon atom shares its 4 valence-shell electrons with 2 oxygen atoms. Oxygen is a very electron-hungry atom and attracts the shared electrons much more strongly than does carbon. However, because the carbon dioxide molecule is linear and perfectly balanced ($O=C=O$), the electron-pulling power of one oxygen atom is offset by that of the other. As a result, the electron pairs are shared equally and orbit the entire molecule, and carbon dioxide is a nonpolar molecule.

In contrast, a water molecule is formed when 2 hydrogen atoms bind covalently to a single oxygen atom. Each hydrogen atom shares an electron pair with the oxygen atom, and again the oxygen has the stronger electron-attracting ability. But in this case, the molecule formed is V-shaped ($^H\diagdown_O\diagup^H$). The 2 hydrogen atoms are located at one end of the molecule, and the oxygen atom is at the other. Consequently, the electron pairs are *not* shared equally and spend more time in the vicinity of the oxygen atom, causing that end of the molecule to become slightly more negative (indicated by S^-) and the hydrogen end to become slightly more positive (indicated by S^+). In other words, a *polar molecule,* a molecule with two charged poles, is formed.

Polar molecules orient themselves toward other polar molecules or charged particles (ions, proteins, and others), and they play an important role in chemical reactions that go on in body cells. Because body tissues are 60% to 80% water, the fact that water is a polar molecule is particularly significant, as described shortly.

HYDROGEN BONDS. Hydrogen bonds are extremely weak bonds formed when a hydrogen atom bound to one electron-hungry nitrogen or oxygen atom is attracted by another electron-hungry atom, and the hydrogen atom forms a "bridge" between them. Hydrogen bonding is common between water molecules (Figure A-8), and is reflected in water's ability to "ball up," or form spheres, when it sits on a surface and in the fact that some insects, such as water striders, can "walk on water," as long as they tread lightly.

Hydrogen bonds are also important *intramolecular bonds*; that is, they help to bind different parts of the *same* molecule together into a special three-dimensional shape. These rather fragile bonds

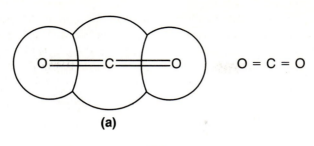

(a)

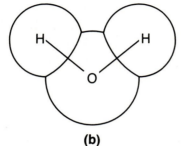

(b)

FIGURE A-7

Nonpolar and polar covalent molecules. **(a)** When carbon dioxide (CO_2) gas is formed by the covalent bonding of carbon to 2 oxygen atoms, a linear molecule is formed. Although oxygen has a greater electron-attracting power than carbon, the electrons are shared equally because the oxygen atoms (as arranged) counterbalance the effects of each other. Thus, carbon dioxide is a nonpolar covalently bonded molecule. **(b)** Water, formed by the combination of 2 hydrogen atoms and 1 oxygen atom, is a polar covalent molecule. The oxygen atom, located at one end of the molecule, strongly attracts the shared electrons, and they spend more time orbiting that end of the molecule. Thus, the oxygen end of the molecule becomes slightly negative (S^-); and the opposite end of the molecule, where the hydrogen atoms are located, becomes slightly positive (S^+). In polar molecules, like water, the charge centers do not coincide.

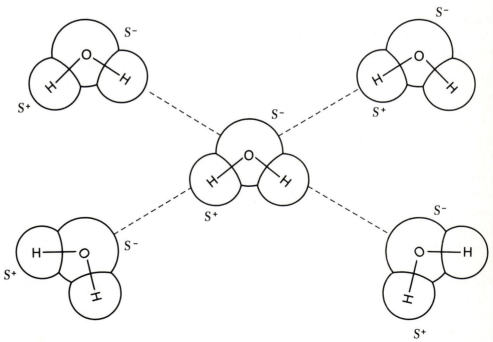

FIGURE A-8

Hydrogen bonding between polar water molecules. The slightly positive ends (indicated by S^+) of some water molecules become aligned with the slightly negative ends (indicated by S^-) of other water molecules. Hydrogen bonding is indicated by the dotted colored lines.

are very important in helping maintain the structure of protein molecules, which are essential functional molecules and body building materials.

Patterns of Chemical Reactions

Chemical reactions involve the making or breaking of bonds between atoms. The total number of atoms remains the same, but the atoms appear in new combinations. Most chemical reactions have one of the three recognizable patterns described below.

SYNTHESIS REACTIONS

A **synthesis reaction** occurs when two or more atoms or molecules combine to form a larger, more complex molecule, which can be simply represented as

$$A + B \rightarrow AB$$

Synthesis reactions always involve bond formation; and because energy must be absorbed to make bonds, synthesis reactions are energy-absorbing reactions.

Synthesis reactions underlie all anabolic (constructive) activities that go on in body cells. They are particularly important for growth, and for repair of worn out or damaged tissues. As shown in Figure A-9a, the formation of proteins by the joining of amino acids into long chains is a synthesis reaction.

DECOMPOSITION REACTIONS

Decomposition reactions occur when a molecule is broken down into smaller molecules, atoms, or ions, and can be indicated by

$$AB \rightarrow A + B$$

Essentially, decomposition reactions are synthesis reactions "in reverse." In these reactions, bonds are always broken, and the products of these reactions are smaller and simpler than the original molecules. As bonds are broken, chemical energy is released.

Decomposition reactions underlie all catabolic (destructive) processes that occur in body cells, that is, they are "molecule-destroying" reactions. Examples of decomposition reactions that occur in the body include digestion of foods to their

building blocks and the breakdown of glycogen (a large carbohydrate molecule stored in the liver) to release glucose (Figure A-9b) when blood sugar levels start to decline.

EXCHANGE REACTIONS

Exchange reactions involve both synthesis and decomposition reactions: Bonds are both made and broken. During exchange reactions, a switch is made between molecule parts (changing partners, so to speak), and different molecules are made. Thus, an exchange reaction can be generally indicated as

$$AB + CD \rightarrow AD + CB$$

As shown in Figure A-9c, an exchange reaction occurs when acids and bases interact to form a salt and water.

BIOCHEMISTRY: THE CHEMICAL COMPOSITION OF LIVING MATTER

All chemicals found in the body fall into one of two major classes of molecules, that is, they are either organic or inorganic compounds. The class of the compound is determined solely by the presence or absence of carbon. **Organic compounds** are carbon-containing compounds. The important organic compounds in the body are *carbohydrates, lipids, proteins,* and *nucleic acids*. All organic compounds are fairly (or very) large covalently bonded molecules. **Inorganic compounds** lack carbon and tend to be simpler, smaller molecules. Examples of inorganic compounds found in the body are *water, salts,* and many (but not all) *acids and bases*.

Organic and inorganic compounds are equally essential for life. Trying to "put a price tag" on which is more valuable can be compared to trying to decide whether the ignition system or the engine is more essential to the operation of a car.

Inorganic Compounds

WATER

Water is the most abundant inorganic compound in the body, and it accounts for about two-thirds

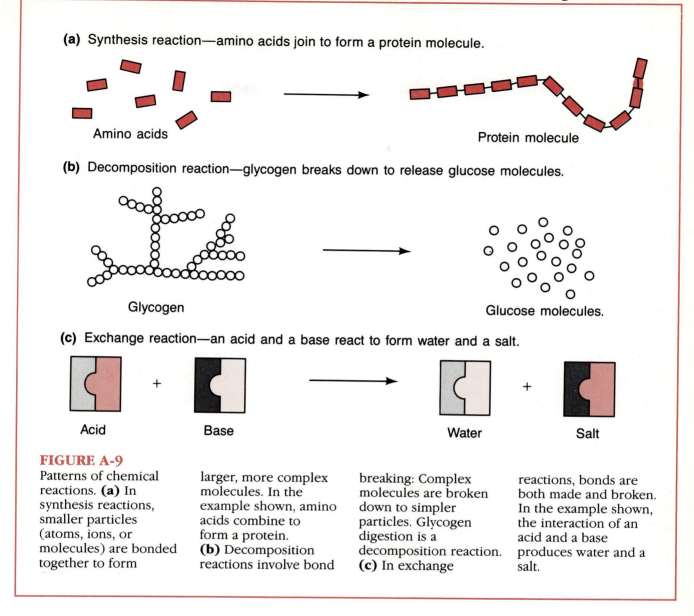

(a) Synthesis reaction—amino acids join to form a protein molecule.

Amino acids   Protein molecule

(b) Decomposition reaction—glycogen breaks down to release glucose molecules.

Glycogen   Glucose molecules.

(c) Exchange reaction—an acid and a base react to form water and a salt.

Acid   Base   Water   Salt

FIGURE A-9

Patterns of chemical reactions. **(a)** In synthesis reactions, smaller particles (atoms, ions, or molecules) are bonded together to form larger, more complex molecules. In the example shown, amino acids combine to form a protein. **(b)** Decomposition reactions involve bond breaking: Complex molecules are broken down to simpler particles. Glycogen digestion is a decomposition reaction. **(c)** In exchange reactions, bonds are both made and broken. In the example shown, the interaction of an acid and a base produces water and a salt.

of body weight. Among the properties which make water so vital are the following:

1. Water absorbs and releases large amounts of heat before changing appreciably in temperature itself. Thus, it prevents the sudden changes in body temperature that might otherwise result from intense sun exposure, chilling winter winds, or internal events, such as vigorous muscle activity, that liberate heat.

2. Because of its polarity, water is an excellent solvent; indeed, it is often called the "universal solvent." A *solvent* is a liquid or gas in which smaller amounts of other substances, called *solutes* (which may be gases, liquids, or solids), are dissolved or suspended. The combination of solvent and solutes is called a *solution* when the solute particles are exceedingly minute and is called a *suspension* when the solute particles are fairly large.

Small reactive chemicals such as salts, acids, and bases dissolve easily in water, and become evenly distributed. Because molecules cannot react chemically unless they are in solution, virtually all chemical reactions that occur in the body depend upon water's solvent properties.

The fact that nutrients, respiratory gases (oxygen and carbon dioxide), and wastes can dissolve in water allows water to act as a transport and exchange medium in the body. For example, all of these substances are carried from one part of the body to another in blood plasma, and are exchanged between the blood and tissue cells by passing through interstitial fluid.

3. Water is an important reactant in some types of chemical reactions. For example, in order to digest foods or break down biological molecules, water molecules are added to the bonds of the larger molecules. Such reactions are called *hydration or hydrolysis reactions*, terms that specifically recognize this role of water.

4. Water serves as the base for all body lubricants. Mucus eases feces along the large intestine; saliva moistens food and prepares it for digestion. Serous fluids reduce friction between internal organs, and synovial fluids "oil" the ends of bones as they move within joint cavities.

5. Water also serves a protective function. In the form of cerebrospinal fluid, water forms a "cushion" around the brain that helps to protect it from physical trauma. Amniotic fluid, which surrounds a developing fetus within the mother's body, plays a similar role in protecting the fetus.

SALTS

The **salts** of many metal elements are commonly found in the body, but the most plentiful salts are those containing calcium and phosphorous, which are found chiefly in bones and teeth. When dissolved in body fluids, salts, which are ionic compounds, easily separate into their ions. This process, called dissociation, occurs rather easily because the ions are already formed. All that remains is to pull the ions apart. This is accomplished by water molecules, which orient themselves toward the ions (their slightly negative ends toward the cations and their slightly positive ends toward the anions) and overcome the attraction between them.

Salts, both in their ionic forms and in combination with other elements, are vital to body functioning. For example, ionic calcium (Ca^{++}) is essential for nerve impulses, and iron forms part of the hemoglobin molecule that transports oxygen within red blood cells.

Because ions are charged particles, all salts are **electrolytes**—substances that conduct an electrical current in solution. When ionic (or electrolyte) balance is severely disturbed, virtually nothing in the body "works." The functions of the elements found in body salts are summarized in Table A-1.

ACIDS AND BASES

Like salts, acids and bases are electrolytes, that is, they ionize and then dissociate in water and conduct an electrical current.

CHARACTERISTICS OF ACIDS.   **Acids** have a sour taste, and can dissolve many metals or "burn" a hole in your rug; but the most useful definition of an acid is that it is a substance that can release hydrogen ions (H^+) in detectable amounts. Because a hydrogen ion is essentially a hydrogen nucleus (a "naked proton"), acids are also defined as *proton donors*.

When acids are dissolved in water, they release hydrogen ions and some anion. The anion is unimportant; it is the release of the protons that determines an acid's effects on the environment. The ionization of hydrochloric acid (an acid produced by stomach cells that aids digestion) is shown in the following equation:

$$HCl \longrightarrow H^+ + Cl^-$$
(hydrochloric acid) (proton) (anion)

Other acids found or produced in the body include acetic acid (commonly called vinegar), sulfuric acid, and carbonic acid.

Acids that ionize completely and liberate all of their protons, as do hydrochloric and sulfuric acid, are called *strong acids*; acids that ionize incompletely (as do acetic and carbonic acid) are called *weak acids*. For example, when carbonic acid dissolves in water, only some of its molecules ionize to liberate H^+:

$$H_2CO_3 \longrightarrow H^+ + HCO_3^- + H_2CO_3$$
(carbonic acid)   (anion)

CHARACTERISTICS OF BASES.   **Bases** have a bitter taste, feel slippery, and are *proton (H^+) acceptors*. The hydroxides are common inorganic bases. Like acids, the hydroxides ionize and dissociate in water; but in this case, the *hydroxyl ion (OH$^-$)* and some cation are released. The ionization of sodium hydroxide (NaOH), commonly known as lye, is shown as:

$$NaOH \longrightarrow Na^+ + OH^-$$
(sodium hydroxide) (cation) (hydroxyl ion)

The hydroxyl ion is an avid proton seeker, and any base containing this ion is considered a strong base.

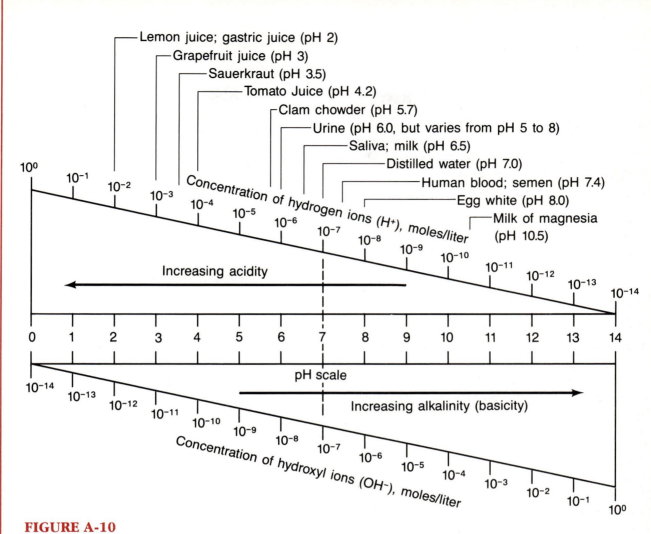

FIGURE A-10

The pH scale and pH values of representative substances. The pH scale is based on the number of hydrogen ions in solution. At a pH of 7, the concentration of hydrogen and hydroxyl ions is equal, and the solution is neutral. As the pH number decreases, the solution becomes increasingly acidic. As the pH number increases, the solution becomes increasingly basic. The pH number is shown in the center of the scale; the actual concentration of hydrogen ions, expressed in moles per liter, is shown on the line above the scale. The corresponding changes in hydroxyl concentration are shown along the line below the scale. The pH values for various body fluids and for substances commonly ingested are provided at the top of the figure.

When acids and bases are mixed, they react with each other (in an exchange reaction) to form water and a salt:

$$HCl + NaOH \rightarrow H_2O + NaCl$$
(acid)   (base)   (water)   (salt)

This type of reaction is more specifically called a *neutralization reaction*.

pH: Acid–Base Concentrations. The relative concentration of hydrogen (and hydroxyl) ions in various body fluids is measured in concentration units called **pH units**. The idea for a pH scale was devised in 1909 by a Danish biologist named Sorenson and is based on the number of protons in solution expressed in terms of moles per liter. (The *mole* is a concentration unit; its precise definition need not concern us here.) The pH scale runs from 0 to 14 (Figure A-10), and each successive change of 1 pH unit represents a 10-fold change in hydrogen-ion concentration.

At a pH of 7, the scale midpoint, the number of hydrogen ions is exactly the same as the number of hydroxyl ions, and the solution is neutral, that is, neither acidic nor basic. Solutions with a pH lower than 7 are acidic: The hydrogen ions outnumber the hydroxyl ions. A solution with a pH of 6 has ten times as many hydrogen ions as a solution with a pH of 7, and a pH of 3 indicates a 10,000-fold ($10 \times 10 \times 10 \times 10$) increase in hydrogen-ion concentration. Solutions with a pH number higher than 7 are alkaline, or basic, and solutions with a pH of 8 and 12 (respectively) have 1/10 and 1/100,000 the number of hydrogen ions present in a solution of pH 7.

Living cells are extraordinarily sensitive to even slight changes in pH; and homeostasis of acid–base balance is carefully regulated by the kidneys, lungs, and a number of chemicals called *buffers,* which are present in body fluids. Because blood comes into close contact with nearly every body cell, regulation of blood pH is especially critical.

Normally, blood pH varies in a narrow range, from 7.35 to 7.45. When blood pH changes more than a few tenths of a pH unit from these limits, death becomes a distinct possibility. Although there are hundreds of examples that could be given to illustrate this point, we will provide just one very important one; that is, when blood pH begins to dip into the acid range, the amount of life-sustaining oxygen that the blood can carry to body cells begins to decline rapidly to "dangerously" low levels. The approximate pH of several body fluids and of a number of commonly ingested substances also appears in Figure A-10.

Organic Compounds

CARBOHYDRATES

Carbohydrates, which include sugars and starches, contain carbon, hydrogen, and oxygen; and (with slight variations) the hydrogen and oxygen atoms appear in the same ratio as in water, that is, 2 hydrogen atoms to 1 oxygen atom. This is reflected in the name carbohydrates, which means "hydrated carbons," and in the molecular formulas of sugars. For example, glucose is $C_6H_{12}O_6$ and ribose is $C_5H_{10}O_5$.

Carbohydrates are classified according to size as monosaccharides, disaccharides, or polysaccharides. Because monosaccharides are joined together to form the molecules of the other two groups, they are the structural units, or building blocks, of carbohydrates.

MONOSACCHARIDES. **Monosaccharide** means one (mono) sugar (saccharide), and thus the monosaccharides are also referred to as *simple sugars.* They are single-chain or ring structures, containing from 3 to 7 carbon atoms (Figure A-11).

The most important monosaccharides in the body are glucose, fructose, galactose, ribose, and deoxyribose. *Glucose,* also called blood sugar, is the universal cellular fuel; *fructose* and *galactose* are converted to glucose for use by body cells. *Ribose* and *deoxyribose* form part of the structure of nucleic acids, another group of organic molecules.

DISACCHARIDES. **Disaccharides**, or *double sugars*, are formed when two simple sugars are joined by a combination reaction known as *dehydration synthesis.* In this reaction, a water molecule is lost as the bond is formed (Figure A-12).

Important disaccharides in the diet are *sucrose* (glucose-fructose), which is cane sugar; *lactose* (glucose-galactose), found in milk; and *maltose* (glucose-glucose), or malt sugar. Because the double sugars are too large to pass through cell membranes, they must be broken down (digested) to their monosaccharide units to be absorbed from the digestive tract into the blood. This is accomplished by *hydrolysis*; as a water molecule is added to each bond, the bond is broken, and the simple sugar units are released (Figure A-12).

POLYSACCHARIDES. **Polysaccharides**—literally, many sugars—are long, branching chains of linked simple sugars (Figure A-11). Because they are large molecules, they are ideal storage products. Another consequence of their large size is that polysaccharides lack the sweetness of the simple and double sugars.

Only two polysaccharides, starch and glycogen, are of major importance to the body. *Starch* is the storage polysaccharide formed by plants. We ingest it in the form of "starchy" foods, such as grain products and root vegetables (potatoes and carrots). *Glycogen* is a slightly smaller, but similar, polysaccharide found in animal tissues (largely in the muscles and the liver). Like starch, it is formed of linked glucose units.

Carbohydrates provide a ready, easily used source of food energy for cells, and glucose is at the top

(a) Simple sugar (monosaccharide)

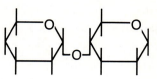
(b) Double sugar (disaccharide)

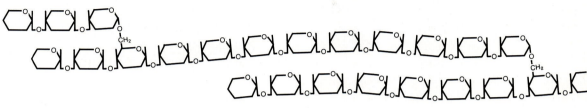
(c) Starch (polysaccharide)

FIGURE A-11

Carbohydrates. **(a)** The generalized structure of a monosaccharide; **(b)** and **(c)** the basic structures of a disaccharide and polysaccharide, respectively.

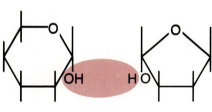

Glucose   Fructose

Dehydration synthesis

Hydrolysis

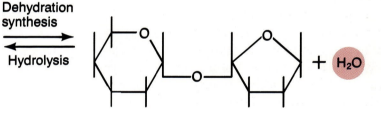

Sucrose   Water

FIGURE A-12

Dehydration synthesis and hydrolysis of a molecule of sucrose. In the reaction going to the right (the dehydration synthesis reaction), glucose and fructose are joined through a process that involves the removal of a water molecule at the bond site. The resulting disaccharide is sucrose. Sucrose is broken down to its simple sugar units when the reaction is reversed (goes to the left). In this hydrolysis reaction, a water molecule must be added to the bond to release the monosaccharides.

of the "cellular menu." When glucose is oxidized (combined with oxygen), in a complex set of chemical reactions, it is broken down to carbon dioxide and water. The energy released as the glucose bonds are broken is trapped in the bonds of high-energy molecules, called ATP molecules, which directly serve as the energy "currency" of all body cells. If not immediately needed for ATP synthesis, dietary carbohydrates are converted to glycogen or fat, and are stored. (The structure of ATP is described later in this appendix.)

Carbohydrates are also used, though to a very small extent, for structural purposes and represent 1% to 2% of cell mass. Some sugars are found in our genes, and others are attached to outer surfaces of cell membranes, where they act as "road signs" to guide cellular interactions.

LIPIDS

Lipids are a large and diverse group of organic compounds (Table A-4) that enter the body in the form of fat-marbled meats, egg yolk, milk products, and oils. The most abundant lipids in the body are neutral fats, phospholipids, and steroids. Like carbohydrates, all lipids contain carbon, hydrogen, and oxygen atoms; but in lipids, carbon and hydrogen atoms far outnumber oxygen atoms, as illustrated by the formula for a typical fat called tristearin: $C_{57}H_{110}O_6$. Most lipids are insoluble in water, but readily dissolve in other lipids and in organic solvents such as alcohol, ether, and acetone.

NEUTRAL FATS. The **neutral fats**, or **triglycerides**, are composed of two types of building blocks, **fatty acids** and **glycerol**; and their synthesis involves the attachment of 3 fatty acids to a single glycerol molecule. The result is an E-shaped molecule that resembles the tines of a fork (Figure A-13a). Although the glycerol backbone is the same in all neutral fats, the fatty-acid chains vary; this results in different kinds of neutral fats.

Neutral fats represent the body's most abundant and concentrated source of usable energy, and when they are oxidized, they yield large amounts of energy. They are stored chiefly in fat deposits beneath the skin and around body organs, where they help insulate the body and protect deeper body tissues from heat loss and bumps.

PHOSPHOLIPIDS. **Phospholipids** are very similar to the neutral fats. They differ in that a phosphorous-containing group is always part of the molecule and takes the place of one of the fatty-acid chains; thus, phospholipids have 2 instead of 3 attached fatty acids (Figure A-13b).

Because the phosphorous-containing portion bears an electrical charge, it gives phospholipids special chemical properties. For example, the charged region attracts and interacts with water and ions, but the fatty-acid chains do not. The presence of phospholipids in cellular boundaries (membranes) allows cells to be selective about what may enter or leave.

STEROIDS. **Steroids** are basically flat molecules formed of 4 interlocking rings (Figure A-13c); thus their structure differs quite a bit from that of fats.

However, like the fats, steroids are made largely of hydrogen and carbon atoms, and are fat soluble.

The single most important steroid molecule is *cholesterol*, which enters the body in animal products such as meat, eggs, and cheese. A certain amount is also made by the liver, regardless of dietary intake. Cholesterol is found in all cell membranes; is particularly abundant in the brain; and is the raw material used to form vitamin D, some hormones (sex hormones and cortisol), and bile salts.

PROTEINS

Proteins account for over 50% of the organic matter in the body, and they have the most varied functions of the organic molecules. Some are construction materials; others play vital roles in cell function. Like carbohydrates and lipids, all proteins contain carbon, oxygen, and hydrogen. In addition, they contain nitrogen and sometimes sulfur atoms as well.

The building blocks of proteins are small molecules called **amino acids**, and approximately 20 common varieties of amino acids are found in proteins. All amino acids have an *amine group,* which gives them basic properties, and an *acid group,* which allows them to act as acids. In fact, all amino acids are identical except for a single group of atoms called their *R-group* (Figure A-14). Hence, it is differences in the R-groups that make each amino acid chemically unique.

Amino acids are joined together in chains to form large complex protein molecules that contain from 50 to thousands of amino acids. (Amino acid chains containing less than 50 amino acids are called *polypeptides.*) Because each type of amino acid has distinct properties, the sequence in which they are bound together produces proteins that vary widely both in structure and function. Perhaps this can be made more understandable if the 20 amino acids are presumed to be a "20-letter alphabet." The letters (amino acids) are then used in specific combinations to form words (a protein). Just as a change in one letter of any word can produce a word that has an entirely different meaning (flour → floor) or is nonsensical (flour → fllur), changes in kinds of amino acids (letters) or in their positions in the protein allow literally thousands of different protein molecules to be made.

TABLE A-4   Representative Lipids Found in the Body

Lipid type	Location/function
NEUTRAL FATS (TRIGLYCERIDES)	Found in fat deposits (subcutaneous tissue and around organs); protect and insulate the body organs; the major source of stored energy in the body
PHOSPHOLIPIDS (LECITHIN AND OTHERS)	Found in all cell membranes; participate in the transport of lipids in plasma; abundant in the brain and the nervous tissue in general, where they help to form insulating white matter
STEROIDS	
Cholesterol	The basis of all body steroids
Bile salts	A breakdown product of cholesterol; released by the liver into the digestive tract, where they aid in fat digestion and absorption
Vitamin D	Produced in the skin, on exposure to UV (ultraviolet) radiation, from a modified cholesterol molecule; necessary for normal bone growth and function
Sex hormones	Estrogen and progesterone (female hormones) and testosterone (male sex hormone) produced from cholesterol; necessary for normal reproductive function; deficits result in sterility
Adrenal cortical hormones	Cortisol, a glucocorticoid—is a long-term antistress hormone that is necessary for life. Aldosterone—helps regulate salt and water balance in body fluids by targeting the kidneys
OTHER LIPOID SUBSTANCES	
Fat-soluble vitamins:	
A	Found in orange-pigmented vegetables (carrots) and fruits (tomatoes); part of the photoreceptor pigment involved in vision
E	Taken in via plant products such as wheat germ and green leafy vegetables; may promote wound healing and contribute to fertility, but not proven in humans; an antioxidant; may help to neutralize free radicals believed to be involved in triggering some types of cancers
K	Made available largely by the action of bacteria resident in the intestine; also prevalent in a wide variety of foods; necessary for proper clotting of blood
Prostaglandins	Fatty-acid derivatives found in all cell membranes; various functions, depending on the specific class, including stimulation of uterine contractions (thus inducing labor and abortions), regulation of blood pressure, and control of stomach secretion and motility of the gastrointestinal tract; believed to be involved as an intermediary in some types of hormone activity

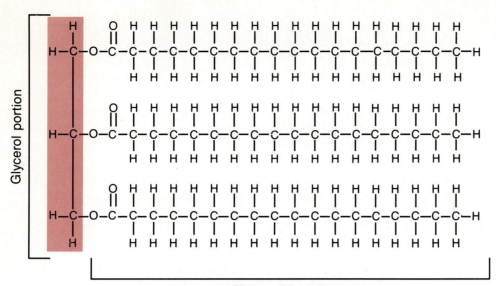

(a)   Fatty-acid portions

$CH_3-CH_2-CH_2-CH_2-CH_2-CH_2-CH_2-CH_2-CH_2-\overset{\displaystyle O}{\overset{\displaystyle \|}{C}}-O-CH_2$ ← Glycerol portion

$CH_3-CH_2-CH_2-CH_2-CH_2-CH_2-CH_2-CH_2-CH_2-\overset{\displaystyle O}{\overset{\displaystyle \|}{C}}-O-CH$

$H_2-C-O-\overset{\displaystyle O^-}{\underset{\displaystyle O^-}{\overset{\displaystyle |}{\underset{\displaystyle |}{P}}}}-O-CH_2-CH_2-\overset{\displaystyle CH_3}{\underset{\displaystyle CH_3}{\overset{\displaystyle |}{\underset{\displaystyle |}{\overset{\displaystyle +}{N}}}}}-CH_3$

Fatty-acid portions
(nonpolar)

Phosphorus-containing group
(polar end)

(b)   Phospholipid molecule

(c)   Steroid nucleus   Cholesterol

FIGURE A-13

Lipids. **(a)** A neutral fat consists of three fatty-acid chains attached to a glycerol backbone. **(b)** A typical phospholipid molecule has two fatty acids and a phosphorous-containing group attached to the glycerol backbone. The presence of the charged phosphorous-containing group makes that end of the molecule polar. **(c)** The steroid nucleus is shown on the left; the generalized structure of cholesterol (the most important body steroid) is shown on the right.

FIGURE A-14
General and specific amino acid structures. **(a)** Generalized structure of all amino acids. As shown, all amino acids have both an amine (basic) group and an acid ($-COOH$) group. They differ only in the atomic makeup of their R-group (shaded). **(b)** through **(e)** Specific structures of 4 amino acids. Possession of an acid group in the R-group makes the amino acid more acidic (for example, aspartic acid); possession of an amine group in the R-group causes the amino acid to become more basic (for example, lysine).

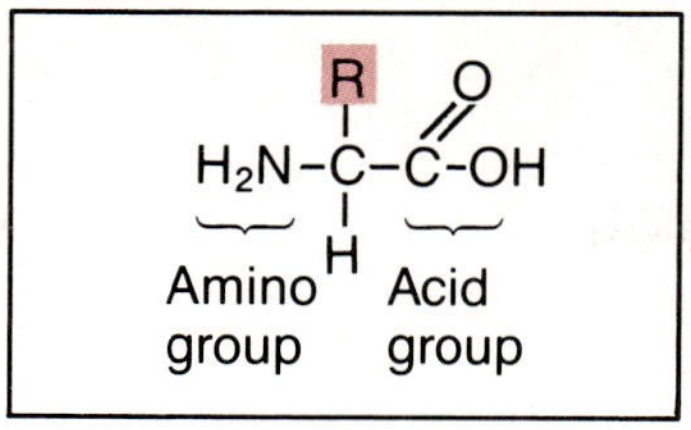

(a) Generalized structure of all amino acids

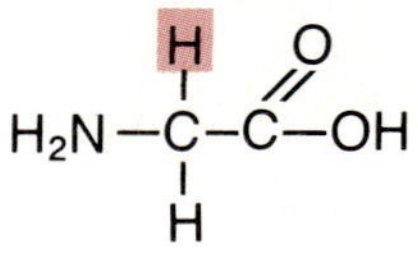

(b) Glycine (the simplest amino acid)

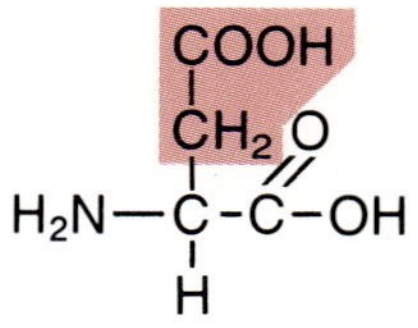

(c) Aspartic acid (an acidic amino acid)

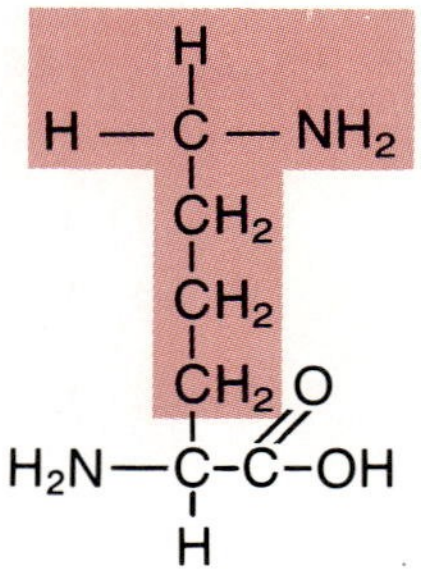

(d) Lysine (a basic amino acid)

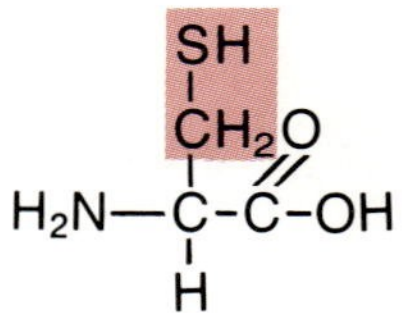

(e) Cysteine (a sulfur-containing amino acid)

STRUCTURAL AND FUNCTIONAL PROTEINS.

Based on their overall shape and structure, proteins are classed as structural or functional proteins. *Structural proteins*, also called *fibrous proteins* because of their strandlike appearance, are most often found comprising body structures. They are very important in binding structures together and for providing strength in certain body tissues. For example, *collagen* is found in bones, cartilages, and tendons, and is the most abundant protein in the body; and *keratin* is the structural protein of hair and nails, and the waterproofing material of the skin.

Functional proteins are mobile, generally spherical molecules that play crucial roles in virtually all biological processes (that is, they *do things* rather than just form structures). As noted in Table A-5, the scope of their activities is remarkable. Some (antibodies) help to provide immunity, others (hormones) help to regulate growth and development, and still others (enzymes) are biological catalysts that regulate essentially every chemical reaction that goes on within the body.

Structural proteins are exceptionally stable, but functional proteins are quite the opposite. Hydrogen bonds are critically important in maintaining their structure, but hydrogen bonds are fragile and are easily broken by heat and excesses of pH.

When their three-dimensional structures are destroyed, functional proteins are said to be *denatured* and can no longer perform their physiologic roles, because their function depends on their specific structure—most importantly on the presence of specific collections of atoms called *active sites* on their surface that "fit," and interact chemically, with other molecules of complementary shape and charge (Figure A-15). As hinted at earlier, hemoglobin becomes totally unable to bind and transport oxygen when blood pH becomes too acidic; and pepsin, a protein-digesting enzyme, is inactivated by alkaline pH. In each case, the struc-

TABLE A-5 Representative Groups of Functional Proteins

Functional group	Role(s) in the body
Antibodies (immunoglobulins)	Highly specialized proteins that recognize, bind with, and inactivate bacteria, toxins, and some viruses; function in the immune response, which helps protect the body from "invading" foreign substances
Hormones	Help to regulate growth and development: Growth hormone—an anabolic hormone necessary for optimal growth during childhood and normal protein metabolism in adults Insulin—regulates blood sugar levels Adrenalin—causes blood vessel constriction and increases the force of heartbeat, thus playing a role in blood pressure regulation Nerve growth factor—guides the growth of neurons in the development of the nervous system
Transport proteins	Hemoglobin, transport of oxygen in the blood; transport by others of iron, cholesterol, and other substances in the plasma
Contractile proteins	Sliding motion of two kinds of muscle proteins (actin and myosin) accomplishes muscle contraction and body movement; on the microscopic level, activity of contractile proteins represented by cell division and sperm propulsion (toward the egg)
Catalysts	Enzymes involved in virtually every biochemical reaction in the body; increase the rates of chemical reactions by at least a millionfold; in their absence (or destruction), biochemical reactions cease

ture needed for function has been destroyed by the improper pH.

Except for enzymes, most important types of functional proteins are described with the organ system or functional process to which they are closely related. For instance, protein hormones are discussed in Chapter 8 (Endocrine System), hemoglobin is considered in Chapter 9 (Blood), and antibodies are described in Chapter 11 (The Immune System). However, enzymes are important in the function of all body cells, and so these incredibly complex molecules are considered here.

ENZYMES AND ENZYME ACTIVITY. **Enzymes** are functional proteins that act as biological catalysts. A *catalyst* is a substance that increases the rate of a chemical reaction without becoming part of the

product or being changed itself. Enzymes accomplish this feat by binding to, and "holding," the reacting molecules in the proper position for chemical interaction. Once the reaction has occurred, the enzyme releases the product. Because enzymes are not changed in "doing their job," they are reusable, and only small amounts of each enzyme are needed by the cells.

Enzymes are capable of catalyzing millions of reactions each minute. However, they do more than just increase the speed of chemical reactions: They also determine just which reactions are possible at a particular time—no enzyme, no reaction. Enzymes can be compared to a bellows used to fan a sluggish fire into flaming activity; without enzymes, biochemical reactions would occur far too slowly to sustain life.

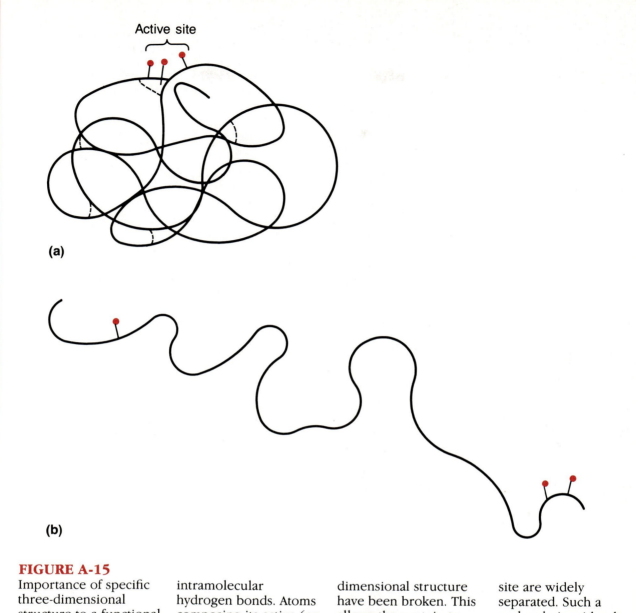

Active site

(a)

(b)

FIGURE A-15

Importance of specific three-dimensional structure to a functional protein, such as an enzyme. In **(a)** the globular structure of the functional protein is maintained by intramolecular hydrogen bonds. Atoms composing its active (or functional site) are shown as stalked particles. In **(b)** the hydrogen bonds maintaining the three-dimensional structure have been broken. This allows the protein to unfold and results in a basically linear molecule in which the atoms formerly constructing the active site are widely separated. Such a molecule is said to be denatured and is nonfunctional.

Although there are hundreds of different kinds of enzymes in body cells, they are very specific in their activity, each controlling only one, or a small group of, chemical reactions and acting only on specific molecules. Most enzymes are named according to that specific type of reaction they catalyze. There are "hydrolases," which add water; "oxidases," which cause oxidation; and so on. (In most cases, an enzyme can be recognized by the suffix **-ase** forming part of its name.)

Many enzymes are produced in an inactive form, which must be activated in some way before they can function. In other cases, enzymes are inactivated immediately after they have performed their catalytic function. Both events are true of enzymes that promote blood clotting when a blood vessel has been damaged. If this were not so, large numbers of unneeded, and potentially lethal, blood clots would be formed.

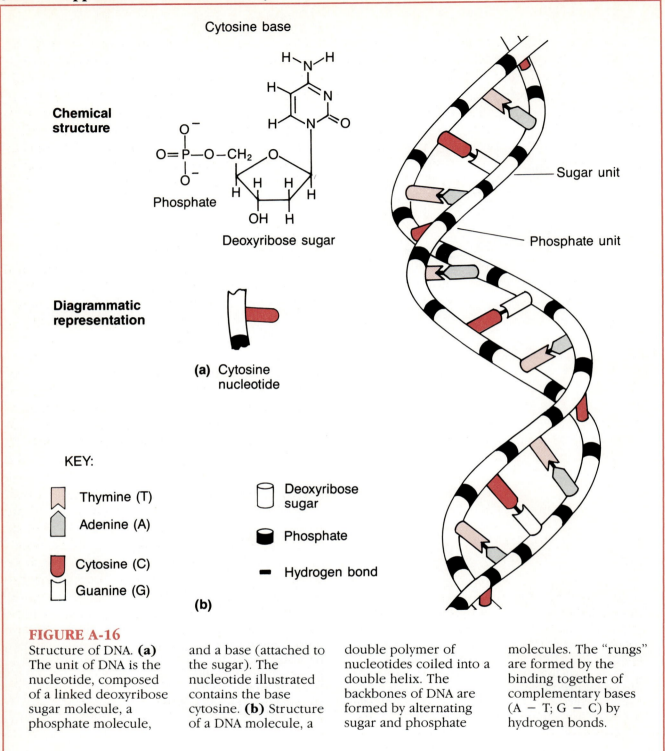

Cytosine base

Chemical structure

Phosphate

Deoxyribose sugar

Sugar unit

Phosphate unit

Diagrammatic representation

(a) Cytosine nucleotide

KEY:

Thymine (T)

Adenine (A)

Cytosine (C)

Guanine (G)

Deoxyribose sugar

Phosphate

— Hydrogen bond

(b)

FIGURE A-16
Structure of DNA. **(a)** The unit of DNA is the nucleotide, composed of a linked deoxyribose sugar molecule, a phosphate molecule, and a base (attached to the sugar). The nucleotide illustrated contains the base cytosine. **(b)** Structure of a DNA molecule, a double polymer of nucleotides coiled into a double helix. The backbones of DNA are formed by alternating sugar and phosphate molecules. The "rungs" are formed by the binding together of complementary bases (A – T; G – C) by hydrogen bonds.

NUCLEIC ACIDS

The role of **nucleic acids** is fundamental: They make up the genes, which provide the basic blue-print of life. Not only do they determine what type of an organism you will be, but they also direct your growth and development. They do this entirely by dictating protein structure. (Remember that enzymes, which catalyze all the chemical reactions that occur in the body, are proteins.)

Nucleic acids, composed of carbon, oxygen, hydrogen, nitrogen, and phosphorous atoms, are the largest biological molecules in the body. Their building blocks, the **nucleotides**, are quite complex. Each consists of three basic parts: (1) a nitrogen-containing base, (2) a pentose sugar, and (3) a phosphate group (Figure A-16a).

The bases come in five varieties: *adenine (A), guanine (G), cytosine (C), thymine (T)*, and *uracil (U)*. A and G are large two-ring bases, whereas the others are smaller single-ring structures. The nucleotides are named according to the base they contain; thus A-containing bases are adenine nucleotides, C-containing bases are cytosine nucleotides, and so on.

The two major kinds of nucleic acid are **deoxyribonucleic acid (DNA)** and **ribonucleic acid (RNA)**. DNA and RNA differ in many respects. DNA is the genetic material found within the cell nucleus (control center of the cell). It has two fundamental roles: (1) it replicates itself exactly before a cell divides, thus ensuring that the genetic information in every body cell is identical, and (2) it provides the instructions for building every protein in the body. RNA is located outside the nucleus and can be considered to be the "molecular slave" of DNA; that is, RNA carries out the "orders" for protein synthesis issued by DNA.

Although both RNA and DNA are formed by the joining together of nucleotides, their final structures are different. As shown in Figure A-16, DNA is a long double chain of nucleotides. Its bases are A, G, T, and C, and its sugar is *deoxyribose*. Its two nucleotide chains are held together by hydrogen bonds between the bases so that a ladderlike molecule is formed. Alternating sugar and phosphate molecules form the "uprights," or backbones, of the ladder, and each "rung" is formed of two joined bases. Binding of the bases is very specific: A always binds to T and G always binds to C. Thus, A and T are said to be *complementary bases,* as are C and G. A base sequence of ATGA on one nucleotide chain would necessarily be bonded to the complementary base sequence TACT on the other nucleotide strand. The whole molecule is then coiled into a spiral staircase–like structure called a *double helix.*

While DNA is double-stranded, RNA molecules are single nucleotide strands. RNA bases include A, G, C, and U (U replaces T found in DNA), and its sugar is *ribose* instead of deoxyribose. Three varieties of RNA exist—*messenger, ribosomal, and transfer*

RNA—and each has a specific role to play in carrying out DNA's instructions. Messenger RNA carries the information for building the protein from the DNA genes to the *ribosomes,* the protein-synthesizing sites, and transfer RNA ferries amino acids to the ribosomes. Ribosomal RNA forms part of the ribosomes, where it acts to oversee the "translation" of the message and the binding together of amino acids to form the proteins.

ADENOSINE TRIPHOSPHATE (ATP)

The synthesis of **adenosine triphosphate**, or **ATP**, is all important because it provides chemical energy in a form that all body cells can use. Without ATP, molecules cannot be made or broken down, cells cannot maintain their boundaries, and all life processes grind to a halt.

Although glucose is the most important "fuel" for body cells, none of the chemical energy contained in its bonds can be used *directly* to power cellular work. Instead, energy released as glucose is catabolized is captured and stored in the bonds of ATP molecules as small "packets" of energy.

Structurally, ATP is a modified nucleotide; it consists of an adenine base, ribose sugar, and three phosphate groups (Figure A-17). The phosphate groups are attached by unique chemical bonds called *high-energy phosphate bonds*. When these bonds are ruptured by hydrolysis, energy that can be used immediately by the cell to power a particular activity—such as synthesizing proteins; moving substances across its membrane; or, in the case of muscle cells, contracting—is liberated. ATP can be compared to a tightly coiled spring that is ready to uncoil with tremendous energy when the "catch" is released. The consequence of cleavage of its terminal phosphate bond can be represented as follows:

$$ATP \longrightarrow ADP + P + E$$

| adenosine | adenosine | inorganic | energy |
| triphosphate | diphosphate | phosphate | |

As ATP is used to provide cellular energy, ADP accumulates, and ATP supplies are replenished by oxidation of food fuels. Essentially, the same amount of energy must be captured and used to reattach a phosphate group to ADP (that is, to reverse the reaction) as is liberated when the terminal phosphate is cleaved off.

FIGURE A-17
Adenosine triphosphate
(ATP). **(a)** The structure
of ATP and **(b)** its
hydrolysis, during
which the end
phosphate group is
cleaved off to produce
adenosine diphosphate
(ADP) and inorganic
phosphate as energy is
liberated. The high
energy phosphate
bonds are symbolized
by ~.

Adenine

(a)   Adenosine triphosphate (ATP)

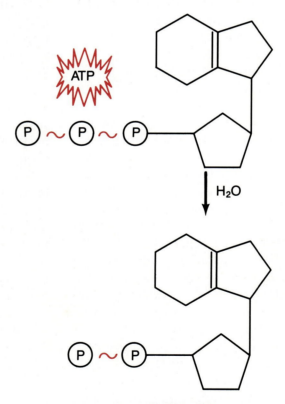

(b)   Adenosine diphosphate (ADP)

+

Inorganic Phosphate

IMPORTANT TERMS*

Matter	**Covalent bond**
Energy	**Hydrogen bond**
Element	**Synthesis reaction**
Atom	**Decomposition reaction**
Atomic symbol	**Exchange reaction**
Proton	**Organic compound**
Neutron	**Inorganic compound**
Electron	**Electrolyte**
Isotope	**Acid**
Radioisotope	**Base**
Radioactivity	**pH**
Molecule	**Carbohydrate**
Compound	**Lipid**
Chemical reaction	**Protein**
Valence shell	**Enzyme**
Ionic bond	**Nucleic acid**
Salt	**Adenosine triphosphate (ATP)**

*For definitions, see Glossary.

SUMMARY

A. DEFINITION OF CONCEPTS: MATTER AND ENERGY

1. Matter
   a. Matter is anything that occupies space and has mass.
   b. Matter exists in three states: gas, liquid, and solid.

2. Energy
   a. Energy is the capacity to do work or to put matter into motion. It exists in kinetic (active) and potential (stored) work capacities.
   b. Energy forms that are important in body functioning include chemical, electrical, mechanical, and radiant.
   c. Energy forms are intraconvertible, but some energy is always unusable (lost as heat) in such transformations.

B. COMPOSITION OF MATTER

1. Elements and atoms
   a. Each element is a unique substance that cannot be decomposed into simpler substances by ordinary chemical methods. A total of 108 elements exist; they differ from one another in their chemical and physical properties.
   b. Four elements (carbon, hydrogen, oxygen, and nitrogen) comprise 96% of living matter. Several other elements are present in small or trace amounts.
   c. The building blocks of elements are atoms. Each atom is described by an atomic symbol consisting of one or two letters.

2. Atomic structure
   a. Atoms are composed of three subatomic particles: protons, electrons, and neutrons. Because all atoms are electrically neutral,

the number of protons in any atom is equal to its number of electrons.

b. The planetary model of the atom portrays all of the mass of the atom (protons and neutrons) as concentrated in a minute central nucleus. Electrons orbit the nucleus within specific areas of space called energy levels.

c. Each atom can be identified by an atomic number, which is equal to the number of protons contained in the atom's nucleus.

d. The atomic mass number is equal to the sum of the protons and neutrons in the atom's nucleus.

e. Isotopes are different atomic forms of the same element; they differ only in the number of neutrons in the nucleus. Many of the heavier isotopes are unstable, and decompose to a more stable form by ejecting particles or energy from the nucleus—a phenomenon called radioactivity. Radioisotopes are useful in medical diagnosis and treatment, and in biochemical research.

f. Atomic weight is approximately equal to the mass number of the most abundant isotope of any element.

C. MOLECULES AND COMPOUNDS

1. A molecule is the smallest unit resulting from the binding of two or more atoms. If the atoms are different, a molecule of a compound is formed.

2. Compounds exhibit properties different from those of the atoms comprising them.

D. CHEMICAL BONDS AND CHEMICAL REACTIONS

1. Bond formation
   a. Chemical bonds are energy relationships. Electrons in the outermost energy level (valence shell) of the reacting atoms are active in the bonding.
   b. Atoms with a full valence shell (2 electrons in shell 1 or 8 in the subsequent shells) are chemically inactive. Those with an incomplete valence shell interact by losing, gaining, or sharing electrons to achieve stability (that is, to fill the valence shell).
   c. Ions are formed when valence-shell electrons are completely transferred from one atom to another; the oppositely charged ions formed attract each other, forming an ionic bond. Ionic bonds are common in salts.

d. Covalent bonds involve the sharing of electron pairs between atoms. If the electrons are shared equally, the molecule is a nonpolar covalent molecule; if the electrons are not equally shared, the molecule is a polar covalent compound. Polarmolecules orient themselves to charged particles.

e. Hydrogen bonds are fragile bonds that bind together different parts of the same molecule (intramolecular bonds). They are common in large complex organic molecules, such as proteins and nucleic acids.

2. Patterns of chemical reactions
   a. Chemical reactions involve the formation or breaking of chemical bonds. They are indicated by the writing of a chemical equation, which provides information about the atomic composition (formula) of the reactant(s) and product(s).
   b. Chemical reactions that result in larger, more complex molecules are synthesis reactions; they involve bond formation.
   c. In decomposition reactions, larger molecules are broken down to simpler molecules or atoms. Bonds are broken.
   d. Exchange reactions involve both bond making and bond breaking. Atoms are replaced by other atoms.

E. BIOCHEMISTRY: THE CHEMICAL COMPOSITION OF LIVING MATTER

1. Inorganic compounds
   a. Inorganic compounds comprising living matter are non–carbon containing and include water, salts, acids, and bases.
   b. Water is the single most abundant compound in the body. It acts as a universal solvent in which electrolytes (salts, acids, and bases) ionize and in which chemical reactions occur. It slowly absorbs and releases heat, thus helping to maintain homeostatic body temperature, and is the basis of transport and lubricating fluids.
   c. Salts in ionic form are involved in nerve transmission, muscle contraction, blood clotting, transport of oxygen by hemoglobin, cell permeability, metabolism, and many other reactions. Additionally, calcium salts (as bone salts) contribute to bone hardness.
   d. Acids are proton donors. When dissolved in water, they release hydrogen ions. Strong acids dissociate completely; weak acids dissociate incompletely.
   e. Bases are proton acceptors. The most important inorganic bases are hydroxides. When

bases and acids interact, neutralization occurs—that is, a salt and water are formed.

f. pH is a measure of the relative concentrations of hydrogen and hydroxyl ions in various body fluids. Each change of one pH unit represents a ten-fold change in hydrogen (or hydroxyl) ion concentration. A pH of 7 is neutral (that is, the concentrations of hydrogen and hydroxyl ions are equal). A pH below 7 is acidic; a pH above 7 is alkaline, or basic.

g. Normal blood pH ranges from 7.35 to 7.45. Slight deviations can be fatal.

2. Organic compounds

a. Organic compounds are those that comprise living matter. They all contain carbon; and carbohydrates, lipids, proteins, and nucleic acids are examples of organic compounds.

b. Carbohydrates contain carbon, hydrogen, and oxygen in the general relationship $(CH_2O)_n$; their building blocks are monosaccharides. (Monosaccharides include glucose, fructose, galactose, deoxyribose, and ribose. Disaccharides include sucrose, maltose, and lactose; and polysaccharides include starch and glycogen.) Carbohydrates are ingested as sugars and starches. Carbohydrates, and in particular glucose, are the major energy source for the formation of ATP.

c. Lipids, also carbon, hydrogen, and oxygen compounds, include the neutral fats (glycerol plus three fatty-acid chains), phospholipids, and steroids (most importantly, cholesterol). Neutral fats are found primarily in adipose tissue, where they serve as insulation and reserve body fuel. Phospholipids and cholesterol are found in all cell membranes. Cholesterol also forms the basis of certain hormones, and of bile salts and vitamin D. Like carbohydrates, the lipids are

degraded by hydrolysis and synthesized by dehydration synthesis.

d. Proteins contain carbon, hydrogen, oxygen, and nitrogen. The building unit of proteins is the amino acid; twenty common types of amino acids are found in the body. Amino-acid sequence determines the proteins constructed. Fibrous, or structural, proteins are the basic structural materials of the body. Globular proteins are functional molecules; examples of these include enzymes, some hormones, and hemoglobin. Disruption of the hydrogen bonds of functional proteins leads to their denaturation and inactivation.

e. Enzymes increase the rates of chemical reactions by combining specifically with the reactants and holding them in the proper position for interaction. They do not become part of the product. Many enzymes are produced in inactive form or are inactivated immediately after use.

f. Nucleic acids are composed of carbon, hydrogen, oxygen, and nitrogen. They include deoxyribonucleic acid (DNA) and ribonucleic acid (RNA). The building unit of nucleic acids is the nucleotide; each nucleotide consists of a nitrogenous base, a sugar (ribose or deoxyribose), and a phosphate group. DNA (the "stuff" of the genes) maintains genetic heritage by replicating itself before cell division and contains the code specifying protein structure. RNA acts in protein synthesis to ensure that the instructions of the DNA are executed.

g. ATP (adenosine triphosphate) is the universal energy compound used by all cells of the body. When energy is liberated by the oxidation of glucose, some of that energy is captured in the high-energy phosphate bonds of ATP molecules and is stored for later use.

REVIEW QUESTIONS

1. Why is a study of basic chemistry essential to understanding human physiology?

2. Matter and your body—how are they interrelated?

3. Matter *occupies space* and *has mass*. Describe how energy *must be* described in terms of these two factors. Then define energy.

4. Identify the energy *form* in use in each of the

following examples:

a. Chewing food.

b. Vision (two types, please—think!).

c. Bending the fingers to make a fist.

d. Breaking the bonds of ATP molecules to energize your muscle cells to make that fist.

5. The statement has been made that "some energy is lost in every energy transformation." Explain the meaning of this statement. (Direct your response to answering the question, Is it really lost? If not, what then?)

6. According to Greek history, a Greek scientist went running through the streets announcing that he had transformed lead into gold. Both lead and gold are elements. On the basis of what you know about the nature of elements, explain *why* his rejoicing was short-lived.

7. What four elements make up the bulk of all living matter? (Provide both their names and their atomic symbols.) Which of these are found primarily in proteins and nucleic acids?

8. List at least six other elements found in the body and describe one way in which each is important to body functioning.

9. What is the relationship of an atom to an element?

10. All atoms are neutral. Explain the basis of this fact.

11. Fill in the table below to fully describe an atom's subatomic particles.

Particle	Position in the atom	Charge	Mass
Proton			
Neutron			
Electron			

12. Define isotope.

13. If an element has three isotopes, which of them (lightest, intermediate mass, or heaviest) is most likely to be a radioisotope and why? Define radioactivity.

14. Complete this statement: Chemical behavior results from interactions of the _____ _____(subatomic-particle type) found in the _____ of the atom.

15. Distinguish between a molecule of an element and a molecule of a compound, and define molecule.

16. Explain the basis of ionic bonding. How do ionic bonds differ from covalent bonds?

17. What are hydrogen bonds, and how are they important in the body?

18. The two oxygen atoms forming molecules of oxygen gas that you breathe are joined by a *polar* covalent bond. Explain why this statement is true or false.

19. Identify each of the following reactions as a synthesis, decomposition, or exchange reaction:

$$2Hg + O_2 \longrightarrow 2HgO$$
$$Fe + CuSO_4 \longrightarrow FeSO_4 + Cu^{++}$$
$$HCl + NaOH \longrightarrow NaCl + H_2O$$
$$HNO_3 \longrightarrow H^+ + NO_3^-$$

20. Distinguish between inorganic and organic compounds, and list the major categories of each in the body.

21. Give at least four reasons a continual intake of water is essential to life.

22. Salts, acids, and bases are electrolytes. What is an electrolyte?

23. Compare and contrast acids and bases.

24. Define pH. State which of the following would be acidic, which basic, and which neutral: pure (distilled) water, vinegar, sodium bicarb, and gastric juice.

25. The pH range of blood is from 7.35 to 7.45. This is slightly _____ (acidic/basic).

26. A pH of 3.3 is _____ (1, 10, 100, 1000) times more acidic than a pH of 4.3.

27. Define monosaccharide, disaccharide, and polysaccharide. List at least two examples of each. Which of these is the building unit of carbohydrates? What is the primary function of carbohydrates in the body?

28. What are the general structures of neutral fats, phospholipids, and steroids? List one or two important uses of each of these lipid types in the body.

29. The building block of proteins is the amino acid. Draw a diagram of the structure of a generalized amino acid. What is the importance of the R-group?

30. Name the two protein classes based on structure and function in the body, and give two examples of each.

31. Define enzyme and describe the mechanism of enzyme activity.

32. Virtually no chemical reaction can occur in the body in the absence of enzymes. How might excessively high body temperature or acidosis (acidic blood pH) interfere with enzyme activity?

33. What is the structural unit of nucleic acids? Name the two major classes of nucleic acid found in the body, and then compare and contrast them in terms of (1) bases and sugar content, (2) three-dimensional structure, and (3) relative functions.

34. What is ATP, and what is its central role in the body?

35. A number of antibiotics act by binding to certain essential enzymes in the target bacteria. How might these antibiotics influence the chemical reaction controlled by the enzyme? What might be the effect on the bacteria? On the person taking the antibiotic prescription?

Glossary

Abdomen *abdō'men* the portion of the body between the diaphragm and the pelvis.

Abduct *ab-dukt'* to move away from the midline of the body.

Abortion *ah-bor'shun* termination of a pregnancy before the embryo or fetus is viable outside the uterus.

Absorption *ab-sorp'shun* passage of a substance into or across a blood vessel or membrane.

Accommodation (1) adaptation in response to differences or changing needs; (2) adjustment of the eye for seeing objects at close range.

Acetabulum *as"ĕ-tab'u-lum* the cuplike cavity on the lateral surface of the hipbone that receives the femur.

Acetylcholine *as"ĕ-til-ko'lēn* A chemical transmitter substance released by nerve endings.

Achilles tendon *ah-kil'ēz ten'don* the tendon that attaches the calf muscles to the heel bone.

Acid a substance that liberates hydrogen ions when in an aqueous solution (compare with *base*).

Acidosis *as"ĭ-do'sis* a condition in which the blood has an excess hydrogen ion concentration and a decreased pH.

Acne inflammatory disease of the skin; infection of the sebaceous glands.

Acromegaly *ak"ro-meg'a-le* an abnormal pattern of bone and connective-tissue growth characterized by enlarged hands, face, and feet and associated with excessive secretion of pituitary growth hormone in the adult.

Acromion *ah-kro'me-on* the outer projection of the spine of the scapula; the highest point of the shoulder.

Acrosome *ak'ro-sōm* an enzyme-containing structure covering the nucleus of the sperm.

Actin *ak'tin* a contractile protein of muscle.

Action potential an event occurring when a stimulus of sufficient intensity is applied to a neuron or muscle cell, allowing sodium ions to move into the cell and reverse the polarity.

Active immunity immunity produced by an encounter with an antigen; provides immunologic memory.

Active transport net movement of a substance across a membrane against a concentration gradient; requires release and use of energy.

Adaptation (1) any change in structure or response to suit a new environment; (2) decline in the transmission of a sensory nerve when a receptor is stimulated continuously and without change in stimulus.

Addison's disease condition resulting from deficient secretion of adrenal cortical hormones.

Adduct *ah-dukt'* to move toward the midline of the body.

Adenosine triphosphate (ATP) *ah-den"o-sin tri-fos'fāt* the compound that is the important intracellular energy source; cellular energy.

Adipose *ad'ĭ-pōs* fatty.

Adrenal glands *ah-dre'nal* hormone-producing glands located superior to the kidneys; each consists of medulla and cortex areas.

Adrenalin *ah-dren"ah-lin* epinephrine.

Adrenergic fibers *ad"ren-er'jik* nerve fibers that release norepinephrine.

Adrenocorticotropic hormone (ACTH) *ad-rē'no-kor"te-ko-trōf'ik* an anterior pituitary hormone that influences the activity of the adrenal cortex.

Aerobic *a-er-o'bik* requiring oxygen to live or grow.

Afferent *af'er-ent* carrying to or toward a center.

Afferent neuron *nu'ron* nerve cell that carries impulses toward the central nervous system.

Agglutination *ah-gloo"tin-a'shun* clumping of (foreign) cells, induced by cross-linking of antigen–antibody complexes.

Agglutinin *ah-gloo'tĭ-nin* an antibody in blood plasma that causes clumping of corpuscles or bacteria.

Agglutinogen *ag"loo-tin'o-jen* (1) an antigen that stimulates the formation of a specific agglutinin; (2) an antigen found on red blood cells that is responsible for determining the ABO blood group classification.

Agonist *ag'o-nist* a muscle that bears the primary responsibility for causing a certain movement.

Albumin *al-bū'min* a protein found in virtually all animal tissue and fluid. The most abundant plasma protein.

Albuminuria *al"bū-mĭ-nu're-ah* presence of albumin in the urine.

Aldosterone *al"do-ster'ōn* a hormone produced by the adrenal cortex that is important in sodium retention and reabsorption by kidney tubules.

Alimentary *al"ĕ-men'tar-e* pertaining to the digestive organs.

Alkalosis *al"kah-lo'sis* a condition in which the blood has a lower hydrogen ion concentration than normal, and an increased pH.

Allergy *al'er-je* overzealous immune response to an otherwise harmless antigen.

Alopecia *al"o-pe'she-ah* baldness, condition of hair loss.

Alveolus *al-ve'o-lus* (1) a general term referring to a small cavity or depression; (2) an air sac in the lungs.

Amino acid *ah-me′no* an organic compound containing nitrogen, carbon, hydrogen, and oxygen; the building block of protein.

Amnion *am′ne-on* the fetal membrane that forms a fluid-filled sac around the embryo.

Amphiarthrosis *am″fe-ar-thro′sis* articulation in which little motion occurs because of fibrocartilaginous connection of the articulating bony surfaces.

Amplitude *am′plĕ-tūd* largeness; fullness or wideness of extent or range.

Anabolism *an-nab′o-lizm* the energy-requiring building phase of metabolism in which simpler substances are combined to form more complex substances.

Anaerobic *an″a-er-o′bik* requiring no oxygen to live or grow.

Anatomy the science of the structure of living organisms.

Androgen *an′dro-jen* a hormone that controls male secondary sex characteristics.

Anemia *ah-ne′me-ah* reduced oxygen-carrying capacity of the blood caused by a decreased number of erythrocytes or decreased percentage of hemoglobin in the blood.

Aneurysm *an′u-rizm* blood-filled sac in an artery wall caused by dilation or weakening of the wall.

Angina pectoris *an-ji′nah pek′to-rus* a severe, suffocating chest pain caused by brief lack of oxygen supply to heart muscle.

Anorexia *an″o-rek′se-ah* loss of appetite or desire for food.

Anoxia *ah-nok′se-ah* a deficiency of oxygen.

Antagonist *an-tag′-o-nist* a muscle that acts in opposition to an agonist or prime mover.

Anterior *an-te′re-or* the front of an organ or part; the ventral surface.

Antibody *an″tĭ-bod″e* a specialized substance produced by the body that can provide immunity against a specific antigen.

Antidiuretic hormone (ADH) *an″tĭ-di″u-ret′ik* a hormone released by the posterior pituitary that promotes the reabsorption of water by the kidney.

Antigen *an′tĭ-jen* any substance—including toxins, foreign proteins, or bacteria—that, when introduced to the body, causes antibody formation.

Anus *a′nus* the distal end of the digestive tract and the outlet of the rectum.

Aorta *a-or′tah* the major systemic artery; arises from the left ventricle of the heart.

Aortic arch *a-or′tik* the portion of the aorta proximal to the heart.

Aortic body a receptor in the aortic arch sensitive to changing oxygen, carbon dioxide, and pH levels of the blood.

Aphasia *ah-fa′ze-ah* a loss of the power of speech, or a defect in speech.

Aplastic anemia *ah-plas′tik* anemia caused by inadequate production of erythrocytes resulting from inhibition or destruction of the red bone marrow.

Apocrine gland *ap′o-krin* the less numerous type of sweat gland. Produces a secretion containing water, salts, and proteins.

Aponeurosis *ap″o-nu-ro′sis* the fibrous or membranous sheet connecting a muscle and the part it moves.

Appendix *ah-pen′diks* a wormlike sac attached to the cecum of the large intestine.

Aqueous humor *a′kwe-us hu′mer* the watery fluid in the anterior chambers of the eye.

Arachnoid *ah-rak′noid* weblike; specifically, the weblike middle layer of the three meninges.

Areola *ah-re′o-lah* the circular, pigmented area surrounding the nipple.

Arrector pili *ah-rek′tor pi′li* tiny, smooth muscles attached to hair follicles, which cause the hair to stand upright when activated.

Arteriole *ar-te′re-ōl* a minute artery.

Arteriosclerosis *ar-te″re-o-sklĕ-ro′sis* any of a number of proliferative and degenerative changes in the arteries leading to their decreased elasticity.

Artery a vessel that carries blood away from the heart.

Arthritis *ar-thri′tis* inflammation of the joints.

Articulate *ar-tik′u-lāt* to join together in such a way as to allow motion between the parts.

Asthma *az′mah* a disease or allergic response characterized by bronchial spasms and difficult breathing.

Astigmatism *ah-stig′mah-tizm* a visual defect resulting from irregularity in the lens or cornea of the eye causing the image to be out of focus.

Atherosclerosis *ath″er-o″skle-ro′sis* changes in the walls of large arteries consisting of lipid deposits on the artery walls. The early stage of arteriosclerosis.

Atlas the first cervical vertebra; articulates with the occipital bone of the skull and the second cervical vertebra (axis).

Atom *at′um* the smallest part of an element, indivisible by ordinary chemical means.

Atomic mass number the sum of the number of protons and neutrons in the nucleus of an atom.

Atomic number the number of protons in an atom.

Atomic symbol a one- or two-letter symbol indicating a particular element.

Atomic weight average of the mass numbers of all of the isotopes of an element.

Atresia *ah-tre′ze-ah* the abnormal closure of a body canal or opening.

Atrioventricular node (AV node) *a″tre-o-ven-trik′u-lar* a specialized mass of conducting cells located at the atrioventricular junction in the heart.

Atrium *a′tre-um* a chamber of the heart receiving blood from the veins.

Atrophy *at′ro-fe* a reduction in size or wasting away of an organ or cell resulting from disease or lack of use.

Auditory *aw′dĭ-to″re* pertaining to the sense of hearing.

Auditory ossicles *os′sĭk′ls* the three tiny bones serving as transmitters of vibrations and located within the middle ear: the malleus, incus, and stapes.

Auditory (eustachian) tube *u"sta'shē-an* tube connecting the middle ear and the pharynx.

Auricle *aw'rĕ-kl* the external ear; pinna.

Auscultation *aws"kul-ta'shun* the act of examination by listening to body sounds.

Autoimmune response *aw"to-im-mūn* the production of antibodies or effector T cells that attack a person's own tissue.

Automaticity *aw-tom"ah-tis'ĭ-te* the ability of a structure, organ, or system to initiate its own activity.

Autonomic *aw"to-nom'ik* self-directed; self-regulating; independent.

Autonomic nervous system the division of the nervous system that functions involuntarily; innervates cardiac muscle, smooth muscle, and glands.

Axial skeleton *ak'se-al* the bones of the skull, vertebral column, thorax, and sternum.

Axilla *ak-sil'ah* armpit.

Axis (1) the second cervical vertebra; has a vertical projection called the dens around which the atlas rotates; (2) the imaginary line about which a joint or structure revolves.

Axon *ak'son* the neuron process that carries impulses away from the nerve cell body; efferent process; the conducting portion of a nerve cell.

B-cells lymphocytes that oversee humoral immunity; their descendants differentiate into antibody-producing plasma cells. Also called B-lymphocytes.

Bacteria any of a large group of microorganisms, generally one-celled; found in humans and other animals, plants, soil, air, and water; have a broad range of functions.

Basal metabolic rate *met"ah-bol'ik* the rate at which energy is expended (heat produced) by the body per unit time under controlled (basal) conditions: 12 hours after a meal, at rest.

Basal nuclei *nu'kle-i* gray matter structures deep inside each of the cerebral hemispheres.

Base a substance that accepts hydrogen ions; capable of uniting with and neutralizing an acid.

Basement membrane a thin layer of substance to which epithelial cells are attached in mucous surfaces.

Basophil *ba'so-fil* white blood cells whose granules stain deep blue with basic dye; have a relatively pale nucleus and granular-appearing cytoplasm.

Benign *be-nīn'* not malignant.

Biceps *bi'seps* two-headed, especially applied to certain muscles.

Bicuspid *bi-kus'pid* having two points or cusps.

Bile a greenish-yellow or brownish fluid produced in and secreted by the liver, stored in the gallbladder, and released into the small intestine.

Biopsy *bi'ŏp-se* the removal and examination of live tissue.

Blastocyst *blas'to-sist* a stage of early embryonic development.

Blood-brain barrier a mechanism that inhibits passage of materials from the blood into brain tissues.

Bolus *bo'lus* a rounded mass of food prepared by the mouth for swallowing.

Bony thorax *bōn'e tho'raks* bones of the thorax, including ribs, sternum, and thoracic vertebrae.

Bowman's capsule *bo'manz* the double-walled cup at the end of a nephron. Encloses a glomerulus.

Brachial *bra'ke-al* pertaining to the arm.

Bradycardia *brad"e-kar'de-ah* slowness of the heart rate, below 60 beats per minute.

Brain stem the portion of the brain consisting of the medulla, pons, midbrain, and diencephalon.

Bronchitis *brong-ki'tis* inflammation of the bronchi.

Bronchus *brong'kus* one of the two large branches of the trachea leading to the lungs.

Buccal *buk'al* pertaining to the cheek.

Buffer a substance or substances that help to stabilize the pH of a solution.

Bundle branch block a blocking of heart action resulting from damage to the bundle of His; delayed contraction of one ventricle.

Bursa *ber'sah* a small sac filled with fluid and located at friction points, especially joints.

Calcitonin *kal"sĭ-to'nin* a hormone released by the thyroid that decreases calcium levels of the blood.

Calculus *kal'ku-lus* a stone formed within various body parts.

Calorie *kal'o-re* a unit of heat; the large calorie is the amount of heat required to raise 1 kg of water 1 C; also used in metabolic and nutrition studies as the unit to measure the energy value of foods.

Calyx *ka'liks* a cuplike extension of the pelvis of the kidney.

Canal a duct or passageway; a tubular structure.

Canaliculus *kan"ah-lik'u-lus* extremely small tubular passage or channel.

Cancer a malignant, invasive cellular neoplasm that has the capability of spreading throughout the body or body parts.

Capillary *kap'ĭ-lar"e* a minute blood vessel connecting arterioles with venules.

Carbohydrate *kar"bo-hi'drāt* organic compound composed of carbon, hydrogen, and oxygen; includes starches, sugars, cellulose.

Carcinogen *kar-sin'o-jen* cancer-causing agent.

Carcinoma *kar"sĭ-no'-mah* cancer; a malignant growth of epithelial cells.

Cardiac *kar'de-ak* pertaining to the heart.

Cardiac muscle specialized muscle of the heart.

Cardiac output the blood volume (in liters) ejected per minute by the left ventricle.

Cardioesophageal sphincter *kar"de-o-ĕ-sof"ah-je'al sfingk'ter* valve between the stomach and esophagus.

Carotid *kah-rot'id* the main artery in the neck.

Carotid body a receptor in the common carotid artery sensitive to changing oxygen, carbon dioxide, and pH levels of the blood.

Carotid sinus *si'nus* a dilation of a common carotid artery; involved in regulation of systemic blood pressure.

Carpal *kar'pal* one of the eight bones of the wrist.

Cartilage *kar'tĭ-lij* white, semiopaque connective tissue.

Catabolism *kah-tab'o-lizm* the process in which living cells break down substances into simpler substances; destructive metabolism.

Cataract *kat'ah-rakt* partial or complete loss of transparency of the crystalline lens of the eye.

Catecholamines *kat"ĕ-kol-am'inz* epinephrine and norepinephrine.

Caudal *kaw'dal* in humans, the inferior portion of the anatomy.

Cecum *se'kum* the blind-end pouch at the beginning of the large intestine.

Cell the basic biological unit of living organisms, containing a nucleus and a variety of organelles; usually enclosed in the membrane.

Cell membrane the selectively permeable membrane forming the outer layer of most animal cells; also called plasma membrane.

Cellular immunity *sel'u-lar ĭ-mu'nĭ-te* immunity mediated by lymphocytes called T-cells. Also referred to as cell-mediated immunity.

Cellulose *sel'u-lōs* a fibrous carbohydrate that is the main structural component of plant tissues.

Cementum *se-men'tum* the bony connective tissue that covers the root of a tooth.

Central nervous system (CNS) the brain and the spinal cord.

Centriole *sen'trĭ-ōl* a minute body found near the nucleus of the cell; active in cell division.

Cerebellum *ser"ĕ-bel'um* part of the hindbrain; controls movement coordination.

Cerebral aqueduct *ser'ĕ-bral ak'we-dukt"* the slender cavity of the midbrain that connects the third and fourth ventricles; also called the aqueduct of Sylvius.

Cerebrospinal fluid the fluid produced in the cerebral ventricles; fills the ventricles and surrounds the central nervous system.

Cerebrum *ser'ĕ-brum* the largest part of the brain; consists of right and left cerebral hemispheres.

Cerumen *sĕ-roo'men* earwax.

Cervical *ser'vĭ-kal* refers to the neck or the necklike portion of an organ or structure.

Cervix *ser'viks* the inferior necklike portion of the uterus leading to the vagina.

Chemical bond an energy relationship holding atoms together; involves the interaction of electrons.

Chemical change a change that alters the basic nature of a substance, resulting in a different substance.

Chemical energy energy form stored in the bonds of chemicals.

Chemical reaction formation or breakage of chemical bonds.

Chemoreceptor *ke"mo-re-sep'tor* receptors sensitive to various chemicals in solution.

Chiasma *ki-as'mah* a crossing or intersection of two structures, such as the optic nerves.

Cholecystectomy *ko"le-sis-tek'to-me* removal of the gallbladder.

Cholecystokinin *ko"le-sis"to-kin'in* an intestinal hormone that stimulates gallbladder contraction and pancreatic juice release.

Cholesterol *ko-les'ter-ol* a steroid found in animal fats and oil as well as in most body tissues, especially bile.

Cholinergic fibers *ko"lin-er'jik* nerve endings that, upon stimulation, release acetylcholine.

Chondrocyte *kon'dro-sīt* a mature cartilage cell.

Chorion *ko're-on* the outermost fetal membrane; forms the placenta.

Choroid *ko'roid* the pigmented nutritive layer of the eye.

Chromatin *kro'mah-tin* the structures in the nucleus that carry the hereditary factors (genes).

Chromosome *kro'mo-sōm* barlike bodies of tightly coiled chromatin; visible during cell division.

Chyme *kīm* the semifluid contents of the stomach consisting of partially digested food and gastric secretions.

Cilia *sil'e-ah* tiny, hairlike projections on cell surfaces that move in a wavelike manner.

Circle of Willis a union of arteries at the base of the brain.

Circumcision *ser"kum-sizh'un* removal of the foreskin of the penis.

Circumduction *ser"kum-duk'shun* circular movement of a body part.

Cirrhosis *sir-ro'sis* a chronic disease, particularly of the liver, characterized by an overgrowth of connective tissue or fibrosis.

Cleavage *klēv'ij* an early embryonic phase consisting of rapid cell divisions without intervening growth periods.

Clitoris *kli'to-ris* a small, erectile structure in the female, homologous to the penis in the male.

Clonal selection *klōn''l* the process during which a B-cell or T-cell becomes sensitized through binding contact with an antigen.

Clone descendants of a single cell.

Cochlea *kok'le-ah* a cavity of the inner ear resembling a snail shell.

Coitus *ko'ĭ-tus* sexual intercourse.

Coma *ko'mah* unconsciousness from which the person cannot be aroused.

Complement group of 11 plasma proteins that normally circulate in their inactive forms; when activated by complement fixation, causes lysis of foreign cells and inflammation, and enhances phagocytosis.

Complement fixation   binding of complement to antigen–antibody complexes; activates complement.

Compound   substance composed of two or more different elements, the atoms of which are chemically united.

Concave   *kon′kāv*   having a curved, depressed surface.

Conductivity   *kon″duk-tiv′ĭ-te*   ability to transmit an electrical impulse.

Condyle   *kon′dīl*   a rounded projection at the end of a bone that articulates with another bone.

Cones   one of the two types of photoreceptor cells in the retina of the eye. Provide for color vision.

Congenital   *kon-jen′ĭ-tal*   existing at birth.

Conjunctiva   *kon″junk-ti′vah*   the thin, protective mucous membrane lining the eyelids and covering the anterior surface of the eye itself.

Conjunctivitis   *kon″junk″tĭ-vi′tis*   an inflammation of the conjunctiva of the eye.

Connective tissue   a primary tissue; form and function vary extensively. Function includes support, storage, and protection.

Contraception   *kon″trah-sep′shun*   the prevention of conception; birth control.

Contraction   *kon-trak′shun*   to shorten or develop tension, an ability highly developed in muscle cells.

Contralateral   *kon″trah-lat′er-al*   opposite; acting in unison with a similar part on the opposite side of the body.

Convergence   *kon-ver′jens*   turning toward a common point from different directions.

Convoluted   *kon′vo-lūted*   rolled, coiled, or twisted.

Cornea   *kor′ne-ah*   the transparent anterior portion of the eyeball.

Corpus   *kor′pus*   body; the major portion of an organ.

Cortex   *kor′teks*   the outer surface layer of an organ.

Corticosteroids   *kor″tĭ-ko-ste′roidz*   the steroid hormones released by the adrenal cortex.

Cortisol   *kor′tĭ-sol*   a glucocorticoid produced by the adrenal cortex.

Costal   *kos′tal*   pertaining to the ribs.

Covalent bond   *ko-va′lent*   a bond involving the sharing of electrons between atoms.

Cramp   a painful, involuntary contraction of a muscle.

Cranial   *kra′ne-al*   pertaining to the skull.

Crenation   *kre-na′shun*   the shriveling of an erythrocyte resulting from loss of water.

Cretinism   *kre′tin-izm*   a severe thyroid deficiency in the young that leads to stunted physical and mental growth.

Cryptorchidism   *krip-tor′kĭ-dizm*   a developmental defect in which the testes fail to descend into the scrotum.

Cubital   *ku′bĭ-tal*   pertaining to the forearm area anterior to the elbow.

Cupula   *ku′pu-lah*   a domelike structure.

Cushing's syndrome   *koosh′ingz sin′drōm*   a disease produced by excess secretion of adrenocortical hormone; characterized by adipose tissue accumulation, weight gain, and osteoporosis.

Cutaneous   *ku-ta′ne-us*   pertaining to the skin.

Cyanosis   *si″ah-no′sis*   a bluish coloration of the mucous membranes and skin caused by deficient oxygenation of the blood.

Cystitis   *sis-ti′tis*   inflammation of the urinary bladder.

Cytology   *si-tol′o-je*   the science concerned with the study of cells.

Cytoplasm   *si′to-plazm″*   the substance of a cell other than that of the nucleus.

Cytotoxic T-cell   see killer T-cell.

Deciduous   *de-sid′u-us*   temporary.

Deciduous (milk) teeth   the 20 temporary teeth replaced by permanent teeth; "baby" teeth.

Decomposition reaction   a destructive chemical reaction in which complex substances are broken down into simpler ones.

Defecation   *def″e-ka′shun*   the elimination of the contents of the bowels (feces).

Deglutition   *deg″loo-tish′un*   the act of swallowing.

Dehydration   *de″hi-dra′shun*   a condition resulting from excessive loss of water.

Dendrite   *den′drīt*   branching; the branching neuron process that transmits the nerve impulse to the cell body; the receptive portion of a nerve cell.

Dentin   *den′tin*   the calcified tissue beneath the enamel forming the major part of a tooth.

Deoxyribonucleic acid (DNA)   *de-ok″se-ri″bo-nu″kle′ik*   a nucleic acid found in all living cells; carries the organism's hereditary information.

Depolarization   *de-po″lar-i-za′shun*   the loss of a state of polarity; the loss of a negative charge inside the cell.

Dermatitis   *der″mah-ti′tis*   an inflammation of the skin; nonspecific skin allergies.

Dermis   *der′mis*   the deep layer of dense, irregular connective tissue of the skin.

Diabetes insipidus   *di″ah-be′tēz in-sip′i-dus*   a disease characterized by passage of a large quantity of dilute urine plus intense thirst and dehydration; a hypothalamic disorder is the cause.

Diabetes mellitus   *mel-li′tus*   a disease caused by deficient insulin release, leading to inability of the body cells to use carbohydrates at a normal rate.

Dialysis   *di-al′ĭ-sis*   diffusion of solute(s) through a semipermeable membrane.

Diapedesis   *di″ah-pĕ-de′sis*   the passage of blood cells through intact vesel walls into the tissues.

Diaphragm   *di′ah-fram*   (1) any partition or wall separating one area from another; (2) a muscle that separates the thoracic cavity from the lower abdominopelvic cavity.

Diarthrosis   *di″ar-thro′sis*   a freely movable joint; a synovial joint.

Diastole   *di-as′to-le*   a period (between contractions) of relaxation of the heart during which it fills with blood.

Diencephalon   *di″en-sef′ah-lon*   that part of the

forebrain between the cerebral hemispheres and the midbrain including the thalamus, the third ventricle, and the hypothalamus.

Diffusion *dĭ-fu'zhun* the spreading of particles in a gas or solution with a movement toward uniform distribution of particles.

Digestion *di-jest'jun* the bodily process of breaking down foods chemically and mechanically.

Dilate *di'lāt* to stretch; to open; to expand.

Disaccharide *di-sak'ĭ-rīd* literally, double sugar; examples include sucrose and lactose.

Distal *dis'tal* farthest from the point of attachment of a limb.

Diverticulum *di"ver-tik'u-lum* a pouch or sac in the walls of a hollow organ or structure.

Dorsal *dor'sal* pertaining to the back; posterior.

Duct *dukt* a canal or passageway.

Duodenum *du"o-de'num* the first part of the small intestine.

Dura mater *du'rah ma'ter* the outermost and toughest of the three membranes (meninges) covering the brain and spinal cord.

Dyspnea *disp'ne-ah* labored, difficult breathing.

Ectopic *ek-top'ik* not in the normal place; for example, in an ectopic pregnancy the egg is implanted at a place other than the uterus.

Edema *ĕ-de'mah* an abnormal accumulation of fluid in body parts or tissues; causes swelling.

Effector *ef-fek'tor* an organ, gland, or muscle capable of being activated by nerve endings.

Efferent *ef'er-ent* carrying away or away from, especially a nerve fiber that carries impulses away from the central nervous system.

Ejaculation *e-jak"u-la'shun* the sudden ejection of semen from the penis.

Electrical energy energy form resulting from the movement of charged particles.

Electrocardiogram (ECG) *e-lek"tro-kar'de-o-gram"* a graphic record of the electrical activity of the heart.

Electroencephalogram (EEG) *e-lek"tro-en-sef'ah-lo-gram"* a graphic record of the electrical activity of nerve cells in the brain.

Electrolyte *e-lek'tro-līt* a substance that breaks down into ions when in solution and is capable of conducting an electric current.

Electron negatively charged subatomic particle; orbits the atomic nucleus.

Element *el'ĕ-ment* any of the building blocks of matter; oxygen, hydrogen, carbon, for example.

Embolism *em'bo-lizm* the obstruction of a blood vessel by a clot floating in the blood; may also be a bubble of air in the vessel (air embolism).

Embryo *em-bre-o* an organism in its early stages of development; in humans, the first two months after conception.

Emesis *em'ĕ-sis* vomiting.

Emphysema *em"fĭ-se'mah* a condition caused by overdistension of the pulmonary alveoli and fibrosis of lung tissue.

Enamel the hard, calcified substance that covers the crown of a tooth.

Endocarditis *en"do-kar-di'tis* an inflammation of the inner lining of the heart.

Endocardium *en"do-kar'de-um* the endothelial membrane lining the interior of the heart.

Endocrine glands *en'do-krīn* ductless glands that empty their hormonal products directly into the blood.

Endometrium *en-do-me'tre-um* the mucous membrane lining of the uterus.

Endomysium *en"do-mis'e-um* the thin connective tissue surrounding each muscle cell.

Endoplasmic reticulum *en"do-plas'mik rĕ-tik'u-lum* a membranous network of tubular or saclike channels in the cytoplasm of a cell.

Endothelium *en"do-the'le-um* the single layer of simple squamous cells that line the walls of the heart and the vessels that carry blood and lymph.

Energy the ability to do work.

Enzyme *en'zīm* a substance formed by living cells that acts as a catalyst in bodily chemical reactions.

Eosinophil *e"o-sin'o-fil* a granular white blood cell whose granules readily take up a stain called eosin.

Epidermis *ep"ĭ-der'mis* the outer layers of the skin; epithelium.

Epididymis *ep"ĭ-did'ĭ-mis* that portion of the male duct system in which sperm mature. Empties into the *ductus* or *vas deferens*.

Epiglottis *ep"ĭ-glot'is* the elastic cartilage at the back of the throat; covers the glottis during swallowing.

Epimysium *ep"ĭ-mis'e-um* the sheath of fibrous connective tissue surrounding a muscle.

Epinephrine *ep"ĭ-nef'rin* the chief hormone of the adrenal medulla.

Epiphysis *ĕ-pif'ĭ-sis* the end of a long bone.

Epithelium *ep"ĭ-the'le-um* one of the primary tissues; covers the surface of the body and lines the body cavities, ducts, and vessels.

Equilibrium *e"kwĭ-lib're-um* balance; a state when opposite reactions or forces counteract each other exactly.

Erythrocyte *ĕ-rith'ro-sīt* red blood cell.

Erythropoiesis *ĕ-rith"ro-poi-e'sis* the process of erythrocyte formation.

Estrogen *es'tro-jen* hormone that stimulates female secondary sex characteristics, female sex hormones.

Eupnea *ūp-ne'ah* easy, normal breathing.

Exchange reaction a chemical reaction in which bonds are both made and broken; atoms become combined with different atoms.

Excretion *eks-kre'shun* the elimination of waste products from the body.

Exocrine glands *ek'so-krin* glands that have ducts through which their secretions are carried to a particular site.

Expiration *eks"pĭ-ra'shun* the act of expelling air from the lungs.

Extracellular *eks"trah-sel'u-lar* outside a cell.

Extracellular fluid fluid within the body but outside the cells.

Fallopian tube *fal-lo'pe-an* the oviduct or uterine tube; the tube through which the ovum is transported to the uterus.

Fascia *fash'e-ah* layers of fibrous tissue covering and separating muscles.

Fascicle *fas'ĭ-k'l* a bundle of nerve or muscle fibers separated by connective tissues.

Fatty acid a building block of fats.

Feces *fe'sēz* material discharged from the bowel composed of food residue, secretions, and bacteria.

Fertilization *fer'tĭ-lĭ-za'shun* fusion of the nuclear material of an egg and a sperm.

Fetus *fe'tus* the unborn young; in humans the period from the third month of development until birth.

Fibrillation *fi-brĭ-la'shun* irregular, uncoordinated contraction of muscle cells, particularly of the heart musculature.

Fibrin *fi'brin* the fibrous insoluble protein formed during the clotting of blood.

Fibrinogen *fi-brin"o-jen* a blood protein that is converted to fibrin during blood clotting.

Filtration *fil-tra'shun* the passage of a solvent and dissolved substances through a membrane or filter.

Fissure *fish'ūr* (1) a groove or cleft; (2) the deepest depressions or inward folds on the brain.

Fistula *fis'tu-lah* an abnormal passageway between organs or between a body cavity and the outside.

Fixator *fiks-a'tor* a muscle acting to immobilize a joint or a bone; fixes the origin of a muscle so that muscle action can be exerted at the insertion.

Flaccid *flak'sid* soft; flabby; relaxed.

Flagella *flah-jel'ah* long, whiplike extensions of the cell membrane of some bacteria and sperm. Serve to propel the cell.

Flexion *flek'shun* bending; the movement that decreases the angle between bones.

Focus *fo'kus* creation of a sharp image by a lens.

Follicle *fol'lĭ-k'l* structure in an ovary. Developing egg surrounded by follicle cells.

Follicle-stimulating hormone (FSH) a hormone produced by the anterior pituitary that stimulates ovarian follicle production in females and sperm production in males.

Fontanels *fon"tah-nelz* the fibrous membranes in the skull where bone has not yet formed; babies' "soft spots."

Foramen *fo-ra'men* a hole or opening in a bone or between body cavities.

Fossa *fos'ah* a depression; often an articular surface.

Fovea *fo-ve'ah* a pit.

Frontal (coronal) plane a longitudinal section that divides the body into anterior and posterior parts.

Fundus *fun'dus* the base of an organ; that part farthest from the opening of the organ.

Gallbladder the sac beneath the right lobe of the liver used for bile storage.

Gallstones particles of hardened cholesterol or calcium salts that are occasionally formed in gallbladder and bile ducts.

Gamete *gam'ēt* male or female reproductive cell (sperm/egg).

Gametogenesis *gam"ē-to-jen'ĕ-sis* the formation of gametes.

Ganglion *gang'gle-on* a group of nerve-cell bodies located in the peripheral nervous system.

Gastrin *gas'trin* a hormone that stimulates gastric secretion, especially hydrochloric acid release.

Gene *jēn* one of the biological units of heredity located in chromatin; transmits hereditary information.

Genetics *je-net'iks* the science of heredity.

Genitalia *jen"ĭ-ta'le-ah* the external sex organs.

Gingiva *jin-jī'vah* the gums.

Gland an organ specialized to secrete or excrete substances for further use in the body or for elimination.

Glaucoma *glaw-ko'mah* an abnormal increase of the pressure within the eye.

Glia *gli'ah* see neuroglia.

Glomerulus *glo-mer'u-lus* a knot of coiled capillaries in the kidney. The "filter."

Glottis *glot'is* the opening between the vocal cords in the larynx.

Glucagon *gloo'kah-gon* a hormone formed by islets of Langerhans in the pancreas; raises the glucose level of blood.

Glucocorticoids *gloo"ko-kor'tĭ-koidz* the adrenal cortex hormones that increase blood glucose levels and aid the body in resisting long-term stress.

Glucose *gloo'kōs* the principal sugar in the blood.

Glycerol *glis'er-ol* a sugar alcohol; one of the building blocks of fats.

Glycogen *gli'ko-jen* the main carbohydrate stored in animal cells; a polysaccharide.

Glycogenesis *gli"ko-jen'ĕ-sis* formation of glycogen from glucose.

Glycogenolysis *gli"ko-jĕ-nol'ĭ-sis* breakdown of glycogen to glucose.

Glycolysis *gli-kol'ĭ-sis* breakdown of glucose into simpler compounds.

Goblet cells individual cells of the respiratory and digestive tracts that produce mucus.

Goiter *goi'ter* a benign enlargement of the thyroid gland.

Gonad *go'nad* an organ producing gametes; an ovary or testis.

Gonadotropins *gon"ah-do-tro'pinz* gonad-stimulating hormones produced by the anterior pituitary.

Graafian follicle *graf'e-an fol'lĭ-k'l* a mature ovarian follicle.

Graded response a response that varies directly with the strength of the stimulus.

Gray matter the gray area of the central nervous system; contains neurons/nerve cell bodies.

Groin the junction of the thigh and the trunk; the inguinal area.

Growth hormone   a hormone that stimulates growth in general; produced in the anterior pituitary; also called somatotropin (STH).

Gustation *gus-ta'shun*   taste.

Gyrus *ji'rus*   an outward fold of the surface of the cerebral cortex.

Hamstring muscles   the posterior thigh muscles: the biceps femoris, semimembranosus, and semitendinosus.

Haustra *haws'trah*   pouches of the colon.

Haversian system or osteon *ha-ver'shan, os'te-on*   a system of interconnecting canals in the microscopic structure of adult compact bone; unit of bone.

Heart block   an impaired transmission of impulses from atrium to ventricle.

Heart murmur   an abnormal heart sound (usually resulting from valve problems).

Heat stroke   the failure of the heat-regulating ability of an individual under heat stress.

Helper T-cell   the type of T-lymphocyte that orchestrates cellular immunity by direct contact with other immune cells and by releasing chemicals called lymphokines; also helps to mediate the humoral response by interacting with B-cells.

Hematocrit *he-mat'o-krit*   the percentage of erythrocytes to total blood volume.

Hemiplegia *hem"e-ple'je-ah*   paralysis of one side of the body.

Hemocytoblasts *he"mo-si"to-blastz*   stem cells that give rise to all the formed elements of the blood.

Hemoglobin *he"mo-glo'bin*   the oxygen-transporting component of erythrocytes.

Hemolysis *he-mol'ĭ-sis*   the rupture of erythrocytes.

Hemophilia *he"mo-fil'e-ah*   an inherited clotting defect caused by absence of a blood-clotting factor.

Hemopoiesis *he"mo-poi-e'sis*   the formation of blood cells.

Hemorrhage *hem'or-ij*   the loss of blood from the vessels by flow through ruptured walls; bleeding.

Heparin *hep'ah-rin*   a substance that prevents clotting.

Hepatic portal system *hĕ-pat'ik*   the circulation in which the hepatic portal vein carries dissolved nutrients to the liver tissues for processing.

Hepatitis *hep"ah-ti'tis*   an inflammation of the liver.

Hilum, hilus *hi"lum, hi'lus*   a depressed area where vessels enter and leave an organ.

Histamine *his'tah-min*   a substance that causes vasodilation and increased vascular permeability.

Histology *his-tol'o-je*   the branch of anatomy dealing with the microscopic structure of tissues.

Homeostasis *ho"me-o-sta'sis*   a state of body equilibrium or stable internal environment of the body.

Homologous *ho-mol'o-gus*   parts or organs corresponding in structure but not necessarily in function.

Hormones *hor'mōnz*   the secretions of endocrine glands; responsible for specific regulatory effects on certain parts or organs.

Humoral immunity *hu'mor-al*   immunity provided by antibodies released by sensitized B-cells and their plasma cell progeny. Also called antibody-mediated immunity.

Hyaline *hy'ah-lin*   glassy; transparent.

Hydrochloric acid *hi"dro-klo'rik*   HC1; aids protein digestion in the stomach; produced by parietal cells.

Hydrogen bond   weak bond in which a hydrogen atom forms a bridge between two electron-hungry atoms. An important intra-molecular bond.

Hydrolysis *hi-drol'ĭ-sis*   the process in which water is used to split a substance into smaller particles.

Hyperopia *hi"per-o'pe-ah*   farsightedness.

Hypertension *hi"per-ten'shun*   high blood pressure.

Hypertonic *hi"per-ton'ik*   excessive, above normal, tone or tension.

Hypertrophy *hi-per'tro-fe*   an increase in the size of a tissue or organ independent of the body's general growth.

Hypothalamus *hi"po-thal'ah-mus*   the region of the diencephalon forming the floor of the third ventricle of the brain.

Hypothermia *hi"po-ther'me-ah*   subnormal body temperature.

Hypotonic *hi-po-ton'ik*   below normal tone or tension.

Hypoxia *hi-pok'se-ah*   a condition in which inadequate oxygen is available to tissues.

Ileum *il'e-um*   the terminal part of the small intestine; between the jejunum and the cecum of the large intestine.

Immunity *ĭ-mu'nĭ-te*   the ability of the body to resist many agents (both living and nonliving) that can cause disease; resistance to disease.

Immunocompetence *im"mu-no-kom'pĕ-tents*   the ability of the body's immune cells to recognize (by binding) specific antigens; reflects the presence of plasma membrane–bound receptors.

Immunodeficiency disease *im"mu-no-de-fish'en-se*   disease resulting from the deficient production or function of immune cells or certain molecules (complement, antibodies, and so on) required for normal immunity.

Immunoglobulin *im"mu-no-glob'u-lin*   a protein molecule, released by plasma cells, that mediates humoral immunity; an antibody.

Infarct *in'farkt*   a region of dead, deteriorating tissue resulting from a lack of blood supply.

Inferior (caudal)   pertaining to a position near the tail end of the long axis of the body.

Inflammation *in"flah-ma'shun*   a physiological response of the body to tissue injury; includes dilation of blood vessels and an increase in vessel permeability.

Inguinal *ing'gwĭ-nal*   pertaining to the groin region.

Inner cell mass   an accumulation of cells in the blastocyst from which the embryo develops.

Innervation   *in"er-va'shun*   the supply of nerves to a body part.

Inorganic compound   a compound that lacks carbon; for example, water.

Insertion   *in-ser'shun*   the movable part or attachment of a muscle as opposed to origin.

Inspiration   *in"spĭ-ra'shun*   the drawing of air into the lungs.

Insulin   *in'su-lin*   the hypoglycemic hormone produced in the pancreas affecting carbohydrate and fat metabolism, blood glucose levels, and other systemic processes.

Integumentary system   *in-teg-u-men'tar-e*   the skin and its accessory structures.

Intercellular   *in"ter-sel'u-lar*   between the body cells.

Intercellular matrix   *ma'triks*   the material between adjoining cells; important in connective tissue.

Internal respiration   the use of oxygen by body cells; synonym: cellular respiration.

Interstitial fluid   *in"ter-stish'al*   the fluid between the cells.

Intervertebral disks   *in"ter-ver'tĕ-bral*   the disks of fibrocartilage between the vertebrae.

Intervertebral foramina   *fo-ram'ĭ-nah*   the openings between the dorsal projections of adjacent vertebrae through which the spinal nerves pass.

Intracellular   *in"trah-sel'u-lar*   within a cell.

Intracellular fluid   fluid within a cell.

Intrinsic factor   *in-trin'sik*   a substance produced by the stomach that is required for vitamin B_{12} absorption.

Invert   to turn inward.

Ion   *i'on*   an atom with a positive or negative electric charge.

Ionic bond   bond formed by the complete transfer of electron(s) from one atom to another (or others). The resulting ions are oppositely charged and attract each other.

Ipsilateral   *ip"sĭ-lat'er-al*   situated on the same side.

Iris   *i'ris*   the pigmented, involuntary muscle that acts as the diaphragm of the eye.

Irritability   *ir"ĭ-tah-bil'ĭ-te*   ability to respond to a stimulus.

Ischemia   *is-ke'me-ah*   a local decrease in blood supply.

Isometric   *i"so-met'rik*   of the same length.

Isotonic   *i"so-ton'ik*   having a uniform tension; of the same tone.

Isotopes   *i'sĭ-top*   different atomic forms of the same element. Isotopes vary only in the number of neutrons they contain.

Jaundice   *jawn'dis*   an accumulation of bile pigments in the blood producing a yellow color of the skin.

Jejunum   *je-joo'num*   the part of the small intestine between the duodenum and the ileum.

Joint   the junction of two or more bones; an articulation.

Keratin   *ker'ah-tin*   an insoluble protein found in tissues such as hair, nails, and epidermis of the skin.

Ketosis   *ke-to'sis*   an abnormal condition during which an excess of ketone bodies are produced.

Killer T-cell   effector T-cell that directly kills foreign cells. Also called a cytotoxic T-cell.

Kinetic energy   the energy of motion.

Kinins   *ki'ninz*   group of polypeptides that dilate arterioles, increase vascular permeability, and induce pain.

Krebs cycle   the citric acid cycle; the series of reactions during which energy is liberated from metabolism of carbohydrates, fats, and amino acids.

Labia   *la'be-ah*   lips.

Lacrimal   *lak'rĭ-mal*   pertaining to tears.

Lactation   *lak-ta'shun*   the production and secretion of milk.

Lacteal   *lak'te-al*   special lymphatic capillaries of the small intestine that take up lipids.

Lactic acid   *lak'tik*   the product of anaerobic metabolism, especially in muscle.

Lacuna   *lah-ku'nah*   a little depression or space; in bone or cartilage, lacunae are occupied by cells.

Lamina   *lam'ĭ-nah*   (1) a thin layer or flat plate; (2) the portion of a vertebra between the transverse process and the spinous process.

Laryngitis   *lar"in-ji'tis*   an inflammation of the larynx.

Larynx   *lar'inks*   the cartilaginous organ located between the trachea and the pharynx. Voice box.

Lateral   *lat'er-al*   away from the midline of the body.

Lens   the elastic, doubly convex structure in the eye that focuses the light entering the eye on the retina.

Lesion   *le'zhun*   a tissue injury or wound.

Leukemia   *lu-ke'me-ah*   a cancerous condition in which there is an excessive production of immature leukocytes.

Leukocyte   *lu-ko'sīt*   a white blood cell.

Ligament   *lig'ah-ment*   a band of fibrous tissue connecting bones.

Lipid   *lip'id*   organic compound formed of carbon, hydrogen, and oxygen; examples are fats and cholesterol.

Lordosis   *lor-do'sis*   an excessive curve in the anterior lumbar spine; otherwise known as swayback.

Lumbar   *lum'ber*   the portion of the back between the thorax and the pelvis.

Lumbar puncture   a procedure involving needle insertion between the third and fourth lumbar vertebrae into the subarachnoid space to withdraw cerebrospinal fluid.

Lumen   *lu'men*   the space inside a tube, blood vessel, or hollow organ.

Luteinizing hormone   *lu'te-in"ĭ-zing*   an anterior pituitary hormone that aids maturation of cells in the ovary and triggers ovulation. Also causes the interstitial cells of the testis to produce testosterone.

Lymph   *limf*   the watery fluid in the lymph vessels collected from the tissue fluids.

Lymph node   a mass of lymphatic tissue

Lymphatic system   *lim-fat'ik*   a system of vessels carrying lymph to the circulatory system and protective lymph nodes.

Lymphocyte   *lim'fo-sīt*   agranular white blood cell formed in the lymphoid tissue.

Lymphokines   *lim'fo-kīnz*   substances involved in cell-mediated immune responses that enhance the basic inflammatory response and subsequent phagocytosis.

Lysosomes   *li'so-sōmz*   organelles that originate from the Golgi apparatus and contain strong digestive enzymes.

Lysozyme   *li'so-zim*   an enzyme capable of destroying certain kinds of bacteria. In lacrimal gland secretion.

Macrophage   *mak'ro-fāj*   a phagocytic cell particularly abundant in lymphatic and connective tissues; important as an antigen-presenter to T-cells and B-cells in the immune response.

Malignant   *mah-lig'nant*   life threatening; pertains to neoplasms that spread and lead to death, such as cancer.

Mammary glands   *mam'er-e*   milk-producing glands of the breasts.

Mastication   *mas"tĭ-ka'shun*   the act of chewing.

Matrix   *ma'triks*   the intercellular substance of any tissue; most often applies to connective tissue.

Matter   anything that occupies space and has mass.

Meatus   *me-a'tus*   the external opening of a canal.

Mechanical energy   energy form directly involved in putting matter into motion.

Mechanoreceptor   *mek"ah-no-re-sep'tor*   a receptor sensitive to mechanical pressures such as touch, sound, or contractions.

Medial   *me'de-al*   toward the midline of the body.

Mediastinum   *me"de-as-ti'num*   the region of the thoracic cavity between the lungs.

Medulla   *mĕ-dul'ah*   the central portion of certain organs.

Meiosis   *mi-o'sis*   the two cell successive cell divisions in gamete formation producing nuclei with half the full number of chromosomes (haploid).

Melanin   *mel'ah-nin*   the dark pigment responsible for skin color.

Melanocyte   a cell that produces melanin.

Memory cells   members of T-cell and B-cell clones that provide for immunologic memory.

Meninges   *mĕ-nin'jēz*   the membranes that cover the brain and spinal cord.

Meningitis   *men"in-ji'tis*   an inflammation of the meninges.

Menopause   *men'o-pawz*   the physiological end of menstrual cycles.

Menses   *men'sēz*   monthly discharge of blood from the uterus. Menstruation.

Menstruation   *men"stroo-a'shun*   the periodic, cyclic discharge of blood, secretions, tissue, and mucus from the mature female uterus in the absence of pregnancy.

Mesenteries   *mes'en-ter"ēz*   the double-layered membranes of the peritoneum that support most organs in the abdominal cavity.

Metabolic rate   *met"ah-bol'ik*   the energy expended by the body per unit time.

Metabolism   *mĕ-tab'o-lizm*   the sum total of the chemical reactions that occur in the body.

Metabolize   *mĕ-tab'o-līz*   to transform substances into energy or materials the body can use or store by means of anabolism or catabolism.

Metacarpals   *met"ah-kar'pal*   one of the five bones of the palm of the hand.

Metastasis   *mĕ-tas'tah-sis*   the spread of cancer from one body part or organ into another not directly connected to it.

Metatarsal   *met"ah-tar'sal*   one of the five bones between the instep and the phalanges of the foot.

Microvilli   *mi"kro-vil'i*   the tiny projections on the free surfaces of some epithelial cells; increase surface area for absorption.

Minerals   the inorganic chemical compounds found in nature; salts.

Mineralocorticoid   *min"er-al-o-kor'tĭ-koid*   an adrenal cortical steroid hormone that regulates mineral metabolism and fluid balance.

Mitochondria   *mi"to-kon'dre-ah*   the rodlike cytoplasmic organelles responsible for ATP generation for cellular activities.

Mitosis   *mi-to'sis*   the division of the cell nucleus; often followed by division of the cytoplasm of a cell.

Mixed nerves   nerves containing the processes of motor and sensory neurons; their impulses travel to and from the central nervous system.

Molecule   *mol'ĕ-kūl*   a very small mass of matter composed of atoms held together as a unit.

Monoclonal antibodies   *mon"o-klōn''l*   pure preparations of identical antibodies that exhibit specificity for a single antigen.

Monocyte   *mon'o-sīt*   a large single-nucleus white blood cell; agranular leukocyte.

Monosaccharide   *mon"o-sak'ĭ-rīd*   literally, one sugar; the building block of carbohydrates; examples include glucose and fructose.

Mons pubis   *monz pu'bis*   the fatty eminence over the pubic symphysis in the female.

Motor unit   a neuron and the muscle cells it supplies.

Mucous membranes   membranes that form the linings of body cavities open to the exterior (digestive, respiratory, urinary, and reproductive tracts).

Mucus   *mu'kus*   a sticky thick fluid, secreted by mucous glands and mucous membranes, that keeps the free surface of membranes moist.

Multiple sclerosis *skle-ro'sis* a chronic condition characterized by destruction of the myelin sheaths of neurons in the spinal cord and the brain.

Muscle fibers muscle cells.

Muscle spindles the complex capsules found in skeletal muscles that are sensitive to stretch.

Muscle twitch a single rapid contraction of a muscle followed by relaxation.

Muscular dystrophy *dis'tro-fe* a progressive disorder marked by atrophy and stiffness of the muscles.

Myelin *mi'ĕ-lin* a white, fatty lipid substance.

Myelinated fibers *mi'ĕ-lĭ-nāt''ed* axons (projections of a nerve cell) covered with myelin.

Myocardial infarction *mi''o-kar'de-al in-fark'shun* a condition characterized by dead tissue areas in the myocardium caused by interruption of blood supply to the area.

Myocardium *mi''o-kar'de-um* the cardiac muscle layer of the wall of the heart.

Myofibril *mi''o-fi'bril* a fiberlike organelle found in the cytoplasm of muscle cells.

Myofilament *mi''o-fil'ah-ment* filaments composing the myofibrils. Of two types: actin and myosin.

Myometrium *mi''o-me'tri-um* the thick uterine musculature.

Myopia *mi-ō'pe-ah* nearsightedness.

Myosin *mi'o-sin* one of the principal proteins found in muscle.

Nares *na'rēz* the nostrils.

Necrosis *nĕ-kro'sis* the death or disintegration of a cell or tissues caused by disease or injury.

Negative feedback feedback that regulates the stimulus.

Neoplasm *ne'o-plazm* an abnormal growth of cells. Sometimes cancerous.

Nephron *nef'ron* the functional unit of the kidney.

Nerve fiber axon or dendrite of a neuron.

Neuroglia *nu-rog'le-ah* the nonneuronal tissue of the central nervous system that performs supportive and other functions; also called glia.

Neuromuscular junction *nu''ro-mus'ku-lar* the region where a motor neuron comes into close contact with a skeletal muscle cell.

Neurons *nu'ronz* the nerve cells that transmit messages throughout the body.

Neutralization *nu''tral-i-za'shun* blockage of the harmful effects of bacterial exotoxins or viruses by the binding of antibodies to their functional sites.

Neutron *nu'tron* uncharged subatomic particle; found in the atomic nucleus.

Neutrophil *nu'tro-fil* the most abundant of the white blood cells.

Nucleic acid *nu-kle'ic* class of organic molecules that includes DNA and RNA.

Nucleoli *nu-kle'o-li* the small spherical bodies in the cell nucleus.

Nucleotide *nu-kle'-o-tīd* the building block of nucleic acids; a complex unit consisting of a sugar, a nitrogen-containing base, and a phosphate group.

Nucleus *nu'kle-us* a dense central body in most cells containing the genetic material of the cell.

Nystagmus *nis-tag'mus* an oscillatory movement of the eyeballs.

Obesity *o-bēs'ĭ-te* a condition of a person being overweight.

Occipital *ok-sip'ĭ-tal* pertaining to area at the back of the head.

Occlusion *ŏ-kloo'zhun* closure or obstruction.

Olfaction *ol-fak'shun* smell.

Oogenesis *o''o-jen'ĕ-sis* the process of formation of the ova.

Ophthalmic *of-thal'mik* pertaining to the eye.

Optic *op'tik* pertaining to the eye.

Optic chiasma *op'tik ki-as'mah* the partial crossover of fibers of the optic nerves.

Oral relating to the mouth.

Organ a part of the body formed of two or more tissues that performs a specialized function.

Organelle *or''gan-el'* a specialized structure in a cell that performs a specific function.

Organic compound a compound containing carbon; examples include proteins, carbohydrates, and fats.

Origin attachment of a muscle that remains relatively fixed during muscular contraction.

Osmoreceptor *oz''mo-re-cep'tor* a structure sensitive to osmotic pressure or concentration of a solution.

Osmosis *oz-mo'sis* the diffusion of a solvent through a membrane from a dilute solution into a more concentrated one.

Ossicles *os'sĭ-k'lz* the three bones of the middle ear: hammer, anvil, and stirrup.

Osteoblasts *os'te-o-blasts''* the bone-forming cells.

Osteoclasts *os'te-o-klasts''* the large cells that reabsorb or break down bone matrix.

Osteocyte *os''te-o-sīt''* a mature bone cell found in each lacuna.

Osteomyelitis *os''te-o-mi'ĕ-li'tis* inflammation of the contents of the marrow cavity, and bone tissue.

Osteoporosis *os''te-o-po-ro'sis* an increased softening of the bone resulting from a gradual decrease in rate of bone formation; a common condition in older people.

Otic *o'tik* pertaining to the ear.

Otitis media *o-ti'tis me'de-ah* middle-ear infection.

Otolith *o'to-lith* one of the small calcified masses in the utricle and saccule of the inner ear.

Ovarian cycle *o-va're-an* the monthly cycle of follicle development, ovulation, and corpus luteum formation in an ovary.

Ovary *o'vah-re* the female sex organ in which ova (eggs) are produced.

Ovulation *o''vu-la'shun* the release of an ovum (or oocyte) from the ovary.

Ovum *o'vum* the female gamete (germ cell); an egg.

Oxidation *ok"sĭ-da'shun* the process of substances combining with oxygen.

Oxygen debt *ok'sĭ-jen* the volume of oxygen required after exercise to oxidize the lactic acid formed during exercise.

Oxyhemoglobin *ok"se-he"mo-glo'bin* hemoglobin combined with oxygen.

Oxytocin *ok"se-to'sin* hormone released by the posterior pituitary that stimulates contraction of the uterus during childbirth and the ejection of milk during nursing.

Palate *pal'at* the roof of the mouth.

Palpation *pal-pa'shun* examination by touch.

Pancreas *pan'kre-as* the gland located behind the stomach, between the spleen and the duodenum, producing both endocrine and exocrine secretions.

Pancreatic juice *pan"kre-at'ik* a secretion of the pancreas containing enzymes for digestion of all food categories.

Pancreatitis *pan"kre-ah-ti'tis* an inflammation of the pancreas.

Papillary muscles *pap'ĭ-ler"e* cone-shaped muscles found in the heart ventricles.

Paralysis *pah-ral'ĭ-sis* the loss of muscle function.

Paraplegia *par"ah-ple'je-ah* paralysis of the lower limbs.

Parasympathetic division *par"ah-sim"pah-thet'ik* a division of the autonomic nervous system; also referred to as the craniosacral division.

Parathyroid glands *par"ah-thi'roid* small endocrine glands located on the posterior aspect of the thyroid gland.

Parathyroid hormone hormone released by the parathyroid glands that regulates blood calcium level.

Parietal *pah-ri'ĕ-tal* pertaining to the walls of a cavity.

Parotid *pah-rot'id* located near the ear.

Passive immunity short-lived immunity resulting from the introduction of "borrowed antibodies" obtained from an immune animal or human donor; immunological memory is not established.

Patella *pah-tel'ah* the kneecap.

Pathogenesis *path"o-jen'ĕ-sis* the development of a disease.

Pectoral *pek'to-ral* pertaining to the chest.

Peduncle *pe-dung'k'l* a stalk of fibers, especially that connecting the cerebellum to the pons, midbrain, and medulla oblongata.

Pelvis *pel'vis* a basin-shaped structure; lower portion of body trunk.

Penis *pe'nis* the male organ of copulation and urination.

Pepsin an enzyme capable of digesting proteins in an acid pH.

Pericardium *per"ĭ-kar'de-um* the membranous sac enveloping the heart.

Perimysium *per"ĭ-mis'e-um* the connective tissue enveloping bundles of muscle fibers.

Perineum *per"ĭ-ne'um* that region of the body extending from the anus to the scrotum in males and from the anus to the vulva in females.

Periosteum *per"e-os'te-um* a double-layered connective tissue that covers and nourishes the bone.

Peripheral nervous system (PNS) *pĕ-rif'er-al* a system of nerves that connects the outlying parts of the body with the central nervous system.

Peripheral resistance the resistance to blood flow offered by the systemic blood vessels.

Peristalsis *per"ĭ-stal'sis* the waves of contraction seen in tubes. Propels substances along the tract.

Peritoneum *per"ĭ-to-ne'um* the serous lining of the interior of the abdominal cavity.

Peritonitis *per"ĭ-to-ni'tis* an inflammation of the peritoneum.

Permeability *per"me-ah-bil'ĭ-te* that property of membranes that permits passage of molecules and ions.

Peroneal *per"o-ne'al* pertaining to the outer side of the leg.

pH the symbol for hydrogen ion concentration; a measure of the relative acid or base level of a substance or solution.

Phagocyte *fag'o-sīt* a cell capable of engulfing and digesting particles or cells harmful to the body.

Phagocytosis *fag"o-si-to'sis* the ingestion of foreign solids by cells.

Phalanges *fah-lan'jēz* the bones of the finger or toe.

Pharynx *far'inks* the muscular tube extending from the posterior of the nasal cavities to the esophagus.

Phospholipid *fos'fo-lip"id* a modified lipid containing phosphorus.

Photoreceptor *fo"to-re-sep'tor* specialized receptor cells that convert light energy into a nerve impulse.

Physical change a change in the physical properties of a substance, for example, in its state, color, or size.

Physiology *fiz"e-ol'o-je* the science of the functioning of living organisms.

Pinna *pin'nah* the irregularly shaped elastic cartilage covered with skin forming the external part of the outer ear; auricle.

Pinocytosis *pi"no-si-to'sis* the engulfing of liquid by cells.

Pituitary gland *pĭ-tu'ĭ-tār"e* the neuroendocrine gland located beneath the brain that serves a variety of functions including regulation of water balance, the gonads, thyroid, adrenal cortex, and other endocrine glands.

Placenta *plah-sen'tah* the temporary organ that provides for nourishment and waste removal of the embryo; has an endocrine function as well.

Plantar *plan'tar* pertaining to the sole of the foot.

Plasma *plaz'mah* the fluid portion of the blood.

Plasma cells members of a B-cell clone; specialized to produce and release antibodies.

Platelet *plāt'let* one of the irregular cell fragments of blood; involved in clotting.

Pleura *ploor'ah*   the serous membrane covering the lungs.

Pleurisy *ploor'ĭ-se*   inflammation of the pleurae, making breathing painful.

Plexus *plek'sus*   a network of interlacing nerves, blood vessels, or lymphatics.

Plica *pli'kah*   a fold.

Pneumothorax *nu"mo-tho'raks*   the presence of air or gas in a pleural cavity.

Podocyte *pod'o-sīt*   an epithelial cell located on the basement membrane of the glomerulus, spreading thin cytoplasmic projections over the membrane. Forms part of the filtration membrane.

Polar body   a minute cell produced during maturation divisions (meiosis) in the ovary.

Polarized *po'lar-īzd*   the state of an unstimulated neuron or muscle cell in which the inside of the cell is relatively negative in comparison to the outside; the resting state.

Polycythemia *pol"e-si-the'me-ah*   presence of an abnormally large number of erythrocytes in the blood.

Polypeptide *pol"e-pep'tīd*   a small chain of amino acids.

Polysaccharide *pol"e-sak'ĭ-rīd*   literally, many sugars; a polymer of linked monosaccharides; examples include starch and glycogen.

Pons   (1) any bridgelike structure or part; (2) the brain area connecting the medulla with the midbrain, providing linkage between upper and lower levels of the central nervous system.

Postganglionic (postsynaptic) neuron *pōst"gang-gle-on'ik*   a neuron of the autonomic nervous system having its cell body in a ganglion and its axon extending to an organ or tissue.

Precipitation   formation of insoluble complexes that settle out of solution.

Preganglionic (presynaptic) neuron *pre"gang-gle-on'ik*   a neuron of the autonomic nervous system having its cell body in the brain or spinal cord and its axon terminating in a ganglion.

Prepuce *pre'pūs*   the loose fold of skin that covers the glans penis or clitoris.

Pressoreceptor *pres"o-re-sep'tor*   a nerve ending in the wall of the carotid sinus and aortic arch sensitive to vessel stretching.

Primary (immune) response   the initial response of the immune system to an antigen; involves clonal selection and establishes immunological memory.

Prime movers   those muscles whose contractions are primarily responsible for a particular movement.

Process   (1) a prominence or projection; (2) a series of actions for a specific purpose.

Progesterone *pro-jes'tĕ-rōn*   a hormone responsible for preparing the uterus for the fertilized ovum.

Pronation *pro-na'shun*   the inward rotation of the forearm causing the radius to cross diagonally over the ulna—palms face posteriorly.

Prone   refers to a body lying horizontally with the face downward.

Proprioceptor *pro"pre-o-sep'tor*   a receptor located in a muscle or tendon; concerned with locomotion and posture.

Protein *pro'te-in*   a complex nitrogenous substance; the main building material of cells.

Proteinuria *pro"te-in-u're-ah*   the passage of proteins in the urine.

Proton *pro'ton*   subatomic particle that bears a positive charge; located in the atomic nucleus.

Proximal *prok'sĭ-mal*   toward the attached end of a limb or the origin of a structure.

Puberty *pu'ber-te*   the period at which reproductive organs become functional.

Pulmonary *pul'mo-ner"e*   pertaining to the lungs.

Pulmonary circuit   the circulatory vessels of the lungs.

Pulmonary edema *ĕ-de'mah*   a leakage of fluid into the air sacs and tissue of the lungs.

Pulse   the rhythmic expansion and recoil of arteries resulting from heart contraction; can be felt from the outside of the body.

Pupil   an opening in the center of the iris through which light enters the eye.

Purkinje fibers *pur-kin'je*   the modified cardiac muscle fibers of the conduction system of the heart.

Pus   the fluid product of inflammation composed of white blood cells, the debris of dead cells, and a thin fluid.

Pyelonephritis *pi"ĕ-lo-nĕ-fri'tis*   an inflammation of the kidney pelvis and surrounding kidney tissues.

Pyloric region *pi-lor'ik*   the final portion of the stomach; joins with the duodenum.

Pyramid   any cone-shaped structure of an organ.

Pyrogen *pi'ro-jen*   an agent or chemical substance that induces fever.

Quadriplegia *kwod"rĭ-ple'je-ah*   the paralysis of all four limbs.

Radiant energy   energy of the electromagnetic spectrum, which includes heat, light, ultraviolet waves, infrared waves, and other forms.

Radiate *ra'de-āt*   diverging from a central point.

Radioactivity   the process of spontaneous decay seen in some of the heavier isotopes, during which particles or energy is emitted from the atomic nucleus; results in the atom becoming more stable.

Radioisotope *ra"de-o-i'sĭ-top*   isotope that exhibits radioactive behavior.

Ramus *ra'mus*   a branch of a nerve, artery, vein, or bone.

Receptor *re-sep'tor*   a peripheral nerve ending specialized for response to particular types of stimuli.

Reduction   restoring broken bone ends (or a dislocated bone) to its original position.

Reflex   automatic reactions to stimuli.

Refract   bend.

Refracting media   substances that bend light.

Refractory period *re-frok'to-re*   the period of unresponsiveness to stimulation immediately after depolarization.

Renal *re'nal*   pertaining to the kidney.

Renal calculus *re'nal kal'ku-lus* a kidney stone.

Renin *re'nin* a substance released by the kidneys; involved with raising blood pressure.

Reticulum *rĕ-tik'u-lum* a fine network.

Retract to draw back, shorten, contract.

Ribonucleic acid (RNA) *ri"bo-nu-kle'ic* the nucleic acid that contains ribose. Acts in protein synthesis.

Ribosomes *ri'bo-sōmz* cytoplasmic organelles at which proteins are synthesized.

Rickets *rik'ets* a disease occurring in infants and young children characterized by softening of the bone; most often resulting from calcium lack.

Rods one of the two types of photosensitive cells in the retina.

Rotate to turn about an axis.

Rugae *roo'je* elevations or ridges, as in the mucosa of the stomach.

Sacral *sa'kral* the lower portion of the back, just superior to the buttocks.

Sagittal plane *saj'ĭ-tal* a longitudinal section that divides the body or any of its parts into right and left portions.

Saliva *sah-li'vah* the secretion of salivary glands ducted into the mouth.

Salt ionic compound that dissociates into charged particles (other than hydrogen or hydroxyl ions) when dissolved in water.

Sclera *skle'rah* the firm white fibrous outer layer of the eyeball; functions to protect and maintain eyeball shape.

Scoliosis *sko"le-o'sis* a lateral curve in the vertebral column.

Scrotum *skro'tum* the external sac enclosing the testes.

Sebaceous glands *se-ba'shus* glands that empty their sebum secretion into hair follicles.

Sebum *se'bum* the oily secretion of sebaceous glands.

Secondary (immune) response second and subsequent responses of the immune system to a previously met antigen; more rapid and more vigorous than the primary response.

Secretion *se-kre'shun* the passage of material formed by a cell to its exterior.

Semen *se'men* the fluid produced by male reproductive structures; contains sperm, nutrients, and mucus.

Semilunar valves *sem"e-lu'nar* valves that prevent blood return to the ventricles after contraction.

Seminiferous tubules *se"mĭ-nif'er-us* highly convoluted tubes within the testes that form sperm.

Sensory nerve a nerve that contains processes of sensory neurons and carries nerve impulses to the central nervous system.

Sensory nerve cell an initiator of nerve impulses following receptor stimulation.

Serous fluid *se'rus* a clear, watery fluid secreted by the cells of a serous membrane.

Sex chromosome chromosome that determines sex of the fertilized egg; X and Y chromosomes.

Sinoatrial node *si"no-a'tre-al* the mass of specialized myocardial cells in the wall of the right atrium. Pacemaker of the heart.

Sinus *si'nus* (1) a mucous-membrane-lined, air-filled cavity in certain cranial bones; (2) a dilated channel for the passage of blood or lymph.

Smooth muscle muscle consisting of spindle-shaped, unstriped (nonstriated) muscle cells; involuntary muscles.

Solute *so'lut* the dissolved substance in a solution.

Solution a homogenous mixture of two or more components.

Somatic nervous system *so-mat'ik* a division of the peripheral nervous system; also called the voluntary nervous system.

Sperm mature male germ cell.

Spermatogenesis *sper"mah-to-jen'ĕ-sis* the process of meiosis (cell division) in the male to produce sperm.

Sphenoid *sfe'noid* wedgelike.

Sphincter *sfingk'ter* a muscle surrounding an opening; acts as a valve.

Sprain the wrenching of a joint, producing stretching or tearing of the ligaments.

Squamous *skwa'mus* pertaining to flat, thin cells that form the free surface of some epithelial tissues.

Stasis *sta'sis* (1) a decrease or stoppage of flow; (2) a state of nonchange.

Static equilibrium *stat'ik e"kwĭ-lib're-um* balance concerned with changes in the position of the head.

Stenosis *stĕ-no'sis* constriction or narrowing.

Steroids *ste'roidz* a specific group of chemical substances including certain hormones and cholesterol.

Stimulus *stim'u-lus* an excitant or irritant; a change in the environment producing a response.

Strabismus *strah-biz'mus* inability to coordinate the movement of the two eyes; crossed eyes.

Stratum *stra'tum* a layer.

Stressor any stimulus that directly or indirectly causes the hypothalamus to initiate stress-reducing responses, such as the fight or flight response.

Striated muscle *stri'āt-ed* muscle consisting of cross-striated (cross-striped) muscle fibers.

Stroke a condition in which brain tissue is deprived of a blood supply, as in blockage of a cerebral blood vessel.

Stroke volume a volume of blood ejected by the left ventricle during a systole.

Sty an inflammation of an oil-secreting gland of the eyelid.

Subcutaneous *sub"ku-ta'ne-us* beneath the skin.

Sublingual *sub-ling'gwal* beneath the tongue.

Sulcus *sul'kus* a furrow on the brain, less deep than a fissure.

Summation *sum-may'-shun* the accumulation of effects, especially those of muscular, sensory, or mental stimuli.

Superficial (external) located close to or on the body surface.

Superior refers to the head or upper body regions.

Supination *su"pĭ-na'shun* the outward rotation of the forearm causing palms to face anteriorly.

Supine refers to a body lying with the face upward.

Suppressor T-cells regulatory T-lymphocytes that suppress the immune response.

Surfactant *sur-fak'tant* a substance on pulmonary alveoli walls that reduces surface tension, thus preventing collapse of the alveoli after expiration.

Suspensory ligament of an eye *sus-pen'so-re* fibrous ligament that holds the lens in place in the eye.

Sutures *su'churz* the immovable joints that connect the bones of the adult skull.

Sweat glands the glands that secrete a saline solution (sudoriferous glands).

Sympathetic division a division of the autonomic nervous system; opposes parasympathetic functions.

Synapse *sin'aps* the region of communication between neurons.

Synaptic cleft *sĭ-nap'tik* the fluid-filled space at a synapse between neurons.

Synarthrosis *sin"ar-thro'sis* a fibrous joint; two types: sutures and syndesmoses. An immovable joint.

Synergist *sin'er-jist* a muscle cooperating with another to produce a desired movement.

Synovial fluid *sĭ-no've-al* a fluid secreted by the synovial membrane; lubricates joint surfaces and nourishes articular cartilages.

Synthesis reaction chemical reaction in which larger molecules are formed from simpler ones.

System a group of organs that function cooperatively to accomplish a common purpose; there are ten major systems in the human body.

Systemic *sis-tem'ik* general; pertaining to the whole body.

Systemic circuit the circulatory vessels of the body.

Systemic edema *ĕ-de'mah* an accumulation of fluid in body organs or tissues.

Systole *sis'to-le* the contraction phase of heart activity.

Systolic pressure *sis-tol'ik* the pressure generated by the left ventricle during systole.

T-cells lymphocytes that mediate cellular immunity; include helper, killer, suppressor, and memory cells. Also called T-lymphocytes.

Tachycardia *tak"e-kar'de-ah* an abnormal, excessively rapid heart beat; over 100 beats per minute.

Tarsal *tahr'sal* one of the seven bones that form the ankle and heel.

Taste buds receptors for taste on the tongue, roof of mouth, pharynx, and larynx.

Tendon *ten'dun* a cord of dense fibrous tissue attaching a muscle to a bone.

Testis *tes'tis* the male primary sex organ that produces sperm.

Tetanus *tet'ah-nus* (1) the tense, contracted state of a muscle; (2) an infectious disease.

Thalamus *thal'a-mus* a mass of gray matter in the diencephalon of the brain.

Thermoreceptor *ther"mo-re-sep'tor* a receptor sensitive to temperature changes.

Thoracic *tho-ras'ik* refers to the chest.

Thorax *tho'raks* that portion of the body trunk above the diaphragm and below the neck.

Threshold the weakest stimulus capable of producing a response in an irritable tissue.

Thrombin *throm'bin* an enzyme that induces clotting by converting fibrinogen to fibrin.

Thrombocyte *throm'bo-sīt* a blood platelet; part of the blood-clotting mechanism.

Thrombophlebitis *throm"bo-fle-bi'tis* an inflammation of a vein associated with blood clot formation.

Thrombus *throm'bus* a clot that is fixed or stuck to a vessel wall.

Thymus gland *thi'mus* an endocrine gland active in immune response.

Thyroid gland *thi'roid* one of the largest of the body's endocrine glands.

Tissue a group of similar cells forming a distinct structure.

Tone *tōn* refers to the state of continuous but staggered muscle stimulation/activity.

Toxic *tok'sik* poisonous.

Trachea *tra'ke-ah* the windpipe; the cartilaginous and membranous tube extending from larynx to bronchi.

Tract a collection of nerve fibers in the CNS having the same origin, termination, and function.

Transverse process the projections that extend laterally from each neural arch of a vertebra.

Trauma *traw'mah* an injury, wound, or shock; usually produced by external forces.

Trochanter *tro-kan'ter* a large, somewhat blunt process.

Trophic *trof'ik* pertains to nutrition

Tubal pregnancy an ecotopic pregnancy that occurs within a uterine tube.

Tubercle *tu'ber-k'l* a nodule or small rounded process.

Tuberosity *tu"bĕ-ros'ĭ-te* a broad process, larger than a tubercle.

Tympanic membrane *tim-pan'ik* the eardrum.

Ulcer *ul'ser* a lesion or erosion of the mucous membrane, such as gastric ulcer of stomach.

Umbilical cord *um-bil'ĭ-kal* a structure bearing arteries and veins connecting the placenta and the fetus.

Umbilicus *um-bil'ĭ-kus* the navel; marks site of the umbilical cord in the fetal stage.

Urea *u-re'ah* the main nitrogen-containing waste excreted in the urine.

Ureter *u-re'ter* the tube that carries urine from kidney to bladder.

Urethra *u-re'thrah* the canal through which urine passes from the bladder to the outside of the body.

Uvula *u'vu-lah* tissue tag hanging from soft palate.

Valence shell *va'lents* the outermost energy level

of an atom that contains electrons; the electrons in the valence shell determine the bonding behavior of the atom.

Varicose vein *var'ĭ-kōs* a dilated, knotted, tortuous vein resulting from incompetent valves.

Vas *vas* a duct; vessel.

Vascular *vas'ku-lar* pertaining to blood channels or vessels.

Vasoconstriction *vas"o-kon-strik'shun* narrowing of blood vessels.

Vasodilation *vas"o-di-la'shun* relaxation of the smooth muscles of the blood vessels producing dilation.

Vasomotor nerve fibers *vas-o-mo'tor* the nerve fibers that regulate the constriction or dilation of blood vessels.

Vein *vān* a vessel carrying blood away from the tissues toward the heart.

Ventral anterior or front.

Ventricle *ven'trĭ-k'l* a discharging chamber of the heart.

Ventricles of the brain cavities within the brain.

Venule *ven'ūl* a small vein.

Vertigo *ver'tĭ-go* dizziness; the feeling of movement such as a sensation that the external environment is revolving.

Viscera *vis'era* the internal organs.

Visceral *vis'er-al* pertaining to the internal part of a structure or the internal organs.

Viscosity *vis-kos'ĭ-te* the state of being sticky or thick.

Visual acuity *ah-ku'ĭ-te* the ability of the eye to distinguish detail.

Vital capacity the volume of air that can be expelled from the lungs by forcible expiration after the deepest inspiration; total exchangeable air.

Vitamins the organic compounds required by the body in minute amounts for physiological maintenance and growth.

Voluntary muscle muscle under control of the will; skeletal muscle.

Vulva *vul'va* female external genitalia.

White matter the white substance of the central nervous system; the myelinated nerve fibers.

Zygote *zi'gōt* the fertilized ovum; produced by union of two gametes.

Index

Acknowledgments

1   R. O. Hynes, MIT.

18   From Spence, A. P., and Mason, E. B. 1983. *Human anatomy and physiology*, 2nd ed., p. 98. Menlo Park, Calif.: Benjamin/Cummings.

39   From Purtilo, D. T. 1978. *A survey of human diseases*, p. 162. Menlo Park, Calif.: Addison-Wesley.

44   From Spence, A. P., and Mason, E. B. 1983. *Human anatomy and physiology*, 2nd ed., p. 95. Menlo Park, Calif.: Benjamin/Cummings.

48   Courtesy, City of Vancouver Health Department.

52   Dr. Robert Chase, Anatomy Dept., Stanford University.

65   From Spence, A. P., and Mason, E. B. 1983. *Human anatomy and physiology*, 2nd ed., p. 146. Menlo Park, Calif.: Benjamin/Cummings.

82   Carolina Biological Supply.

87   Clara Franzini-Armstrong, University of Pennsylvania.

115 Ed Reschke.

130 William Thompson, RN, Limited Horizons © 1982.

162 Michael Coppinger, CRA/J.M.C. Eye Photography.

179 From Kimball, J. W. 1978. *Biology*, 4th ed. Reading, Mass.: Addison-Wesley.

197 W. Rosenberg, Iona College/BPS.

205 Manfred Kage/Peter Arnold Inc.

269 K. E. Muse, Duke University Medical Center/BPS.

273 From Saxton, D. F. et al. (eds.). 1983. *The Addison-Wesley manual of nursing practice*, p. 1041. Menlo Park, Calif.: Addison-Wesley.

313 From Purtilo, D. T. 1978. *A survey of human diseases*, p. 314. Menlo Park, Calif.: Addison-Wesley.

337 R. Yanagimachi, School of Medicine, University of Hawaii at Manoa/BPS.

342 R. Yanagimachi, School of Medicine, University of Hawaii at Manoa/BPS.

354 Drs. Robert Rugh and Landrum Shettles.